Laser Interaction
and Related Plasma Phenomena
Volume 2

Laser Interaction
and Related Plasma Phenomena

Volume 2

Proceedings of the Second Workshop, held at Rensselaer Polytechnic Institute, Hartford Graduate Center, Hartford, Connecticut, August 30–September 3, 1971

Edited by
Helmut J. Schwarz
Professor of Physics
Rensselaer Polytechnic Institute

and
Heinrich Hora
Adjunct Associate Professor of Physics
Rensselaer Polytechnic Institute
and
Max-Planck-Institut für Plasmaphysik
Garching, Germany

℗ *PLENUM PRESS • NEW YORK-LONDON • 1972*

ADVISORY BOARD

N. G. BASOV
*P. N. Lebedev Physical
Institute of the Academy
of Sciences of the USSR*

N. BLOEMBERGEN
Harvard University

J. M. DAWSON
Princeton University

A. H. GUENTHER
Kirtland Air Force Base

P. HARTECK
Rensselaer Polytechnic Institute

ISBN 978-1-4684-7740-5   ISBN 978-1-4684-7738-2 (eBook)
DOI: 10.1007/ 978-1-4684-7738-2
Library of Congress Catalog Card Number 79-135851

© 1972 Plenum Press, New York
Softcover reprint of the hardcover 1st edition 1972
A Division of Plenum Publishing Corporation
227 West 17th Street, New York, N.Y. 10011

United Kingdom edition published by Plenum Press, London
A Division of Plenum Publishing Company, Ltd.
Davis House (4th Floor), 8 Scrubs Lane, Harlesden, London, NW10
6 SE, England

All rights reserved

No part of this publication may be reproduced in any form without
written permission from the publisher

FOREWORD

Paul Harteck

Rensselaer Polytechnic Institute

Troy, New York

When the Maser and the Laser were discovered, people were speculating if this was the beginning of a new page, or even a new chapter, in the Book of Physics. The Second Workshop on "Laser Interaction and Related Plasma Phenomena" held in Hartford made it clear that the perspective had changed, that people now question if the consequences of these discoveries constitute a new chapter, or possibly a new era in Physics.

While the papers presented were all stimulating and of outstanding quality, of special interest were the experiments which demonstrated that triggering of thermonuclear fusion by Laser techniques is indeed in the realm of the possible. Along these lines, I enjoy recalling an anecdote concerning the late F. G. Houtermans. I think that all who knew him will agree that he was an unusual genius and at the same time a very amusing colleague. He had, in fact, decades ago considered the possibility of Laser type effects. Unfortunately, his experimental efforts ended with the destruction of the transformers of his Institute. It happened - so goes the anecdote - that a few years ago, there was a meeting where the possibilities of thermonuclear fusion were discussed. After the meeting a newspaper correspondent asked Houtermans if he thought it would some day be possible to use thermonuclear fusion for energy production. Houtermans answered that he did not really understand this question because old Jehovah had shown that thermonuclear fusion works. The curious newspaperman asked: "Where is the laboratory of Professor Jehovah and how long has he been performing this experiment?" Houtermans answered that he thought the title Professor was not quite adequate, that the laboratory is the entire universe, and that these experiments had been in progress for billions of years!

You are all familiar with the brilliant names in Physics which are connected with the discovery of the Maser and the Laser. In recent years we have learned that old Jehovah's name must certainly stand at the head of the list. Laser or, more precisely, Maser signals have been found emanating from certain distant cosmic areas, carrying messages concerning molecular makeup, temperature, and density across many light-years of space.

Our knowledge of Laser phenomena is indeed increasing and speculation on the future may be boundless. It appears that the "Laser Conference" in Hartford will be a continuing series for some time to come.

PREFACE

The First Workshop on "Laser Interaction and Related Plasma Phenomena" held at Rensselaer Polytechnic Institute, Hartford Graduate Center, in 1969, brought out the complexity of the results obtained by different and differing theoretical and experimental groups. The Second Workshop in 1971 corroborated this complexity as more and more unexpected observations were reported indicating the possible existence of non-equilibrium and non-linear plasma properties. Tremendous advances were made in the interval between the First and Second Workshops in the research of plasmas produced by laser irradiation as well as in the development of very high power lasers.

This significant progress was indicated by the increased number of highly qualified scientists participating in the Second Workshop. Their lectures and discussions provided up-to-date knowledge and stimulated further fundamental research, development of new techniques, and possible industrial applications.

As in the First Workshop, many of the presentations gave valuable fundamental insights for those starting in the field, while at the same time bringing into focus very advanced material. In addition to the invited lectures, attendees were given the opportunity to present short papers on their own research, and several of their contributions are included in these Proceedings.

As suggested by C. G. Young in his Review of the Proceedings of the First Workshop ("Laser Focus", June 1971, p. 45), the editors have attempted to include some of the discussions of the presentations (for example, nuclear fusion plasmas), although these comments may not reflect the exact view of the participants involved.

The first chapter contains results of the development of high intensity lasers. Although at this time, one cannot tell in what direction the research will go, the major results of experts in the field were presented which indicated what one might expect from future developments for plasma applications. Beaulieu, the pioneer in CO_2-TEA-lasers reports on his present work and future plans, while

DeMaria reviews the electric discharge lasers in more general terms. Vlases presents many details of gasdynamic lasers. Miley reports on how electrons and other particles of about MeV-energies originating from nuclear radiation influenced the discharge of CO_2 lasers. Chemical lasers are reviewed by Bronfin; Hohla and Kompa supplement this review with special results on a chemical low-cost 5-Joule 10-nsec laser system. The most advanced neodymium glass laser system is described in a presentation by Basov, Krokhin and Sklizkov, which is included in their paper in Chapter VI. From this system, the next major results in laser produced fusion plasmas can be expected. Gerstmayr, Harteck and Reeves describe the techniques of multiphoton absorption in gases, which are important for many gas lasers.

The second chapter is devoted to laser produced gas breakdown. Our First Workshop did not deal in depth with the complex problems of this field. However, due to the advances made, more detailed information could be presented at the Second Workshop. Papoular gives some unexpected experimental results. Although the question of the "first electron" is still not answered, possible directions for the solution of this problem are indicated. Guenther's and Pendleton's contribution reveals many detailed measurements of laser produced gas breakdown. Hull, Lencioni and Marquet report measurements for the decrease of the breakdown threshold by a CO_2-TEA-laser focused in air, when NaCl dust particles of known size are present. Bradley discusses switching of megavolt with subnanosecond jitter applying laser produced gas breakdown. Hereby he develops a new technique of studying the streamer processes. Alcock's advanced diagnostic techniques for measuring the self-focusing in laser produced gas breakdown shows numerous details of the self-focusing filament, its inner hole in electron density, time behavior, and dependence on the molecular state of the gas. Further observations on self-focusing and gas breakdown are given by Tomlinson.

In the third chapter, plasma diagnostics and special interaction processes are considered. The scattering of laser radiation, first successfully measured by Fünfer, Kronast and Kunze in 1963, is a very important method for the diagnosis of plasmas of medium density. Now, Kronast gives a report on recent developments in this area, especially dealing with the micro-state of the plasma. Experimental and new theoretical results of scattering of laser photons by free electrons in vacuum (Kapitza-Dirac Effect) and in solids (Quantum Mechanical Modulation) are reported by Schwarz. Probe measurements in the diagnosis of plasmas produced by laser irradiation of solid targets in vacuum are sometimes difficult to interpret. This leads to a presentation by Segall on the treatment of cylindrical Langmuir probes. Yablonovich gives results on his research of CO_2 laser produced breakdown in alkali-halides. Büchl's contribution deals with an advanced study of the plasma created by CO_2-TEA-laser pulses from solid hydrogen. Siller, Büchl and Hora discuss

PREFACE

latest theoretical aspects of the anomalously high electron emission from laser produced plasmas.

The fourth chapter deals with the interaction of laser produced plasmas with gases and magnetic fields. Stamper, Dean and McLean discuss unexpected interactions of laser produced plasmas with background gas and the creation of magnetic fields within plasmas that expand into the field free space. Vlases' paper in Chapter I also touches the question of heating magnetically confined thermonuclear plasmas of medium density by intense CO_2 lasers. He also indicates that the Lawson criterion of thermonuclear fusion by magnetic confinement of a laser produced plasma might be reached by injecting a neutral beam. Haught elaborates on this concept. Bhadra theorizes on the behavior of a laser heated plasma in a magnetic field. Schwarz reports measurements of the increase of ion energies in laser produced plasmas due to magnetic fields, and Hora interprets these results with his nonlinear theory of ponderomotive forces.

The fifth chapter covers the theory of high intensity laser interaction with plasmas. The gasdynamic or thermokinetic theory as discussed in the First Workshop and supplemented in the meantime by a paper of R. G. Rehm (Phys. Fluids $\underline{13}$, 921 (1970)) seems to have been substantiated. Therefore, the main interest in this Workshop was directed towards instabilities and nonlinear effects. Dawson and coworkers were among the first in this specific field of laser produced plasmas. A review of the present status is given in the paper by Kruer and Dawson. Godwin comments by presenting a classical mechanism of anomalous absorption. Hora reviews his concepts of nonlinear forces by bringing his previous contribution in the First Workshop up-to-date. Palmer reports on stimulated scattering and self-focusing processes in dense plasmas. New results of numerical dynamic treatments with the nonlinear ponderomotive force are given by Green and Mulser.

The sixth chapter focuses on the possibility of achieving nuclear fusion with laser irradiated high density plasmas. Included in the Second Workshop were all groups that had reported fusion neutrons up to August 1971. The first fusion neutrons were detected from LiD targets irradiated by 10 psec laser pulses (Basov et al 1968). Basov, Krokhin and Sklizkov now report longer laser pulses (in the order of 1 nsec) which led to up to 10^5 neutrons per pulse. Immediately after our First Workshop (see Appendix of Floux' presentation, Proceedings of First Workshop), Floux et al applied this technique successfully for the first time and measured an increase of the neutron production higher by orders of magnitude in relation to the first psec pulses. Floux, Bernard, Cognard and Saleres give an extensive description of their measurements performed under varying conditions. Lubin's work on laser produced neutrons concentrates on laser pulses of a little less than one nsec duration applying tailored prepulses; he justifies his techniques in a theoretical analysis. Jones, Gobeli and Olsen contribute with measurements related to carefully generated laser pulses selected from prepulses, which indicate the problems involved in such techniques. Yamanaka points out new aspects of neutron production using high power neodymium glass lasers. In an Appendix to his paper, there is a valuable investigation of the damages occurring in glass lasers used for such high power applications. Büchl, Eidmann, Mulser, Salzmann, and Sigel report on refined diagnostic techniques of glass laser produced plasmas for the detection of neutrons, x-rays, and reflectivity properties. Hora and Pfirsch contribute with numerical treatments on the fusion gain of laser produced plasmas. Finally, Shatas gives calculations on laser assisted dense plasma focus whereby fusion neutrons and soft x-rays are generated.

The success of the Second Laser Workshop was mainly due to the high caliber of the lecturers and participants, who also enthusiastically contributed to the discussions. At the writing of this Preface, there are indications that one of the main objectives, namely to initiate and stimulate further work, has been accomplished. The organizers (the editors) gratefully acknowledge not only the excellent presentations of the invited speakers but also those who contributed with short presentations.

We should like to thank the following Advisors of the Second Workshop for their counsel: Professor N. G. Basov (P. N. Lebedev Physical Institute, Moscow, USSR), Professor N. Bloembergen (Harvard University), Professor John Dawson (Princeton University), Dr. A. H. Guenther (Kirtland Air Force Base) and Professor P. Harteck (Rensselaer Polytechnic Institute).

We are indebted to Rensselaer Polytechnic Institute of Connecticut for partial financial support, and to Vice President Warren C. Stoker, Dean Harry Kraus, and Director of Special Programs Preston T. Reed.

We wish to acknowledge permission of the American Institute of Physics, the IEEE, and the American Chemical Society to print figures and texts as indicated by the references.

Again, as in the First Workshop, Mrs. Lucy Myshrall of Rensselaer Polytechnic Institute, Hartford Graduate Center, was an invaluable assistant in the organization of the Workshop and for the preparation of these Proceedings. Her care and interest were an important contribution to its success.

January 1972   Helmut Schwarz

                                                     Heinrich Hora

CONTENTS

I. HIGH INTENSITY LASERS

High Power CO_2 Lasers
    A. J. Beaulieu 1

Survey of Electric-Discharge CO_2 Lasers
    A. J. DeMaria 9

Thermokinetic Theories of Gasdynamic Lasers and Application to Controlled Thermonuclear Fusion
    G. C. Vlases and A. L. Hoffman 25

Nuclear Radiation Effects on Gas Lasers
    G. H. Miley 43

Chemical Molecular Lasers (Abstract Only)
    B. R. Bronfin 59

Photochemical Iodine Laser - A High Power Gas Laser
    K. Hohla, P. Gensel, and K. L. Kompa 61

Photodissociation of N_2O by Pulsed Laser Light at 6943Å
    J. Gerstmayr, P. Harteck, and R. Reeves 67

Summary of Discussion 77

II. LASER INDUCED GAS BREAKDOWN

The Initial Stages of Laser-Induced Gas Breakdown
    P. Papoular 79

Laser-Produced Gaseous Deuterium Plasmas
    A. H. Guenther and W. K. Pendleton 97

Influence of Particles on Laser Induced
Air Breakdown
    R. J. Hull, D. E. Lencioni, and L. C. Marquet . . .   147

Laser Produced Plasma-Streamer Interaction
(Abstract Only)
    L. P. Bradley   153

Experiments on Self-Focusing in Laser Produced
Plasmas
    A. J. Alcock   155

Scattering and Beam Trapping Laser-Produced
Plasmas in Gases (Abstract Only)
    R. G. Tomlinson   177

Summary of Discussion   179

III. PLASMA DIAGNOSTICS AND SPECIAL INTERACTION PROCESSES

Recent Developments in Light Scattering
Experiments on Laboratory Plasmas
    B. Kronast   181

Electron-Photon Interaction
    H. Schwarz   209

Interpretation of Cylindrical Langmuir Probe
Signals from Streaming Laser-Produced Plasmas
    S. B. Segall   227

The Dielectric Strength of Alkali-Halide
Crystals at Optical Frequencies (Abstract Only)
    E. Yablonovitch   243

Plasma Production by CO_2 Laser
    K. P. Büchl   245

Intense Electron Emission from Laser
Produced Plasmas
    G. Siller, K. Büchl, and H. Hora   253

Summary of Discussion   271

IV. LASER PRODUCED PLASMAS INTERACTING WITH GASES AND MAGNETIC FIELDS

Interactions of Laser-Produced Plasmas with
Background Gases
    J. A. Stamper, S. O. Dean, and E. A. McLean   273

CONTENTS

Magnetic Field Confinement of Laser Irradiated
Solid Particle Plasmas (Abstract Only)
A. F. Haught 289

Dynamics of a Resistive Plasmoid in a
Magnetic Field
D. K. Bhadra 291

Magnetic Field Enhanced Energy Increase of
Ions Emitted from Laser Irradiated Solid
Targets
H. Schwarz and H. Hora 301

Theoretical Aspects of Ion Energy Increase in
Laser Produced Plasmas Due to Static
Magnetic Fields
H. Hora 307

Summary of Discussion 315

V. THEORY OF HIGH INTENSITY LASER INTERACTION WITH PLASMAS

Anomalous Absorption of Intense Radiation
W. L. Kruer and J. M. Dawson 317

A Classical Resonance in the Optics of Thin
Films at the Plasma Frequency
R. P. Godwin 339

Nonlinear Forces in Laser Produced Plasmas
H. Hora 341

Stimulated Scattering and Self-Focusing
Processes in Dense Plasmas
A. J. Palmer 367

A Self Consistent Calculation of Ponderomotive
Forces in the Laser Plasma Interaction
B. J. Green and P. Mulser 381

Summary of Discussion 387

VI. FUSION NEUTRONS FROM LASER IRRADIATED HIGH
DENSITY PLASMAS

Heating of Laser Plasmas for Thermonuclear Fusion
N. G. Basov, O. N. Krokhin, and G. V. Sklizkov . . . 389

Nuclear DD Reactions in Solid Deuterium
Laser Created Plasma
    F. Floux, J. F. Benard, D. Cognard,
    and A. Saleres 409

Laser Heated Overdense Plasmas for
Thermonuclear Fusion
    M. Lubin, J. Soures, E. Goldman,
    T. Bristow, and W. Leising 433

Nanosecond and Picosecond Laser Irradiation
of Solid Targets
    E. D. Jones, G. W. Gobeli, and J. N. Olsen 469

Thermonuclear Fusion Plasma by Lasers
    C. Yamanaka 481

Appendix to Preceding Paper: Investigation of
Damages in Laser Glasses
    C. Yamanaka 495

Plasma Production with a Nd Laser and
Non-Thermal Effects
    K. Büchl, K. Eidmann, P. Mulser,
    H. Salzmann, and R. Sigel 503

Influence of Fast Ion Losses in Inertially
Confined Nuclear Fusion Plasma
    H. Hora and D. Pfirsch 515

Fusion Neutron and Soft X-ray Generation in
Laser Assisted Dense Plasma Focus
    R. A. Shatas, T. G. Roberts, H. J. Mayer,
    and J. D. Stettler 527

Summary of Discussion 545

LIST OF CONTRIBUTORS AND ATTENDEES 547

AUTHOR INDEX 551

SUBJECT INDEX 571

HIGH POWER CO_2 LASERS *

A. Jacques Beaulieu

GEN-TEC (1969) INC.

INTRODUCTION

Until very recently, the study of high energy laser interaction and related plasma phenomena has been done almost exclusively with solid state lasers. For many years, CO_2 gas lasers have been recognized as the highest average power sources and their efficiency is considerably higher than for solid state lasers. However, until the TEA laser was developed, the very high peak powers required for plasma interaction studies were not considered possible with gas lasers because of the very low density of laser molecules compared to solid state lasers. The atmospheric pressure operation of CO_2 lasers which was pioneered at the Defence Research Establishment Valcartier has changed this situation and the ease with which gas lasers can be increased in size indicates that gas lasers may be the best high energy sources of the future.

Before going into the potential and limitations of future large gas lasers, I would like to review briefly the present developments in high pressure CO_2 lasers.

REVIEW OF STATE OF THE ART IN TEA LASERS

The foremost problem in developing atmospheric pressure gas lasers is to excite a sufficiently large volume of gas

---

*Presented at the Second Workshop on "Laser Interaction and Related Plasma Phenomena" at Rensselaer Polytechnic Institue, Hartford Graduate Center, August 30- September 3, 1971.

to achieve appreciable gains. The required excitation per unit volume increases rapidly with pressure. Obviously, the greater number of molecules to be excited per unit volume, requires that the excitation energy increases linearly with operating pressure. However, due to the shortening lifetime of the excited level, this energy must be delivered more rapidly as pressure increases. Hence the excitation power must increase as the square of the operating pressure to achieve the same gain. To achieve maximum efficiency in Q switched or giant pulses, the excitation time must be short with respect to the lifetime of the upper laser level which for CO_2 lasers at atmospheric pressure is of the order of 10 microseconds. Excitation pulses of the order of one microsecond or less are desirable and this consideration has led to the use of transverse excitation which offers a much lower impedance than the more conventional longitudinal discharges and greatly simplifies the electrical discharge system which can operate in the kilovolt rather than the megavolt range.

At atmospheric pressure one can obtain a 1 microsecond uniform discharge between two long electrodes only if the energy of the discharge is very low. For the desired energies, the discharge becomes rapidly unstable and takes the form of a single bright arc where most of the energy is delivered, producing excessive heating and causing a major scattering center for the infrared radiation due to the lesser refraction index of the hot gas. The first simple solution to this problem has been the use of a large number of resistively loaded cathodes to increase the stability of the discharge. It has been found that by increasing the applied voltage and decreasing the discharge time, the discharges from the multiple pins were more diffused, providing better excitation volume and smaller thermal gradients thus reducing the scattering. In the limit of short pulses, it is even possible to eliminate completely the resistive elements and still have a well distributed excitation discharge at energies compatible with laser operation. Such a resistanceless pin electrode structure has yielded better than 15% efficiency with energies in excess of 2 joules from a laser 1 metre long, using a single row of pins spaced 5 mm apart and an electrode cathode spacing of 3 cm. Voltages of 50 KV and 0.01 MFarad condensers were used. The total estimated discharge volume was between 100 and 150 cc (the exact volume being difficult to determine more accurately because of ill defined discharge section). Thus laser energies from 15 to 20 joules per litre have been obtained.

For larger lasers with better mode purity, the multipe pin structure is neither elegant nor practical. Discharge systems providing more uniform discharge distributions and which can more readily expanded to very large volume lasers

have been investigated by different groups. Laflamme at DREV
developed the grid-laser or double discharge laser where the
anode consists of a grid of wires with an insulated trigger
electrode immediately behind it. The trigger electrode being
connected to the cathode, upon application of the voltage
pulse, a low energy preionizing cathode to trigger discharge
first occurs which creates initial conditions for the main
anode to cathode discharge that greatly improves its stability
and uniformity. Using this technique, 4 joules were obtained
from a 2.5 X 2.5 X 100 cm discharge volume with 8% efficiency.
Similar figures of approximately 5 joules/litre were obtained
with lasers up to 5 X 5 cm cross-section. Similar work has
been done by Dumanchin et al at the C.G.E. using a cathode
made of parallel fins and insulated trigger wires running
between the fins. They have reported 20 joules/litre of laser
energy in their latest systems and have achieved a 140 joule
laser which is 3 metres long with cross-section of 10 X 10 cm.
Lamberton et al at SERL has also announced energies of 20
Joules/litre in 100 cc laser using two shaped electrode and a
separately excited trigger wire. Other techniques are also
being investigated which are aimed at creating a uniform pre-
ionization to improve both the uniformity and the stability
of the pumping discharge. Such methods are preionization by
U.V. flash tubes or by a high energy electron beam. Unfortunately, I have no information yet about the performance that
have been achieved with these techniques.

FUNDAMENTAL LIMITATIONS OF TEA LASERS

After this review of the rapid progress in atmospheric
pressure CO_2 lasers, it is natural to wonder where will it
end and what are the true limitations of this technique. To
try and answer this question it is necessary to look at the
basic properties of the CO_2 molecular laser. When the excitation discharge is very rapid (less than 1 µsecond) the low
gain per unit length of the laser causes the electromagnetic
field to build up rather slowly so that by the time the field
intensity in the laser approaches the saturation value, the
gain is completely built up and a giant pulse results which
is essentially identical to a Q-switch pulse and has a duration of the order of 100 nanosecond. Only half the available
laser energy stored in the V_3 vibration of the CO_2 molecules
is extracted during this giant pulse because of the equalization
of the upper and lower laser level population during this giant
pulse. Since in the usual CO_2 laser gas mixes the lifetime
of the lower laser lever is appreciably shorter than that of
the upper laser level, a certain amount of population inversion will reappear and if the optical cavity has sufficiently
small losses a second laser pulse will appear during which

most of the remaining vibrational energy of the V_3 vibrations can ultimately be extracted. In fact, this second pulse is seldom observed in pure CO_2 or CO_2-He gas mixes. When nitrogen is added to the gas mix the stored vibrational energy is shared between the CO_2 and the N_2 molecules. During the giant pulse, only half the energy stored in the CO_2 moleculer is extracted. However at the end of the giant pulse, population inversion is rebuilt not only by the decay of the V_1 population but also by reexcitation of the CO_2 moleculer to the V_3 level by resonnant exhange with the excited N_2. The second pulse becomes much more important. Thus for example in a gas mix with equal amounts of CO_2 and N_2, the giant pulse energy can be as little as 1/4 of the total pulse energy and the importance of the second pulse can become even greater for gas mixes with large amount of N_2, this second pulse being as long as 5 microseconds. The most efficient TEA lasers use substantial amount of N_2 such that their output is usually in a rather long pulse without appreciably increasing the peak power over other systems with low N_2 concentrations which are less efficient. The use of N_2 is still advantageous however from the peak power point of view as it is more readily excited than CO_2 by the electrical discharge and does increase somewhat the net CO_2 excitation.

How much energy can ultimately be expected per unit volume of discharge? In a litre of pure CO_2 at atmospheric pressure one could extract 500 joules if all the molecules were excited to the V_3 level. In fact this is a very optimistic figure and it is difficult to see how more than 50% of the molecules can be in the upper level. Approximately 20% of the available molecules in the V_3 energy level is a more realistic figure for CO_2 - N_2 gas mixes and 10% for pure CO_2. Thus the best figures would be about 25 Joules/litre for the giant pulse and from 20 to 75 Joules/litre for the second pulse depending on the amount of N_2 in the gas mix. Furthermore, all TEA lasers at atmospheric pressure have used substantial amounts of He primarily for assisting in producing uniform and well diffused discharges. From 70% to 90% helium concentration are usual which further cuts the realistic estimates to 5 Joules/litre for the giant pulse and from 5 to 20 Joules/litre for the second pulse. Higher energy values can be achieved if the electrical discharge is longer than one microsecond because repumping of the active molecules after the initation of the laser pulse is present but this does not bring any increase in peak power, only pulse length.

Thus for the design of high peak power lasers, one should consider that 5 Joules/litre or 50 MWatts/litre are practical figures for a well optimized system. Higher peak power figures are possible if shorter pulses are amplified. Mode locking of of TEA lasers have produced pulses shorter than one nanosecond and similar short pulses could be achieved using pulse shortening techniques in saturated amplifiers. However for pulses shorter than one nanosecond one cannot expect to be able to extract the full 5 Joules/litre because of relaxation time between the different rotational levels of the V_3 vibration is of the order of one nanosecond and only the energy in one rotational line would be availabe for very short pulses, which can represent only a few percent of the overall energy of the V_3 vibration.

HIGH PEAK POWER SYSTEMS

For the design of very high power lasers with a uniform intensity distribution, the best approach is probably the same as that evolved to solid state lasers, i.e. to use a relatively small and well controlled oscillator laser followed by many stages of amplifiers using increasing cross section. Some relatively crude measurements have indicated that most material surfaces, either window, lenses or mirrors will tolerate up to approximately 5 Joules/cm^2 for a giant pulse at 10.6 u. Thus the cross section of the last stage of amplification will be determined by the ratio of the desired output power and this maximum energy density that can be handled. This implies that the optimum amplifier length should be of the order of 10 metres. A maximum efficiency amplifier should have a gain of approximately 10 if the preliminary figures of .5 Joules/cm^2 from extracting all the availble energy in an amplifier is valid. Such amplifiers would have small signal gains of 50 to 60 db which means that they should not be superradiant but great care will be needed to prevent self oscillation. Individual stages would have to be decoupled to prevent superradiance and the use of bleachable absorbing gasses seems a promising technique.

The design of such amplifiers is not yet proven however and considerable work is still required before reliable design parameters are available. The main unknown factor is the magnitude of the scattering losses. When a laser pulse propagates through an amplifier medium of infinite length, a limit in the peak power will be obtained when the amount of energy lost by scattering by unit length is equal to the energy extracted from the excited molecules in the same length. In solid state lasers the scattering losses are primarily due to imperfections in the crystal structure and impurities which are always present to some extent. In gas lasers, such scattering factors should be extremely small. However the uniformity of the excitation discharge becomes extremely important. A non uniform discharge will produce a non uniform gain but

even more important, it will lead to temperature gradients
and shock waves that give rise to a variation in the refractive index of the amplifier which will cause appreciable scattering. The multipe pin electrode system is very likely to be
unacceptable from the uniformity point of view. One such amplifier 5 metre long has been studied by Robinson at DREV.
The output energy was observed to be linear with the length
of the excited region up to levels of 2 Joules/cm^2. Distortion
of the output beam was observed but the exact amount is very
difficult to assess as the oscillator laser was affected by
the reflection of the measuring equipment which was amplified.
Recent measurements at GEN-TEC indicate that such a structure
corresponds to about 5% scattering losses per metre. Since
only 150 millijoules/cm^2 could be extracted over a one meter
length, this corresponds to a maximum energy density output
of 3.0 joule/cm^2 for an infinitely long amplifier. These numbers are of coure quite dependent on the electrical discharge
characteristics and vary from one system to the next. The
much greater uniformity of the double-discharge laser excitation leads to believe that an order of magnitude less scattering losses can be expected from these structures and output energy densities of 5 Joules/cm^2 should be efficiently
achieved. A 15 metre long laser test facility is presently
under construction at DREV and much more reliable figures
on the scattering losses should be available for different
types of structures within a year. To consider a typical
example, a laser system capable of generating 1000 joules in
a Q-switched pulse, would have a final amplifier stage of
200 cm^2 cross-section and a length of 10 metres. It would require a 100 joule input which could be produced by a laser
5 X 5 cm cross-section and 10 metre long which could be used
to produce a fundamental mode output.

    The length of the output laser pulse is difficult to as-
sess because some pulse shortening will occur, specially if
bleachable absorbing cells are used to prevent self-oscilla-
tion and superradiance. In the absence of non-linear scattering losses the pulse could shorten to a time period of the
order of magnitude of the rotational relaxation time, i.e.
approximately 1 nanosecond. This may not be achievable at the
highest energy densities because of field-induced variations
in the refractive index of the gas could lead to focussing and
eventually breakdown of the gas. The limit of peak power density remains to be established but based on breakdown characteristics at the focal point of a lense, it is probably within
an order of magnitude of one gigawatt/cm^2.

LASER OPERATION ABOVE ATMOSPHERIC PRESSURE

    The most common consideration of scientists when they
realize the great advantages obtained by increasing the op-

perating pressure of gas lasers is to explore the possibility
of working well above atmospheric pressure. The first advantage of above atmospheric operation is an increase in possible
energy per unit volume. Because all relaxation rates are faster,
the second pulse would be shorter and much more powerfull but
the peak power would still appear during the first giant pulse
which would have the same duration because gain does not vary
with pressure unless one considers operation well above 10
atmospheres where the collision broadening would begin to
cause an overlap of the amplification bands of the different
lines of the P branch. On the other hand, the cross section
of high power amplifiers would still be as large or even larger
because the optical components cannot stand any more power
density but the scattering losses would increase due to the
higher refractive index of the gas and the breakdown treshold
decreases with increasing pressure. The laser length could
be reduced but the energy discharge system could not and its
size is roughly equivalent to that of the laser at atmospheric
pressure. Considering that electrical discharges become more
unstable with pressure and that the required applied field
is higher certainly increases the electrical engineering task.
The increased complexity of constructing high pressure containers and finding high pressure windows really requires more
important advantages than a simple reduction in length to
deserve a serious amount of development efforts. The only important advantage may be for the generation of sub-nanosecond
pulses because the bandwidth of the gain characteristics
would increase with pressure.

CONCLUSIONS

The development of TEA lasers has established that gas
lasers can compete advantageously with solid state lasers
for the production of high peak power pulses. While further
developments are still needed to produce reliable uniform
excitation systems, it is only a question of time before 10.6
micrometer, 1000 joule lasers with 1 nsec output are
available. These lasers will certainly be larger than solid
state lasers of equivalent output but their power supplies
may actually be smaller because of the higher efficiency of
gas lasers. Another factor of importance is that these lasers
can operate easily at 1 p.p.s. with a slow gas flow and above
1000 p.p.s. with rapid gas flow. The cost of going to much
higher energies will certainly be considerably less than it
would be for solid state lasers and the development of 100,000
joule lasers with an amplifier output stage 10 metre long
and a 2 X 2 meters cross section would certainly represent
and appreciable amount of engineering efforts but which may be
well worthwhile in the development of controlled thermonuclear
fusion.

SURVEY OF ELECTRIC DISCHARGE CO$_2$ LASERS*

A. J. DeMaria

United Aircraft Research Laboratories

East Hartford, Connecticut 06108

ABSTRACT

Based on research performed with sealed-off discharge tubes and with slow flowing gas lasers, most researchers believed, up until several years ago, that an upper limit of approximately 100 watts per meter of discharge length was obtainable from CO$_2$ lasers. In 1967, a power output of 800 watts in a 30 cm long CO$_2$:N$_2$:He discharge was obtained at United Aircraft Research Laboratories by convectively cooling the CO$_2$ laser medium by rapidly flowing the gas through the discharge region. In the fall of 1968 and in the spring of 1969, the power was pushed up to 9 kW and 11.5 kW per meter respectively. As a result of this work, the merits of the convectively cooling process in electrical discharge CO$_2$ lasers has now become well known[1-10].

ENERGY LEVELS OF CO$_2$

The CO$_2$ is a linear symmetric molecule which has an axis of symmetry "C_∞" and a plane of symmetry perpendicular to the C_∞ axis (see Fig. 1). There are three normal modes of vibration ν_1, ν_2, and ν_3, each of which is associated with one of the species, Σ_g,

---

*Presented at the Second Workshop on "Laser Interaction and Related Plasma Phenomena" at Rensselaer Polytechnic Institute, Hartford Graduate Center, August 30-September 3, 1971.

π_u, and Σ_u respectively. The designations of the resonances for the CO_2 molecule are chosen in the usual way as for the electronic states of homonuclear diatomic molecules. The species π_u corresponds to $\ell = 2$ and represents a double degenerate vibration, usually indicated by ν_{2a} and ν_{2b}, which occurs with equal frequency both in the plane and perpendicular to the plane of the paper as illustrated by Fig. 1.

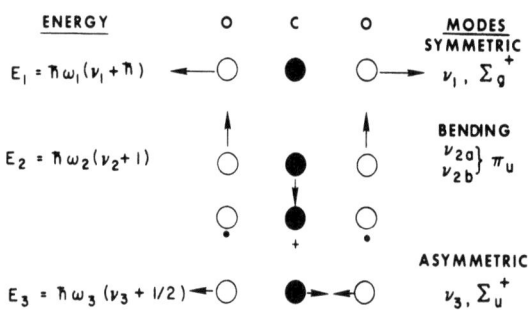

Figure 1. Normal Vibrational Modes of the CO_2 Molecule

The energy levels (E_i) and the vibrating frequencies (ω_i) are given on Fig. 1 for the CO_2 molecule where ν_1, ν_2, and $\nu_3 = 0, 1, 2...$, K_1 and K_δ = valence force constants, M_O and M_C = mass of the oxygen and carbon atoms, and ℓ = the inter-nuclear distance between the carbon and the oxygen atom.

Figure 2 illustrates a simplified energy level diagram of the CO_2 molecule. For the sake of simplicity, the fine structure due to the rotational levels is not shown. Also illustrated in Fig. 2 is the position of the $v = 1$ level of the N_2 molecule. Notice that there is only an energy difference of $\Delta E = 18$ cm^{-1} between the $v = 1$ level of N_2 and the $00°1$ level of CO_2. The lifetime of the $v = 1$ level of N_2 is extremely long. As a result, if one runs an electrical discharge in N_2, the $v = 1$ level will be heavily populated. If one adds CO_2 molecules to the discharge, collisions between CO_2 ground state molecules and N_2 (v = 1) excited molecules will heavily overpopulate the $00°1$ level of CO_2 by the following process:

$$N_2^* \ (v = 1) + CO_2 \ (00°1) \rightarrow \\ N_2 \ (v = 0) + CO_2 \ (00°1) - 18 \ cm^{-1} \quad (1)$$

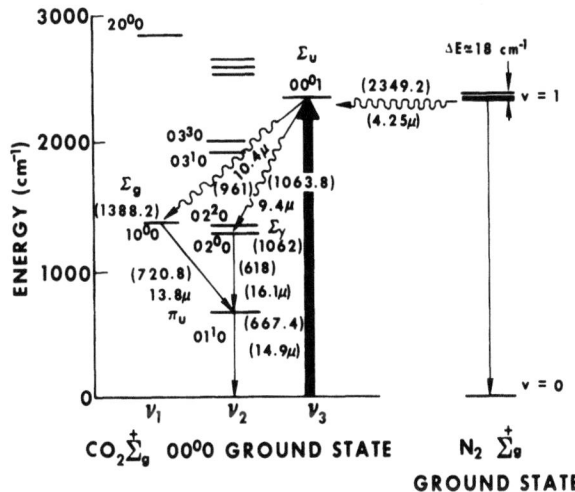

Figure 2. Simplified Energy Level Diagram of the CO_2 Molecule

Experimental data gathered within our Laboratories as well as elsewhere indicate that the lifetime of the $00°1$ to $00°0$ CO_2 transition is approximately 3.5 microsec/atms pressure at $300°$ K and approximately 1 microsec/atms pressure at $500°$ K. Since the lifetime of both the $10°0$ to $01'0$ and the $02°0$ to $01'0$ transitions are at least an order of magnitude faster, an overpopulation between these levels and the $00°1$ level occurs. Laser action can, therefore, take place at either 10.6 micron (i.e., the $00°1$ to $10°0$ transition) or at 9.4 micron (i.e., the $00°1$ to $02°0$ transition). Unfortunately, the lifetime of the $01'0$ to $00°0$ transition has a lifetime of approximately 6.5 microsec/atms pressure at $500°$ K. At high laser power outputs, the relatively slow $01'0$ to $00°0$ CO_2 transition presents a substantial bottleneck, and its population increases markedly. An increase population in the $01'0$ level produces an increase in the population of both the $10^\circ 0$ and the $02^\circ 0$ lower laser levels. This in turn reduces the population difference between these levels and the $00^\circ 1$ upper laser level. A reduction in the over population difference of the two laser level results in a decrease in the laser output power.

EFFECT OF CONVECTIVE COOLING ON CO_2 LASERS

In CO_2 lasers employing sealed-off tubes or slow flow velocities (i.e., the characteristic flow transit times are much less than the characteristic times for the diffusion of the component species to the walls), one has to rely on the cooling effects from the walls of the tube for reducing the population of the $01'0$ CO_2 level. At high electrical drive power, the walls are not effective in cooling the gas in the large diameter tubes required for the high power. On the other hand, if the gas flow transit is much faster than the characteristic times for the diffusion to the walls of the laser channel, the excess heat absorbed by the gas from the discharge is swept away by the flowing gas. In this manner the laser medium is cooled convectively rather than by diffusion to the walls.

In a diffusion cooled laser, waste energy is rejected in a characteristic time approximately that of the diffusion time (τ_D). If an electrical discharge column of diameter D is assumed, the number of mean free paths during which the energy diffuses is given by D/λ where λ is the mean free path of the CO_2 molecules in the gas mixture. The mean free time between collision is given by λ/v_t where v_t is the thermal molecular speed. Since diffusion is a random walk process, τ_D is equal to:

$$\tau_D = D^2/\lambda\, v_t \quad . \tag{2}$$

If the gas is moved at a speed v in a flowing gas system, waste energy is rejected in a characteristic time (τ_F) equal to:

$$\tau_F = D/v \quad . \tag{3}$$

Figure 3 illustrated the fractional population variation of the various levels of the CO_2 molecule as a function of temperature. The assumptions are that all levels except the $00^\circ 1$ are in thermal equilibrium with the translational temperature of the gas and that

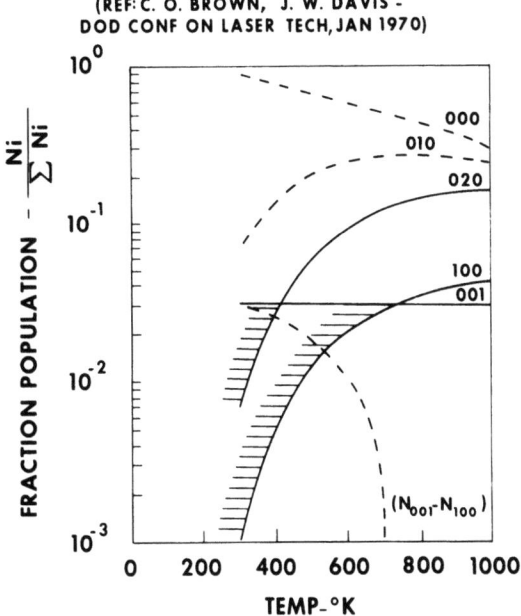

Figure 3. Fractional Population Variation of the Various Levels of the CO_2 Molecule as a Function of Temperature

3 percent of all the CO_2 molecules are maintained in the $00°1$ level by collisional excitation with excited N_2 (v = 1) molecules. The figure shows how the relative population between the lower laser levels (i.e., both the lower $10°0$ and the $02°0$ levels responsible for the 10.6 micron and the 0.96 micron transitions, respectively) and the upper laser level, decreases as a function of increasing gas temperature. For the assumption chosen, the overpopulation between the 10.6 micron transitions (i.e., the $00'1$ and the $10°0$ levels) goes to zero (i.e., laser action becomes impossible) when the temperature of the gas rises up to $680°K$. The overpopulation between the 9.6 micron transitions (i.e., the $00°2$ and the $02°0$ levels) goes to zero when the temperature of the gas rises up to $400°K$. The higher population differences maintained by the 10.6 micron transition throughout the temperature range clearly show why the 10.6 micron transition is dominant over the 9.6 micron transition.

Assume we have a volume of gas of cross-sectional area A and thickness x. We dissipate P_E watts of electrical power into the volume. We extract P_L watts of laser power from the volume and P_H

watts of heat by convectively flowing the gas through the volume. If the laser extraction efficiency is η, then

$$P_L = \eta \, P_E \quad , \tag{4}$$

and

$$P_H = \left(\frac{1-\eta}{\eta}\right) P_L = (1-\eta) \, P_E \quad . \tag{5}$$

Since

$$P_H = \rho \, c_p \, v \, \frac{\Delta T}{\Delta X} \, V \quad , \tag{6}$$

where ρ is the gas density, c_p is the specific heat of the gas, v is the velocity of the flow, and V is the volume. Substituting Eq. (6) into Eq. (5) we obtain:

$$P_L = \rho \, c_p \, v \, \Delta T \, A \left(\frac{\eta}{1-\eta}\right) \quad . \tag{7}$$

For a temperature rise of $300°K$ and an extraction efficiency of 15 percent, we obtain:

$$P_L \simeq 2.5 \times 10^{-2} \text{ torr (cfm)} \quad , \tag{8}$$

or for a pressure of 100 torr, a few watts per cfm of gas flow may be obtained from a convectively cooled, electrically driven CO_2 laser.

In summary, it is obvious that for high power, compact, low weight, CO_2 lasers, convective cooling of the gas medium is desirable. Heating of the gas causes degradation of the population inversion, primarily because of filling up of the lower states and eventual blockading of the lower laser level. In the case of cooling by diffusion, the maximum power output per unit volume is proportional to:

$$P_L \propto \rho \, \lambda \, v_t / D^2 \quad , \tag{9}$$

where λ is the mean free path, v_t is the thermal velocity, and D is the diameter of the discharge region. Since the mean free path is inversely proportional to the gas density, the positive effect on the output power obtained by increasing ρ is concealed by the negative effect of a decreasing λ when the pressure is increased. The available power output per unit volume decreases with increasing tube diameter at the same rate at which the volume itself is increasing. This leaves the power per unit length unchanged. The only alternative means for increasing the power output in diffusion-cooled lasers is to increase the tube length. Such tube scaling leads to 200 meter long CO_2 lasers having output powers in the few kilowatt range.

The essential desirable features obtained with a high volume flowing gas laser are: 1) the convective cooling reduces the thermal population of the lower laser level, 2) the laser pumping mechanism can be increased because of the more effective cooling and 3) the flow of new material into the laser active region increases the potential laser power that can be obtained from the device.

The qualitative effects of convection can be examined in more detail by using the rate equations. For the two level system illustrated in Fig. 4, the population of the upper and lower levels

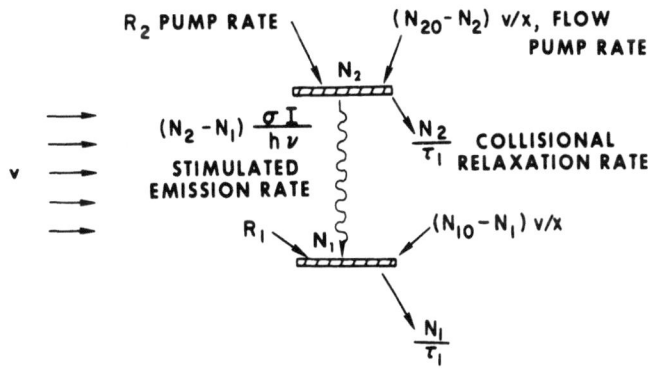

$$\frac{\partial N_2}{\partial t} = R_2 - \frac{N_2}{\tau_2} - (N_2 - N_1)\frac{\sigma I}{h\nu} + \frac{(N_{20} - N_2)}{\tau_F}$$

$$\frac{\partial N_1}{\partial t} = R_1 - \frac{N_1}{\tau_1} + (N_2 - N_1)\frac{\sigma I}{h\nu} + \frac{(N_{10} - N_1)}{\tau_F}$$

WHERE $\tau_F = \frac{x}{v}$ = GAS FLOW TRANSIENT TIME

Figure 4. Two Level System and Rate Equations

are given by:

$$\frac{\partial N_2}{\partial t} = R_2 - \frac{N_2}{\tau_2} - (N_2 - N_1)\frac{\sigma I}{h\nu} + \frac{(N_{20} - N_2)}{\tau_F} \qquad (10)$$

and

$$\frac{\partial N_1}{\partial t} = R_1 - \frac{N_1}{\tau_1} + (N_2 - N_1)\frac{\sigma I}{h\nu} \quad \frac{(N_{10} - N_1)}{\tau_F} \qquad (11)$$

where N_2 is the upper level population, N_{20} is the upper level population in the absence of stimulated emission, R_2 is the upper level volume pumping rate, τ_2 is the upper level collisional relaxation rate, $\tau_F = X/v$, v is the velocity of the gas in the X dimension, N_1 is the lower level population, N_{10} is the lower level population in the absence of stimulated emission, R_1 is the lower level volume pumping rate, τ_1 is the lower level collisional relaxation time, I is the laser beam intensity, σ is the cross section for stimulated emission, and $h\nu$ is the photon energy. Under equilibrium conditions, the net rate of change of the upper and lower level population is equal to zero, and the gain coefficient (α) of the laser can be obtained from the equilibrium solution of Eqs. (10) and (11).

$$\alpha = \sigma(N_2 - N_1) = \frac{\sigma(R_2 \tau_2 - R_1 \tau_1)}{1 + I\frac{\sigma}{h\nu}\left[\frac{\tau_2 \tau_F}{\tau_2 + \tau_F} + \frac{\tau_1 \tau_F}{\tau_1 + \tau_F}\right]} \qquad (12)$$

The cw laser is usually characterized by two parameters; the small signal gain coefficient (α_0) at the peak of the spectral line, and the saturated intensity I_S. The small signal gain coefficient (α_0) is defined as the gain coefficient with no stimulated emission. From Eq. (12) we obtain:

$$\alpha_0 = \sigma(R_2 \tau_2 - R_1 \tau_1) \qquad (13)$$

The saturated intensity (I_S) is defined as the intensity required to reduce the laser gain by a factor of 2. Using this definition and Eq. (12), the saturated density is:

$$I_S = \frac{h\nu}{\sigma}\frac{1}{\left[\frac{\tau_2 \tau_F}{\tau_2 + \tau_F} + \frac{\tau_1 \tau_F}{\tau_1 + \tau_F}\right]} \qquad (14)$$

Substituting Eqs. (13) and (14) into Eq. (12) yields the well-known relationship:

$$\alpha = \frac{\alpha_0}{1 + I/I_S} \quad . \tag{15}$$

For large flow velocities, $\tau_F \ll \tau_2, \tau_1$:

$$\alpha \simeq \frac{\alpha_0}{1 + \frac{2\sigma I}{h\nu} \tau_F} \quad \alpha \ \nu \quad , \tag{16}$$

and

$$I_S = \frac{h\nu}{2\sigma \tau_F} \ \alpha \ \nu \quad . \tag{17}$$

Equations (16) and (17) indicate that both the gain coefficient and the saturated intensity increase proportionately with flow velocity. For small flow velocities, $\tau_F \gg \tau_2, \tau_1$,

$$\alpha \simeq \frac{\alpha_0}{1 + \frac{\sigma I}{h\nu} (\tau_2 + \tau_1)} \quad , \tag{18}$$

and

$$I_S = \frac{h\nu}{\sigma (\tau_2 + \tau_1)} \quad . \tag{19}$$

Since $\tau_2 \gg \tau_1$, Eqs. (18) and (19) can be simplified slightly. Figure 5 illustrates the variation of the gain and saturation intensity as a function of the inverse of the transient time (i.e., the time required for the flowing gas to move across the discharge region). The curve clearly shows the two regions; for slow velocity, the gain coefficient and saturation intensity are constant. For fast flow velocities, the gain coefficient and saturation intensity

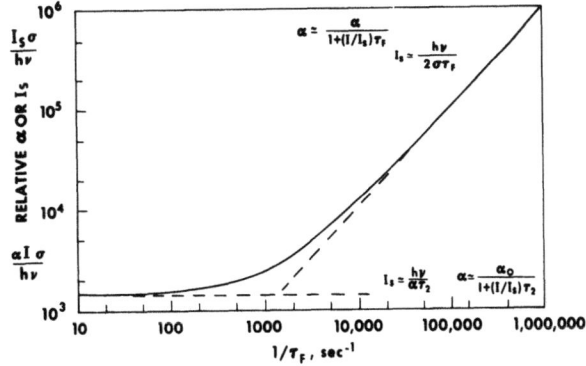

Figure 5. Variation of Gain and Saturation Intensity as a Function of the Flow Velocity

increase linearly with velocities. Since the relaxation rates τ_1 and τ_2 vary with pressure and gas composition, a family of such curves can be plotted for various values of τ_1 and τ_2.

The effects of convective cooling can also be shown by considering the balance under steady state conditions between ohmic heating, convective and conductive cooling:

$$JE = \rho\, c_p\, v\, \frac{\partial T}{\partial x} - \frac{K}{r}\frac{\partial}{\partial r}\left(v\frac{\partial T}{\partial r}\right) \qquad (20)$$

where r is the discharge radius, K is the thermal conductivity, J is the current density, and E is the electric field strength. This equation determines the background gas temperature if we assume that the heat contained in vibrationally excited states of the gas is negligible. In addition, we have neglected the additional cooling effect which results by diffusion of excited states to the laser tube wall combined with wall recombinations. Such diffusion losses are important for small diameter tube lasers. The order of magnitude of the flow velocity needed for convection heat transfer to be comparable or greater than the thermal conduction heat transfer is:

$$v \geq \frac{Kx}{\rho\, c_p\, a^2} \quad , \qquad (21)$$

where a is the coordinate in the direction of thermal conduction heat transfer.

TYPES OF FLOWING GAS LASERS

In describing the various types of flowing gas lasers, we shall accept rectangular coordinates as a means to specify 1) the direction of the optical axis (O), 2) the discharge current (I), 3) the gas flow velocity vector (V), and 4) the magnetic field vector (B) ---if a magnetic field is utilized. In discussion of the various types of lasers, the optical axis is maintained as the reference axis and the current, gas flow velocity, and magnetic field vectors will be described with respect to the optic axis of the laser.

Figure 6 is a schematic representation of a laser in which the optical axis, current flow, and gas flow are in the same direction or coaxial. This laser will be called the coaxial laser or the $(OIV)_x$ laser. Figure 7 is a schematic representation of a laser in

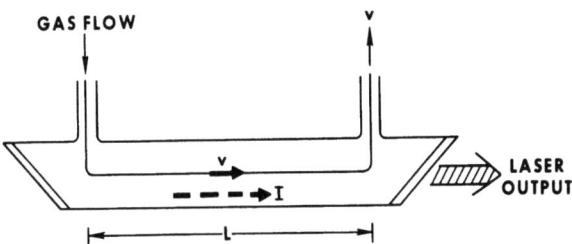

Figure 6. Schematic Representation of a Co-Axial Laser $(OIV)_x$

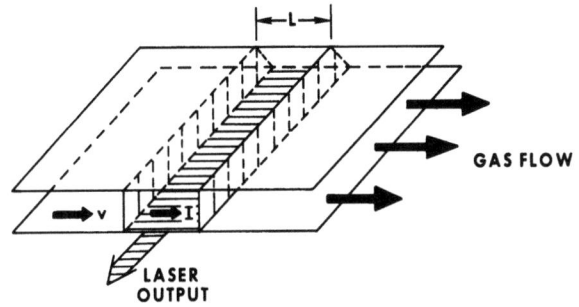

Figure 7. Schematic Representation of a Transverse Flow Discharge Laser (O_x, IV_y)

which the gas flow and discharge currents are perpendicular to the optic axis or (O_x, IV_y). This laser is called the transverse flow discharge laser. A slightly different laser is shown by Fig. 8. This laser is called a transverse flow, cross-excited laser or $(O_x, V_y - I_z)$.

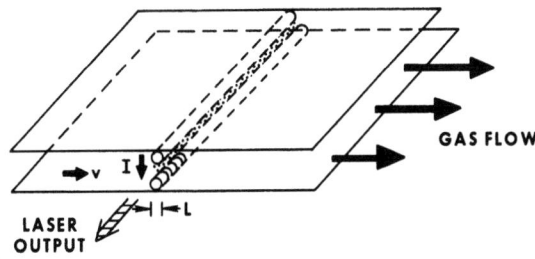

Figure 8. Schematic Representation of a Transverse Flow Cross Excited Laser $(O_x V_y I_z)$

Another variation of the transverse flow laser is represented by Fig. 9. This laser is called the transverse flow, magnetic stabilized laser or (OI_x, V_y, B_z).

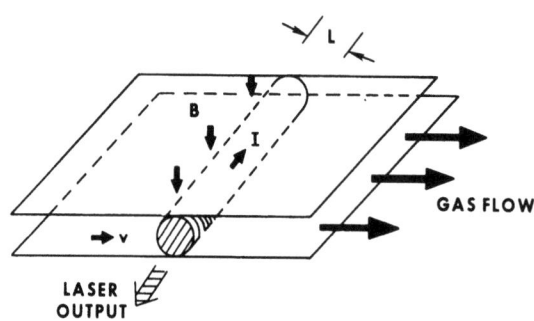

Figure 9. Schematic Representation of a Transverse Flow, Magnetic Stabilized Laser (OI_x, V_y, B_z)

Figure 10 illustrates a schematic representation of a flow mixing laser or $(O_x V_y)$. Notice a current specification is not given for the mixing laser. For the case of a CO_2 mixing laser, the nitrogen is excited by some means upstream of the laser's optical axis. It can be excited by a discharge, or by a heater and then subsequently expanded, etc. At some point, cold CO_2 is injected into the excited nitrogen stream and mixed with the nitrogen. The upper laser level of CO_2 is populated by collision with the nitrogen, and subsequent laser action results. Researchers at UARL have obtained up to 11.2 kW of power from a mixing laser device.

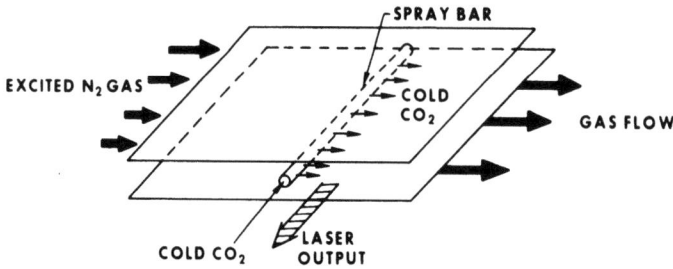

Figure 10. Schematic Representation of a Flowing Mixing Laser

The coaxial laser is the simplest of the five different types of flow lasers. The convective cooling is inversely proportional to the time spent by the CO_2 molecules in the discharge region. Since the length of discharge (L) through which the CO_2 molecules must pass is relatively long in the coaxial laser, the gas flow velocity must be relatively large in order to minimize the transit time. High flow velocity means high pressure drops which in turn requires more power driving the gas pumping systems. Continuous powers in the range of 8 kW have been obtained with such lasers at UARL. Due to the high flow velocities required for good convective cooling, the coaxial flow laser is not well suited for most airborne applications.

The distance (L) in the transverse flow discharge laser is relatively short so that relatively slow flow velocities are required for good convective heating. Laseroscillators prefer to operate with gaussian beam cross section. Since the gain region of the transverse flow discharge laser is rectangular in cross section, a good geometrical match is not obtained with conventional optical cavities. As a result, this laser appears to be ideally suited as a power amplifier for airborne applications requiring more than a few thousand watts of power. One can take more than one pass through the rectangular gain region with a gaussian cylindrical laser beam. Another advantage of this laser is the relatively low voltage required across the electrodes arising from the nearness of the electrodes. This type of laser has also been operated up in the kilowatt range at UARL and up to 18.1 kW as an amplifier by A. Hill of the Air Force.

The transverse flow, cross-excited laser also has the convective cooling and low voltage advantages of the transverse flow, discharge laser. As a result of the flowing or bowing of the discharge column downstream by the gas flow, the laser does not operate very well as a continuous oscillator. It does operate well as a high peak power. pulsed laser. This type of laser has operated up to a pressure of an atmosphere with a peak power in excess of one megawatt at UARL. The closed spacing of the electrodes enable one to obtain high electric fields to pressure ratios across the discharge.

The transverse flow, magnetic stabilized laser also has a relatively short transit time through the plasma discharge so that relatively slow flow velocities are required for good convective cooling. The discharge geometry is tubular in shape and can be made of an optical cavity. This type of laser is thus well suited as a relatively high power master oscillator or a low noise amplifier.

The transverse flow, mixing laser has the best convective cooling of all the five types of flowing gas lasers. Unfortunately, the mixing laser is not well suited for closed-cycle operation due to the fact that its operation requires separate input ports for the N_2, He and CO_2 gas.

CLOSED-CYCLE LASERS

Convectively cooled, electrically driven, CO_2 lasers basically consist of the following components: 1) a laser discharge channel, 2) heat exchangers, 3) gas recirculating equipment, 4) ducting connecting the mechanical components, 5) optical feedback cavity, and 6) an electrical dc power supply. A block diagram of a closed-cycle convectively cooled, electrical discharge gas laser is given by Fig. 11. Two heat exchangers are illustrated in the figure: one for cooling the gas before entering the circulating gas pump but after it is passed through the laser discharge region, and the other heat exchanger for removing the heat added to the gas by the circulating gas pump. For well-designed systems, the second heat exchanger can be eliminated. Due to the dissociation by the electrical discharge of the molecular species of the continuously recirculating gas and subsequent chemical reaction of the chemical activated species, the gas composition of the closed-cycle laser may change with time. As a result, one may wish to inject a very small amount of make-up gas into the system as shown. This will

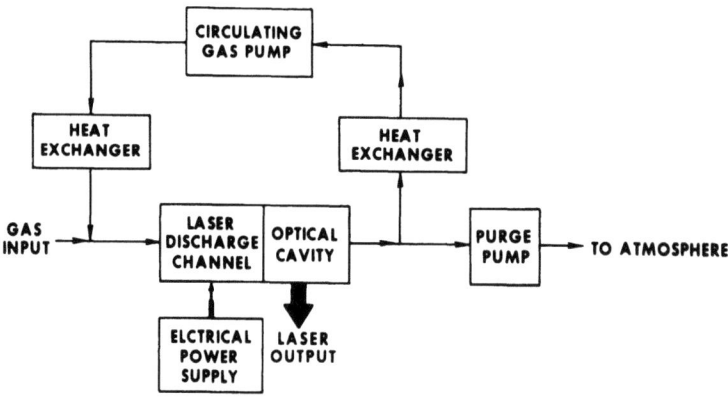

Figure 11. Block Diagram of a Closed-Cycle Flowing Gas Laser

necessitate the removal of an equal amount of gas from the system by a small purge pump. With proper design and the use of proper chemical catalyst, the requirement of the make-up gas, and the purge pump can be eliminated such as has already been accomplished with sealed-off, low power gas tube-type lasers.

REFERENCES

1. I. R. Hurle and A. Hertzberg, Phys. Fluids $\underline{8}$, 1601 (1965).
2. T. A. Cool and J. A. Shirley, Appl. Phys. Letters $\underline{14}$, 70 (1969).
3. T. F. Deutsch, F. A. Harrigan and R. I. Mudko, Appl. Phys. Letters $\underline{15}$, 88 (1969).
4. A. P. Walch, J. B. Burnham and J. W. Davis, "A High Power Electric-Discharge Mixing Laser," Proc. Fourth DoD Conf. on Laser Technology.
5. C. O. Brown and J. W. Davis, "The Electric-Discharge Laser," Proc. Fourth DoD Conf. on Laser Technology.
6. A. C. Eckbreth, J. W. Davis and E. A. Pinsley, "Investigation of a CO_2 Laser Pulse Amplifier," Proc. Fourth DoD Conf. on Laser Technology.
7. R. Targ and W. B. Tiffany, "Compact CO_2 Oscillator-Amplifier," Proc. Fourth DoD Conf. on Laser Technology.
8. W. B. Tiffany, R. Targ, and J. D. Foster, Appl. Phys. Letters $\underline{15}$, 91 (1969).
9. R. Targ and W. B. Tiffany, Appl. Phys. Letters $\underline{15}$, 302 (1969).
10. D. C. Smith and J. W. McCoy, Appl. Phys. Letters $\underline{15}$, 289 (1969).

THERMOKINETIC THEORIES OF GASDYNAMIC LASERS AND APPLICATION TO

CONTROLLED THERMONUCLEAR FUSION*

George C. Vlases

University of Washington and Mathematical Sciences
Northwest, Inc.

Alan L. Hoffman

Mathematical Sciences Northwest, Inc.

ABSTRACT

A simple model is used to illustrate the kinetics of the inversion and energy extraction (lasing) processes in gasdynamic lasers. Particular emphasis is placed upon the range of pulse lengths that can be obtained in pulsed gasdynamic lasers by suitable variation of mixture and pressure. The model yields analytical results for the case of small CO_2 concentration which can be readily interpreted physically. It is shown that the pulsed CO_2 laser is very well suited to the problem of heating magnetically confined plasmas to thermonuclear temperatures.

I. INTRODUCTION

In this paper we discuss the operation of the $CO_2 - N_2$ gasdynamic laser from the point of view of the operating conditions set by the molecular kinetics. Particular emphasis is placed upon pulsed gasdynamic lasers, although continuous or quasi-cw operation is included as a special case. It is shown that a simple kinetics model can be used to analytically predict scaling for both the inversion and the energy extraction processes. Since the scaling for inversion has been treated by other authors [1-3] it is reviewed quickly and the bulk of the discussion centers on the kinetics of energy extraction. It is seen that pulsed g.d.l.'s can provide, in principal, energetic pulses with

---

*Presented at the Second Workshop on "Laser Interaction and Related Plasma Phenomena" at Rensselaer Polytechnic Institute, Hartford Graduate Center, August 30-September 3, 1971.

durations from about 10 milliseconds down to the submicrosecond range.

Fig. 1 is a simplified schematic diagram showing the elements of a gasdynamic laser. At the left, upstream of a converging diverging supersonic nozzle, is a region of nearly stagnated CO_2 -N_2 catalyst gas mixture at a temperature of from 1300° - 2000° K, under which conditions the upper laser level of CO_2 has its maximum population. While higher energy storage can be achieved by raising the temperature further, other considerations make it ineffective to operate above this range. This stagnation region may be produced either by a reflected shock wave[2], as shown in the figure, or by combustion occuring in an upstream burner.[1] The purpose of heating the mixture is solely to deposit energy in the nitrogen and CO_2 asymmetric vibrational modes. The lower level population of CO_2 is of course higher than that of the upper level in the stagnation region, since thermodynamic equilibrium exists there. It has been pointed out by several authors over the last decade that an inversion can be obtained by rapidly cooling the heated gas by expanding it in a supersonic nozzle.[4,5,6] In passing through the nozzle the translational temperature changes from its stagnation value to about 300°K. The lower level population depletes rapidly to a new equilibrium value at the lower temperature, but the upper level and the nitrogen relax much more slowly, thereby creating an inversion. The inversion thus arises by preferential de-excitation of the lower level, rather than by a combination of this effect plus preferential excitation of the upper level as in electrical CO_2 - N_2 lasers.

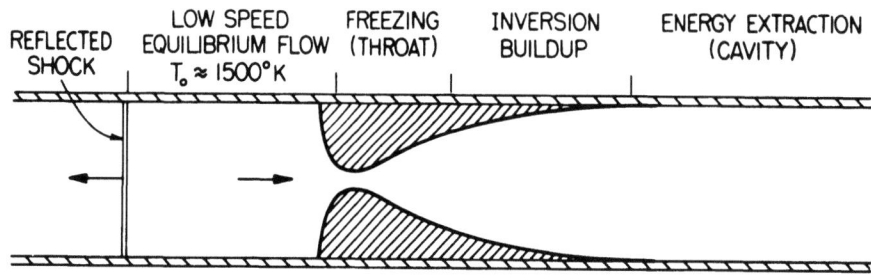

Fig. 1 Elements of a Shock Tube Gasdynamic Laser

When conditions are chosen properly for the N_2 vibrational energy to "freeze" near the nozzle throat, there results a relatively long distance downstream of the nozzle in which the inversion persists because the temperature and pressure, and therefore the collision rates, decrease by orders of magnitude downstream of the throat.

This region of inverted flow is called the energy extraction or cavity region. There are three principal ways in which energy can be extracted as coherent radiation. First, the inverted mixture can be passed continuously through a transverse beam of radiation in either an amplifier or an oscillator configuration. The streamwise length of the radiation field or cavity must be sufficient to remove all the vibrational energy. The duration of the pulse then simply equals the duration of the flow through the nozzle. This may be continuous, as in the case of a combustion driven GDL, or may be on the order of one to ten milliseconds for shock tube lasers.

Secondly, in a mode called "natural time" operation, the inverted gas mixture may be used to fill a chamber whose length is the flow speed times the vibrational energy characteristic decay time, typically a meter. This volume of inverted gas is then made to lase either as an oscillator by Q switching the cavity, or as an amplifier by illuminating it with an intense beam from outside. The time for energy extraction in this case is determined solely by the vibrational relaxation times and can be varied from about 10μsec to 100μsec at a pressure of 1 atm by suitable choice of CO_2 - N_2 ratio and catalyst; it is inversely proportional to pressure for a given mixture.

Finally, very short pulses can in principle be extracted by a method due to Christiansen[2] termed "pulse compression." In this method a short pulse is passed through the box and extracts the energy stored in the CO_2 only. This pulse is optically delayed external to the box for a time long enough for the CO_2 to be replenished by transfer from the nitrogen, and then passes through the box again. The process is repeated until all the nitrogen energy has been depleted; the number of passes is approximately equal to the ratio of N_2 to CO_2 concentrations. The minimum pulse length is determined only by the gain bandwidth.

II. KINETICS MODEL

Derivation of the kinetics model used in the following sections is discussed in detail in a paper by Hoffman and Vlases[7] and will only be summarized here. Fig. 2 shows a simplified energy-level diagram of the CO_2 - N_2 system. The analysis used here is restricted to the use of Helium as a catalyst. We shall be concerned with only three collisional transfer rates[8]:

$$N_2(v=1) + CO_2 \rightleftarrows N_2 + CO_2^*(\nu_3) \qquad P_4^{1,0}(N_2^{1,0}; \nu_3^{0,1}) \qquad (1)$$

$$CO_2^*(\nu_2) + M \rightleftarrows CO_2 + M + KE \qquad P_1^{1,0}(\nu_2 : M) \qquad (2)$$

$$CO_2^*(\nu_3) + M \rightleftarrows CO_2^*(\nu_2) + M + KE \qquad P_8(\nu_3^1; \nu_2^0 : M) \qquad (3)$$

where M is the collision partner and may be CO_2, N_2, or He. These reactions are shown on Fig. 2 and are designated by characteristic times τ_4, τ_1, and τ_8 which are derived from the probabilities P_4, P_1, P_8 and the collision frequencies as in reference 7, e.g.

$$\tau_1 = (\Psi_{He} K_{1_{He}} + \Psi_{N_2} K_{1_{N_2}} + \Psi_{CO_2} K_{1_{CO_2}})^{-1}$$

where the Ψ_M are species mole fractions, the K_{1_M} are given by $Z_{CO_2,M} P_{1_M} (1-e^{-\theta_2/T})$

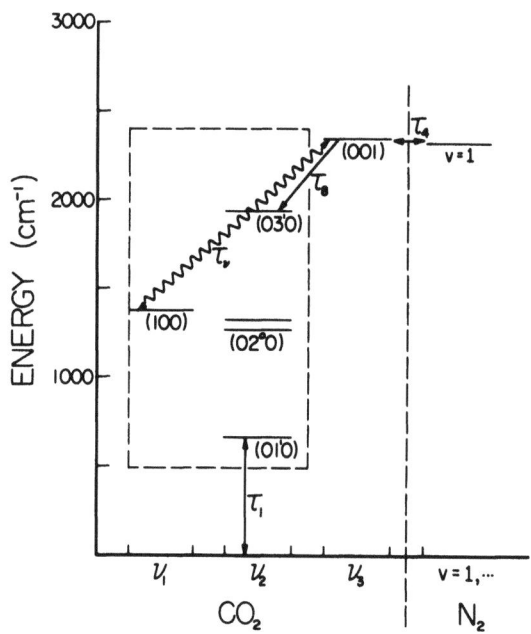

Fig. 2 Simplified CO_2-N_2 Energy Level Diagram

and the $Z_{CO_2,M}$ are the collision frequencies of CO_2 with collision partners M. Helium has a very large influence on τ_1 but does not greatly affect τ_8. If H_2O, which occurs naturally in some combustion driven lasers were included, then its effect on both depopulation of $CO_2^*(\nu_3)$ (τ_8) and direct depopulation of N_2 (v=1) would have to be considered. The effect of CO_2 on τ_8 can in some cases be large[8], but for the small CO_2 concentrations to be discussed here will be negligible. Therefore, in summary, τ_1 can be altered greatly by adjusting the concentration of He, but τ_8 and τ_4 are relatively independent of mixture for $\psi_{CO_2} \ll 1$. This is shown in Table I which was constructed using rate constants from Ref. 8.

It is assumed that v-v intramode collisions are sufficiently fast that the levels within each mode are Boltzmann distributed at a mode vibrational temperature, and further that the coupling between modes 1 and 2 is fast enough that they have a common vibrational temperature. The energy of each of the modes CO_2 (ν_1, ν_2, ν_3) and N_2 (v=1) is related to its vibrational temperature by

$$E_\nu = N\psi_{CO_2,N_2} g_\nu (k\theta_\nu)(e^{\theta_\nu/T}-1)^{-1} \qquad (4)$$

where N is the total number density, g_ν is the mode degeneracy and is equal to 1 for modes ν_1, ν_3 and N_2(v=1) and 2 for ν_2, k is Boltzmann's constant, and the θ_ν are the characteristic mode vibrational temperatures whose values are taken to be 1920°K, 960°K, 3380°K and 3355°K for modes ν_1, ν_2, ν_3 and N_2(v=1) respectively. The lower laser level characteristic temperature θ_L may be slightly different from θ_1, but the ratio θ_L/θ_1 has been set equal to 1 for simplification.

ψ_{He}	250°K	350°K	500°K	250°K	350°K	500°K	250°K	350°K	500°K
0	19	11	5.8	.058	.100	.182	14.4	10.0	4.8
.1	3.3	2.5	1.8	.064	.110	.202	14.6	10.2	4.9
.2	1.8	1.4	1.1	.073	.125	.228	14.8	10.4	5.1
.3	1.2	.97	.76	.083	.143	.260	15.0	10.6	5.3
.4	.95	.75	.59	.097	.167	.303	15.3	10.9	5.5
.5	.76	.60	.48	.116	.200	.364	15.5	11.1	5.7
	$p\tau_1$ (atm-μsec)			$p\tau_4$ (atm-μsec)			$p\tau_8$ (atm-μsec)		

TABLE I. BASIC RELAXATION RATES FOR $\psi_{CO_2} \ll 1$

In addition to the collisional energy transfer already described, energy exchange between modes 1 and 3 takes place by the mechanisms of absorption and stimulated emission between levels 100 and 001. The characteristic time for radiative transfer is called τ_ν and is defined by

$$\tau_\nu = \frac{k(\theta_3-\theta_L)Q'}{(\frac{G_0}{N})I} \qquad (5)$$

where Q' is the reduced partition function, Q/ε_3, (G_o/N) the absorption cross section times the rotational occupancy factor, and I the cavity intensity. The appropriate definitions are:

$$Q = (1-e^{-\theta_1/T_1})^{-1} (1-e^{-\theta_2/T_2})^{-2} (1-e^{-\theta_3/T_3})^{-1} \qquad (6)$$

$$\varepsilon_3 = (1 - e^{-\theta_3/T_3}). \qquad (7)$$

G_o/N depends on the particular rotational line on which the laser operates, as well as on the temperature, pressure and mixture, and for lean CO_2 mixtures is given approximately by $718/NT$ cm^2 for the rotational line of maximum gain.[7] The gain coefficient is then given by $g = (G_o/N)(n(001)-n(100))$. We now define the dimensionless energies by

$$X = \frac{E_n}{Nk\theta_3}, \quad Y = \frac{E_3}{Nk\theta_3}, \quad Z = \frac{E_{12}}{Nk\theta_1} \qquad (8)$$

where $E_{12} = E_1 + E_2$. The rates of change of these mode energies resulting from the three collisional reactions (Equations 1-3) and the radiation transfer are described by a set of coupled first order equations similar to those developed by Basov et. al.[6] Under a set of non-restrictive assumptions, they become[7]

$$\frac{dX}{dt} = \frac{\psi}{\tau_4} [X - \psi^{-1}Y] \qquad (9)$$

$$\frac{dY}{dt} = \frac{1}{\tau_4} [\psi X - Y] - \frac{1}{\tau_8} [Y - \bar{Y}] - \frac{1}{\tau_\nu} [Y - \delta Z] \qquad (10)$$

$$\frac{dZ}{dt} = -\frac{1}{\tau_1} [Z - \bar{Z}] + \frac{1}{\tau_8} \frac{3\theta_2}{\theta_1} [Y - \bar{Y}] + \frac{\theta_L}{\theta_1 \tau_\nu} [Y - \delta Z] \qquad (11)$$

The bars denote equilibrium values, $\Psi = (\Psi_{CO_2}/\Psi_{N_2})$ and $\delta = \varepsilon_1/\varepsilon_3$ is approximately the ratio of the number of molecules in the lower laser level (100) to those in mode 2. ε_1 is given by:

$$\varepsilon_1 = e^{-\frac{\theta_L}{T_2}} \left((e^{\theta_1/T_2}-1)^{-1} + \frac{2\theta_2}{\theta_1} (e^{\theta_2/T_2}-1) \right)^{-1} \qquad (12)$$

and the gain coefficient becomes

$$g = \frac{G_o}{Q'} (Y - \delta Z). \qquad (13)$$

Equations (9) – (11) are weakly non-linear in the energies X, Y, Z since δ and Q' are functions primarily of the bending mode temperature and hence of Z. δ can vary significantly during a lasing pulse, when the bending mode temperature may rise. However, the equations become linear if δ and Q' are taken to be constant and I is assumed to be a specified parameter, and therefore can be solved analytically. The solutions can be readily interpreted physically and provided much insight into the design considerations for gasdynamic lasers. Moreover, the approximate solutions are very close to those obtained by numerical integration with varying Q' and δ for trial cases.[7]

III. THE INVERSION PROCESS

As a model for the inversion process we assume that the gas is in thermal equilibrium at an initial temperature ($T_o \approx 1300°K$) and is cooled instantaneously to a new translational temperature on the order of 300°K. The radiation intensity is zero, i.e. $\tau_\nu = \infty$. In general, the set of equations (9) – (11) possesses (constant) particular solution, corresponding to the long time ($t \to \infty$) solution, plus a homogeneous solution consisting of three decaying exponential terms. For $\tau_\nu = \infty$ equations (9) and (10) are uncoupled from (11) and the three roots can be written down explicitly. For $\Psi \ll 1$ they simplify to

$$\tau_\alpha = \Psi^{-1} \tau_8 = \tau_p \tag{14}$$

$$\tau_\beta = \tau_4 \tag{15}$$

$$\tau_\gamma = \tau_1 \tag{16}$$

The solution is

$$X = X' + X_p$$
$$Y = Y' + Y_p$$
$$Z = Z' + Z_p$$

where the particular solution is $X_p = \Psi^{-1} Y_p = \Psi^{-1} \bar{Y}$, $Z = \bar{Z}$, and the homogeneous solution becomes

$$X' = X'_0 \left\{ \left(1 + \Psi \frac{\tau_4}{\tau_8}\right) e^{-\frac{t}{\Psi^{-1}\tau_8}} - \Psi \frac{\tau_4}{\tau_8} e^{-t/\tau_4} \right\} \tag{17}$$

$$Y' = \psi X_0' \left\{ \left(1 - \frac{\tau_4}{\tau_8}\right) e^{-\frac{t}{\psi^{-1}\tau_8}} + \frac{\tau_4}{\tau_8} e^{-t/\tau_4} \right\} \quad (18)$$

$$Z' = \left\{ Z_0' - \frac{3}{2} \psi X_0' \left(\frac{\tau_1}{\tau_8} + \frac{(\tau_4/\tau_8)^2}{1-\tau_4/\tau_1} \right) \right\} e^{-t/\tau_1} \quad (19)$$

$$+ \frac{3}{2} \psi X_0' \left\{ \frac{\tau_1}{\tau_8} e^{-\frac{t}{\psi^{-1}\tau_8}} + \frac{(\tau_4/\tau_8)^2}{1-\tau_4/\tau_1} e^{-t/\tau_4} \right\}$$

Here the subscript 0 denotes initial values and the overbar denotes final (low temperature) equilibrium values.

A plot of equations (17) - (19) is shown on Fig. 3. The upper level CO_2 (Y) adjusts to a value slightly below ψX on a time scale τ_4 and then both ψX and Y relax on a single time scale $\tau_p = \psi^{-1}\tau_8$, termed the pumpout time by Christiansen.[2] The lower level behavior is dominated by a rapid decay to very near its final equilibrium value on a time scale τ_1 followed by a final approach to equilibrium on a time scale τ_p.

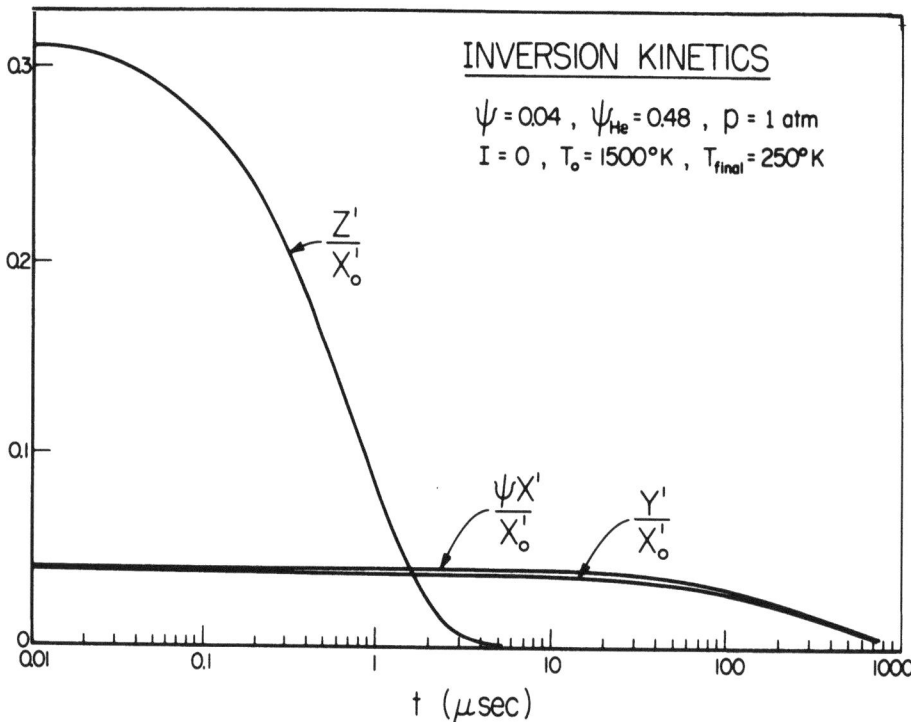

Fig. 3 Decay of Mode Energies in Response to Instantaneous Cooling

The pumpout time scale τ_p is important for two reasons. First, it determines scaling laws for the expansion. The characteristic flow time through the nozzle is of the order of h/a^* where h is the throat height and a^* is the sonic speed at the throat. Therefore, for freezing to occur, we require that

$$(\psi^{-1}\tau_8)_{throat} > h/a^* \qquad (20)$$

Since conditions in the throat and stagnation region are simply related, this can be written $\psi^{-1}\tau_8 = $ const. h. The vibrational energy flux per unit area is directly proportional to the sum of CO_2 and N_2 partial pressures, and it is thus advantageous to operate at high pressures. In order to satisfy the freezing condition, Eq. (20), one finds then that $\psi_c P_o h/\psi_{N_2}$ must be held constant when individual parameters are varied. In order to operate at reasonably high pressures, most GDL's employ a grid of small nozzles (h = 0 (1 mm)) and mixtures with $\psi_{CO_2} \simeq 0.1$.[1] In order to increase the pressure further, ψ_c must be decreased as $1/P_o$. These arguments were first put forth by Christiansen[2] who further shows that the (pressure broadened) gain coefficient is reduced as a consequence according to the scaling rule g = const. $\psi_c \sim 1/P_o$ for fixed ψ_{N_2}, h. The pumpout time is also important in that it determines the time scale for decay of vibrational energy in the cavity region and hence the distance over which an inversion persists; i.e. the length of box that can be filled. Since the freezing condition requires $(\psi^{-1}\tau_8) U_{flow} > h$ at the throat where the pressure and temperature are high and the velocity relatively low, the corresponding length scale in the cavity region, $\ell = [(\psi^{-1}\tau_8) U]_{cavity}$, is on the order of 10^3 times greater, or 1 meter.

The scaling laws given above have led to intensive investigation of high mass flow grid nozzle configurations for proper matching of shock tube capabilities to kinetic requirements, as reported in References 2 and 9.

IV. KINETICS OF ENERGY EXTRACTION

Assume now that there exists an inversion in the cavity and that at some time t = 0 the gas is subjected to a constant radiation intensity I. We wish to determine the time required for the vibrational energy to be depleted from the upper level CO_2 and nitrogen in the presence of the given radiation field. This gives the time scale on which the energy can be extracted from a filled box or alternatively, the length of cavity required for a quasi CW or CW operation. We shall also be concerned with the efficiency of conversion of vibrational energy into radiation, and with the concepts of gain saturation and saturation intensity as applied to a pulsed GDL.

In the presence of a radiation field equations (9) - (11) are fully coupled and the characteristic equation for the roots is a cubic which in general cannot be solved analytically. The particular solution corresponding to $t \to \infty$ is

$$X_p = \psi^{-1} Y_p = \frac{\psi^{-1}}{1 + \tau_\nu/\tau_8} [\delta Z_p + \frac{\tau_\nu}{\tau_8} \bar{Y}] \tag{21}$$

$$Z_p = \left[\frac{\bar{Z}}{\left(1 - \frac{\delta \tau_1}{2(\tau_\nu + \tau_8)}\right)}\right] \left[1 - \frac{\tau_1}{2(\tau_8 + \tau_\nu)} \frac{\bar{Y}}{\bar{Z}}\right] \tag{22}$$

As $I \to 0$ ($\tau_\nu \to \infty$), we recover our earlier result that $Y_p \to \bar{Y}$, and $Z_p \to \bar{Z}$. For $I \to \infty$ ($\tau_\nu \to 0$), we obtain the "fully bleached" condition corresponding to equal populations of $CO_2(001)$ and $CO_2(100)$, $Y_p \to \delta Z_p \simeq \delta \bar{Z}$ for $\delta \tau_1 \ll \tau_8$.

In the limit where $I \to \infty$, or more precisely, $\tau_\nu \ll \delta \tau_1, \tau_4$, and τ_8, an analytic solution for the three roots can be found. The upper and lower laser levels first approach each other on a time scale

$$\tau_{bleach} = \tau_\nu (1 + \delta) \tag{23}$$

and the remaining two roots can be found by putting $Y \simeq \delta Z$ in Equations (9) - (11) to give, for $\psi \ll 1$,

$$\tau_{nat} = \psi^{-1} (\delta \tau_1 + \tau_4) \tag{24}$$

$$\tau_{\beta 1} = \frac{\tau_4 \tau_1 (1 + \delta)}{\delta \tau_1 + \tau_4} \tag{25}$$

The homogenous solution for the decay of energy levels from an initial inversion state for $I \to \infty$ is then:

$$X' = X'_0 e^{-t/\tau_{nat}} \tag{26}$$

$$Y' = \psi X'_0 \left\{ \frac{\delta \tau_1}{\delta \tau_1 + \tau_4} e^{-t/\tau_{nat}} + \left[\frac{\delta}{1+\delta}\left(1 + \frac{Z'_0}{\psi X'_0}\right) - \frac{\delta \tau_1}{\delta \tau_1 + \tau_4}\right] e^{-t/\tau_{\beta}'} \right.$$

$$\left. + \frac{1}{1+\delta}\left(1 - \delta \frac{Z'_0}{\psi X'_0}\right) e^{-t/\tau_{bleach}} \right\} \tag{27}$$

$$Z' = \psi X_0' \left\{ \frac{\tau_1}{\delta\tau_1 + \tau_4} e^{-t/\tau_{nat}} + \left[\frac{1}{1+\delta} \left(1 + \frac{Z_0'}{\psi X_0'} \right) - \frac{\tau_1}{\delta\tau_1 + \tau_4} \right] e^{-t/\tau_\beta'} \right.$$
$$\left. - \frac{1}{1+\delta} \left(1 - \delta \frac{Z_0'}{\psi X_0'} \right) e^{-t/\tau_{bleach}} \right\} \quad (28)$$

This solution is plotted in Fig. 4. Following the initial rapid bleaching, the laser levels $Y \simeq \delta Z$ adjust on a time scale τ_β, to new levels given with respect to ψX by the ratio $\delta\tau_1/(\delta\tau_1+\tau_4)$, after which the entire mixture relaxes on the time scale τ_{nat}. This time scale τ_{nat} is thus seen to determine the shortest Q-switched pulse length available or the shortest permissible streamwise cavity dimension for full energy extraction.

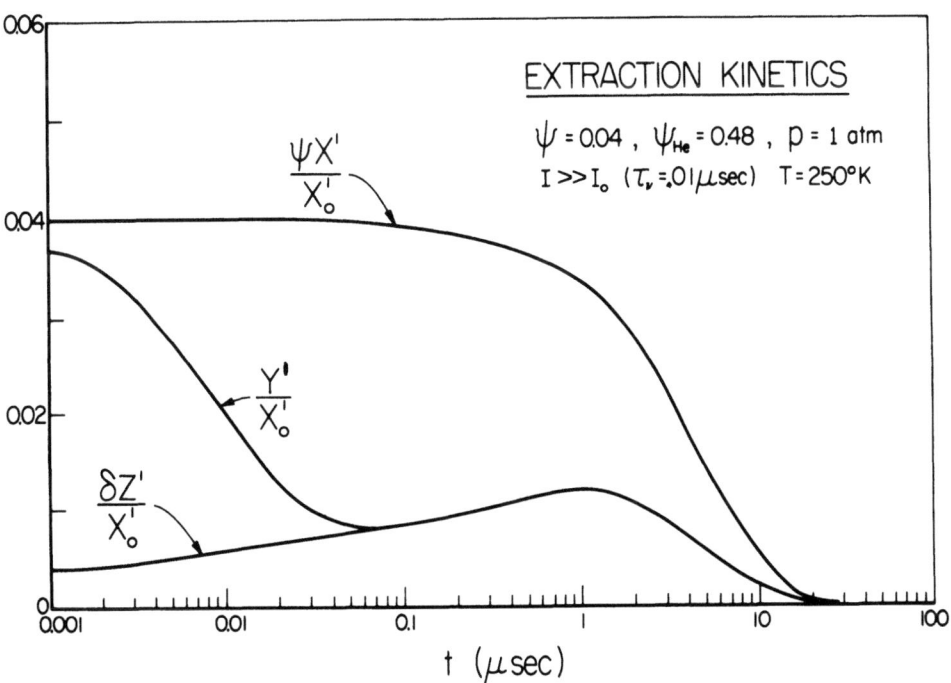

Fig. 4 Decay of Mode Energies in a Strong Radiation Field

Fig. 5 shows the variation of τ_{nat} with Helium concentration for a 1 percent CO_2 mixture and a cavity pressure of 1 atm; τ_{nat} scales as $p^{-1/2}$. It is seen that the minimum pulse length can be varied considerably at fixed ψ and p simply by varying the helium concentration; this results principally from the strong effect of He on τ_1 (Table I). Since the inversion level produced is relatively independent of ψ_{He} this quantity can be chosen in a shock tube GDL to produce the desired $p\tau_{nat}$.

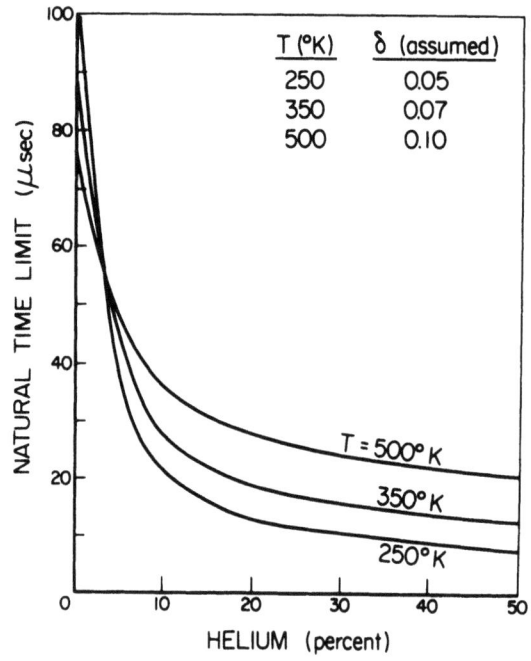

Fig. 5
Variation of Natural Time with ψ_{He} for $\psi_{CO_2} = 0.01$ and p = 1 atm

In the general case when τ_ν is neither zero nor infinite the roots are more complicated. The fact that in both limits the mixture relaxes with a single time scale, however, suggests that this should also be true for arbitrary intensity. The slowest root can be found analytically and is

$$\tau_{pulse} = \psi^{-1} \frac{(\delta\tau_1 + \tau_4 + \tau_\nu)}{(1 + \tau_\nu/\tau_8)} \quad (29)$$

which reduces to τ_p and τ_{Nat} for the two limiting cases $\tau_\nu = \infty$ and o. Equation (29) thus gives the time scale for energy extraction for arbitrary intensity.

The gain coefficient can now be calculated from Equations (13) and (26-28). For efficient operation the radiative transfer from 001 to 100 should exceed the collisional transfer, which implies $\tau_\nu \ll \tau_8$. In addition, for lasing mixtures of interest it can be seen from Table I that τ_1 and τ_4 are much less than τ_8. Then, neglecting terms of order τ_1/τ_8, τ_4/τ_8, τ_ν/τ_8, the gain becomes

$$g = \frac{G_o}{Q'} \frac{\psi X'}{1+\left(\frac{\delta\tau_1+\tau_4}{\tau_\nu}\right)} = \frac{G_o}{Q'} \frac{\psi X_o' e^{-t/\tau_{pulse}}}{1+\left(\frac{\delta\tau_1+\tau_4}{\tau_\nu}\right)} = \frac{g_o e^{-t/\tau_{pulse}}}{1+\left(\frac{\delta\tau_1+\tau_4}{\tau_\nu}\right)} \quad (30)$$

where g_o is the small signal gain at $t = 0$. This equation naturally defines a saturation intensity I_o such that

$$(\tau_\nu)_{I=I_o} = \delta\tau_1 + \tau_4, \quad \text{or from equation (5)}$$

$$I_o = \frac{Nk(\theta_3-\theta_L)Q'}{G_o(\delta\tau_1+\tau_4)} \quad (31)$$

The gain is thus

$$g = \frac{g_o e^{-t/\tau_{pulse}}}{1 + I/I_o}$$

where

$$\tau_{pulse} \simeq \psi^{-1}(\delta\tau_1 + \tau_4 + \tau_\nu) = \tau_{nat}(1 + I_o/I)$$

In the presence of a radiation field, the gain is reduced by a factor of $1 + I/I_o$ from that which it would be for low intensity at a given inversion level $\psi X'$. In general it takes a much higher intensity to produce given gain saturation in a g.d.l. than in a steadily pumped laser (e.g. an electric discharge laser), where the appropriate saturation intensity is that for which $\tau_\nu = \tau_8$.

From the foregoing discussion it can be seen that operating the cavity at a radiation intensity of I_o as defined by equation (31) represents a good choice. The gain is reduced by two from the small signal level and the pulse length is twice the shortest possible. At higher intensities the gain becomes too low, while at lower I collisional losses reduce efficiency.

The results of the approximate model described above can be used to design oscillators by matching saturated gain to losses in the usual manner, both for Q switched and quasi-cw operation. Such calculations have been reported in Ref. 7, where comparison between the approximate solution and exact integration of the governing equations indicates very good agreement. This enables one to use the simple approximate results with confidence, subject to the restriction $\psi \ll 1$, to evaluate the performance of various amplifier and oscillator configurations. In addition to providing analytical simplification, the approximate method yields physical insight into the relaxation processes of the $CO_2 - N_2$ system with and without radiation.

V. APPLICATION TO CONTROLLED THERMONUCLEAR FUSION

In keeping with the spirit of this workshop on laser-matter interaction, we turn now to one of the most interesting potential applications of pulsed gasdynamic lasers, that of heating a magnetically confined plasma to thermonuclear temperatures.

It has been pointed out by Dawson[10,11] and his collaborators that the CO_2 laser is uniquely suited to the heating of magnetically confined plasmas by virtue of its wavelength, its efficiency, and its capability of being scaled to large energies.

A conceptual diagram of a laser heated magnetically confined plasma is shown in Fig. 6. The plasma column is contained within a cylindrical magnet structure and the radiation propagates along the axis. The magnetic field may be longitudinal, as in a θ-pinch or superconducting solenoid arrangement, or may be azimuthal external to the plasma with a longitudinal field within the plasma, as in the case of a stabilized Z pinch. Particular problems associated with laser heating of θ and Z pinches have been studied by Vlases, Ahlstrom, and Steinhauer.[12-14]

It was shown by Lawson[15] in 1957 that in order to create a thermonuclear reaction which produces net useful energy two conditions must be met. First, for a 50 percent-50 percent mixture of deuterium and tritium, the ion temperature must exceed approximately 5 kev. Secondly, the plasma must be contained at least for a time $\tau_c = (10^{14}/n_e)$ seconds, for a pulsed reactor with a conversion efficiency of 33 percent, where n_e is the electron density (cm^{-3}).

For a simple linear system such as the Z pinch or θ pinch with uninhibited end losses, the confinement time τ_c is the time required for the plasma to flow out the end. Thus $\tau_c = (L/2a)$, where L is the length and a is the speed of sound, which at a temperature of 10 kev is on the order of 1 meter/μsec. Thus the higher the density, the shorter the reactor. Conventional θ pinches produce temperatures sufficiently high for thermonuclear reactions, but at densities of around 10^{16}, which requires a coil too long to be of practical interest.

The length of coil required to satisfy Lawson's conditions can clearly be reduced by increasing the density while maintaining the temperature. The upper density limit is set by magnet strength considerations. To confine a plasma at a density of 5×10^{17} and 10 kev requires a field of 630 kg, which is about the limit of current pulsed magnet technology. If we accept a field strength at the coil wall of 500 kg, the length of a θ pinch reactor satisfying Lawson's criteria comes out to be about 800 meters[11] and the density is 3×10^{17}. Methods for inhibiting end losses have been suggested which could reduce the length by a factor of five or more.[11]

Increasing the density from the currently available 10^{16} to 3×10^{17}, without reducing the temperature, requires a prohibitively large and high voltage capacitor bank if conventional pinch techniques are used. Moreover, with conventional techniques the pinched column occupies only 1-5 percent of the coil volume, so that the ratio of plasma to magnetic field energy, a measure of efficiency, is correspondingly low.

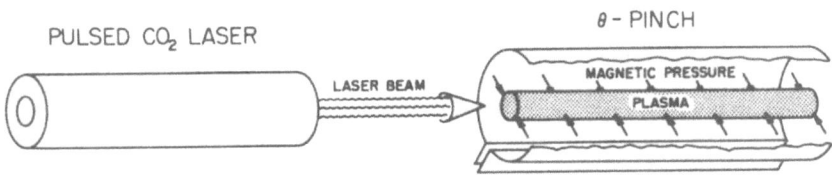

Fig. 6 Schematic Diagram of Laser Heated Pinch

The use of a CO_2 laser to heat the plasma overcomes both of these difficulties. First, the laser can be made to supply nearly any desired portion of the total energy required, and secondly, the plasma density and temperature can be controlled independently so that the plasma can be made to occupy an arbitrary fraction of the coil volume. In addition to making high density high temperature operation possible, the laser makes it practical by reducing the total energy required for a given plasma energy by a large factor.

It has been assumed in the preceeding paragraphs that the laser energy can be efficiently absorbed in the long, slender plasma column. A plasma absorbs 10.6μ radiation at electron densities n_e up to 10^{19} cm   efficient absorption takes place in the range of $10^{17} - 10^{19}$. For the 500 kg reactor example given above, the absorption length is .8 times the reactor length at 10 kev, so that all the radiation will be absorbed. Refraction of the beam is a problem of some concern, since transverse e-m waves propagating through a plasma are bent towards regions of lower electron density. Thus it is necessary to operate with a density minimum on axis. This occurs naturally, however, in the dynamic phase of pinches and the addition of heat on the axis will maintain the required minimum. Details of the rays and wavefronts are given by Steinhauer and Ahlstrom.[13]

Time scales for laser energy deposition should be less than the plasma confinement time by about one order of magnitude. For the 500 kg reactor the confinement time is approximately 400μsec, so that the laser pulse width should be 40μsec. or less. From section IV, however, it is seen that this is precisely within the range of natural pulse times available from high pressure gasdynamic lasers. Finally, we note that the efficiencies of open cycle g.d.l.'s are considerably lower than that assumed in Lawson's analysis. However, proposals for closed cycle systems have recently been advanced[16] which offer the prospect of efficiencies at least as high as the 33 percent figure required for $n \tau_c = 10^{14}$ to be sufficient.

In conclusion, pulsed CO_2 lasers in general and pulsed CO_2 gasdynamic lasers in particular appear to be naturally matched, from the standpoints of absorption, efficiency, and pulse length, to the problem of producing a controlled thermonuclear reaction in a magnetically confined plasma column.

VI. ACKNOWLEDGEMENTS

The authors wish to thank W. Christiansen for stimulating the development of the kinetics model and for many interesting discussions. Helpful conversations with S. Byron, H. Ahlstrom and A. Hertzberg are also gratefully acknowledged. The work was supported jointly by AFWL, Air Force Systems Command, United States Air Force, Kirtland Air Force Base, Contract F 29601-70-C-0033, by NASA Grant NGL 48-002-044, and by NSF Grant GK 28562.

REFERENCES

1. E. T. Gerry, Bull. Am. Phys. Soc. 15, 563 (1970); I.E.E.E. Spectrum (USA) 7, 51 (1970)
2. W. H. Christiansen, A.I.A.A. Paper 71-572 (1971)
3. W. H. Christiansen and G. A. Tsongas, Phys. Fluids (to be published)
4. I. R. Hurle and A. Hertzberg, Phys. Fluids, 8, 1601 (1965)
5. V. K. Konyukhov and A. M. Prokhorov, J.E.T.P. Letters, 3, 786 (1966)
6. N. G. Basov, A. N. Oraevskii, and V.A. Shcheglov, Sov. Phys. Technical Physics 12, 243 (1967)
7. A. L. Hoffman and G. Vlases, I.E.E.E. Jour. Quantum Elec. (to be published)
8. R. L. Taylor and S. Bitterman, Rev. Mod. Phys. 41, 26 (1969)
9. D. A. Russell, W. H. Christiansen, and A. Hertzberg, Proc. 8th International Shock Tube Symposium, London. (June 1971)
10. J. M. Dawson, A.I.A.A. Paper, 70-779 (1970)
11. J. M. Dawson, A. Hertzberg, R. Kidder, G. Vlases, H. Ahlstrom, and L. Steinhauer, Paper CN/28 D 13, I.A.E.A. Fourth International Fusion Conference, Madison, Wisconsin (June 1971)
12. G. C. Vlases and H. G. Ahlstrom, Bull. Am. Phys. Soc. 14, 1022 (1969)

13. L. Steinhauer and H. G. Ahlstrom, Phys. Fl. 14, 1109 (1971)
14. G. C. Vlases, Physics of Fluids, 14, 1287 (1971)
15. J. D. Lawson, Proc. Roy. Soc. B 70, 6 (1957)
16. A. Hertzberg, E. Johnston, H. Ahlstrom, A.I.A.A. Paper 71-106 (Jan. 1971)

NUCLEAR RADIATION EFFECTS ON GAS LASERS*

George H. Miley

Nuclear Engineering Program, University of Illinois

Urbana, Illinois   61801

INTRODUCTION

The irradiation of a gas laser via nuclear radiation (e.g., using γ-radiation, high energy ions from nuclear reactions, or emissions from radioisotopes) provides an important and unique means of energy input. As such, potential applications range from total pumping by the radiation to the use of relatively small radiation fluxes to enhance the performance of a laser operating primarily by electrical pumping. The former (direct pumping) is still somewhat speculative, but conclusive evidence for enhancement of CO_2 laser operation has been reported. Both are briefly reviewed here and some recent results from CO_2 enhancement experiments at the University of Illinois are presented.

DIRECT PUMPING

The status of direct pumping has been discussed in some detail by the author and his associates[1,2], and also by Thom and Schneider[3], hence only a few highlights are noted here.

Characteristically, high radiation fluxes are required to achieve inversion. One of the most convenient methods of achieving such fluxes is to use neutrons from a nuclear reactor to initiate a nuclear reaction which releases high-energy charged-

---

*Presented at the Second Workshop on "Laser Interaction and Related Plasma Phenomena" at Rensselaer Polytechnic Institute, Hartford Graduate Center, August 30 - September 3, 1971.

particles in the gas. A common approach is to coat the inside of the laser tube with boron (enriched in B-10) to obtain the neutron induced reaction:

$$^{1}_{0}n + ^{10}_{5}B \rightarrow ^{7}_{3}Li + ^{4}_{2}He + 2.79 \text{ MeV}$$

The MeV lithium and alpha ions escaping from the coating ionize and excite the gas, in effect creating a "radiation-induced" plasma. While there are other possible neutron reactions [$^{3}He(n,p)^{3}H$; $^{235}U(n)ff$; etc.], the fluxes required to deposit a given amount of energy in the gas are not too different from the boron case. Thus several investigators have estimated the neutron flux required to cause inversion in the gas assuming a boron lining. Some typical results are shown in Table I.

Table I. Some Threshold Estimates

[For pressures and cavities typical of electrical pumping]

Laser System	Threshold Flux n/cm^2-sec on B^{10}	Investigator
$CO_2 - N_2 - H_e$ (10.8μ)	2×10^9	DeShong[4]
He - Ne (3.39μ)	3×10^{12}	Guyot[5]; Herwig[6]; and DeShong[4]
Ar (neutral; new lines)	$\approx 10^{15}$	Russell[7]
(ion; uv, blue lines)	$\approx 10^{18}$	Ganley[8]
Ne - O_2 (0.84μ)	3×10^{16}	Rusk, et al.[9]

Two points should be stressed: first, most of the fluxes indicated are accessible with a reactor such as the U. of Illinois pulsed TRIGA [peak thermal-neutron fluxes during a pulse of about 5×10^{16} n/(cm^2 sec)]; and second, these calculations are quite approximate and open to questions. Most sutdies assume that once the charged-particle energy is deposited in the gas, the subsequent splitting into various states and levels follows the same route as in the conventional electron pumping case. There is, indeed, mounting experimental evidence that this is a poor assumption (see References 10-12 and the following section). Russell's calculations are the most complete, but still he was forced to make key assumptions about excited state formation during recombination and also about the relative contribution from the primary ions and their δ-rays.

In summary, an adequate treatment of the non-equilibrium character of the radiation-induced plasma (caused by the superposition of the high-energy particles on a cold-background electron gas) represents an unsloved theoretical problem. As a result the primary thrust in nuclear pumping must rest on experimental studies.

Experiments in Direct Pumping

Experimental studies of direct pumping have been centered at four locations: at Moscow State University in Russia[13,14]; at the University of Illinois[1,10]; and, prior to 1970, at the Northrup Corporation[9] and also at Argonne National Laboratory[4,15]. All have used a boron or an uranium lining scheme such as discussed above, or in several cases ^{3}He has been substituted for normal helium in the laser mixture. Three positive results have been reported, but in the author's opinion, no completely conclusive proof of direct pumping has been achieved to date. The Northrup studies[9] originally reported enchanced fluorescence "or lasing" in a Ne-O_2 mixture and also at 6684 Å in Argon. However, no test of coherence nor studies of the dependence of the output on mirror alignment, etc. were reported.

Guyot, et al.[1], reported a gain at 3.39μ in a He-Ne-air mixture for very short neutron pulse times, and more recently he[10] extrapolated steady state measurements in pure He-Ne to predict gain for fluxes several orders of magnitude above those indicated in Table 1.

Andriakhin's[14] recent studies of ^{3}He-Hg are perhaps most promising since quite strong outputs were observed (see Fig.1,next page).

The signal amplitude (≈2V) is estimated to correspond to an optical radiation power of ≈10 mW which the investigators note is much higher than the spontaneous-radiation power obtainable with their geometry even if all the nuclear reaction energy were converted to radiation. They did not, however, report tests for coherence, nor are data about pressure effects, threshold flux requirements, etc., available.

Some negative results are also clear. Corresponging to Table 1, CO_2 should be quite easy to pump. However, attempts by DeShong[15], Rusk, et al.[9], and also Ganley[16] have failed to date. Guyot's studies[10] seem to indicate that the threshold estimates for He-Ne in Table I are in error. The reasons for these problems are not entirely clear, but they in general seem to follow from mechanisims associated with the unique non-equilibrium character of the radiation-induced plasma.

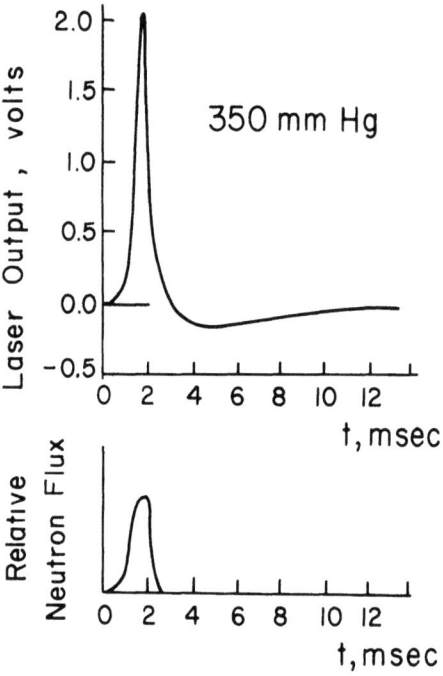

Fig. 1
Output signal from a Hg-^{3}He laser during neutron bombardment as reported by Andriakhin, et al.[14]. The lower trace shows the neutron pulse, the peak flux being about 5×10^{16} n/(cm^2-sec).

Potential Applications of Direct Pumping

L. Herwig[6] first stressed that nuclear radiation pumping might lead to unique laser characteristics, e.g., new lines, frequency control, etc. High-power systems are also envisioned by some[4,6]. However, to achieve this, high-pressure (atmospheric or above) operation is necessary to obtain efficient absorption of the high-energy nuclear radiation. There is no experimental evidence yet available which would verify that this is feasible, although spectroscopic studies[17,18] of various noble gases irradiated by alpha particles from a polonium source have been undertaken to investigate the basic mechanisms involved at higher pressures ($\bar{<}$ 1 atm.).

Another potential application, discussed by Thom and Schneider[3], is to extract power from a "plasma-core" nuclear reactor. These authors have also discussed several coupled laser-reactor concepts where a conventional reactor is used to irradiate specially designed laser tubes.

Admittedly, all such applications must be considered to be quite speculative at this point. However, direct pumping has truly a basic character in that it extends pumping from electron or photon excitation to ion excitation. In this sense it opens a whole new area. There will clearly be some unique features associated with ion pumping and only time will tell how these features can be exploited and utilized.

ENHANCEMENT

Several experiments[11,13,19-22] have demonstrated that a small nuclear radiation input (< 1% of the electrical power input) can be used to increase the power and efficiency of low-pressure CO_2 lasers, and recent results reported here also indicate an improvement in high-pressure TEA CO_2 laser performance. These results are attributed to a tailoring of the electron energy-distribution such that it overlaps the excitation cross section better. Thus, the effect is a function of the detailed energy dependence of the excitation cross section, and as such it would not be expected to occur in all gases. In fact, experiments with other systems have not been reported to date so the present discussion will be restricted to CO_2.

Low-Pressure CO_2 Lasers

Andriakhin, at al.[13] first reported experiments that indicated enhancement by nuclear radiations was possible. They injected 5 to 7 µA of 2.8-MeV protons from an accelerator along the axis of an operating CO_2 laser. The proton beam deposited ≈ 10 watts in the gas equal to about 10% of the electrical power. Increases in the laser output up to a factor of 3 were observed as a result of the proton beam.

Later Allario and Schneider[19] observed small changes in output when a CO_2 (N_2-He) laser with ^{3}He substituted for He was irradiated by low-level neutron fluxes. [Energy is deposited in the gas by the recoil proton and triton from the reaction ^{3}He (n,p)^{3}H].

Ganley, Verdeyen, and Miley[11,20] first reported sizable changes when a low-pressure dc excited CO_2 laser with a boron-lined discharge tube was irradiated by neutrons from a nuclear reactor. Some typical results are reproduced in Fig. 2.

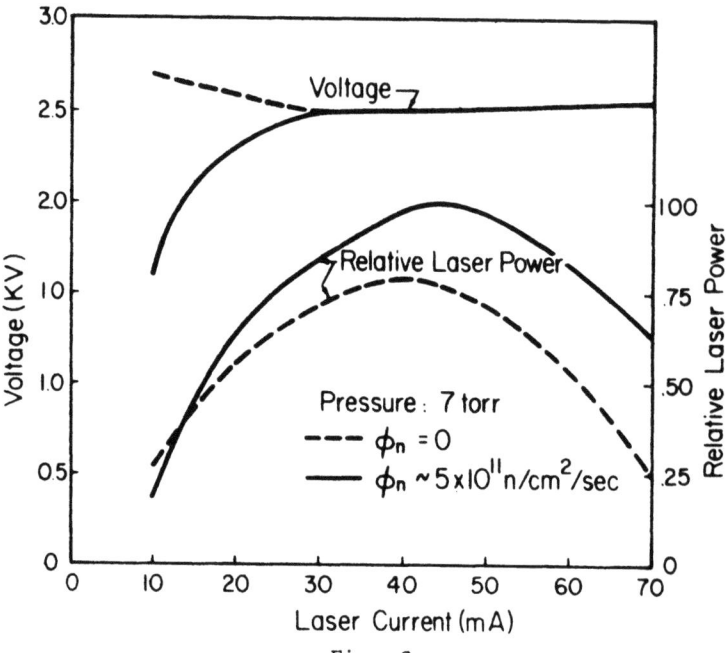

Fig. 2
Changes in low-pressure CO_2 laser output under neutron irradiation, (From Ref. 11).

For discharge currents above about 30 mA, the voltage with and without the reactor "on" was essentially the same. Thus the electrical power input remained constant, but the laser output increased with the reactor on, corresponding to neutron fluxes of order of 10^{11} n/cm²sec.) or about 10^{10} ions/cm²sec/ from the boron surface. This enhancement varies from about 20% at the peak in the putput (corresponding to ≈ 40mA current) to a factor of two or so at higher discharge currents. Note, however, that both outputs decrease at high currents where the increased input-power cause gas heating in the low-flow laser system employed here. Thus, the interpretation of the results in this region is somewhat ambiguous. While it is of interest to extrapolate these data to other systems such as high-flow gas-dynamic lasers, predictions are risky, since heating effects should be subtracted out.

These results differ somewhat from Andriakhin's in that the radiation energy inputs are always much less than 1% of the electrical input as compared to ≈ 10% in the Russian work. In fact, the present studies show that an increased radiation input results in such a high electrical conductivity that the voltage drops, automatically reducing both the electrical input and the laser output. (Note the region below 13 mA in Fig. 2).

As described in more detail in Refs. 11 and 22, the catalytic effect of the radiation is attributed, in this case, to a shift in the electron energy-distribution in the discharge so that it overlaps better with the cross sections responsible for "pumping" the laser. In the normal discharge, the energy distribution must produce sufficient ionization to sustain the discharge. Only the electrons in the high-energy tail actually contribute to the ionization, and the energy actually going into ionization is only a small fraction of the total input (less than a watt out of 125 W input in the present device). Small amounts of nuclear radiation are able to produce comparable ionization rates, and this in turn permits the electron energy-distribution associated with the electrical discharge to relax to a shape having fewer electrons in the high-energy tail.

In order to better understand the mechanisims involved in this enhancement, Ganley, et al.[22] recently performed a series of gain measurements. To do this a boron coated discharge tube was placed in the reactor as illustrated in Fig. 3.

The output of a CO_2 reference laser was directed into a sodium chloride beam splitter. Part of the beam was reflected into a reference detector, D1, while part was passed through the beam

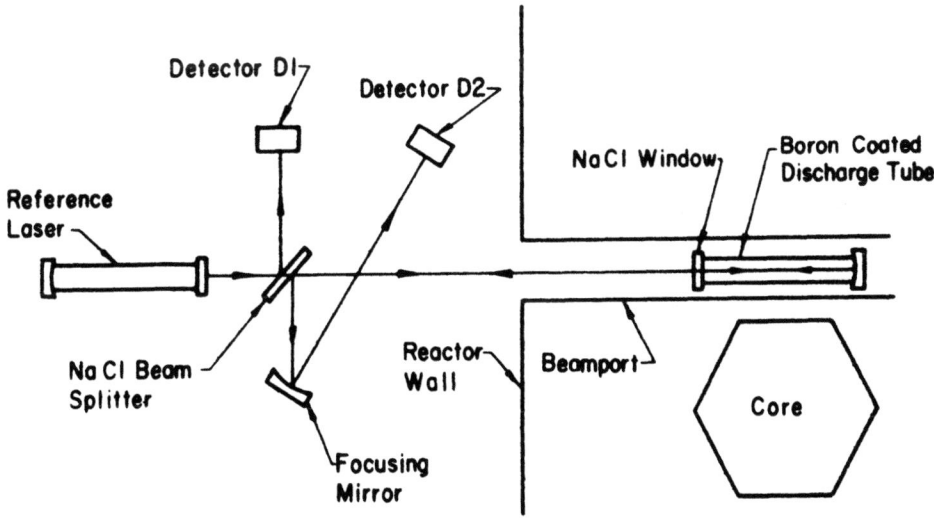

Fig. 3
Schematic of the gain experiment.

splitter and into the boron-coated discharge tube. After passing through this tube, the laser light was reflected back out toward the beam splitter by a gold-coated spherical mirror. At the beam splitter, part of this light from the discharge tube was reflected onto the output detector, D2. Since the beam entering D2 made a double pass through the discharge tube, the magnitude of the signal from D2 was dependent on the optical gain of the medium in the discharge tube as well as on the intensity of the light coming from the reference laser. Variations in the reference laser could be subtracted out by making use of the signal from D1. Thus, by comparing the relative magnitude of the signals from the two detectors, it was possible to determine the percentage increase (gain) of the laser light passing through the discharge tube. The gain coefficient is directly proportional to the population inversion in the medium. Therefore, the change in gain when the reactor is turned "on", $\Delta\alpha$, gives a measure of the relative change in population inversion that is produced as a result of the external ionization.

The pressure range from 5 to 20 Torr has been explored for values of discharge current from 5 to 45 mA. The value of the change in the gain coefficient with the reactor on relative to the normal gain is plotted as a function of discharge current for various pressures in Fig. 4, along with the corresponding voltage-current curves.

These results appear to be consistent with Fig. 2, for example, for a given pressure an increased gain occurs at higher currents. This increase in gain scales with pressure although a saturation appears to start at $\approx$ 17 torr, corresponding to the highest pressure compatible with the power supplies employed here.

Since the reference laser was carefully controlled outside of the radiation field, these results confirm that the enhancement effect is not due to mode switching, changes in lasing transitions, gas lens effects, etc. Thus while this is not entirely conclusive, the hypothesis of a change in electron energy distribution remains the most likely explanation. Experiments are now in progress to measure the distribution directly using probe techniques.

These results have been partly confirmed in recent experiments reported by Rhoads and Schneider[21]. With a similar CO_2 laser, but using ^{3}He instead of a boron lining, they roughly doubled the laser output and efficiency using neutron fluxes of order of 10^{10} n/cm^2-sec. This corresponded to a nuclear energy input of 1 mW as compared to a 100 to 200 W electrical input. In contrast to Fig. 2, however, they find the maximum enhancement occurs at low currents where they were unable to sustain a discharge without the external radiation.

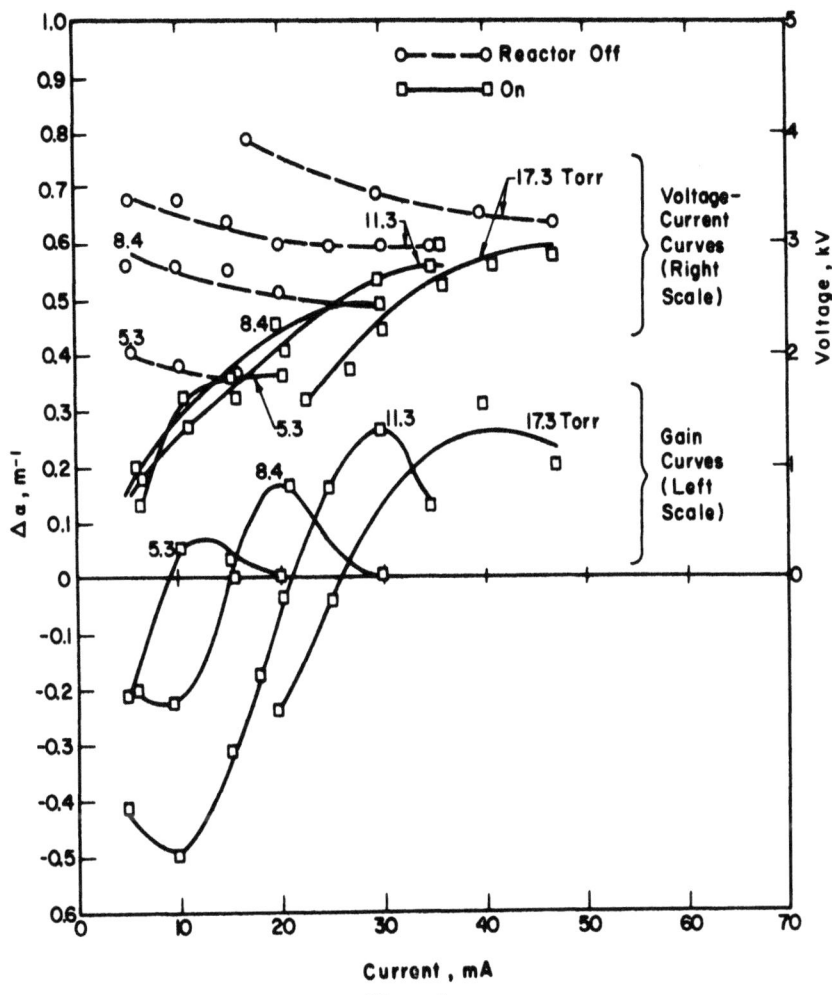

Fig. 4
Change in gain of a boron-lined CO_2 discharge due to a neutron flux of $\approx 5 \times 10^{11}$ n/(cm^2 sec).

Low-Pressure Pulsed CO_2 Results

Recent studies at Illinois have extended the dc discharge results to a pulsed low-pressure CO_2 discharge. Current pulses were applied to the same laser as used by Ganley, et al.[11] in the earlier dc experiments. The current pulse was obtained by discharging a 0.015 µf capacitor through the laser tube with a thyratron pulser, giving a 4-5 A pulse of roughly 10 µsec. The resulting laser output was delayed from the current pulse by about 75 µsec. This time lag is not completely understood, but it could be due to the transfer time associated with N_2 pumping of CO_2, or alternately it

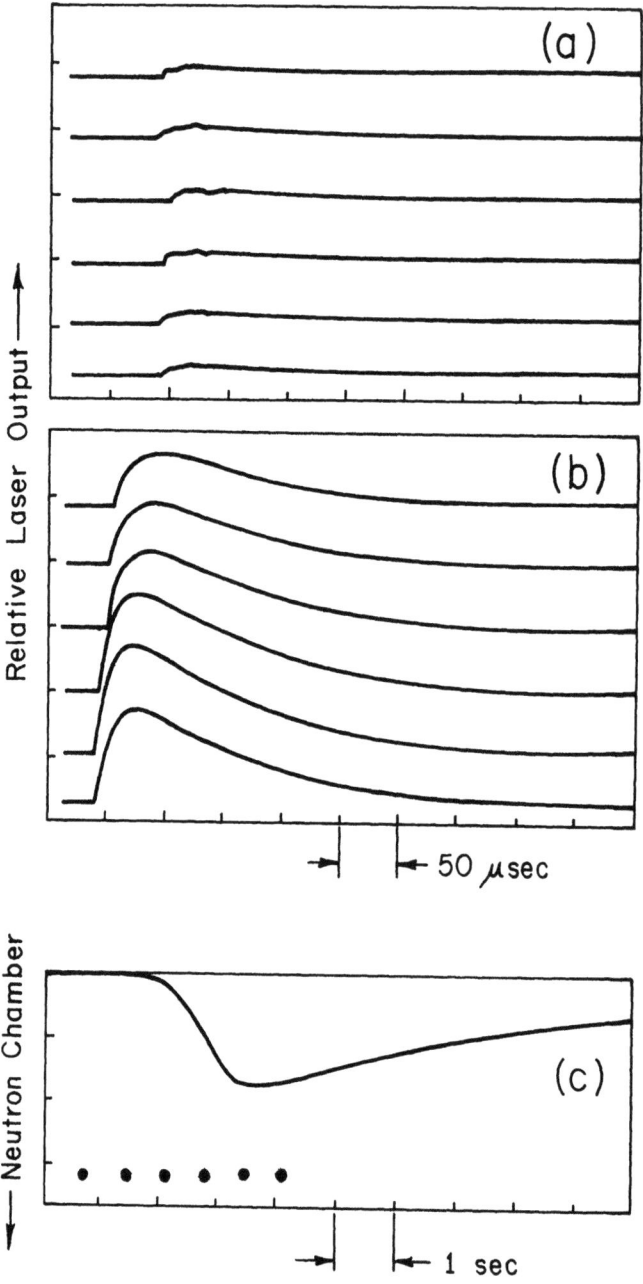

Fig. 5
Effect of nuclear radiation on a pulsed low-pressure CO_2 laser: (a) neutron flux equals zero, (b) laser output corresponding to the neutron flux shown in (c). Counting from the top, each trace in (b) corresponds to a dot in (c).

may be indicative of the dissociation of CO_2 followed by lasing upon recombination[23].

The effects of the nuclear ionization source on the output of this pulsed laser are shown in Fig. 5. The top figure (a) is a reproduction of a scope trace showing a series of laser pulses taken with no external ionization. The laser was pulsed at intervals of 0.8 seconds, starting with the top trace. The second trace (b) shows the same pulsing operation, with the same discharge conditions, for various levels of nuclear radiation. Figure 5 (c) shows the radiation levels corresponding to the laser pulses in 5 (b). The dots correspond to the times at which the laser was pulsed.

Even at the start of the pulse corresponding to very low levels of incident radiation, the favorable effects on laser output are quite pronounced. As the radiation level increases, the laser power also increases. Increased outputs as much as 10 times the no radiation case are observed; however, a point is eventually reached where the radiation effect tends to saturate. Also, notice that in addition to increasing the output of the laser, the external ionization causes the laser pulse to follow the current pulse more closely.

The results shown in Fig. 5 are surprisingly large. They are quite reproducible, and similar results have been obtained under a variety of conditions during experiments carried on over a period of several weeks. A complete explanation cannot be given at this time. A combination of effects may be involved--better pumping efficiency, larger active volume, and increased uniformity of the discharge.

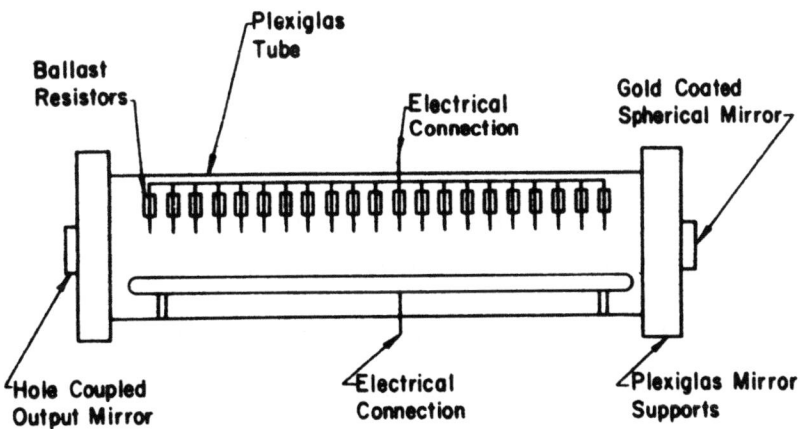

Fig. 6

Construction of the TEA laser. The 1/2-in. cylindrical anode (lower electrode in the drawing) was coated with boron.

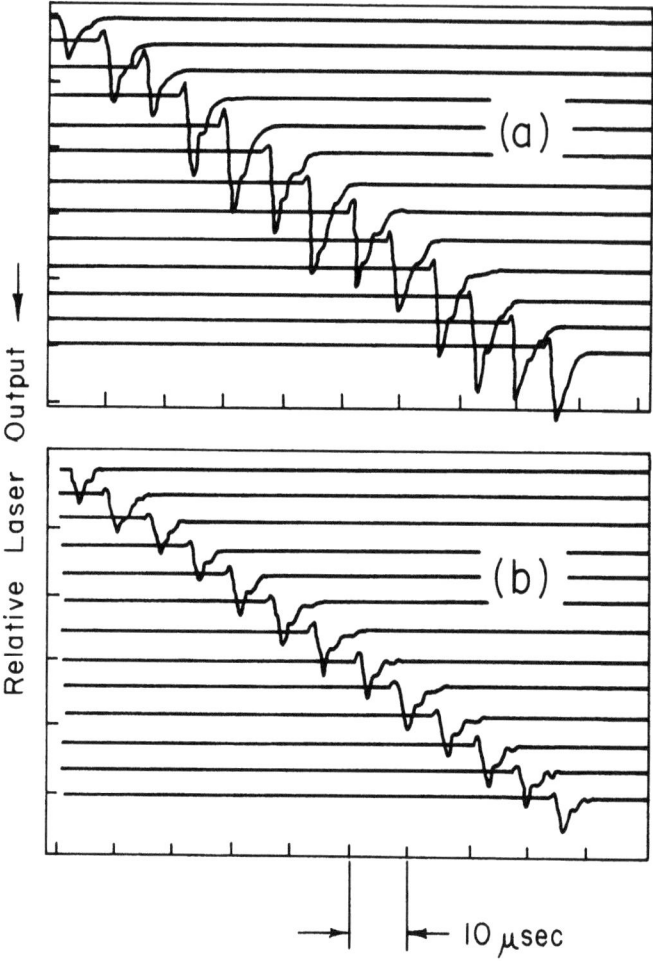

Fig. 7
Output of a 1/2-Atm TEA-laser (a) during a reactor pulse such as shown in Fig. 5; and (b) immediately after the reactor pulse.

High-Pressure (TEA Laser)

The results obtained in experiments with a high-pressure, transversely pulsed (TEA) laser (Fig. 6) are qualitatively the same as those discussed above for the low pressure pulsed discharge.

The laser used here is of the conventional linear TEA design except for several minor modifications. The anode was coated with boron to provide the external radiation source. Also, fairly large (10K) resistors were employed and they were potted in silicone compound to prevent spurious internal arcing which might be enhanced by the reactor radiation.

Figure 7 shows typical increases in output power which were achieved by neutron irradiation.

A raster display was employed so that each laser pulse (separated by 1 sec; the traces shown were inverted by the scope) appears as a separate trace. The first pulse at the top corresponds to the start of the reactor pulse [see 5(c)] and the 6th pulse corresponds to the peak reactor power. The laser pulses were displaced to the right in time by the raster for clarity. A maximum enhancement of a factor of 2 or 3 times is observed. Again, quite modest neutron fluxes, of the order of 10^{11} n/(cm^2 sec), were found to be effective.

The peak output of the TEA laser was of the order of 10 KW. It is hoped that the effect observed here can be scaled-up to higher power systems; however, this depends on the detailed mechanism responsible for the change, and considerable more study is required to positively identify this. Theoretically, the added radiation should provide a more uniform discharge and prevent formation of the "white arcs" that plague TEA lasers. Simultaneously, some tailoring of the electron energy distribution may occur such as to improve the pumping efficiency. However, the possibilities of mode jumping, transition switches, internal gas lens effects, etc. cannot be ruled out as contributing to these early results.

It should be noted that in both the low- and high-pressure pulsed discharges, the presence of the external ionization source makes it possible to achieve lasing as higher pressures for a given breakdown voltage. This is, in itself, an important result in light of the attempts now being made to pump CO_2 lasers pressures above 1 Atm where breakdown with obtainable power supplies becomes a problem.

Potential Applications of Enhancement

Although encouraging initial results have been obtained, these studies are only in their infancy and any comments about potential applications are purely speculative. However, several points of interest should be stressed. As brought out in other papers in this conference, the CO_2 laser offers an attractive approach to the high-power systems sought for laser-effect studies, laser-induced fusion, etc. To do this, it is clearly desirable to attempt operation of TEA-type lasers at pressures well above an atmosphere. However, this introduces problems relative to obtaining an uniform discharge or even the initial breakdown. Auxiliary ionization by an external source such as an electron beam as suggested in Ref. 24 or by nuclear radiation such as suggested here afford one approach. Nuclear radiation has an advantage at ultra-high pressures since a volume source can still be obtained via the neutron-^{3}He reaction or, alternately, by using gamma rays from the reactor, possibly supplemented by neutron bombardment of a cadmium liner to produce capture gammas. When the size, cost, and power level of some of the projected CO_2 lasers are considered, the use of a nuclear reactor as an auxiliary radiation source may not be at all unrealistic. However, it should be noted that the radiation intensities involved in the studies reported here are relatively low by reactor standards [10^{11} n/(cm^2 sec) vs. peak fluxes of 10^{16} from the Illinois TRIGA during a pulse.] The resulting ion currents are in fact only an order of magnitude above that obtainable with advanced radioisotopic sources. If future developments could close this gap, extramely attractive and simple systems employing radioisotope sources could be envisioned.

ACKNOWLEDGEMENTS

The work described at the University of Illinois represents studies by J. T. Verdeyen, J. C. Guyot, T. Ganley, and P. Thiess, and the author. This research is supported by grant AT (11-1) 2007 from the AEC Research Division. Continued support by the Illinois TRIGA staff is also gratefully acknowledged.

REFERENCES

1. J. C. Guyot, G. H. Miley, J. T. Verdeyen, and T. Ganley, "On Gas Laser Pumping via Nuclear Radiations," Symposium on Uranium Plasmas, U. of Fla., 1970; SP236. NASA Sci. and Tech. Inf. Div., Washington, D.C. (1971). pp 357-368.
2. G. H. Miley, J. T. Verdeyen, T. Ganley, J. Guyot, and P. Thiess, "Pumping and Enhancement of Gas Lasers via Ion Beams," 11th Int. Symp. on Electron, Ion, and Laser Beam Technology, U. of Colo., 1971 (in press, San Francisco Press).
3. K. Thom and R. T. Schneider, "Nuclear Pumped Gas Lasers," 9th Aerospace Sciences Meeting, N.Y.C., Jan. 1971, AIAA paper 71-110.
4. J. DeShong, Jr., "Optimum Design of High-Pressure Large-Diameter, Direct-Nuclear-Pumped, Gas Lasers," ANL-7030, Argonne Natl. Lab, Argonne, Ill. (June 1965).
5. J. C. Guyot, Nucl. Eng. Program, U. of Ill. private communication, August, 1965.
6. L. O. Herwig, "Summary of Preliminary Studies Concerning Nuclear Pumping of Gas Laser Systems," Rept. C-110053-5, United Aircraft, East Hartford, Conn. (Feb. 1964).
7. G. R. Russell, "Feasibility of a Nuclear Laser Excited by Fission Fragments Produced in a Pulsed Nuclear Reactor," Symposium on Uranium Plasmas, U. of Fla., 1970, SP236, NASA Sci. and Tech. Inf. Div., Washington, D.C. (1971). pp 53-62.

8. T. Ganley, Nucl. Eng. Program, U. of Ill., Urbana, Ill., private communication, August 1969.
9. J. R. Rusk, R. D. Cook, J. W. Eerkins, J. A. DeJuren, and B. T. Davis, "Research on Direct Nuclear Pumping of Gas Lasers," Rept. AFAL-TR-68-256, Northrop Labs, Hawthorne, Calif. (Dec. 1968).
10. J. C. Guyot, G. H. Miley, and J. T. Verdeyen, J. Appl. Phys. (to appear, Nov. 1971).
11. T. Ganley, J. T. Verdeyen, G. H. Miley, Appl. Phys. Ltrs., 18, 568 (1971).
12. P. E. Thiess and G. H. Miley, Trans. Am. Nucl. Soc., 14, 134 (1971).
13. V. M. Andriakhin, E. P. Velikhov, S. A. Golubev, S. S. Krasil'nikov, A. M. Prokhorov, V. D. Pis'mennyi, and A. T. Rakhimiov, Sov. Phys. JEPT Ltrs., 8, 214 (1968).
14. V. M. Andriakhin, V. V. Vasil'nov, S. S. Krasil'nikov, V. D. Pis'mennyi, and V. E. Khvostionov, ZHETF, Pis. Red., 12, 83 (1970).
15. J. A. DeShong, Jr., "Summary of Model-I Nuclear Pumped Gas Laser Experiments," Internal Argonne Nat. Lab, Rept., Argonne, Ill. (April, 1967).
16. T. Ganley, Nuclear Eng. Program, Univ. of Illinois, Urbana, Ill., unpublished results.
17. P. E. Thiess and G. H. Miley, "Spectroscopic and Probe Measurements of Excited State and Electron Densities in Radiation Induced Plasmas," 2nd Symposium on Uranium Plasmas, Georgia Tech, Nov. 1971 (to be published, AIAA).
18. P. E. Thiess and G. H. Miley, Trans, Am. Soc., 14, 131 (1971).
19. F. Allario, R. T. Schneider. R. A. Lucht, and R. V. Hess, "Enhancement of Laser Output by Nuclear Radiations," Symposium on Uranium Plasmas, U. of Fla., 1970; SP236, NASA Sci. and Tech, Inf. Div., Washington, D.C. (1971), pp. 397-400.
20. T. Ganley, J. T. Verdeyen, and G. H. Miley, Trans, Am. Nucl. Soc., 14, 133 (1971).
21. H. S. Rhoads and R. T. Schneider, Trans. Am. Nucl. Soc., 14, 429 (1971).
22. T. Ganley, J. T. Verdeyen, and G. H. Miley, "Nuclear Radiation Enhancement of CO_2 Laser Performance," 2nd Symposium on Uranium Plasmas, Georgia Tech, Nov., 1971 (in press, AIAA).
23. C. J. Chen, J. Appl. Phy., 42, 1016 (1971).
24. C. A. Fenstermacher, M. J. Nutter, W. T. Leland, and K. Boyer, "An Electron Beam Controlled Electrical Discharge as a Method of Pumping Large Volumes of CO_2 Laser Media at High Pressure," in press, Appl. Phys. Ltrs.

CHEMICAL MOLECULAR LASERS*

B. R. Bronfin

United Aircraft Research Laboratories

East Hartford, Connecticut 06108

ABSTRACT

Gaseous molecular media, which can be pumped via chemical reaction or selective molecular energy transfer processes, exhibit interesting characteristics which afford high power and efficiency. Within this category are the thermally-pumped gas dynamic lasers (CO_2 laser transitions) and the bimolecular exchange chemically pumped lasers (HF and CO laser transitions). Experiments in which these exemplifying lasers are established shall be reviewed with special attention being given to the mechanisms leading to population inversion.

---

*Presented at the Second Workshop on "Laser Interaction and Related Plasma Phenomena" at Rensselaer Polytechnic Institute, Hartford Graduate Center, August 30 - September 3, 1971.

PHOTOCHEMICAL IODINE LASER+

A HIGH POWER GAS LASER

K. Hohla, P. Gensel, K.L. Kompa

Max-Planck-Institut für Plasmaphysik

Euratom Association, Garching, Germany

For high power lasers there is a general tendency presently to explore the potential of gas lasers. Gas lasers have potential advantages over solid state laser materials. One can achieve larger geometrical dimensions and the active material can easily and cheaply be replaced if necessary. In addition, scattering losses will normally be lower and the breakdown field strength can be higher than for instance in glass lasers. Diffraction limited beam quality is obtained without problems and high repetition rates are possible. On the other hand there are two kinds of limitations peculiar to gas lasers: Firstly the cross section for stimulated emission σ is comparatively large due to the small emission line widths $\Delta\nu$ of gases. This causes limitations to the storage of energy due to superradiance and parasitic oscillations unless the line can be broadened by suitable means. The second disadvantage is collisional quenching which is usually quite efficient especially at higher pressures in gas laser operating on vibrational excitation.

We wish to report here some data on the high power operation of the photochemical iodine laser. This laser

---

+ Presented at the Second Workshop on "Laser Interaction and Related Plasma Phenomena" at Rensselaer Polytechnic Institute, Hartford Graduate Center, August 3o - September 3, 1971

was described first by Kasper and Pimentel[1] in 1964. It uses the photodissociation of alkyl iodides, mostly trifluoromethyl iodide CF_3I, to generate electronically excited iodine atoms in the $5^2P_{1/2}$ state (hereafter I*) which will lase to the ground state $5^2P_{3/2}$. Lasing occurs on the magnetic dipole transition at 1.315 μ ($h\nu_{laser}$ = .95 eV). While breaking the CF_3-I bond requires ~ 2.5 eV, the photolysis is accomplished via an absorption band of the CF_3I molecule around 2700 Å ($h\nu_{flash} \approx$ 4.75 eV) giving ample excess energy for the excitation of the products.

Photolysis $CF_3I + h\nu_{flash} \longrightarrow CF_3 + I^*$

Laser $I^* \longrightarrow I + h\nu_{laser}$

A variety of secondary chemical processes changes the concentrations of I* and I during and after the photolysis flash. A more complete picture of the relevant chemical reactions is given in the level scheme of fig. 1

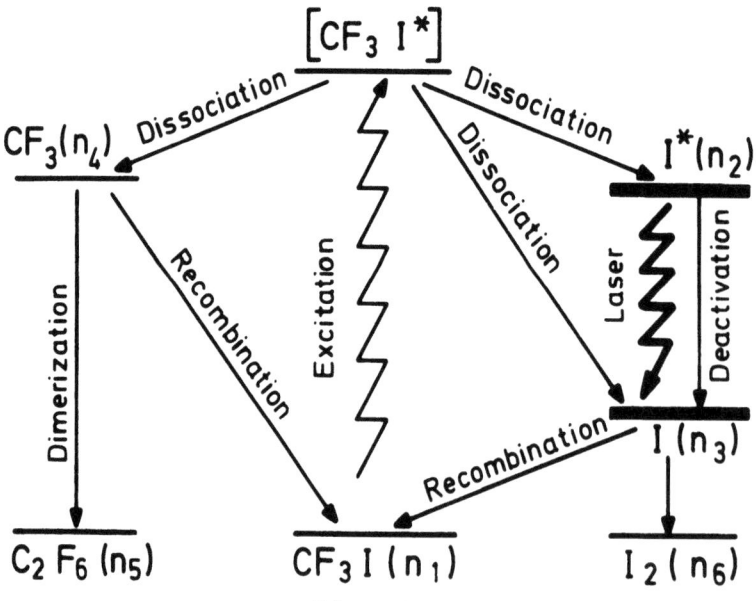

Fig. 1

Level scheme of the species n_1, ... n_6 present in the photochemical iodine laser. The relevant radiative and collisional processes are indicated.

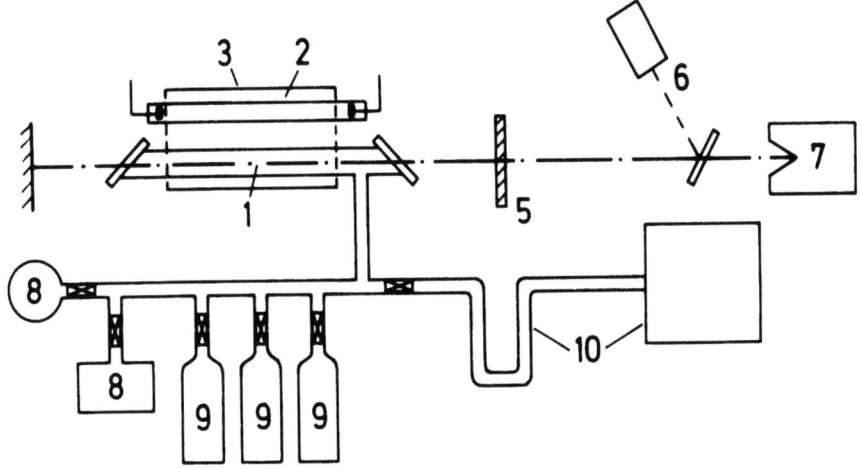

Fig. 2

Principal parts of the iodine laser oscillator.
(1) Laser tube with brewster angle windows,
(2) xenon flash tube, (3) aluminum foil reflector, (4) totally reflecting mirror, (5) partially transmitting mirror, (6) take-off plate and photodiode for control of the pulse shape, (7) conus calorimeter for measurement of the laser energy, (8) manometers and vacuum gauges, (9) bulbs for storage and purification of chemicals connected to a vacuum line, (1o) liquid nitrogen trap and oil diffusion pump.

Figure 2 shows a simple laser oscillator employing conventional flash photolysis. By using flash lamps of known light output and spectral energy distribution, the efficiency of the laser was found to exceed .5 % of the electrical energy input. However, no attempts were made so far to optimize the experimental parameters to further increase the radiation yield. In these experiments the output energy was 1o J.

Any giant pulse laser operation involves storing energy for some time to permit control of the pulse shape as for instance by Q-switching. Therefore, radiational and collisional loss processes which may severely limit the accumulation and storage of energy in the medium have to be investigated to explore the potential of a gas laser. From the scheme of fig. 1 a set of rate equations may be derived which includes the chemical reactions and the energy transfer processes. Using known rate constants or estimated values for all processes, numerical calculations were carried out to predict the decay of I* as function of time after the photolytic excitation. Time-dependent gain measurements were then chosen as the diagnostic method to check these calculations. A laser oscillator amplifier arrangement was used in these experiments. Since the gain in an amplifier disappears as the population inversion $\Delta N = N_{I*} - N_I/2$ decays gain measurements yield direct information on the concentration of excited iodine atoms as a function of time. If losses in the active medium can be ignored the following energy gain formula applies:

$$E_\ell = \frac{1}{1.5\sigma E_o} \ln(1 + (\exp 1.56 E_o - 1)\exp \sigma N\ell)$$

Here E_o is the input signal and E_ℓ is the amplified signal after a single pass through the amplifier length ℓ. The two unknown parameters are the inversion ΔN and the cross section of stimulated emission σ. Both can be determined by measuring the amplification for different input signals E_o. A severe problem were inhomogeneities caused by the formation of pressure waves in the photolysis mixture. These were checked by observing the diffraction of a He-Ne laser beam passing axially through the laser tube and were still weakly observable at pressures as low as 5 torr CF_3I.

The value of the stimulated emission cross section was found to be $\sigma = 6 \times 10^{-18}$ cm^2 at 20 torr CF_3I and 2×10^{-18} cm^2 at 100 torr. These data were confirmed by investigating σ as function of the CF_3I pressure independently in oscillator experiments. Another result of these measurements is that at CF_3I pressure of up to 100 torr the energy can be stored for several msec. Thus chemical limitations to the operation of this laser do not exist in this pressure range. The storage of energy is ultimately limited by the accumulation of molecular iodine which is a very efficient quencher. This however is formed only slowly by a three body recombination reaction.

The comparatively large value of σ gives rise to a very high gain. It is simply seen from the Schawlow Townes threshold condition $R_1 R_2 T^2 V^2 = 1$ ($R_{1,2}$ = mirror loss, T = transmission loss of the medium; $V = \exp \sigma \Delta N \ell$ (optical gain), that unwanted oscillations can hardly be prevented for any reasonable inversion density ΔN. Therefore σ has to be reduced if higher I^* densities are to be controlled. This is accomplished to some extent by foreign gas pressure broadening of the transition line. More effectively, however, inhomogeneous magnetic fields can be used. The effect of an inhomogeneous magnetic field is not only a splitting of the degenerate upper and lower energy levels of the iodine atom but additionally a Zeeman shift of the lines which varies along the axis of the laser tube according to the magnetic field gradient. The principle is shown in fig. 3

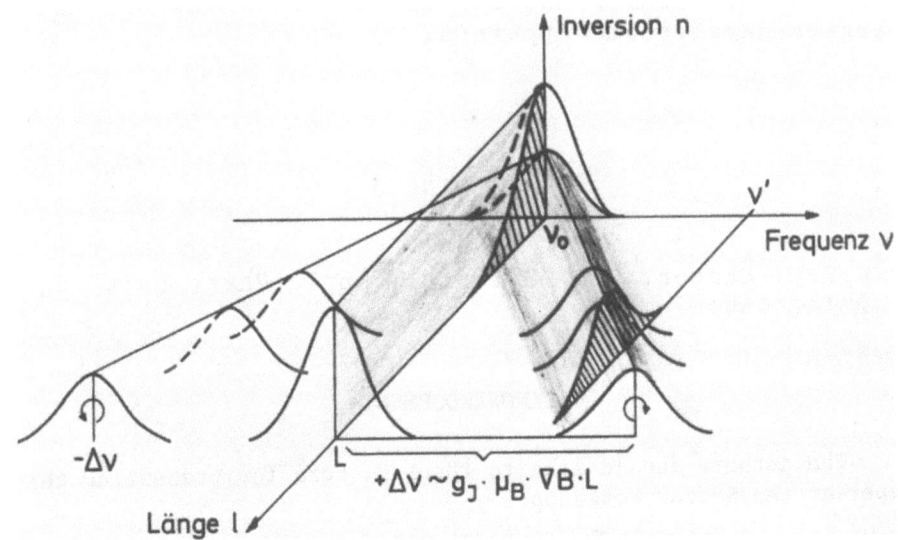

Fig. 3

Line broadening in the iodine laser by an inhomogeneous magnetic field.

It has been possible in this way to decrease the stimulated emission cross section to practically any desired value. With a magnetic field of 14 kG the cross section was found to be $\sigma = 8 \times 10^{-20}$ cm^2 and with 150 kG a value of $\sigma \sim 10^{-20}$ can be expected. This then is in the same range as for solid state laser materials. Accordingly the laser energy that can be stored is ~ 100 J cm^{-2} under these conditions, permitting full optical control of excited state densities up to 5 torr of I^*.

Preliminary experiments to release the stored energy in short pulses have led to an output of 4o MW with 15 nsec halfwidth. Besides using Pockels cell switches it has been found advantageous to use the principle of gain switching to obtain a small pulse width without switching elements in the cavity. This is possible if the build-up time for the photons in the cavity is longer than the pumping pulse. Operation in the lowest order transverse mode has been possible for small laser oscillators. Experiments are now in progress to generate giant pulses in the 1o J range using the principles of operation described here. A storable energy of > 5o J/liter will permit the construction of high power systems with reasonable dimensions. So far no limitations for large dimensions and energies have become apparent.

REFERENCES

1. J.V.V. Kasper, G.C. Pimentel, Appl. Phys. Lett. $\underline{5}$, 231 (1964)

ACKNOWLEDGEMENT

The authors should like to thank H. Hora for presenting this paper at the Second Workshop.

PHOTODISSOCIATION OF NO_2 BY PULSED LASER LIGHT AT 6943A[*]

John Gerstmayr, Paul Harteck & Robert Reeves

Chemistry Department, Rensselaer Polytechnic Institute, Troy, New York 12181

ABSTRACT

Nitrogen dioxide was photodissociated using a pulsed ruby laser at 6943A. The energy of a single photon at this wavelength was equivalent to only 57% of the dissociation energy. The mechanism proposed to account for the results was the consecutive absorption of two photons, the first resulting in a short-lived excited state. The second photon is then absorbed by the excited species resulting in dissociation.

[*]Presented at the Second Workshop on "Laser Interaction and Related Plasma Phenomena" at Rensselaer Polytechnic Institute, Hartford Graduate Center, August 30-September 3, 1971. Reprinted by permission from the Journal of Physical Chemistry, February 14, 1972.

INTRODUCTION

In work done previously in this laboratory it was demonstrated that two-photon emission was present in the reaction SO + O:

$$SO + O \rightarrow SO_2^* + h\nu$$
$$SO_2^* \rightarrow SO_2 + h\nu$$

The time delay between the two emissions was found to be of the order of 30 nanoseconds.[1] (Smith observed emission corresponding to a lifetime of SO_2^* of 12 nanoseconds in pulse electron beam studies.[2])

The purpose of the present work has been to investigate the possibility of a reverse mechanism of this type occurring in the dissociation of NO_2. Photodissociation becomes energetically possible at wavelengths below about 3945A.[3] Some dissociation still occurs around 4070A due to the availability of the vibrational and rotational energy of the molecule. At 4358A, however, no dissociation is found.[4]

At higher wavelengths, up to approximately 7900A, the combined energy of two photons would once again make the dissociation reaction possible. The following three mechanisms are considered for discussion:

(1) $NO_2 + h\nu \rightarrow NO_2^*$
    $NO_2^* + h\nu \rightarrow NO + O$

(2) $NO_2 + 2h\nu \rightarrow NO + O$

(3) $NO_2 + h\nu \rightarrow NO_2^*$
    $NO_2^* + NO_2^* \rightarrow NO_2 + NO + O$

A fourth possible mechanism would be:

(4) $NO_2^* + NO_2 \rightarrow 2NO + O_2$

The reaction of ground state molecules is 26 kilocalories endothermic or for the excited molecule it would need the equivalent energy from photons at wavelengths of about 10,000A or less. However, such a reaction has not been observed as noted above. The net result of any of these mechanisms is the production of oxygen when the fast reaction of NO_2 with O-atoms is included:

(5) $NO_2 + O \rightarrow NO + O_2$

The simultaneous absorption of two photons has been observed by several investigators. Pao and Rentzepis were the first to report a multiphoton process terminating in a specific chemical reaction: the photo-initiation of the polymerization of styrene and of P-isopropylstyrene.[5] G. Porter has reported the initiation of the explosive reaction of H_2 and Cl_2 by a two photon absorption at 6943A leading to dissociation of the Cl_2.[6] Speiser has used a Q-switched ruby laser to achieve a two photon absorption in iodoform, followed by the liberation of iodine.[7]

The absorption in some cases may have been, at least in part, consecutive rather than simultaneous, with a short lived excited intermediate absorbing the second photon. Porter has observed this consecutive two-photon absorption in the photodissociation of phthalocyanine.[8]

EXPERIMENTAL

A Korad K-1QP laser system was used in these experiments. The 9/16" ruby rod was operated with a passive Q-switch containing cryptocyanine dye to obtain single pulses of 1-2 joules energy and 10 nanosecond duration at 6943A. The energy of the laser output was verified using a Korad KJ-2 calorimeter.

The gases used were obtained from the Matheson Company, East Rutherford, New Jersey. The argon was supplied at 99.995% purity and was used without further purification. The nitrogen dioxide was further purified until it was better than 99.99% purity. Several mixtures of NO_2 and Argon were prepared and gas analyses were performed on a CEC 21-130 mass spectrometer.

The fluorescent lights in the laboratory were found to cause some dissociation of the NO_2, therefore the storage vessels containing the gas mixtures were covered with black cloth and the laboratory was in virtually total darkness at all times.

A cylindrical quartz cell of 9.5cm path length was filled to the desired pressure with the mixture to be irradiated and placed in the path of the laser beam. The laser was fired five times at two minute intervals. Each flash was monitored to ensure that the laser had produced only a single pulse, using a RCA-1P21 photomultiplier. The output signal of the phototube was recorded on film by a Tektronix 545-A oscilloscope fitted with a camera. The NO_2 was frozen out of the sample and the amount of O_2 present was measured against the argon standard on the mass spectrometer. Each set of experiments was run in one time span so that effects of variables related to laser operation, room temperature, etc. would be minimized.

Samples of NO_2 were repeatedly exposed to only the light from the Xenon flash lamp of the laser under normal experimental conditions to ensure that no oxygen was being produced in this way. The results consistently showed no detectable formation of oxygen (i.e., <.01%).

The dimerization which occurs in NO_2 ($2NO_2 \rightleftarrows N_2O_4$) had to be considered in these experiments. The true pressures of NO_2 were calculated for a series of gas pressures at 20°C using the equilibrium constants of Harris and Churney.[9] The results are shown in Fig. (1). All pressures of NO_2 cited in this paper refer to true pressures of NO_2 in the equilibrium mixtures.

The absorption coefficient of NO_2 was measured in the region of 6943Å using a Beckman DK-2 spectrophotometer. The experimental value of $\alpha = .15$ cm^{-1}atm^{-1} was in good agreement with the value found by Dixon in this region of the spectrum.[10] Dixon also reports that Beer's Law is valid in the pressure region of these experiments. Due to the fine structure of the NO_2 bands, the actual absorption coefficient for the very narrow laser line may be somewhat different.

Results

Using various pressures of NO_2 in the cell, oxygen was produced by firing the laser five times for each data point. Fig. (2) depicts the O_2 production curve in a mixture of 47% argon and 53% ($NO_2 + N_2O_4$), which is representative of our observations below 15mm of NO_2. Both O_2 as a percentage of total gas pressure and the number of O_2 molecules formed are plotted against the pressure of NO_2.

In order to have sufficient amounts of O_2 produced for purposes of gas analysis it was desirable to irradiate each sample 5 times. Calculations showed that the fractional loss of O_2 via the back reaction: $2NO + O_2 \rightarrow 2NO_2$, was negligible during the 20 min. interval between the first laser pulse and the analysis.

PHOTODISSOCIATION OF N₂O BY PULSED LASER LIGHT AT 6943A

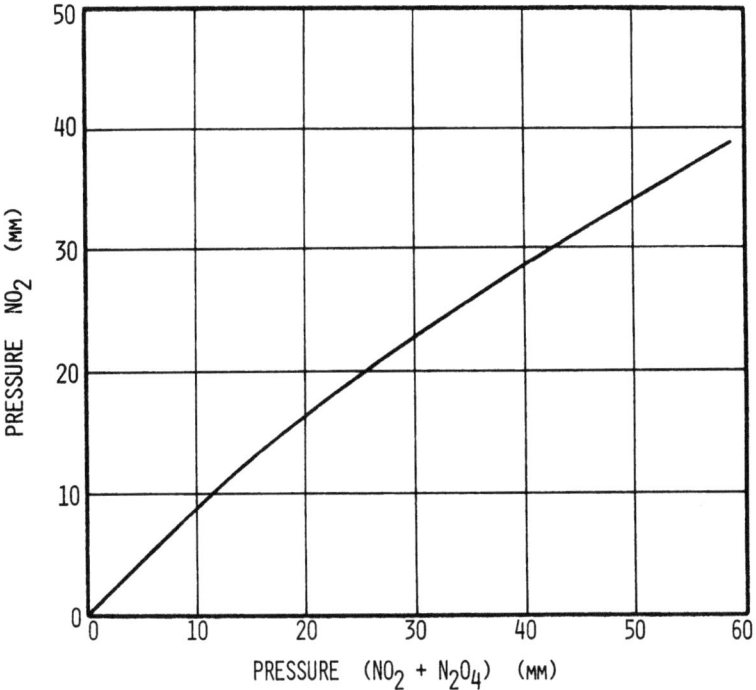

Figure 1 - Partial pressure of NO_2 vs. pressure of ($NO_2 + N_2O_4$) equilibrium mixture at 20°C.

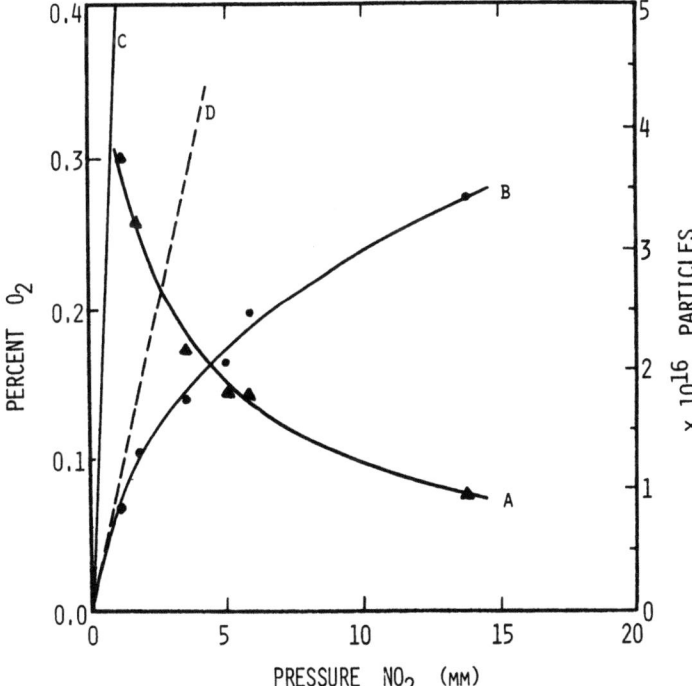

Figure 2 - O_2 formed by five pulses in a mixture of 53% NO_2 and 47% Argon: (A) as percent of total gas pressure and (B) as number of particles vs. pressure of NO_2; (C) number of NO_2 formed by five pulses vs. pressure of NO_2; and (D) tangent drawn to production curve at lowest pressures of NO_2.

Discussion

The formation of oxygen as shown in Figure 2 cannot be explained by ordinary photochemical mechanisms. As mentioned earlier, the absorption of light above ~4300A does not result in dissociation, but rather the formation of an excited state which eventually fluoresces or is quenched by collision. The dissociation energy of NO_2 into NO and an O-atom is put at 71.8 kcal/mole, while an einstein of light quanta at the laser wavelength was calculated to supply only 41 kcal. The results suggest that the energies of two photons are combining to cause dissociation of the molecule. Several mechanisms are possible: the simultaneous absorption of two photons, the consecutive absorption of two photons, and the collisional interaction of two singly excited species.

The results appear to be consistent with the consecutive absorption mechanism represented by:

$$NO_2 + h\nu_1 \rightarrow NO_2^*$$
$$NO_2^* + h\nu_2 \rightarrow NO + O.$$

(followed by $NO_2 + O \rightarrow NO + O_2$)

From the known absorption coefficient at 6943A, and expected radiative and collisional lifetimes, a significant amount of oxygen should be formed, assuming a reasonable coefficient for the second absorption to the continuum, resulting in dissociation of the NO_2.

The simultaneous process is expected to produce only small amounts of product such as mentioned by Porter in the initiation of the $H_2 - Cl_2$ reaction[6], much less than those observed. These would also follow a different reaction pattern because the results of Porter should be independent of quenching. The amounts of O_2 observed are also far in excess of those permitted by the low probability of the collision of two excited species before collisional deactivation occurs.

A primary consideration in discussing the dissociation of NO_2 by the consecutive absorption process is the production of the NO_2^* state by the laser light. Calculations made from Beer's Law using an average photon flux of 5×10^{18} photons/pulse (calculated from pulse energy measurements) showed that the number of excited molecules created during each pulse:

$$NO_2^* = 1 \times 10^{16} \text{ (particles/mm)} P(NO_2)$$

or for the five pulses:

$$NO_2^* = 5 \times 10^{16} \text{ (particles/mm)} P(NO_2)$$

where $P(NO_2)$ is the pressure of NO_2 in millimeters. This relationship is plotted along with the O_2 production curve in Fig. (2). Collisional deactivation at very low pressures becomes negligible during the pulse time of 10^{-8} sec., and a comparison with a tangent drawn to the experimental O_2 production curve at the lowest pressures suggests that about one in five of the NO_2^* produced eventually absorbs a second photon leading to dissociation.

We believe that the consecutive absorption process can be described by three rate equations:

$$+ \frac{d(NO_2^*)}{dt} = \gamma_1 (NO_2) \qquad (1)$$

$$- \frac{d(NO_2^*)}{dt} = \gamma_2 (NO_2^*) = + \frac{d(O_2)}{dt} \qquad (2)$$

$$- \frac{d(NO_2^*)}{dt} = (\lambda_1 + \lambda_2 + \ldots) (NO_2^*) \qquad (3)$$

where equation (1) governs the production of the singly excited species; equation (2) governs the loss of the NO_2^* via the absorption of the second photon; hence, also the production of O_2. Equation (3) governs the loss of the NO_2^* due to collisional deactivation with λ_1 λ_2, etc. relating to the quenching effects of the various components of the gas mixture. Since the duration of the laser pulse is much shorter than the radiative lifetime of 4×10^{-5} sec,[11] loss by fluorescence is considered to be negligible. The coefficients $\gamma_1, \gamma_2, \lambda_1, \lambda_2$, all have dimensions of reciprocal time. The rate coefficient for production of the NO_2^* was calculated from Beer's Law as $\gamma_1 = 2.0 \times 10^6$ sec^{-1}.[12] The general form of λ_1, λ_2, etc. is $\lambda = Qk(M)$, where Q is the quenching efficiency of a gas component, $k = 2 \times 10^{-10}$ particles^{-1} sec^{-1} and (M) is the number of particles of that particular gas component in the reaction volume. The only approximation which has been made is the assumption of a constant light intensity for the duration of the pulse. Solving these equations for the net O_2 production during the laser pulse time ($T = 10^{-8}$ sec.) yields the expression:

$$(O_2) = \frac{\gamma_2 \gamma_1 (NO_2) T}{(\gamma_2 + \lambda_1 + \lambda_2 + \ldots)} \times \left[1 + \frac{1}{(\gamma_2 + \lambda_1 + \lambda_2 + \ldots)T} \left(e^{-(\gamma_2 + \lambda_1 + \lambda_2 + \ldots)T} - 1 \right) \right] \quad (4)$$

If collisional quenching of the reaction is set equal to zero, the fraction of the NO_2^* which eventually absorbs a second photon and dissociates, is given by:

$$\frac{NO_2^* \to O_2}{NO_2^*} = \frac{O_2}{\gamma_1 (NO_2) T} = \left[1 + \frac{1}{\gamma_2 T} \left(e^{-\gamma_2 T} - 1 \right) \right] \quad (5)$$

This expression was calculated for several values of γ_2. Graphical analysis showed that our experimental observation of the dissociation of one in five singly excited molecules corresponds to $\gamma_2 = 4.6 \times 10^7$ sec^{-1}. Calculating again from Beer's Law (using the reaction volume, $V_R = 15$cc) yields the coefficient for the second absorption, $\alpha_2 = 3.5$ atm^{-1} cm^{-1}. The value of expression (5) is plotted against different values of the second absorption coefficient in Figure 3.

In order to test the validity of the theory presented, O_2 production curves were calculated for a mixture containing 50% ($NO_2 + N_2O_4$) and 50% Argon (a mixture similar to the experimental mixture of 53% ($NO_2 + N_2O_4$) and 47% Argon which is reported). The coefficient for the second absorption, α_2, was taken as 3.5 atm^{-1} cm^{-1} and quenching efficiencies were estimated from the results of Myers as .5 for NO_2, .1 for Ar, and 1.0 for N_2O_4.[13] If the dimerization is ignored, oxygen production levels off at higher pressures. When the quenching of the N_2O_4 molecule is taken into account, oxygen production reaches a maximum and begins to diminish. These calculated curves are shown in Figure - 4.

Preliminary experiments have verified this diminishing oxygen yield at higher pressures and have suggested a possible quenching efficiency for the N_2O_4 molecule that exceeds gas kinetic expectations. We are continuing our investigations in the higher pressure

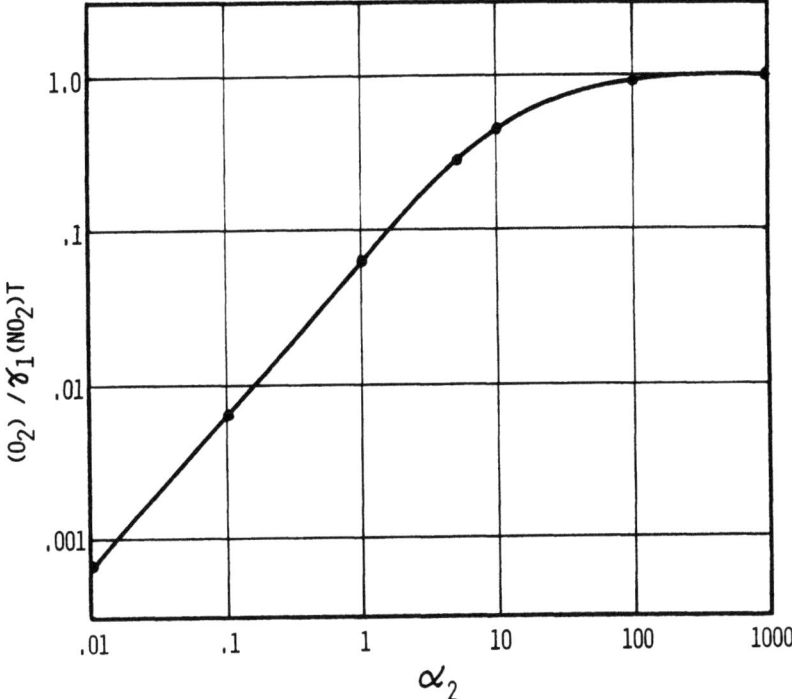

Figure 3 - Fraction of NO_2^* which absorbs a second photon with zero quenching vs. the second absorption coefficient, α_2.

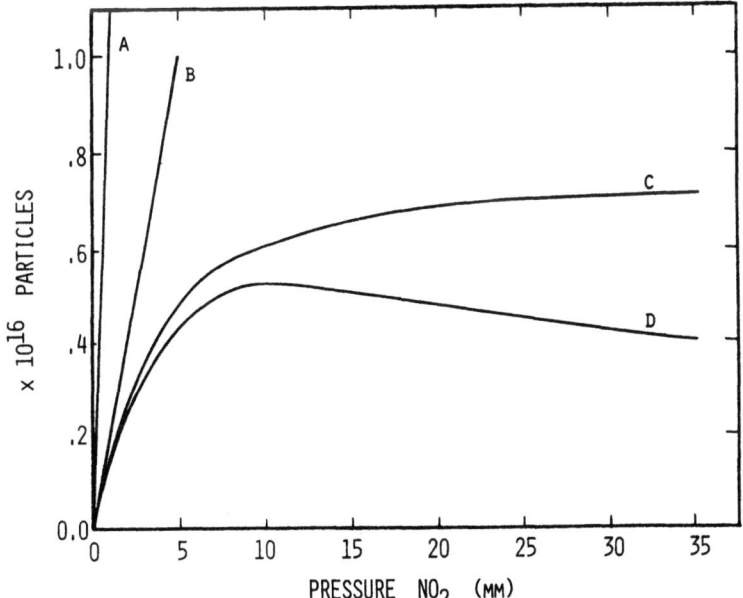

Figure 4 - Calculated values for a mixture of 50% ($NO_2 + N_2O_4$) and 50% Argon: (A) total number of NO_2^* formed per pulse vs. pressure of NO_2; O_2 production curves for (B) zero quenching, (C) quenching but no dimerization, (D) quenching and dimerization of NO_2 as shown in Fig. 1.

region where the deactivation processes dominate and are studying the effects of various other gases, such as CO_2, on the overall reaction.

This work was supported by a grant from the National Aeronautics and Space Administration NGL 33-018-007.

REFERENCES

1. J. A. Emerson, Ph.D. Dissertation, Rensselaer Polytechnic Institute, Troy, New York, (1969).
2. W. H. Smith, J. Chem. Phys. $\underline{51}$, 3410 (1969).
3. J. G. Calvert and J. M. Pitts, Jr., Photochemistry, John Wiley and Sons, Inc., New York (1966), pp. 217, 219.
4. P. A. Leighton, Photochemistry of Air Pollution, Academic Press, New York (1961), P. 47.
5. Y. H. Pao and P. M. Rentzepis, Appl. Phys. Letters $\underline{6}$, 93 (1965).
6. G. Porter, Nature $\underline{215}$, 502 (1967).
7. S. Speiser and S. Kimel, J. Chem. Phys. $\underline{51}$, 5614 (1969).
8. G. Porter and J. I. Steinfeld, J. Chem. Phys. $\underline{45}$, 3456 (1966).
9. L. Harris and K. L. Churney, J. Chem. Phys. $\underline{47}$, 1703 (1967).
10. J. K. Dixon, J. Chem. Phys. $\underline{8}$, 157 (1940).
11. D. Neuberger and A.B.F. Duncan, J. Chem. Phys. $\underline{22}$, 1693 (1954).
12. Applying the approximation that, for small values of y,

    $e^{-y} = 1 - y$ to Beer's Law results in the following expression:

    $$\Delta I = \#NO_2^* = \left[\frac{I_o \alpha_1 X}{\left(\frac{\#part.}{cc \cdot mm}\right) \cdot V_R} \right] (NO_2)$$

    where I_o is the average photon flux, α is the coefficient for the first absorption, X is the path length of the cell, V_R is the volume of gas exposed to the laser light (V_R = 15cc), and $\frac{\#part}{cc \cdot mm}$ is a conversion factor from pressure to particles. The term in brackets is equal to γ_1.
13. G. H. Myers, D. M. Silver and F. Kaufman, J. Chem. Phys. $\underline{44}$, 718 (1966).

SUMMARY OF DISCUSSION

(I. High Intensity Lasers)

The question arose as to what direction laser technology should go to develop laser pulses of some megajoule energy and about one nanosecond length in order to obtain inertially confined nuclear fusion plasmas[1] or to trigger a controlled nuclear explosion[2] for the purpose of generating controlled fusion energy. The general concensus of opinion was to continue building large neodymium glass laser systems as the best means for studying laser plasma interactions. However, the conclusion was reached that the neodymium glass laser will not be the final answer. Chemical lasers may be preferable because of their high energy concentration. For example, H. Hora brings out that the combustion energy of one gallon (about 3kg) of gasoline exceeds 10^8 joule. The report by B. R. Bronfin of a transfer of 60-70% classical energy into initial vibrational energy sounds encouraging.

As to what energy per pulse the TEA-laser can be developed, A. J. Beaulieu answered 10^5 joule. He indicated the limitations of such a system by the 5 J/cm^2 damage threshold of the materials used for the tube, mirrors, and other precision elements. C. Yamanaka in Chapter VI (see Appendix of his presentation) gives a value for the volume of damage threshold in platinum-free crown glass as 400 J/cm^2, whereas he finds a surface damage threshold of 150 J/cm^2 after treating the surface for 10 minutes with 10% HF.

References
1. I. J. Spalding, Culham Report CLM-R 109(1970); N. G. Basov, Kvantova Elektronika $\underline{1}$, No. 3 (1971); see also H. Hora and D. Pfirsch, these Proceedings.
2. J. G. Linhart, Nuclear Fusion $\underline{10}$, 211 (1970).

THE INITIAL STAGES OF LASER-INDUCED GAS BREAKDOWN*

Renaud Papoular

Association Euratom-CEA, DPh PFC,

B.P. N°6,92-Fontenay-aux-Roses (France)

ABSTRACT

Instruments and methods are described which have been used to detect and count free electrons released in a gas by laser radiation at power levels below breakdown threshold. The experimental results are analyzed and their possible relations to known elementary phenomena are discussed.

INTRODUCTION

Gas breakdown is defined as the sudden onset of a high electrical conductivity in a normally non-conducting gas. This, of course, is due to the appearance of free electrons in the medium and is generally accompanied by the emission of a bright light and, in the case of laser-induced gas breakdown, by a strong absorption of the incident laser light.

In fact, breakdown is a composite and dynamic phenomenon, extending over a measurable time and going through different stages, starting from the absence of free electrical charges and ending possibly in complete

---

*Presented at the Second Workshop on "Laser Interaction and Related Plasma Phenomena" at Rensselaer Polytechnic Institute, Hartford Graduate Center, August 30-September 3, 1971.

ionization of the gas, not to speak of the following after glow and recombination.

This paper is concerned with <u>the first stages of gas-radiation interaction</u> and with the mechanisms by which electrons are set free in the gas. Eight years after the discovery of laser-induced breakdown[1], these mechanisms are by no means completely clarified, probably because of the smallness of both space and time scales in the relevant experiments. The following pages must therefore only be considered as an attempt to describe experimental observations and their possible relations to already known elementary phenomena.

EXPERIMENTAL OBSERVATIONS

Fig. 1 represents the incident, unperturbed laser pulse (P_o), the light pulse (P) transmitted <u>through</u> the gas, and the light emission <u>from</u> the gas (E) <u>in the</u> visible band. The term "breakdown" is often used to designate the time when P decreases abruptly while e_λ increases rapidly. <u>In this paper, we are interested in the events occuring before that time</u>. Indeed, the laser

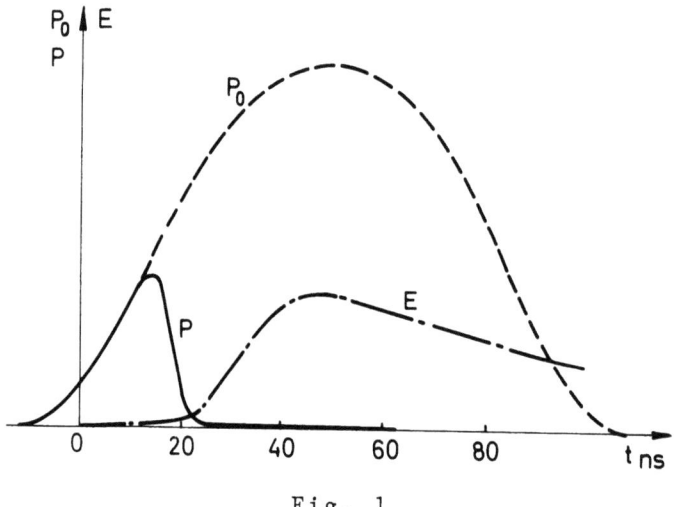

Fig. 1

Gas absorption and emission on breakdown.
P_o : incident laser light ; P : transmitted laser light, E^o : light emitted by the gas in the visible band.

pulse itself has a bell-shaped time profile and starts much earlier than the instant of breakdown (a few tens of nanoseconds). As a consequence, and although nothing particular seems to occur, on Fig. 1, before breakdown, it is possible, even then, to detect and count free electrons, if only sensitive instruments are used, such as cloud chambers photomultipliers and proportional counters.

Cloud Chambers[2]

When free charges appear at some point in a supersaturated vapor with enough density, the vapor condenses locally at that point and may be detected photographically as a small cloud or droplet. Naiman et al. used a cloud chamber based on this principle to detect charges formed at the focus of a laser beam. They used methanol or dimethylmethylphosphonate as condensible target vapors. The sensitivity of their chamber was 5000 ion pairs/cm^3, enough to detect minimum ionization cosmic-ray tracks. No such tracks were normally observed passing through the focal volume when the laser was fired. A clearing electric field (100 v/cm) was applied to sweep away any electron originating from other parts of chamber. Thus, any charge detected on firing the laser could only be due to ionization of the vapor by the laser itself.

Using a Q-switched ruby laser which could deliver a maximum of 0,2 J in 30 ns (i.e. about 10MW), they could exceed the breakdown threshold. A bluish plasma was then formed at the focus, giving rise to profuse swirling clouds that expanded throughout the chamber and persisted for a few minutes.

At lower powers, however, the passage of the laser beam resulted in a cluster of droplets, ranging in diameter from 0.1 to 0.5 cm, which fell to the bottom of the chamber at about the same rate as the background mist.

This proves that the laser could release a number of charges without necessarily ending in complete breakdown (ionization) of the vapor : this is called prebreakdown ionization.

Photomultipliers[3]

Let the light beam from a Q-switched laser be focused by a lens onto the gas in a pressure chamber. In the particular experiment to be described, the laser had an Nd : glass rod and delivered 40 nsec pulses of 25 MW peak power, and the lens had a 100 mm focal length. The light emitted by the focal region at a mean angle of 90° relative to the laser beam direction was collected by a field lens of 25 mm focal length and detected by a photomultiplier type Radiotechnique 56 TVP (S20 cathode, 14 dynodes, 2-ns rise-time). The anode signal was displayed on a Tektronix oscilloscope type 585 A (4.5 nsec rise-time)[4]. The stray laser light is trapped by a system of black-paper stops and a high-pass filter in front of the photocathode.

For a given gas and a given set of pressure and laser power values, oscillograms are taken with different achromatic attenuators in front of the photocathode. The smaller the attenuation, the earlier one can follow the development of the discharge ; from the corresponding partial curves (shown in fig. 2), an overall curve of brightness (I) against time (t) is drawn for each set of physical conditions.

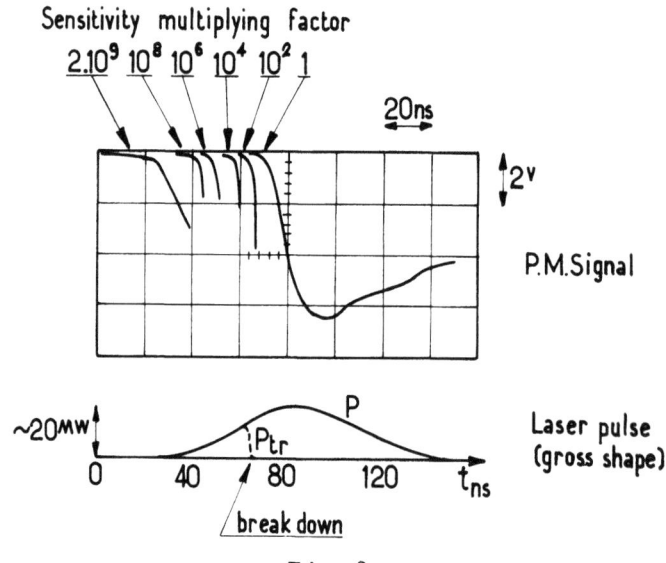

Fig.2

Raw photomultiplier data ; gas : argon, 1 atm; laser energy, $\varepsilon \simeq 1$ J.

At low attenuator settings the photomultiplier is strongly saturated during most of the time, but this does not cause any permanent damage ; it is enough to overlook anode signals higher than about 100 mA.

The experimental results are summarized in the following figures. Fig. 3 shows the result of <u>spectral analysis</u> of I(t) by means of a premonochromator with a spectral resolution of about 200 Å at λ = 5000 Å.

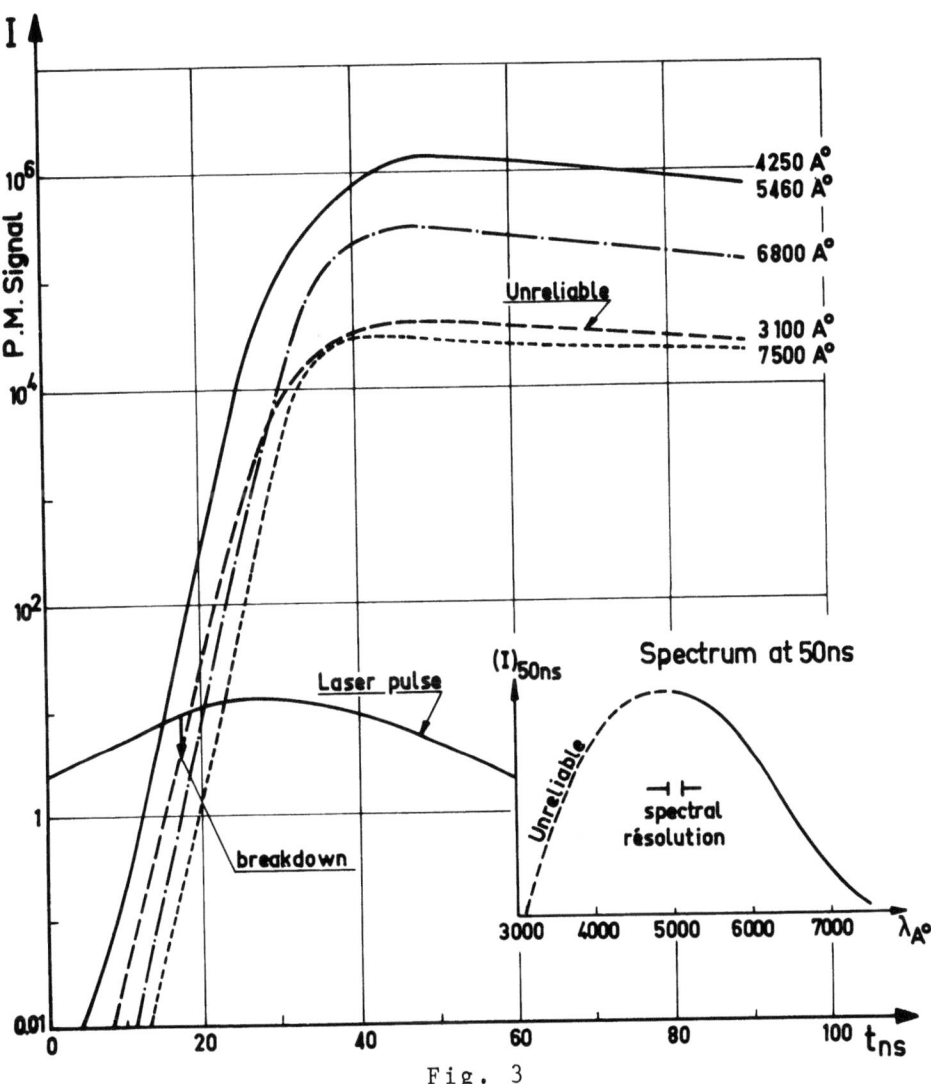

Fig. 3

I(t) for a few wavelengths (vertical scale arbitrary), Argon, 1 atm. $\varepsilon \simeq$ 1 J. Laser pulse shape also shown ; the vertical arrow shows time of breakdown.

Each curve corresponds to one wavelength and the inset shows the spectral profile at one instant near maximum brightness. These very crude measurements are an indication of <u>the predominance of the continuum over the line emission</u> during the period of interest.

Fig. 4 shows <u>I(t) for different pressures</u>. Notice that, when the pressure increases, the curves are displaced towards earlier times, their gross shape remaining unchanged in that part which occurs before breakdown (the latter is represented by a vertical arrow for each pressure). Light emission from the gas is clearly detected as early as 20 to 30 nsec before breakdown. <u>At such times, the laser power is less than 1/10 th of the threshold</u>, i.e. the minimum peak power to cause breakdown, which is indicated for 1 atm. by the horizontal dashed line, marked $(P_{th})_{1 atm}$.

Around the time of breakdown, <u>light emission increases approximately exponentially</u> during 20 to 30 nsec. In order to interpret this observation in terms of electron density, it should be noted that, before breakdown, the degree of ionization of the gas is still low and the light must be emitted mainly by free-free bremsstrahlung due to electron-neutral collisions ; I(t) should therefore be proportional to the free electron density in the observed volume. At breakdown and immediately after wards, electron-ion collisions are prevailing and I(t) should be proportional to the square of electron density.

Fig. 5 shows <u>the influence of laser energy</u> (ε) on the early gas emission. An increase of ε from 0,6 J up to 1J does not alter very much the gross shape of I(t) but shifts this curve to earlier times through about 60 ns. It has not been possible to draw a reproducible brightness curve for $\varepsilon < 0,6$ J.

Fig. 6 shows <u>I(t) for different gases</u> at 1 atm and $\varepsilon \simeq 1$ J.

Finally, Fig. 7 summarizes the <u>general behaviour</u> of the brightness I(t) in correspondance with the laser power P(t). The scale for I and P is arbitrary. If it is assumed that ionization is complete at maximum brightness, the free electron density, N, is then about 10^{19} cm^3 ; if, furthermore, I is taken to be proportional to N, as the case should be for emission by electron-neutral bremsstrahlung at constant temperature, then,

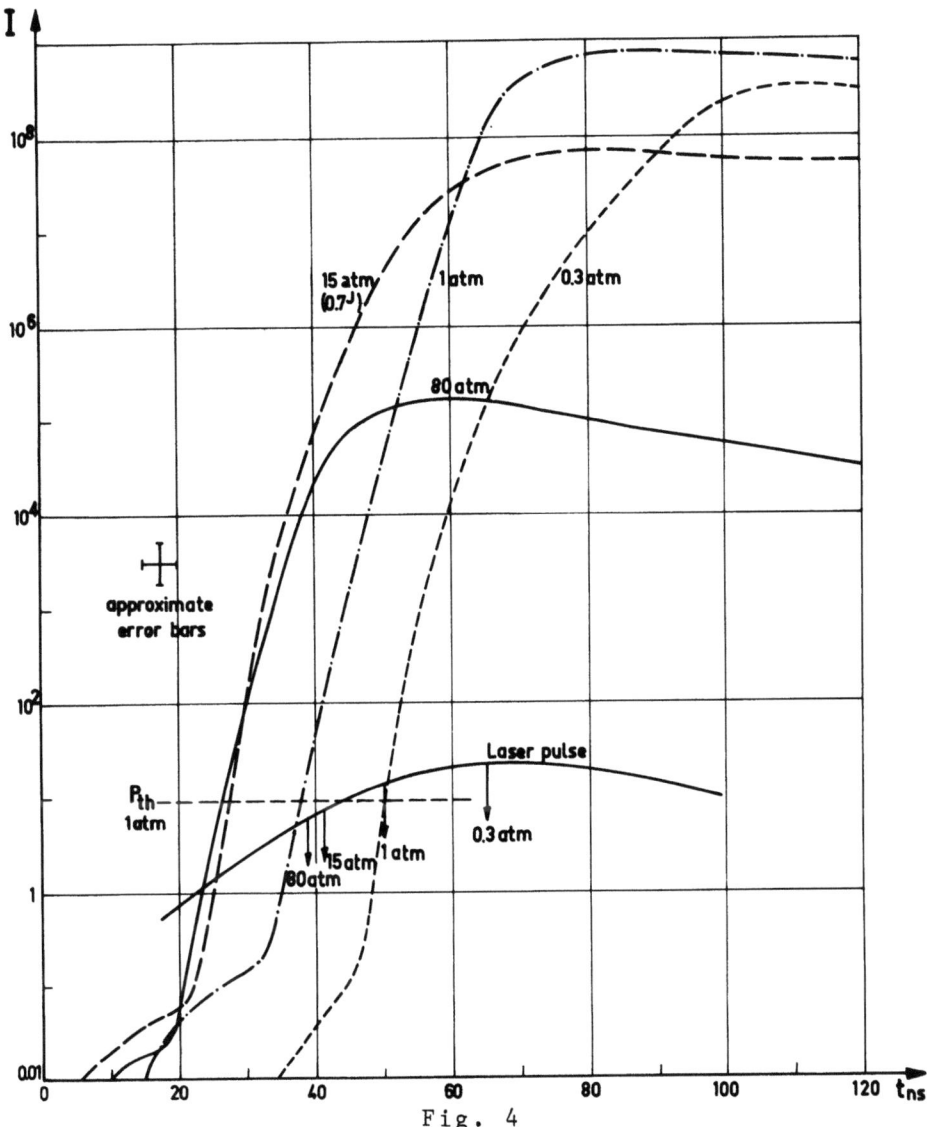

Fig. 4

I(t) for various argon pressures. $\varepsilon \simeq 1$ J.

at the foot of I(t), N should range between 10^9 and 10^{12} cm^{-3}, corresponding to 10^4 to 10^7 electrons in the focal volume (10^{-5} cm^3). If account were taken of electron-ion collisions beyond breakdown, and also of temperature variations, then still larger numbers would have been found for the minimum observable N. This is an indication that the photomultiplier method is not

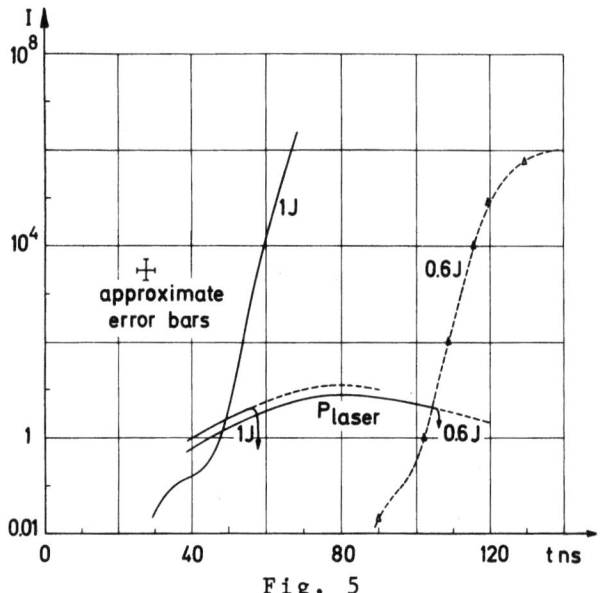

Fig. 5

I(t) for argon at 1 Atm. and two different laser energies (0,6 and 1 J) ; the corresponding laser pulses are also shown (P_{laser}).

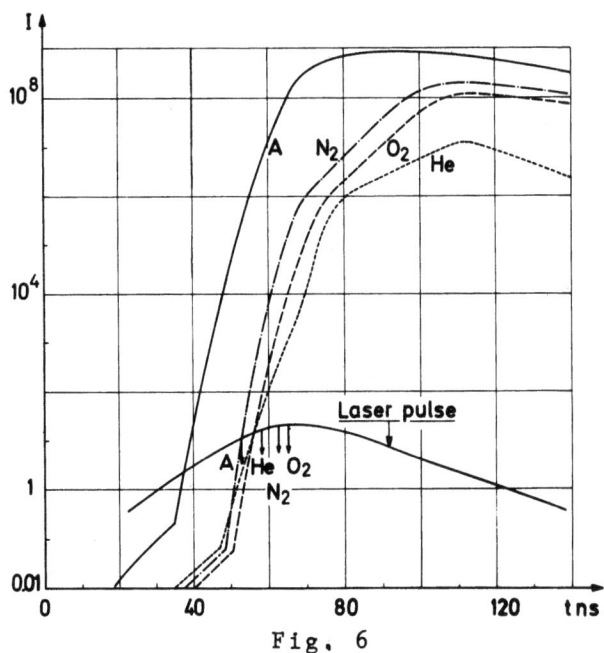

Fig. 6

I(t) for different gases at 1 atm and $\varepsilon \simeq 1$ J.

THE INITIAL STAGES OF LASER-INDUCED GAS BREAKDOWN

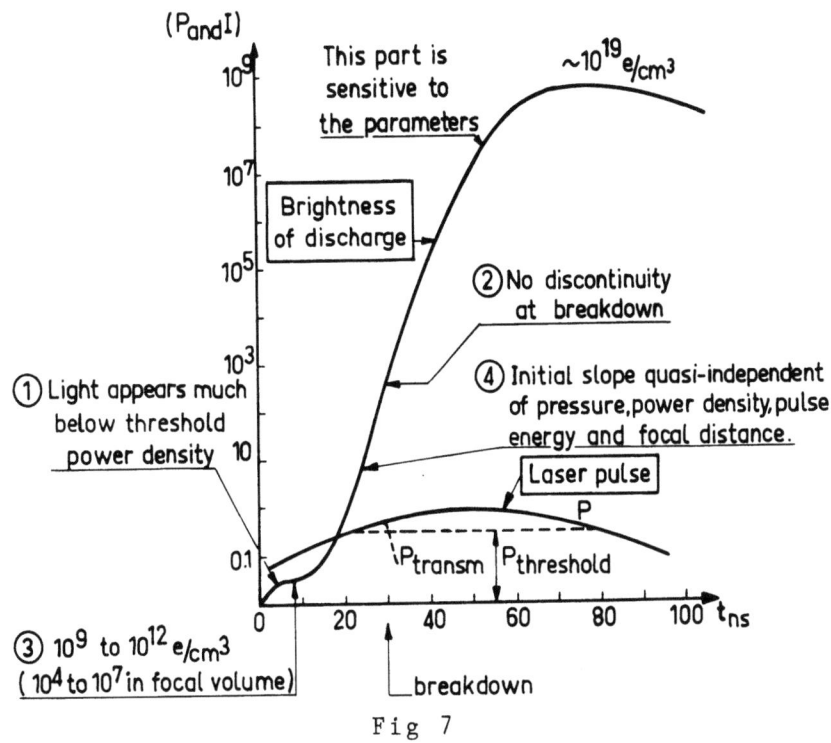

Fig 7

Summary of observations for a pressure of order 1 atm.
Vertical scale for I and P is arbitrary.

enough to detect the <u>initial</u> or <u>priming</u> electrons
for a definition, see "Discussion" below). Probably, when
meaningful signals start to be detected by this method,
the breakdown process has already developed into the
<u>avalanche</u> stage, (which is treated in detail elsewhere).
This could provide a method of studying the growth of
<u>cascade ionization</u> ; there is, however, a difficulty
here, in that it is not easy to account for the role of
temperature in the light emission from the focal volume.

Proportional counters[5,6,7]

Fig. 8 is a sketch of a measuring device[6], showing
the laser beam traversing a pyrex chamber containing
two spherical bronze electrodes, 50 mm in diameter, to
which is applied a dc potential difference varying between 0 and 40 Kv. The laser beam crosses the common
axis of the electrodes, a few millimeters above the
surface of the lower one (cathode). The n_t electrons
which are released by the laser pulse in this region are
accelerated by the dc electric field and thus can ionize,

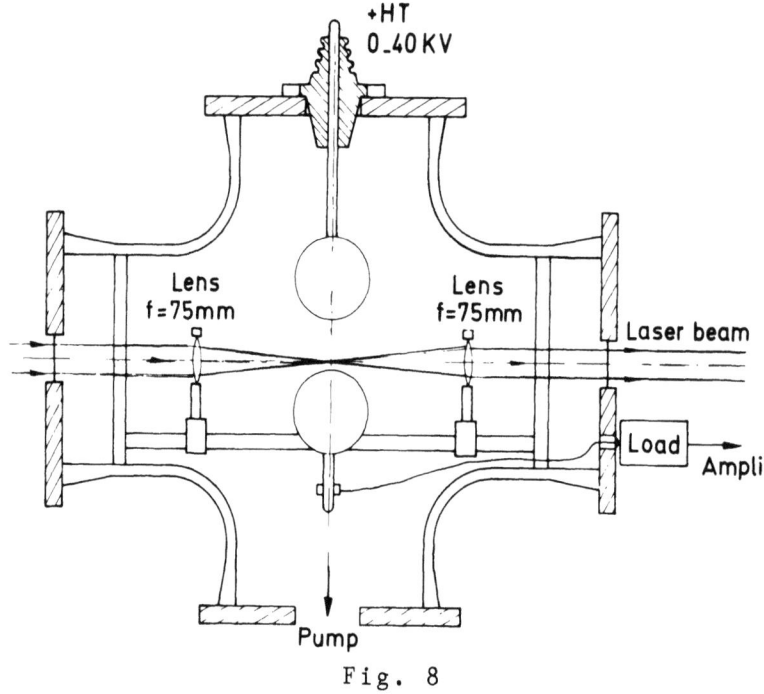

Fig. 8

Proportional counter with spherical electrode.

by collision, the molecules of the ambiant gas. The resulting free electrons are also accelerated and the cascade process goes on, giving rise to a growing avalanche of charges which can easily be detected when they reach the anode. If the total space charge, n, does not exceed the critical value (10^8 elementary charges) leading to the formation of a streamer[8], then the amplitude of the voltage pulse in the high voltage circuit is proportional to the initial number of free electrons in the focal region, assuming the time constant of the load is large enough (a few tens of μsec). It is clear that the maximum amplification factor $M = n/n_t$, is of the order of 10^8; but since 10^5 charges is already enough to give a detectable pulse of 1 mV (if the load capacity is of the order of 10 pF), an amplification of only 10^5 will allow one to detect a <u>single</u> initial electron.

Of course, the amplification factor, M, must be decreased as the expected number of initial electrons increases, so as not to exceed the streamer limit. This is done by dereasing the high voltage, for given gas

and pressure. Because of the stochastic nature of
the cascade process, M only represents the average value
of a number of measurements, the dispersion of which
decreases as n_t increases. M is determined either by
direct calibration, using a known value of n_t, or by
computation, using the known values of Townsend and
attachment coefficients.

The analysis of experimental results often requires
a knowledge of the volume in which every free electron
gives rise to an avalanche. In the case of fig. 8, this
is grossly the volume limited by the surface of the
spheres and the circumscribed cylinder, and the amplification factor is function of the initial position of the
priming electron. In practice, one can take M to be
constant over about 20 mm along the laser beam axis. The
cylindrical counter[7] with coaxial electrodes is superior
in this respect, because M is not sensitive to the initial position inside the counter (see fig. 9)

Fig. 10 shows experimental results obtained with a
counter of the type in fig. 8 ; curves of n_t (number of
electrons released by the laser pulse) against Pm (peak
laser power) are drawn for CO_2 at different pressures
and a focusing lens of 75 mm focal length (f_1), using
an Nd : glass laser with 40-nsec. pulses. The vertical
bars define the dispersion of measurements. For each
pressure, P_m has a minimum value, P_i, below which
no electrons are detected. This value is to be distinguished from the laser-induced gas breakdown threshold,
P_c, which is much higher (>20 MW for p<200 torr).

In between P_i and P_c, the curves exhibit three
distinct regions :

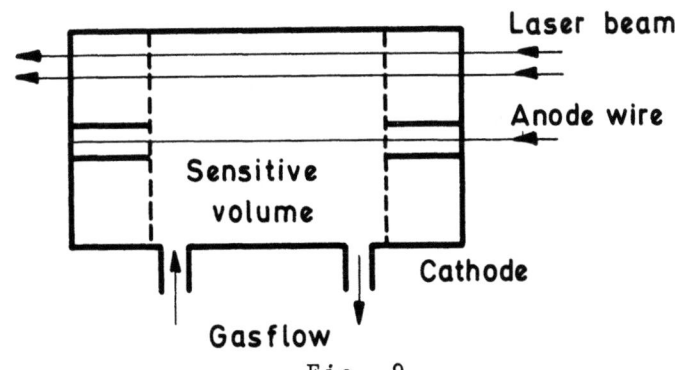

Fig. 9

Proportional counter with cylindrical electrodes.

a) an <u>initial abrupt rise</u> of n_t from zero to 10 or 100;

b) an <u>intermediate</u> region where $n_t \sim P_m^k$, with $k \simeq 2$;

c) an approximately <u>exponential</u> region, ending up in complete laser-induced gas breakdown.

The initial abrupt part is probably to be identified with results obtained with a cylindrical counter[7]; apparently, the range of powers used there was not sufficient to cover the three regions observed here (except, perhaps, in the case of CO_2).

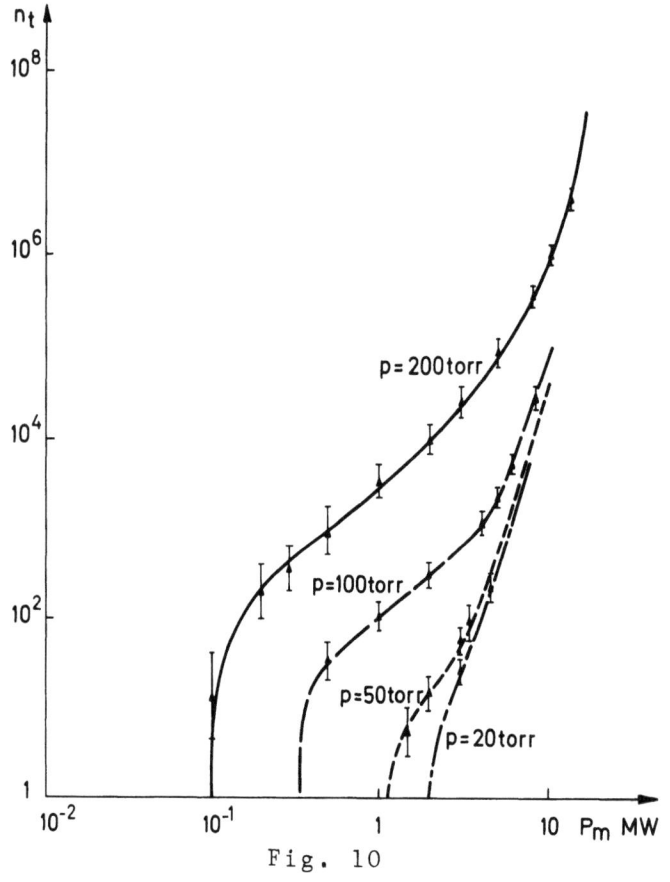

Fig. 10

Number of priming electrons, n_t, as a function of peak laser power, P_m; f_1 = 75mm is the focal length of the focusing lens; gas: CO_2; laser wavelength : 1,06μ

As regards stage (C), one is tempted to link it up with the exponential part of the brightness curve (fig.4) in the sense that, in both cases, the breakdown process has developed to the point where the laser light gives rise to cascade ionization of the gas. It must be kept in mind, however, that the counter <u>integrates</u> the electron production over the pulse duration while the photomultipliers respondes to the <u>instantaneous</u> population of electrons.

Fig. 11 shows typical curves of the minimum power, P_i, as a function of pressure, p, for a couple of focal distances, f_1=75 and 2500 mm. Below a pressure of some tens of torrs, P_i scarcely depends on p while, at higher pressures, the slope of the curves lies between -1 and -2. Similar results have been obtained for molecular hydrogen and for a ruby laser.[6]

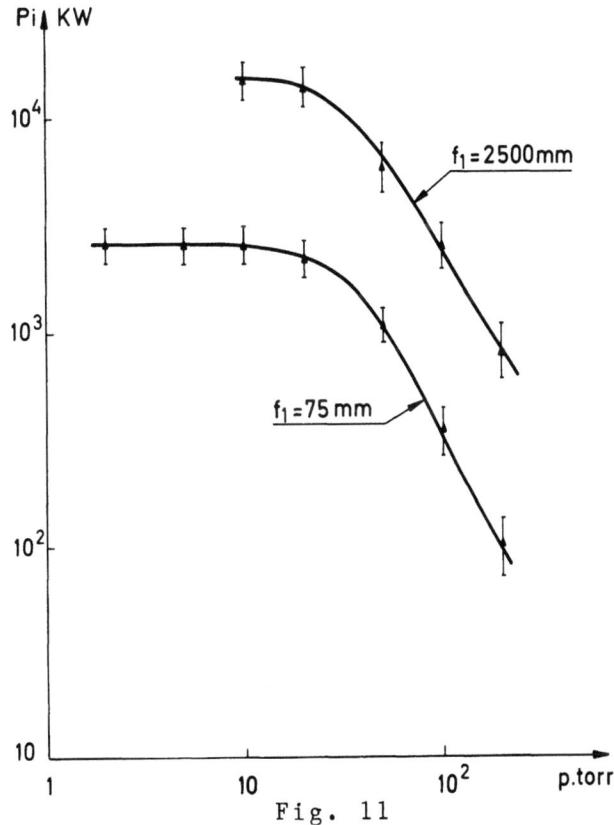

Fig. 11

Minimum power, P_i, as a function of pressure, p, for different focal distances ; gas : CO_2 ; laser wavelength 1,06µ.

As regards absolute values, P_i is of the order of a megawatt at pressures of about 100 torrs ; this value is not very sensitive to the nature of the gas (CO_2, H_2, N_2, A, He, Kr) nor to the difference in wavelength between ruby and neodymium glass ; it only increases by a factor of 10 when focal distance increases from 75 to 2500 mm.

Finally, the light flux corresponding to P_i is about 10^9 to 10^{10} W/cm^2, for f_1=75mm. This result is in order of magnitude agreement with results on CO_2, obtained in reference 7.

DISCUSSION[9,10,11].

In order to explain the production of free electrons during the early stages of breakdown, which we are interested in, one can think of two different mechanisms :

a) <u>photoionization</u> of neutral particles by one or more (say k) photons "acting together" so as to climb the potential barrier[12]: this is the <u>photoelectric effect of order k</u> and gives rise to the so-called <u>initial</u> or <u>priming</u> electrons ;

b) <u>cascade ionization</u> : this is due to free-free photon absorption by electron-neutral collisions, leading to an increase of electron energy, followed by neutral ionization by electrons (like in microwave breakdown) ; at least one priming electron is required to start this process.

(It is generally admitted that cosmic or other high energy radiation do not produce more than 10 electron-ion pairs per cm^3 and sec, and that the equilibrium concentration of negative-positive ion pairs does not exceed 10^3 cm^{-3} at NTP[13]. This, together with direct experimental evidence (e.g. reference 2), shows that background sources of electrons are not to be worried about).

The evolution of the free electron density N, in a gas of neutral particles of density N_o, under the action of a photon flux F, is then given by

$$\frac{dN}{dt} = a_k N_o F^k + b N_o N F \qquad (1)$$

where a_k and b are constant coefficients, and k is the smallest integer such that k times the photon energy exceeds the ionization potential (photoelectric effect of order k) ; for simplicity eq. (1) overlook loss terms (attachment, recombination and diffusion) and electron-ion collisions (the ion density being negligible in early stages).

The photons will be considered to be delivered in a pulse of duration τ and amplitude $F = P_m/s$, where τ is the half-power width of the laser pulse and P_m its peak amplitude, and s the area of the focal spot. By integration of (1) with suitable initial and boundary conditions, and neglecting the second term on the r.h.s, the minimum flux necessary to produce one free electron in the volume v under observation, is then found to be

$$F_i = P_i/s = (a_k N_o \tau v)^{-1/k} \quad (2)$$

which is proportional to $p^{-1/k}$, since $N_o \sim p$

When $F > F_i$, it is found that the total number of electrons set free in the volume v at the end of the pulse is :

$$n_t \simeq a_k F^k N_o \tau v \quad (3)$$

When the photon flux is much larger than F_i,

$$n_t \simeq \frac{a_k}{b} F^{k-1} e^{bN_o F\tau} \cdot v \quad (4)$$

which is exponential in F when $bN_o F\tau$ is large enough.

If the photon flux is not spatially homogeneous, the v in the above formulae is an "effective" volume over the flux distribution.

Let us now proceed to compare these equations with the experimental results, fig. 10 and 11. From (3) and (4), the slope $\alpha = d(\log n_t)/d(\log P_m)$ is equal to k or $k-1 + bN_o\tau F$, respectively. Thus, α should always be greater than k and increase with F steadily. Also, for a given F, α should increase with p (i.e N_o). Neither of these predictions is verified experimentally <u>in the range of parameters considered here</u>.

Also, according to ref. 2, the slope $\beta = d(\log F_i)/d(\log p) = d(\log F_i)/d(\log \cdot N_o)$ should be equal to $-1/k$. For the gases and photons employed here, k is about 10 to 20 , so that β should be very small, which fig. 11, for instance, shows not to be the case, <u>except at pressure lower than a few tens of torrs</u>.

Finally, one can compare experimental values of F_i with the values deduced from eq.2 by injecting the coefficients a_k given by measurements of the multiphoton effect[14], done at low pressures ($<2\times10^{-3}$ torr) so that cascade ionization is absent. It appears that multiphoton ionization of the <u>main</u> gas (CO_2, H_2, N_e, H_e, etc.) is hardly probable in the present experiments (at p>50 torrs) because the photon fluxes involved are much too low.

One would then conjecture that <u>impurities</u> of low ionization potential are present among the main gas molecules and give rise to the priming electrons[11,15]. However, the concentration of impurities should be independant of pressure, which make it difficult to explain why doubling the pressure can result in a increase of the electron yield by a factor larger than 10, at constant light flux (see fig. 10). The fact that the slope β (fig. 11) is of the order of -1 or less is also hardly accounted for by the presence of impurities. The above considerations apply all the more to breakdown induced by CO_2 lasers[16], since the photons, there, are ten times less energetic than those of Nd : glass lasers

Another conjecture could be that, at pressures higher than about 10 torrs, some sort of permanent, absorbing[17] or transient, "macromolecules" are formed with the molecules of the main gas, their size and/or concentration depending on pressure. For a given pulse duration, a minimum power would be required to ionize these particles (viz P_i). Would this critical value be reached, a large number of electrons could be liberated simultaneously, but once the particles were blown off, the only remaining mechanism for the production of free electrons would be cascade ionization.

This rough scheme is compatible a) with the sharp initial rise of $N_t(P)$ for low P_i's, followed by a bend going into an exponential, and b) with the strong inverse dependance of P_i on p.

On top of any such mechanism, which remains to be assessed, multiphoton ionization of impurities could also occur, and it would be useful to estimate the relative importance of the two phenomena. One way to do so is to analyze the products of ionization by means of a mass spectrograph and hence determine the concentration of impurities in the main gas.

In conclusion, it may be noted that a large amount of effort has been spent to build a theory of the avalanche process [10,11,18]. The present theories appear to be quite successful in predicting breakdown thresholds. This, however, is only a global test of the theory, and it is very much desirable that a direct experimental study of the cascade be carried out (e.g. the e-folding time as a function of pressure, gas and light flux). In principle, the proportional counter could be employed for this purpose, by measuring n_t as a function of the variables.

REFERENCES.

1. P.D. Maker, R.W. Terhune and C.M. Savage, Proc. 3rd Int. Quantum Electr. Conf. Paris (1963) ; Dunod (Paris) vol. II, 1559 (1964).

   E. K. Damon and R.G. Tomlinson, Appl. Opt. $\underline{2}$, 546 (1963).

   R.G. Meyerand and A.F. Haught, Phys. Rev. Lett. $\underline{11}$, 401 (1963)

2. C.S. Naiman et al, Phys. Rev., $\underline{146}$, 133 (1966)

3. V. Chalmeton and P. Papoular, C.R. Acad. Sc. (Paris), $\underline{264}$, 213 (1967) ; Meeting of the Plasma Physics Division of the A.P.S., Boston 1966, N° 3C2 ;
   V. Chamelton, Thèse (Paris) 1969, Chap. II.

4. The time response of a chain comprising a photomultiplier and an oscilloscope has been studied by R. Papoular, in Rev. Phys. Appl., $\underline{3}$, 169 (1968) for different time-profiles of the light signal.

5. W.R. Pendelton and A.H. Guenther, Rev. Sc. Inst., $\underline{36}$ 1546 (1965).

6. V. Chalmeton and R. Papoular, Phys. Lett. $\underline{26\ A}$ (11), 579 (1968) ,
   V. Chalmeton, J. de Phys. $\underline{30}$, 687 (1969) - Thèse (Paris, 1969), Chap. III-IV

7. A. Blanc and D. Guyot, Int. Conf. Phys. Ionized Gases Bucharest 1969, p.35, and private communications.

8. P. Raether. Electron Avalanches and Brakdown in Gases, London, Butterworth's, 1964.

9. V. Chalmeton, Thèse (Paris) 1969, Chap. V-VI

10. F.V. Bunkin and A.M. Prokhorov, Sov. Phys. JETP, 25, 1072, (1967)
    Ya. B Zeldovich and Yu. P. Raïzer, Sov. Phys. JETP, 20, 772 (1965).

11. A.V. Phelps, in Physcs of Quantum Electronics, 538, Mc. Graw-Hill, (1966).
    M. Young and M. Hercher, J. Appl. Phys, 38, 4393 (1967).

12. L.V. Keldysh, Sov. Phys. JETP, 20, 1307 (1965)

13. In Am. Inst. of Phys. Handbook, 5-278 (Mc Graw-Hill, 2nd ed.)

14. N.K. Berejetskaya et al, 9th Inter. Conf. on Phys. of Ionized Gases, Bucharest, 40 and 43 (1969).
    P. Agostini et al., C.R. Acad. Sc., 270B, 1566(1970)

15. S.L. Chin, Canad. J. Phys., 48, 1314 (1970)

16. N.A. Generalov, et al. JETP Lett., 11 (7), 228(1970)

17. Non-resonant absorption of light by gases at medium pressures has been observed by C. Bordier et al., C.R. Acad. Sc., 262 B, 1389 (1966) ; 263B, 619 (1966) and N.R. Isenor and M.C. Richardson, Appl. Phys. Lett. (1971)

18. F. Morgan et al, J. Phys. D, 4, 225 (1971)

LASER-PRODUCED GASEOUS DEUTERIUM PLASMAS*

Arthur H. Guenther and Winston K. Pendleton

Air Force Weapons Laboratory

Kirtland AFB, Albuquerque, NMex 87117

ABSTRACT

A brief summary of the theory of laser-produced plasmas is presented. Experimental measurements of the properties of laser-produced, low pressure (100-600 Torr) deuterium plasmas are discussed and interpreted using the best available theories. These measurements include, laser intensity breakdown threshold versus pressure, plasma growth rate, laser-plasma absorption and boundary interactions, spatial and temporal electron density distribution, and temporal electron temperature. A Time Variable Reflectivity (TVR) ruby laser having an output of up to 2 joules in 4 nanoseconds (FWHM) was used in this investigation.

INTRODUCTION

The production of opaque, highly ionized, high-temperature plasmas by laser beam focused into normally transparent gases has aroused considerable interest since its first observation in 1963.[1] The bright flash and sharp, audible crack resembled a small "spark discharge" at the laser focus and was immediately characterized as a region of highly ionized gas. Additional research has shown these "sparks" to be of high temperature, greater than 10^6°K in some gases. These characteristics, along with their explosive nature, have prompted the term "fireball" in their description.[2] In many laser interaction experiments, plasma generation is an

---

*Presented at the Second Workshop on "Laser Interaction and Related Plasma Phenomena" at Rensselaer Polytechnic Institute, Hartford Graduate Center, August 30 - September 3, 1971.

annoying limitation; however, the study of these laser produced plasmas has developed into a fertile area of research of its own in recent years. The Air Force Weapons Laboratory's interest in laser-induced breakdown relates to studies of the transport of high-power laser beams through the atmosphere as well as to the use of laser-produced plasmas in nuclear effects simulation programs. In addition, there is considerable interest in using extremely high-power lasers to enhance existing plasma generating devices such as the dense-plasma focus (DPF). Deuterium gas (D_2) has, to date, been of primary interest in this laboratory because of its general utility in numerous plasma devices (e.g., DPF), and because the simple atomic and molecular structure of deuterium facilitates the theoretical/experimental analysis necessary in the development of a comprehensive theory of gas breakdown. Also, certain necessary assumptions made in the experimental design and in the interpretation of the results are more valid for deuterium than other less simple gases. From an engineering standpoint it is essential that the fundamental parameters affecting the attainment of breakdown and the resulting plasma properties be understood so that scaling toward more powerful lasers and more intense plasmas can be achieved. In addition, it is believed that knowledge of the electron density and temperature in gaseous deuterium will contribute to a better understanding of research now being done elsewhere on the use of lasers to initiate a controlled thermonuclear reaction. An excellent review article by DeMichelis[3] summarizes the majority of papers published on laser induced gas breakdown in the open literature up to May 1968. The section on theory which follows is a qualitative discussion of the theories currently used to describe laser initiated gas breakdown. A comprehensive presentation of existing theoretical and experimental results is beyond the scope of this paper and is not necessary for its intent, that is, the presentation of laser-plasma experimental techniques and the plasma characteristics which are evidenced.
These measurements should form the basis for validation of various proposed theories.

THEORY

The process of gas breakdown at optical frequencies can be divided into five stages: initiation, growth, plasma development, extinction (recombination), and shock-wave propagation. Each state is somewhat distinct, dominated by a particular interaction or process. Each will be discussed briefly in turn according to the most generally accepted theory. This is done in order that the experimental results to be presented later are placed in their proper context.

Initiation

The initiatory stage is the time between arrival of the first laser photon at the lens focal region and the appearance of the first free electron-ion pair in the region. Recent observations of the breakdown with a picosecond ruby pulse[4] implies virtually instantaneous initiation. There are primarily two mechanisms which have been proposed for initiation: optical tunneling and multiphoton absorption. Tunneling, related to field emission, results when a bound electron passes through the Coulomb potential barrier to the free state under the extremely high electric field of the focused laser. Multiphoton absorption results, for example, from the simultaneous absorption of at least nine ruby photons ($h\nu$ = 1.78 eV) to surmount the 15.4 eV hydrogen molecular barrier.[5] Keldysh[6] has shown by a single, general formula that the nature of these two effects is essentially the same. Tunneling is the best physical model for high fields of low frequency, and the multiphoton picture is best for low fields of high frequency. The high fields at high frequency in focused laser beams fall between the two extremes.

Initiation has been carefully studied experimentally by Chalmeton[7] and by Blanc and Guyot[8]. In each case free-electron formation was studied at low pressures by careful charge amplification and collection. Referenced results show that free electron initiation occurs at laser intensity values at least two orders of magnitude below that needed to obtain a visible breakdown at higher pressure. Such initiation experiments are difficult at pressures where visible breakdown occurs, because of the uncertainty in accounting for electron losses in the neutral gas.

Growth

The formative growth stage is that period from initiation to "breakdown". Breakdown is an ill-defined term and at least one author has arbitrarily defined it as the attainment of an electron concentration greater than 10^{15} electrons/cm^3.[9] The electron multiplication (or avalanche) process characteristic of this stage is best described by an extension of the microwave gas breakdown model to optical frequencies. Growth of electron population is by cascade ionization whereby a low energy free electron absorbes photons by neutral atom and ion collisions (inverse bremsstrahlung or free-free absorption). After acquiring an energy greater than the ionization potential, the electron further ionizes the gas by impact. This results in two electrons of lower energy which start a new generation, causing the population to grow exponentially. The rate equations for this process were derived by Zel'dovich and Raizer[10] and have recently been solved[11] for deuterium and the experimental conditions of this research. The results indicate that a quasi-equilibrium electron distribution is established very early and is maintained throughout the growth stage.

The influence of diffusion and recombination electron losses and laser pulse duration upon attainment of breakdown (i.e., electron density $> 10^{15}/cm^3$) has been determined by Morgan, etal.[6] They have shown the most probable breakdown mechanism to be primarily pressure dependent under any given experimental conditions. They have classified the breakdown phenomenon into three pressure domains by solving the following electron-production equation assuming no electron drift or attachment:

$$\frac{\partial N_e}{\partial t} = D\nabla^2 N_e + N_e \nu_{ie} - R'N_e^2 \qquad (1)$$

where N_e is the electron density
  D is the diffusion rate coefficient
  ν_{ie} is the cross section for collisional ionization
  R' is the recombination rate coefficient.

The low, intermediate, and high-pressure regions are respectively dominated by the first, second, and third terms on the right-hand side. This equation was solved in cylindrical geometry for a triangular laser pulse of half width, T, and peak laser power, P_L. Threshold was defined as $N_e > 10^{15}/cm^3$. At low pressures where electron mean free paths are long, attainment of threshold is governed by electron diffusion; J_{th} (threshold intensity) is given by

$$J_{th} = \frac{\phi \ln(\beta P)}{PT} + \frac{\eta}{P^2 T} \qquad (2)$$

where ϕ, β, η are functions independent of pressure (P) and laser duration (T). At high pressure, electron diffusion losses are negligible; recombination losses dominate. The appropriate threshold expression in this region is

$$J_{th} = \frac{\phi \ln(\beta P)}{PT} + \xi \frac{1}{\sqrt{PJ_{th}}} \qquad (3)$$

where ϕ, β, ξ are again functions independent of pressure and laser duration. Between these limiting cases of very low and very high pressures there is an intermediate region in which the duration T of the laser becomes the dominant factor in the onset of breakdown. Ignoring both diffusion and recombination, the appropriate threshold expression is

$$J_{th} = \frac{\phi \ln(\beta P)}{PT} \qquad (4)$$

Examination of equations (2), (3), and (4) shows that for sufficiently large T the slopes of $\ln J_{th}$ versus $\ln P$ are expected to be (-2) for the very low (diffusion limited) pressure, (-1) for the intermediate (pulse limited), and (-1/3) for very high recombination limited) pressure regions. Direct comparison between these predicted slopes and experiment will be made in the next section for the pressure region from 150 to 600 torr. The measured slope will be shown to be -0.87 $\pm$ 0.08 which is very nearly the value -1.0 applicable in the pulse duration limited region.

Plasma Development

The plasma development stage begins at the moment of "breakdown" and continues until the end of the laser pulse. The higher electron density during this stage inhances the absorption of laser energy by the plasma. This increased energy content in the plasma in turn results in increased free electron concentrations, higher temperatures and plasma expansion. It is the very rapid plasma expansion, preferentially upstream toward the laser which has been the main subject of interest during this stage. Three different and independent mechanisms have been proposed for this very rapid axial growth of the plasma during laser irradiation[12]: breakdown wave, radiative transport wave, and radiation-supported shock wave.

A breakdown wave results when the local intensity of the focused laser beam reaches the threshold value. For a sharply rising laser pulse the location of threshold intensity moves toward regions of the focusing cone (caustic) having a larger cross section. To date, detailed calculations for the rate of propagation of this breakdown threshold wave have been restricted to the incoming caustic and have taken a geometric optics approach to the description of the laser intensity[12]. This approach was adopted because strong absorption of the laser by the plasma was generally observed and because of the extreme difficulty in incorporating the spatial microstructure of the focused intensity for multimode lasers. Under the conditions of low absorption evident in the present research, the propagation of the threshold wave down the exit caustic becomes important. The geometric calculation with low plasma absorption leads to a prediction of a very nearly symmetric threshold wave propagation about the plasma center. Observations shows the upstream growth velocity to be about 40 times greater than that downstream to 600 torr. The breakdown research of Alcock et al.[13] using single-mode lasers has simplified the description of the focused laser intensity distribution, however, as shown by Evans and Morgan[14] there is still considerable asymmetry in the focal region even for a well-behaved, Gaussian laser beam. Their calculated isointensity profiles show that local intensity maximum exist upstream of the geometric focus and that the output caustic intensity is much more uniform, having peak intensity values much lower than those upstream. The incorporation of these consideration into the breakdown wave theory may well yield the best physical description of plasma development under the present experimental conditions.

The radiative-transport wave theory of plasma development assumes that at a given time after breakdown the existing plasma heats and ionizes a layer in the surrounding gas, making further absorption of the incident laser radiation more likely. As suggested by Korobkin et al.[15], the plasma emission may be instantaneously absorbed in a layer of gas approximately equal in thickness to a mean free path for absorption of plasma radiation in the cold gas. The outer edge of this layer would then become absorbing, showing a jump in the location of breakdown. The focused beam is strongly absorbed at the shock wave boundary further increasing the local pressure and temperature toward the laser. The velocity of the detonation front can be derived from Chapman-Jouget theory[16] making use of the absorbed laser energy just behind the expanding shock front. It has been shown that once breakdown occurs, the radiation intensity required to support a detonation wave is at least two orders of magnitude lower than that corresponding to the breakdown threshold [17]. Such a breakdown mechanism predicts a smooth spatial asymmetric growth of the plasma. The observations of plasma symmetry, lack of evidence of a boundary shock, and low values of absorption are not consistent with the radiative-transport or radiation-driven shock wave theories at these pressures.

Extinction

After the laser pulse, the total energy of the plasma must decrease. Recombination of charge species, thermal radiation and conduction, and plasma expansion all contribute to this drop in energy as the localized breakdown region returns to equilibrium with its surroundings. The extinction phase lasts until all electron ion pairs have recombined and the temperature of the gas drops to a value where excited-state populations are negligible. In addition, any shock-wave ionization and excitation must have subsided. Experimentally, this stage is characterized as the period of decreasing plasma luminosity. Typically, depending on the gas and its pressure, it lasts for a few hundred nanoseconds after breakdown. During this stage the luminous region may grow to several times the volume attained during the development phase. This stage of hydrogen breakdown has been studied spectroscopically in detail by Litvak and Edwards[18] under conditions when local thermodynamic equilibrium is reasonably well established. Their work illustrates the extreme care which must be taken to obtain and analyze spectroscopic data from high-density, high-temperature plasmas.

Shock Wave Expansion

The final remnant of a laser-produced plasma is the hydrodynamic shock wave produced during the development stage. During the extinction phase the compressive shock wave leaves the luminous region and propagates into the surrounding gas. The theory of shock wave propagation in un-ionized gases is very well established, as a result of approximate solutions of the equations of conservation of mass, energy, and momentum at the shock front. The best calculations, good to second order in reciprocal Mach number, are those of Sakurai [19]. His results extend the hydrodynamic theory to weak shocks with Mach number less than two. First order theory (Taylor-Sedov) is generally useful only for strong shocks (> Mach 4). Laser-induced hydrodynamic effects last until the shock wave is dissipated at the walls of the experimental chamber or the walls of the laboratory. If there is no boundary close to the breakdown region, the shock wave is audible as a sharp crack.

Virtually all the experimental results now to be presented were obtained during the plasma development stage, i.e., during the time from "breakdown" to the end of the laser pulse. The results are in four parts. The first part deals with those measurements which yield gross properties of the plasma: threshold laser intensity, plasma growth rates, plasma/laser absorption, and boundary interactions. The second part treats the spatial and temporal electron density and the third part the temporal electron temperature. The final set of results were obtained during the shock wave stage of the breakdown. The results are in air, however, they illustrate the nature of this stage and the general utility of pulsed holography in the study of shock phenomenon.

PLASMA DYNAMICS

The laser-produced plasma characteristics classified under the heading plasma dynamics include, threshold, plasma transmission, plasma growth rate and laser/plasma boundary interactions. These characteristic are related to the macroscopic properties of the plasma and are normally determined prior to experiments on the plasma in detail. It is from these types of plasma properties that the theories discussed earlier were derived. The interpretaion of the electron density and temperature results is supported by the understanding obtained from such preliminary investigations. Figure 1 is a schematic diagram of the apparatus used to characterize the plasma dynamics.

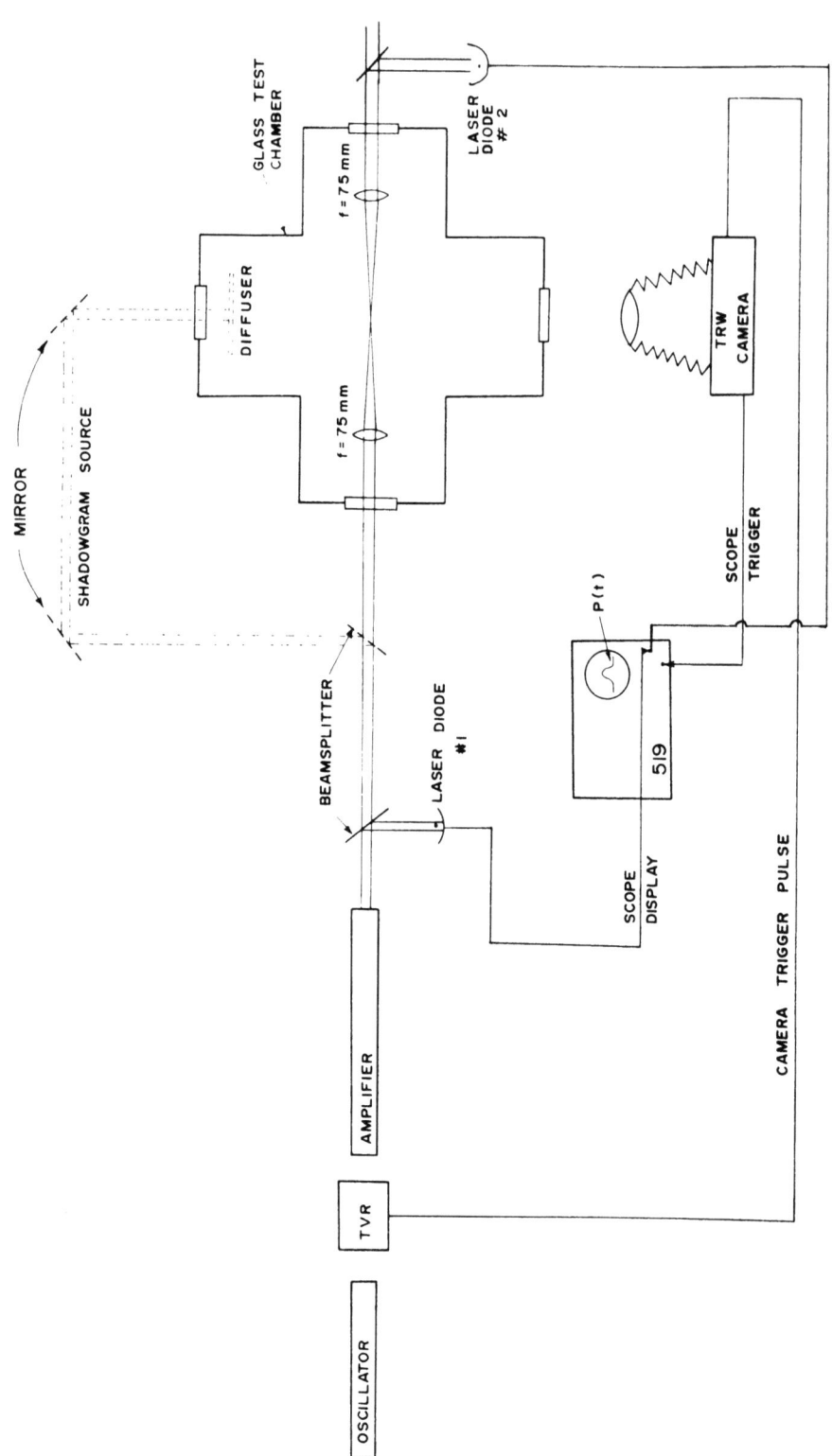

Figure 1. Plasma Dynamics Experimental Apparatus

Laser

The laser system used throughout this study was a Spacerays, Model OA-100/TVR (time-variable reflectivity) ruby oscillator/amplifier. Oscillations are achieved between a five-element resonant reflector in the rear and a time-variable reflectivity unit (sometimes referred to as pulse transmission mode, PTM) in front. The TVR unit contains a pulsed, deuterated potassium dihydrogen phosphate (KD*P) polarization rotator, a calcite polarizing prism and a nearly 100-percent reflectivity dielectric mirror in the configuration shown in Figure 2. The TVR unit behaves like a front reflector having either zero or 100% reflectivity depending on whether a bias voltage is applied to the KD*P. During flashlamp pump prior to achieving maximum population inversion the reflectivity is zero. The cavity is then Q-spoiled by applying the correct voltage to the KD*P to rotate the polarization by 90°. 100% reflectivity is then achieved by total internal reflection in the calcite prism to the dielectric mirror. At the preset time when maximum photon density in the cavity is achieved the bias voltage is removed and the cavity energy is transmitted through the calcite prism, amplified by a second ruby rod and focused into the deuterium gas. Further details on this particular TVR technique can be obtained from reference 20. The emitted laser pulse is nearly triangular with a width (full width at half-maximum intensity) of 4 nanoseconds and having total energy of 2 joules. This results in a peak power of up to 500 megawatts. Associated with TVR operation is an electrical spark gap signal which occurs $120 \pm .5$ nanoseconds before oscillator cavity dump. This feature gives sub-nanosecond timing capability. The output is monitored with a beam splitter and a fast, ITT bi-planar photodiode and Techtronix 519 oscilloscope.

The laser beam enters the breakdown chamber through a quartz window and is focused with a fused silica, plano-convex lens of 75-mm focal length. The convex face was toward the laser to prevent feedback of any specular reflections to the laser as well as to minimize spherical aberrations produced in the focal region by the lens. Confocal with the focusing lens was an identical lens to recollimate the transmitted light.

Threshold

Threshold has previously been defined as an attainment of 10^{15} electrons/cm^3 throughout the focal volume. Such a criteria is precise but highly impractical for experimental purposes. Therefore, as has been done traditionally, threshold is taken as that level of ionization detectable by the human eye. The value of $10^{15}/cm^3$ is actually an estimate of that density which is "seeable"

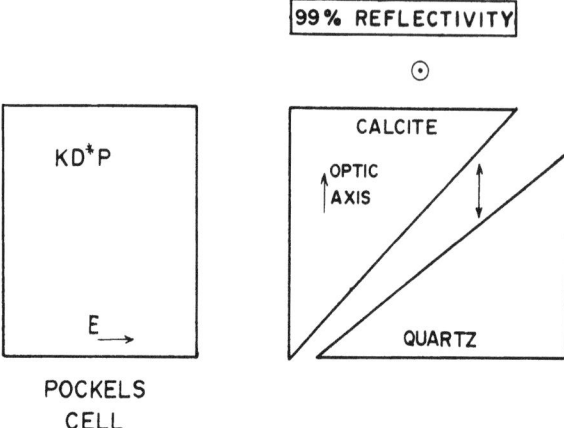

Figure 2. Time-Variable Reflectivity Q-Switch Device

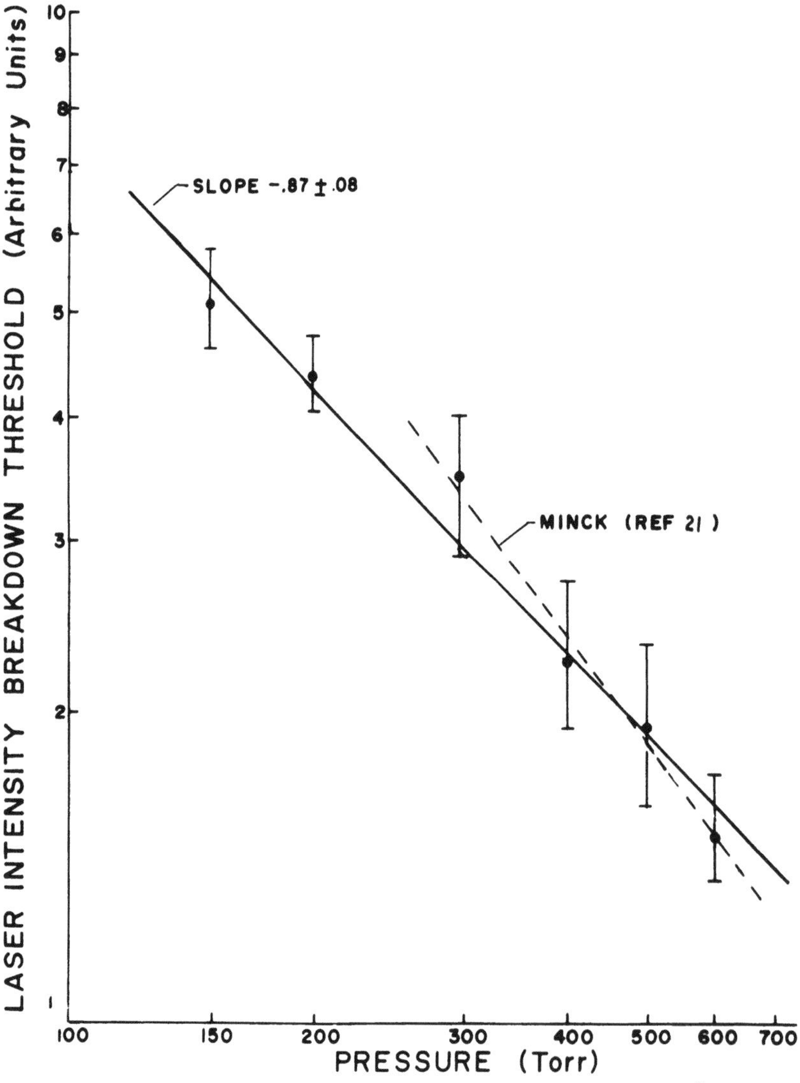

Figure 3. Deuterium Breakdown Threshold vs Pressure

by the eye under the conditions normally present in these types of experiments. Admittedly this is a subjective observation but it has proved useful because breakdown is highly dependent on laser intensity with breakdown first occuring over a range of laser intensity smaller than measurement resolution. The breakdown region was observed through a Nikon 35-mm single lens reflex camera placed in the TRW camera position shown in Figure 1. A series of laser shots was taken until the minimum power to produce observable breakdown was determined. These values are plotted in Figure 3 as a function of deuterium pressure from 150 to 600 torr on log-log scales. The ordinate is shown as normalized focused intensity averaged over the focal spot. Uncertainty in accurately determining the smallest spot size and the poor spatial uniformity of the beam did not justify the calculation of laser intensity on an absolute scale. For the purpose of determining the pressure dependence the normalized values given and the subjective detection scheme are sufficient. The data were least-squares fitted with a straight line having a slope of $-0.87 \pm .08$. This compares with a value of -1.0 predicted by Morgan, etal.[9] for the threshold dominated by the laser pulse length. The dashed curve is hydrogen data due to Minck and Rado[21].

Plasma Transmission

The transmission properties of the plasma were determined by simultaneously monitoring the laser pulse before and after it enters the plasma. As shown in Figure 1 the outputs of laser diodes 1 and 2 are combined by an impedence matching tee and displayed on a 519 oscilloscope. The chamber was evacuated to 0.2 torr (far below minimum breakdown pressure) and the photodiode gains adjusted to give pulses of approximately equal amplitude. In this way all reflection losses at windows and lenses were compensated for. The pressure in the chamber was raised and numerous traces of the transmitted pulse obtained. The insert in Figure 4 is a typical scope trace at 600 torr showing the incident and transmitted pulses. The equivalent vacuum pulse is shown as a dashed curve in the insert. Taking the ratio of transmitted/incident energy (the integral of the power trace) at various pressures results in Figure 4. Even at the highest pressure studied the transmission was 82 percent. This leaves at most 18 percent of the energy for producing and heating the plasma. Such low coupling of energy from the laser to the plasma has important consequences related to achieving the kinds of densities and temperatures deemed necessary for thermonuclear initiation or nuclear effects simulation. The values of absorption quoted here will be compared later with calculated absorption coefficients based on measured electron density and temperature. Also shown are results in D_2 of Bobin, etal.[22] and results in H_2 of Wick[23]. Because of the experimental conditions manifest in this work the results show the smallest observed

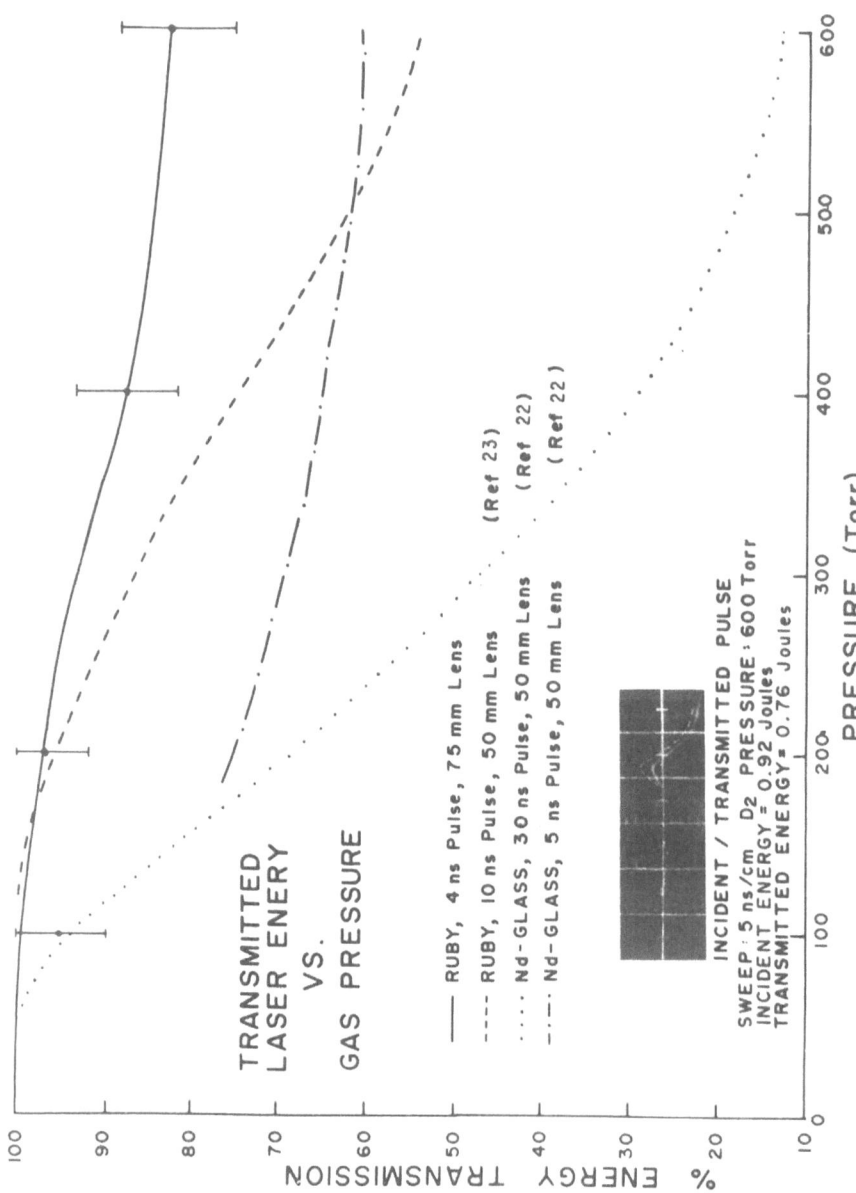

Figure 4. Transmitted Laser Energy vs Gas Pressure

absorption. It is not known definitely whether this is due to pulse length, wavelength or laser intensity. Based on the importance of pulse length demonstrated by the threshold pressure dependence it is believed that pulse duration plays a major role in the degree of absorption.

Plasma Development Rate

The determination of plasma development rate was made with the experimental arrangement of Figure 1, ignoring the dotted path titled Shadowgram Source. The diffuser A was replaced by a black velvet cloth as background for photography. High-speed streak photography was done using an image converter camera with a calibrated writing speed of 0.87 mm/ns $\pm$ 2%. Synchronization between the camera and laser was obtained by triggering the camera with the TVR prelaser signal. The camera monitor, which indicates the precise time the camera begins its streak, in turn triggers the 519 scope displaying the output of laser diode number 1. Subnanosecond variation in time correlation was routine and points out the value to TVR operation for these kinds of experiments.

Figure 5 shows several typical streak pictures. The directions of the incident laser and time axis are indicated. Streak 5a is the total emitted radiation of the D_2 plasma at 600 torr. Breakdown begins at the focal point and propagates upstream at an initial rate of 4×10^7 cm/sec until the peak of the laser pulse is reached at which time the velocity drops to about 3×10^5 cm/sec. Streak 5b was taken through a 0.6943 μm narrow band-pass filter. Regions of intense ruby scattering are shown to be at the plasma boundary. The discontinous nature of the breakdown is shown best in this record. Streak 5c was taken at 200 torr. Only the total plasma radiation is shown because the intensity of the scattered ruby radiation was too weak to detect at these pressures. The initial growth velocity upstream in this case is about 2×10^7 cm/sec during the rising portion of the laser pulse. The growth is seen to be smoother, not having the irregularities evident in streak records 5a and 5b. The asymmetric growth is notable since it does not match that predicted by existing plasma growth models at low absorption values.

Laser/Plasma Boundary Interactions

Two-dimensional profiles of the plasma were obtained by the shadowgram method; figure 1 indicates the arrangement. About 10 percent of the incident ruby pulse was diverted by means of a beam splitter and two mirrors onto a ground glass diffuser behind the plasma. The streak head on the camera was replaced with a 5-nanosecond exposure time framing head. Synchronization was

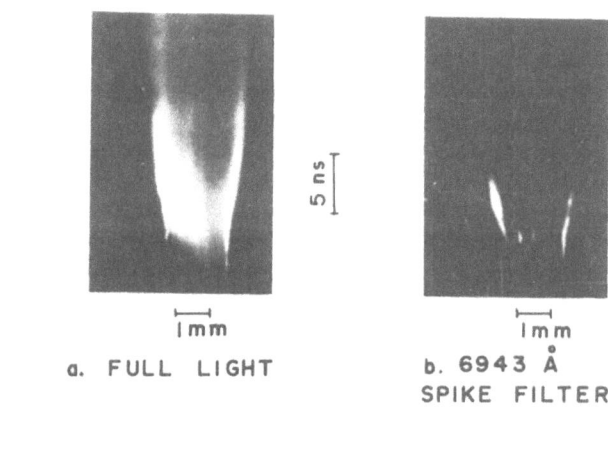
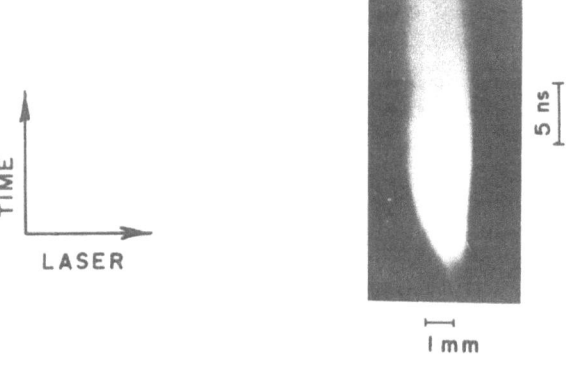

Figure 5. High-Speed Streak Photographs of Plasma Expansion at 200 and 600 Torr

obtained as before. A 0.69μm pass filter was placed before the camera so that only ruby light was detected. Figure 6 shows two typical shadowgrams taken at 600 torr. Shots 6a and 6b were taken on different plasmas and at a time difference of 1 nanosecond relative to the start of the laser pulse. In addition to the plasma shadow there are bright areas at the front and back of 6a, and the front only in 6b corresponding to areas of interaction between the laser and plasma boundary. The radius of the laser spot entering the plasma in 6b is about 100 μm. The actual dimensions are smaller because the exposure was taken over a 5-nanosecond interval when the plasma is expanding at about 10μm/ns. Thus, the spot is really about 50 μm in radius, very close to the estimated focal radius. These bright spots are conclusively due to reflections at the plasma boundary as calculations of refraction and scattering at 90° will show.

ELECTRON DENSITY

The electron density was determined by double exposure, holographic interferometry, using the same ruby pulse to produce as well as to diagnose the plasma. Interferometry, in general, is a technique for detecting wave-front distortions in space due to an object. Double-exposure, holographic interferometry permits the observation, of phase distortions in a given region of space which develop during the interval between the two exposure. The phase of an electromagnetic wave is determined by the phase index of refraction of the media through which the wave propagates. The measurement of electron density by this technique hinges on the fact that the plasma has a different index of refraction than the ambient, un-ionized gas. In particular, the free electrons in the plasma contribute a unique change in the phase refractive index during breakdown. This was first recognized experimentally in 1953 in the work of Dolgov and Mandel'shtam[24]. The application of optical interferometry to the quantitative study of the electron concentration in a plasma was worked out in detail by Alpher and White in 1958[25].

The dispersion formula for free electrons is derived from the solution of Maxwell's equations for sinusoidal wave propagation in a free electron gas with no magnetic field or resistivity[26]. The phase index of refraction, due to the electrons is given by:

$$\eta_e = 1 - \left(\frac{\omega_p}{\omega}\right)^2 \qquad (5)$$

where ω_p, electron plasma angular frequency, is defined as

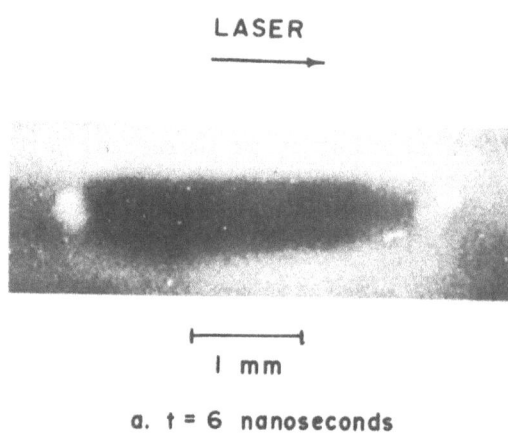

a. t = 6 nanoseconds
PRESSURE = 600 Torr

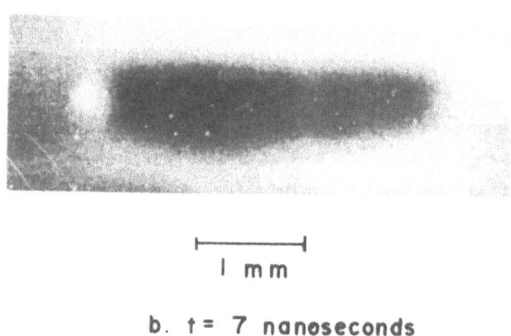

b. t = 7 nanoseconds
PRESSURE = 600 Torr

LASER POWER = 170 MW

Figure 6. Shadowgrams of 600 Torr Deuterium Breakdown Plasmas

LASER-PRODUCED GASEOUS DEUTERIUM PLASMAS

$$\omega_p = \left[\frac{4 N_e e^2}{M_e}\right]^{1/2} \cdot \left[1 + Z \frac{M_e}{M_i}\right]^{1/2} \quad (6)$$

Ignoring $M_e/M_i \ll 1$ and substituting in constants,

$$\omega_p = 5.63 \times 10^4 N_e^{1/2} \quad [N_e \text{ in electrons/cm}^3].$$

For the case of ruby, $\omega = 2.72 \times 10^{15}$ radians/sec and for $N_e \leqslant 3 \times 10^{19}/\text{cm}^3$,

$$\left(\frac{\omega_p}{\omega}\right)^2 \leqslant 0.013$$

giving equation (5) approximately as

$$(\eta - 1)_e \approx -2.150 \times 10^{-22} N_e \quad (7)$$

As can be seen the electron contribution to the phase index of refraction is negative. The index of refraction of molecular deuterium is given by

$$(\eta - 1)_{D_2} = 5.06 \times 10^{-24} N_{D_2} \quad [27] \quad (8)$$

and for atomic deuterium (assumed equal to ordinary atomic hydrogen)

$$(\eta - 1)_D = 4.18 \times 10^{-24} N_D \quad [28] \quad (9)$$

Thus, the change in index of refraction of the breakdown region due to the plasma is

$$\eta_p - \eta_{D_2}^\circ = -2.15 \times 10^{-22} N_e + 5.06 \times 10^{-24} N_{D_2}$$
$$+ 4.18 \times 10^{-24} N_D - 5.06 \times 10^{-24} N_{D_2}^\circ \quad (10)$$

where $N_{D_2}^\circ$ is the original concentration of molecular deuterium. Equation (10) can only be solved for N_e by eliminating N_D and N_{D_2}. This was done by assuming all deuterium atoms were equally divided between atoms and molecules. (The number of atoms is twice the number of molecules.) The relationships between original molecular density, electron density, atomic and molecular densities are

$$N_D = \frac{2N_{D_2}^\circ - N_e}{2} \quad (11)$$

$$N_{D_2} = \frac{N_D}{2} \quad (12)$$

This assumption (at 10 percent ionization) results in an error in N_e of $\pm$ 3 percent. Substituting equations (11) and (12) into equation (10) gives the final expression for the index of refraction change of a given point in the breakdown region as

$$\eta_p - \eta_{D_2}^\circ = -2.18 \times 10^{-22} N_e + 1.65 \times 10^{-24} N_{D_2} \quad (13)$$

Serious underestimation in the value of N_e would result if any high positive index species were present. Values of N_e will be shown to be consistent with other observations given support to the validity of equation (13).

Experimental Apparatus

Figure 7 is a schematic diagram of the experimental apparatus used to determine the electron density and its distribution. The optical path from the laser to the chamber was lengthened as shown to enable holograms to be taken as early as 4 nanoseconds before the peak of the plasma-producing pulse reached the focal volume. Two, spatially equivalent holographic beams were produced by the wedge and apertures. The scene beam was directed through the breakdown region and onto a 55-mm focal length achromatic lens located such that the breakdown region was focused on the holographic film. The reference beam was routed around the breakdown region by a prism and inverted by a 100-mm lens. This lens was located such that the reference beam spot size on the film was the same size as the scene beam. The path length difference between the two beams was less than 3 centimeters. This arrangement minimizes the requirement for spatial and temporal coherence which are notoriously poor for high-power, multimode ruby lasers. More on the theory and technique of pulsed, interferometric holography can be found in references 29 and 30. The holograms were made in two exposures. For the first exposure, the plasma producing pulse was blocked to prevent breakdown. The reference beam mirror was then rotated about 4 milliradius around its vertical axis. The plasma was produced during the second exposure. Rotation of the reference mirror has the effect of imposing background fringes on the reconstructed image, the spacing of the fringes being controlled by the angle of rotation. Figure 8 shows the wavefront reconstruction scheme. A spatially filtered, collimated Spectra-Physics 125 He-Ne beam was focused through the hologram. Real images were formed

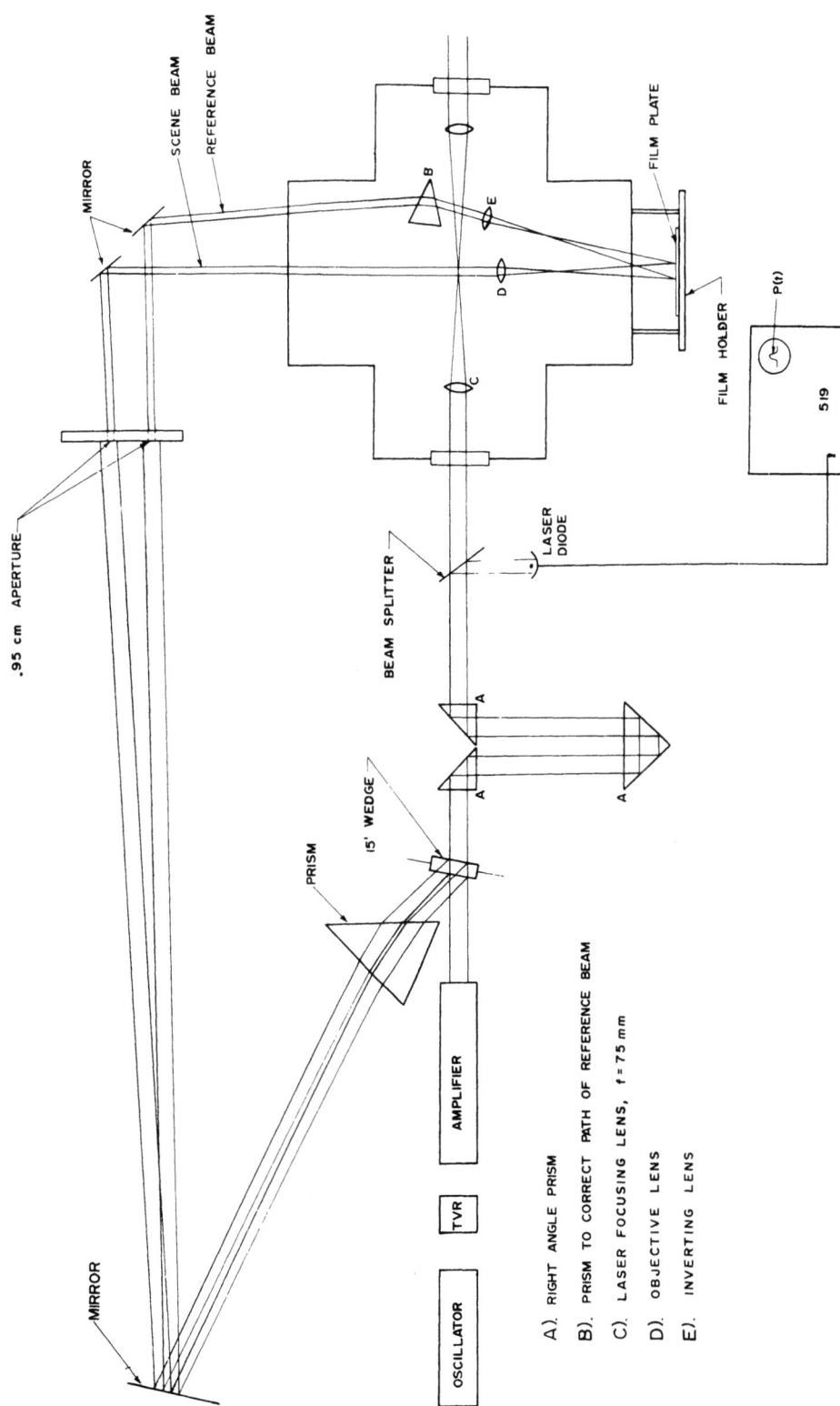

Figure 7. Holographic Experimental Apparatus

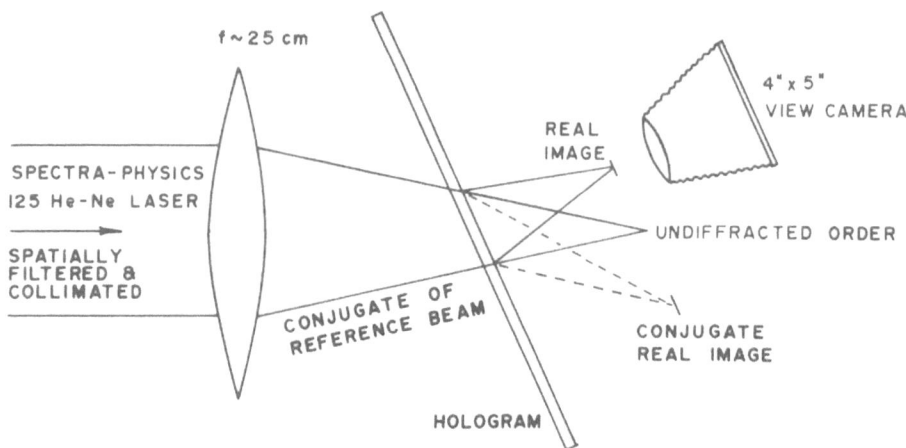

Figure 8. Wavefront Reconstruction Scheme

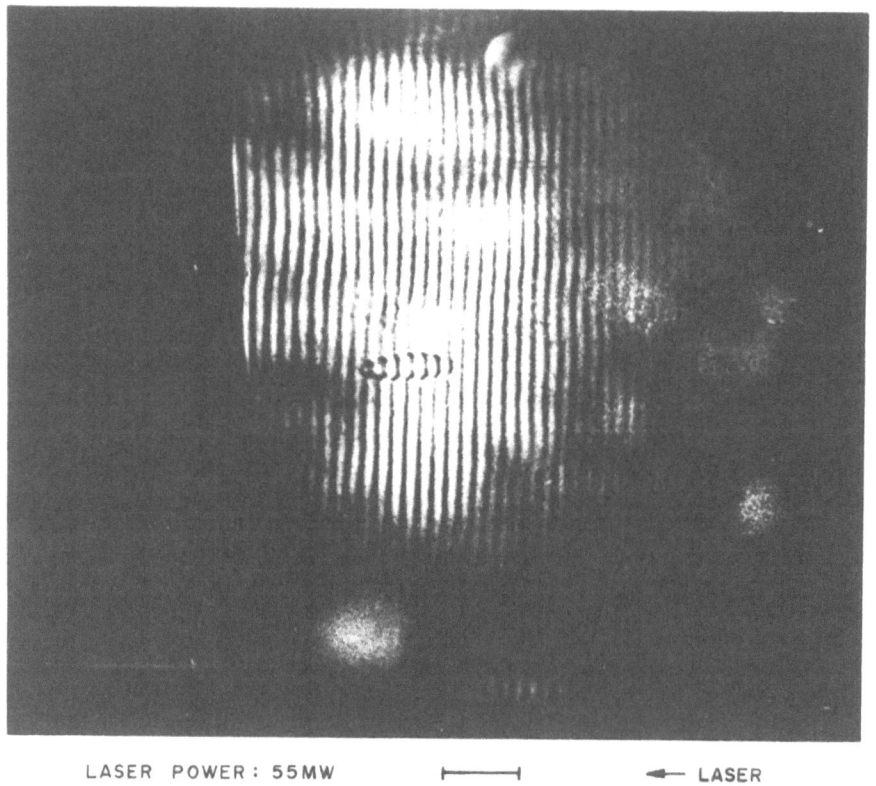

Figure 9. Photograph of Reconstructed Interferogram

and photographed as indicated. Two needles, 0.32 centimeters apart, placed below the breakdown region served as fiducial marks.

Data & Results

Figure 9 is a typical hologram, taken 3 nanoseconds after the peak of the plasma-producing laser pulse. The fringes are relatively straight outside the breakdown region and bent to the right within the plasma indicating a lowering of the index of refraction of the plasma. This was easily verified by making a hologram of a thin glass wedge (index 1.5) in the plasma position. The fringes in this case were bent to the left. Figure 9 contains the necessary information to determine the electron density within the plasma. In addition to the assumptions related to plasmas composition an addition assumption of cylindrical symmetry is necessary in order to extract the radial electron distribution of a given cross section of the plasma. Figure 10 shows the geometry employed in interpreting the fringe shift data.

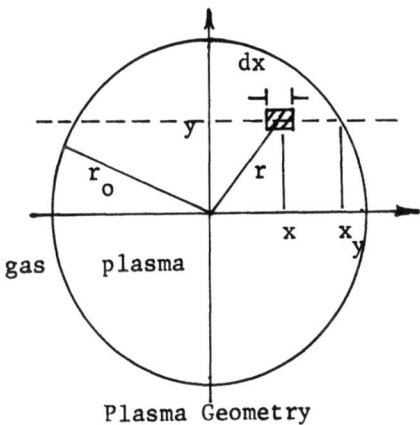

Plasma Geometry

Figure 10

The change in phase of a light ray along the dashed path due to the presence of the plasma is

$$\Delta\phi(y) = 2 \int_0^{x_y} [k_p(r) - k_g] \, dx \qquad (14)$$

where
$$k_p(r) = \frac{2\pi n_p(r)}{\lambda_o} \qquad (15)$$

is the magnitude of the wave vector in the plasma and k_g is that in the gas,

and $\chi = \sqrt{r^2 - y^2}$ (16)

and $dx = \dfrac{rdr}{\sqrt{r^2 - y^2}}$ (17)

Each adjacent fringe represents a phase change of 2π, therefore, equation (14) in terms of fractions of a fringe shift is

$$\Delta f(y) = \frac{\Delta\phi(y)}{2\pi} = \frac{2}{\lambda_o} \int_y^{r_o} \frac{[n_p(r) - n_g]}{\sqrt{r^2 - y^2}} rdr \qquad (18)$$

This form of the Volterra integral equation was first derived by Abel and now bears his name. The solution is

$$n_p(r) - n_g = -\frac{\lambda_o}{\pi} \int_r^{r_o} \frac{\Delta f'(y)}{\sqrt{y^2 - r^2}} dy \qquad (19)$$

as can be verified by substituting equation (19) into equation (18) or by consulting references 31 and 32. The function $\Delta f(y)$ is read directly from the photograph of the reconstructed hologram with an x - y opaque film reader. These data is then fitted with a fourth-degree symmetric polynomial using the method of least square. The resulting polynomial is substituted into equation (19) which has an exact integral. The justification for using a polynomial of fourth degree in the case of these interferograms is based on results of several polynomial fits and using physical arguments to select the most meaningful form. Attempting to use higher order fits resulted in radial profiles with large gyrations and in some cases physically unattainable electron densities. Figure 11 is an example of a particular fringe taken from Figure 9; the raw data and the least-squares fit are shown. The limitations of the results are primarily in the spatial resolution of the radial density profile. One should not infer that the smooth curves in Figures 12 to 15 are necessarily that uniform over distances less than about 25 μm. Variations in refractive index over distances of this dimension would appear as noise on the interferogram. This noise has been deleted from the data by the selected fitting scheme. The dashed curves in Figure 12 represent errors resulting from systematic uncertainty in the fringe positions. Figure 13 is a graph of the center density as a function of axial position relative to

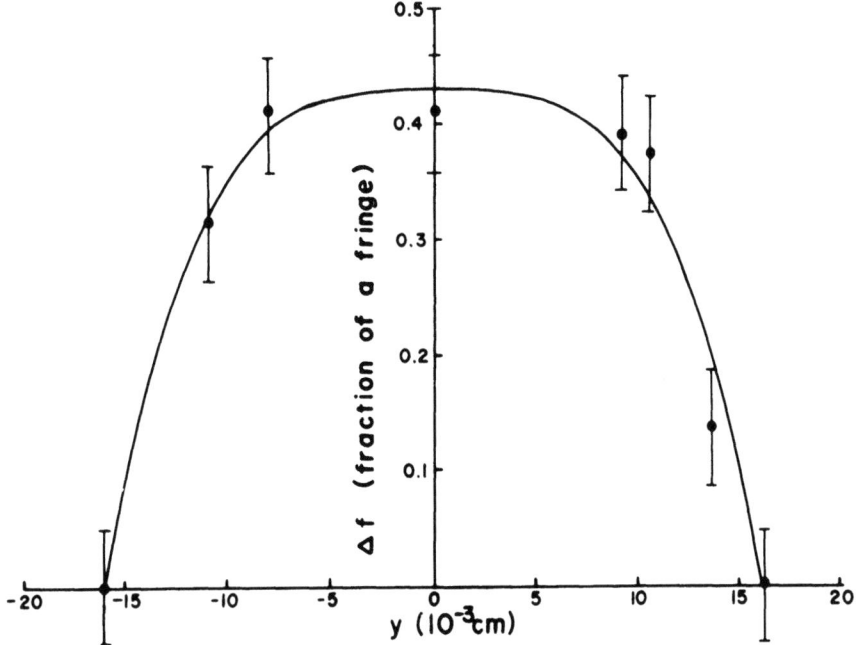

Figure 11. Typical Fringe Shift Data and Fitted Profile

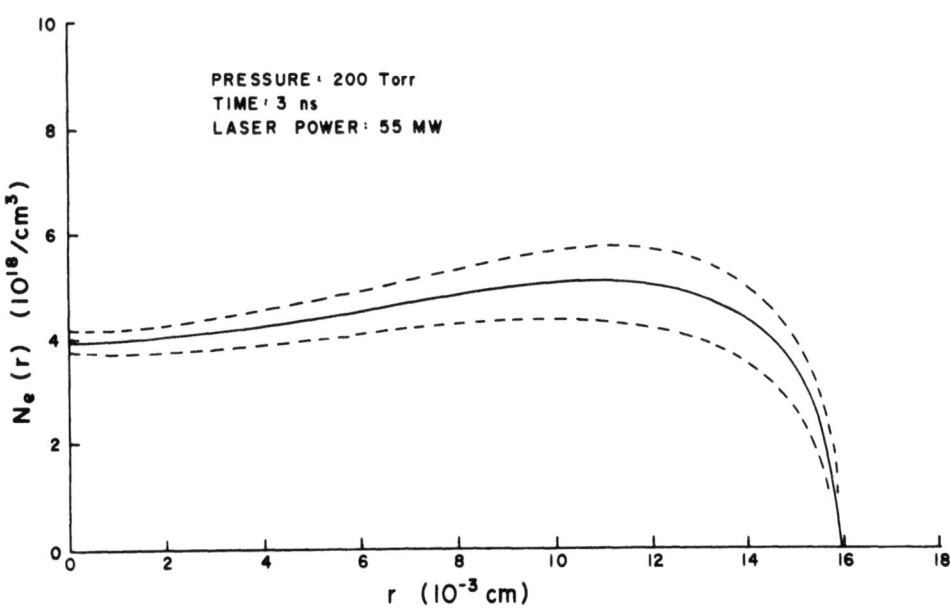

Figure 12. Radial Electron Density Distribution

the focal point of the laser. The electron density is again seen
to be fairly constant except at the boundary. It is more difficult
to analyze the fringes near the ends of the plasma because the
cylindrical geometry is lost. With the fringes spaced closer, the
results would allow more spatial resolution in the axial direction.
Figure 14 shows a sequence of radial density profiles taken at
various times (1, 4, and 7 nanoseconds) after breakdown at the same
axial position. The pressure is 300 torr and the average laser
power for the three shots was about 120 megawatts. The plasma
radius increases with time at a rate of about 1.4×10^6 cm/sec.
The peak electron density increases by about 30 percent during this
time. Figure 15 is a comparison of the radial profiles for two
different pressures taken at a delay time of 5 nanoseconds. The
radial dimensions are virtually the same; the electron density is
four times greater at 600 torr than at 100 torr.

ELECTRON TEMPERATURE

The state of a plasma is essentially specified if the composition of the plasma is known and if the energy of each constituent can be determined. On a microscopic scale this means knowing where each particle is and its velocity. Clearly, this is impossible on an individual particle basis since the number is so large. Therefore, one describes the plasma in terms of particle concentration and the particle velocity distribution. For laser-produced plasmas the electron properties has been chosen as the most important plasma constituent. First, energy is preferentially coupled by inverse bremsstrahlung to the electrons by the laser electric field, because of their mobility and long-range Coulomb forces. Second, the radiative processes in these plasmas are almost always more characteristic of the electron properties than those of any other specie. The electron density was discussed in the previous section; the temperature is now discussed. The use of the term "temperature" places a very restrictive condition on the electron velocity distribution, i.e., it must be Maxwellian. It is well know that any given velocity distribution of particles will eventually thermalize (become Maxwellian) if unperturbed. The time required for this to occur depends on the particle concentration and the temperature. From estimates of the magnitude of densities and temperatures of laser-produced plasmas the electron thermalization times are of the order 10^{-12} sec.[26] On a nanosecond time scale (10^{-9} sec) the electrons have had sufficient time to thermalize. Such a condition allows one to assign a single parameter (temperature) to the electrons which then describes the Maxwellian distribution. Very often one multiplies the temperature (T_e in units of °K) by Boltzmann's constant (k) to obtain the parameter kT_e which has units of energy, (e.g. electron volts, eV). This product is commonly called the temperature and will be so done from here out.

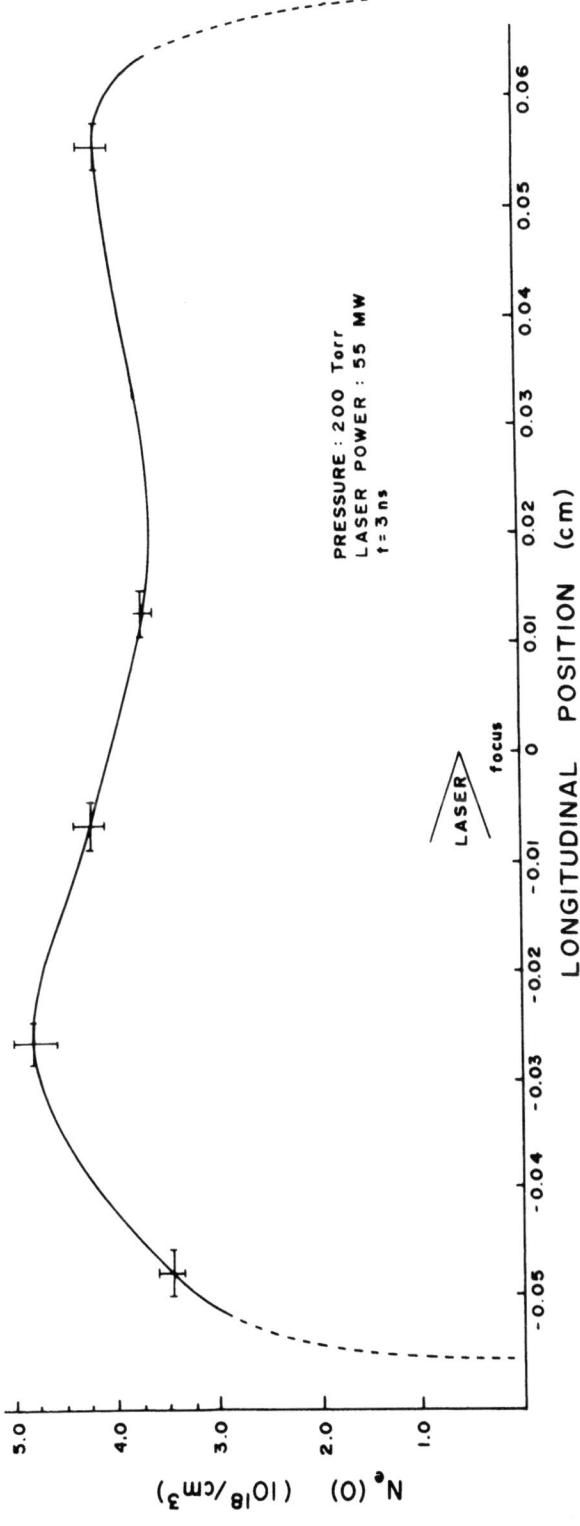

Figure 13. Axial Electron Density Distribution

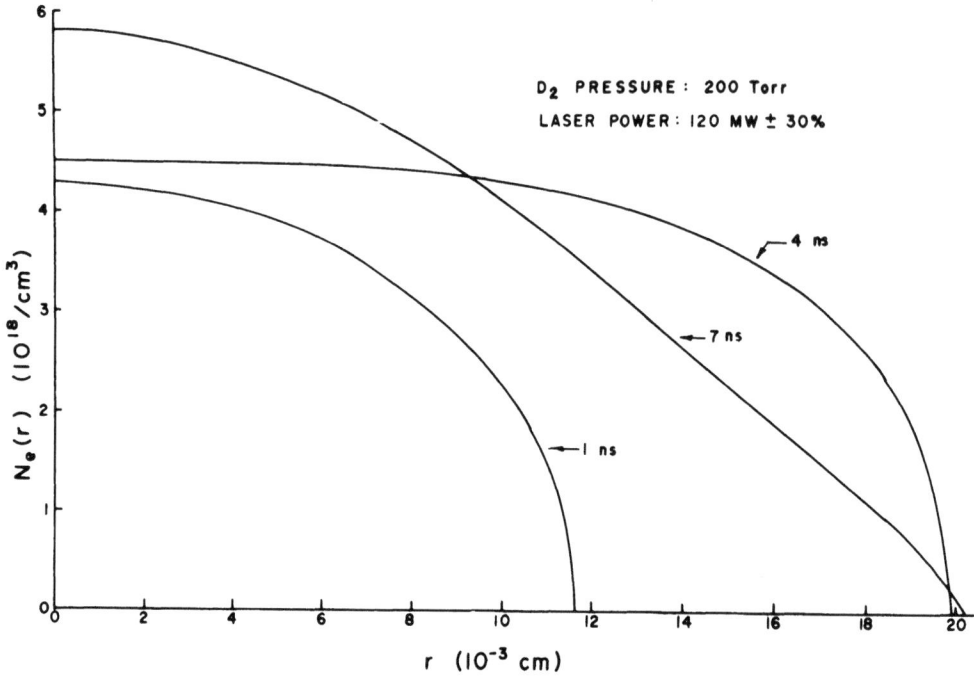

Figure 14. Temporal Sequence of Radial Electron Density Distribution for Constant D_2 Pressure

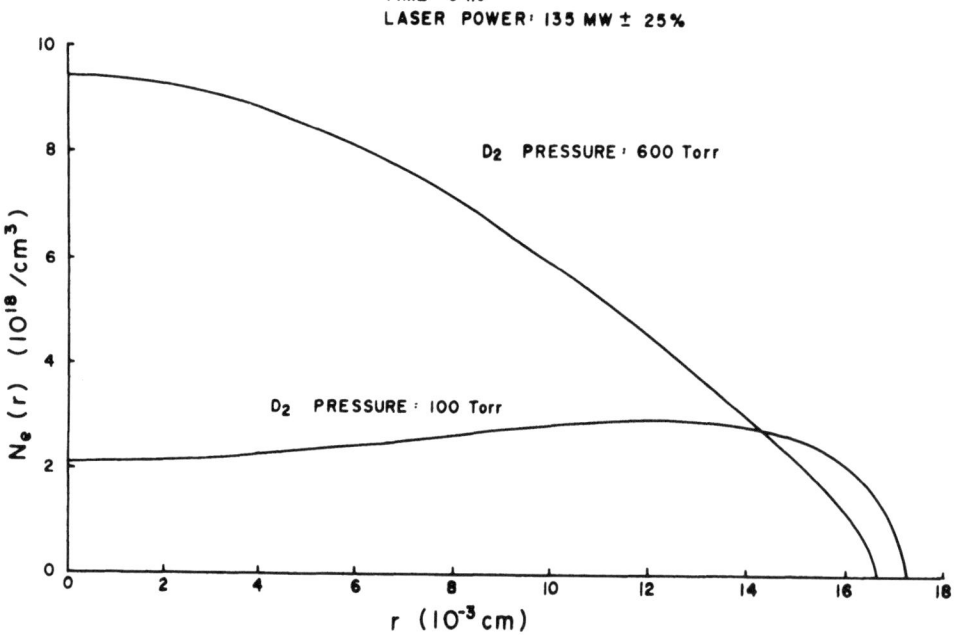

Figure 15. Radial Electron Density Distributions for 100 and 600 Torr D_2 Pressures

A great deal is known about plasma composed of Maxwellian electrons. In particular, the radiation produced by such a plasma is well known. At temperatures greater than about 10 eV plasma radiation is primarily of two types: free-free (bremsstrahlung) and free-bound (recombination). Free-free transitions results when energy is lost from a free electron as it is accelerated by the Coulomb field of a positively charged nucleus or ion. The free electron remains free but with less energy. The lost energy is radiated away as electromagnetic energy. Free-bound transitions result when a free electron is captured into a bound quantum state of an ion. The difference in energy between the two states is likewise radiated. For the case of hydrogen and a Maxwellian free electron distribution of temperature kT_e the combined spectrum is calculable exactly and the emission intensity per unit wavelength interval is

$$\frac{dI}{d\lambda} = \frac{C_1 N_e^2}{\lambda^2} \left(\frac{X_H}{kT_e}\right)^{1/2} e^{-hc/k\lambda T_e}$$ [33] (20)

$$+ 2 \frac{C_1 N_e^2}{\lambda^2} \left(\frac{X_H}{kT_e}\right)^{3/2} e^{-[(\frac{hc}{\lambda} - X_H)/kT_e]}$$

where X_H is the first ionization potential of hydrogen and $C_1 = 5.4 \times 10^{-16}$ ergs - cm^5/A-sec. The first term in the brackets is that due to bremsstrahlung and the second term that due to recombination into ground state only. At increasing temperature, bremsstrahlung becomes more dominant. Equation (20) is plotted in Figure 16 for a value of kT_e = 80eV. This is the spectrum emitted by an 80 - eV D_2 plasma assuming no impurities are present. The expression is plotted over a wavelength range just beyond the hydrogen ionization potential. In this wavelength region the spectrum is strongly temperature dependent and its analysis is thus especially suited to the determination of the plasma electron temperature. The spectrum is determined by simultaneously measuring the transmitted energy through two thicknesses of Beryllium foil. The ratio of transmitted energy is a unique function of the temperature of the incident radiation.

Experimental Apparatus

Figure 17 depicts the experimental arrangement employed in the temperature measurement. The laser pulse is monitored with a Techtronix 519 oscilloscope as before; in turn, the delayed-gate output of this scope triggers two Hewlett-Packard Model 183 oscilloscope employed to record the photomultiplier signals. The X-ray signal generated by the plasma is detected through nominally

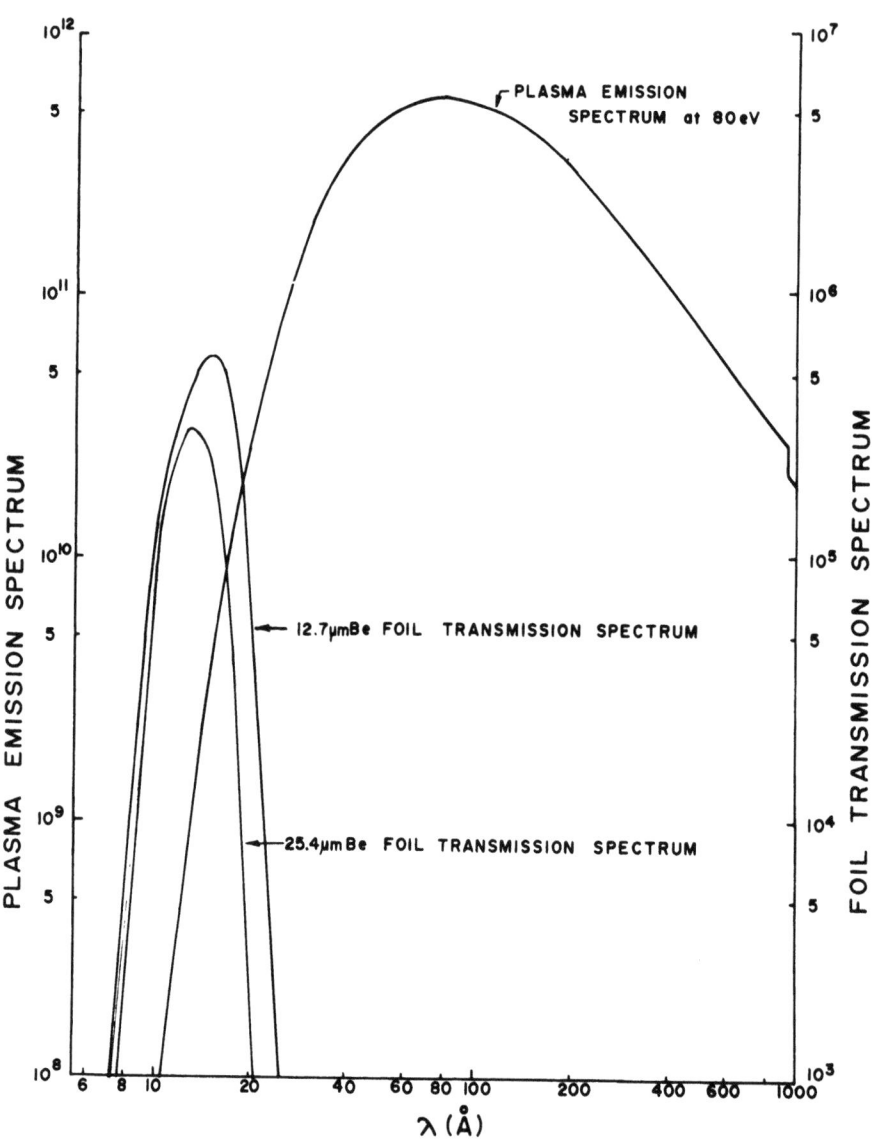

Figure 16. Plasma Emission/Foil Transmission Spectra for 80-eV Deuterium Plasma

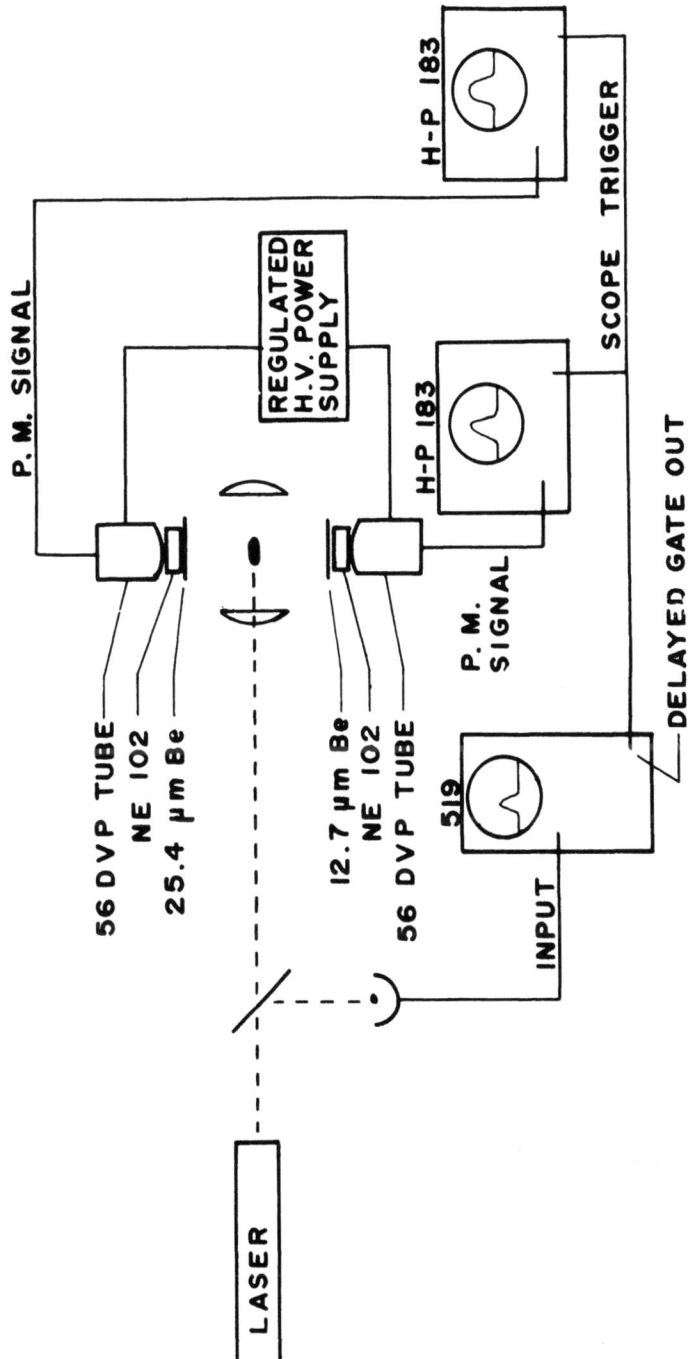

Figure 17. Electron Temperature Experimental Apparatus

0.5 and 1.0 mil (0.001 in.) beryllium foils. The transmitted x-rays are absorbed in NE 102 scintillation plastic and the resulting scintillations monitored by two Amprex 56 DVP photomultiplier tubes. Each signal is recorded on a separate oscilloscope. The two foil-scintillator-PM tube combinations were located in a plane normal to the incident laser pulse and at 90° to one another. This design was based on the assumption that the plasma radiation was isotropic. The measured radial electron density profile supported this assumption and the tedious procedure of direct verification was not deemed necessary.

Data Analysis

The data analysis procedure is two-fold. First, the ratio of foil transmissions must be calculated based on measured foil characteristics (absorption cross section & thickness) and geometrical configuration and second, the ration must be measured taking into account the PM tube gain and scope calibration.

The problem of theoretical calculation of the ratio of transmitted energy through the two foils is partly geometrical and partly radiation transport through the foils. The geometry is easily handled with a high-speed digital computer. Calculation of radiation transport through the foil required knowledge of the absorption coefficient of the foil and the area density (units of milligrams/cm^2) of each foil. The cross sections were those compiled by Henke[34] in the range from 2 to 200Å He has estimated that for beryllium the "best fit" function is likely to be in error by as much as 10 percent. The area density was determined by accrate measurement of the foil area and weight employing respectively a travelling microscope and microbalance. Combining these physical parameters with the cross sections (nominal and nominal ± 10%) yielded the three ratio curves of Figure 18. Absorption in the D_2 gas between the plasma and the foil was ignored because D_2 cross sections are at least an order of magnitude smaller than beryllium and the effective area-mass is likewise small. In the authors' opinion the primary limit of accuracy of these electron temperature measurements is the uncertainty in beryllium cross section. The experimental determination of foil transmission resulted from simutaneously recording each photomultiplier signal on a separate oscilloscope. A typical set of signals is shown in Figure 19 along with the corresponding transmission ratio versus time graph for this measurement taken at one nanosecond intervals. The foil transmission ratio is obtained from the photomultiplier signals with the relationship

$$\frac{T_1}{T_2} = \frac{V_1}{G_1} \bigg/ \frac{V_2}{G_2} \tag{21}$$

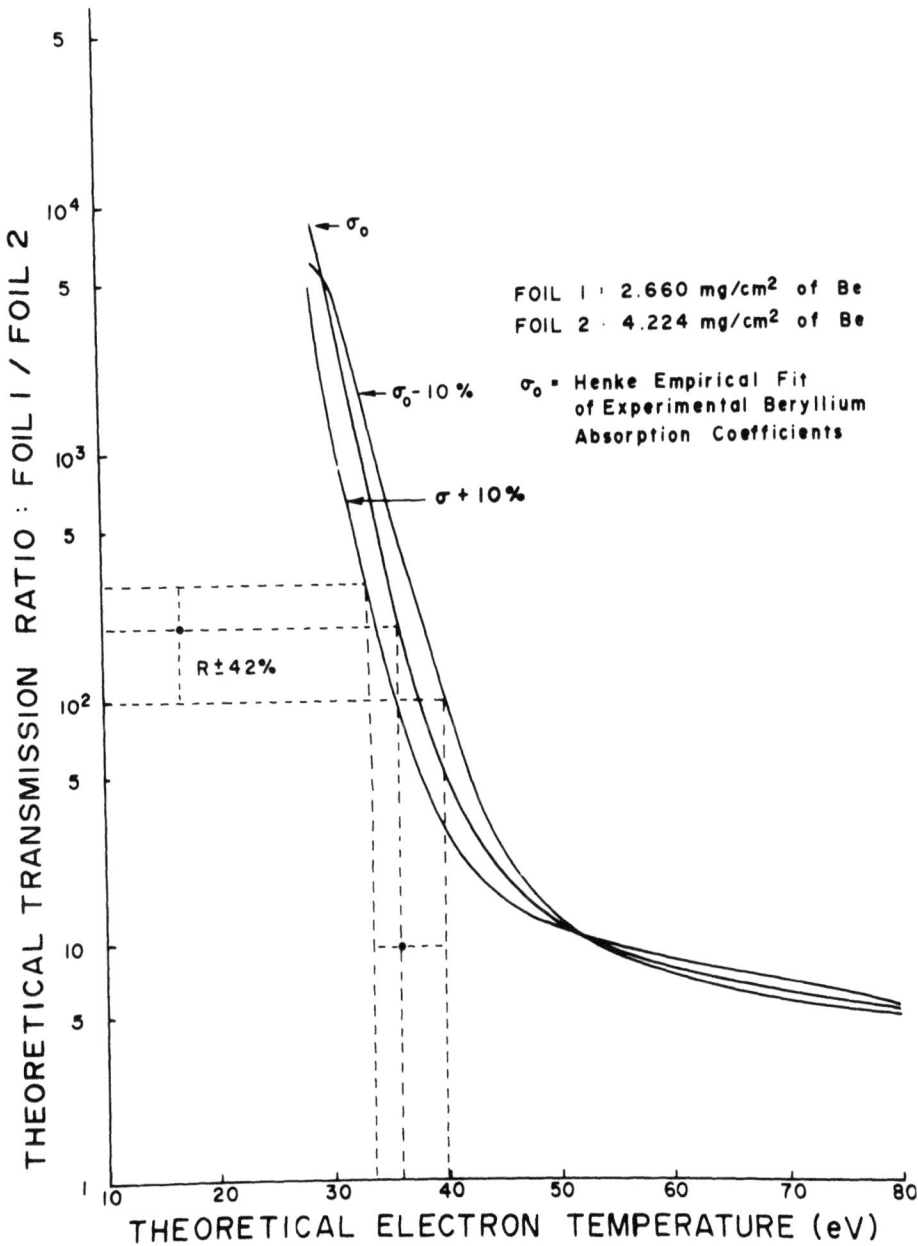

Figure 18. Theoretical Foil Transmission Ratio vs Electron Temperature

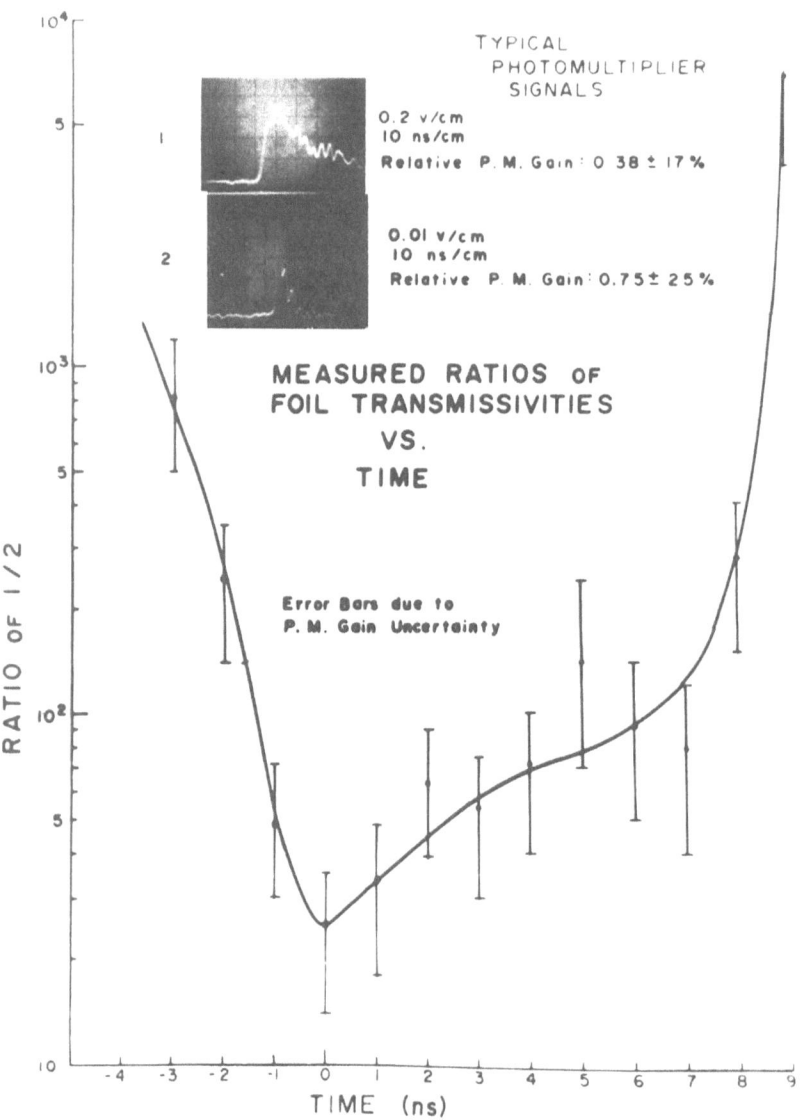

Figure 19. Experimental Foil Transmission Ratio vs Time

where V_1 and V_2 are the voltage deflections recorded by the scopes from the PM tubes behind the 0.5 mil and 1.0 mil foils respectively. G_1 and G_2 are the photomultiplier gains, respectively. G_1 and G_2 were determined under pulsed conditions by detecting the scattered 0.69 μm ruby pulse through a small light leak in the chamber introduced by removing a light shield around the beryllium. Correlation between PM output and incident laser intensity for each tube gave the relative gain <u>at this wavelength</u>. The relative gain throughout the visible spectrum was found <u>statically</u> with a monochromator and the gain variation was within the experimental limits established by the first method. A better method is to reverse the roles of the PM tubes in the actual experiment on a series of shots keeping the plasma as nearly as possible the same. From a series of shots the relative gain and its variation can be established with greater confidence. Due to the very steep dependence of signal ratio with temperature below 40eV improvement in experimental ratio determination yields a small reduction in temperature error.

Time correlation between the two foil-transmission signals was obtained by assuming that the respective peak amplitudes occured at the same time. This is a correct assumption because the transmitted power is a monatonically increasing function of electron temperature Thus, if one signal begins to decrease the other must also. The zero time for these measurements was taken as the time fo peak x-ray production. This generally corresponded to the time of peak electron temperature and normally occurred less than 1 nanosecond after the peak of the laser pulse. Extreme difficulty was encountered in obtaining these data. The signals were of small amplitude (millivolts) and varied over a large range. Electrical noise from the laser had to be shielded from the scopes by double-braided cable and carefully made connections.

Results

Figure 18 illustrates the method employed to obtain the electron temperature and errors from the measured ratio at a given instant. The ratio is obtained from the scope traces as explatined previously Reading to the right and down from the σ_0 curve gives the nominal value of electron temperature. Using a value of experimental error in the ratio of±42% sets the upper and lower limits on the temperature as shown.

Figure 20 is a comparison of the temporal electron density at 200 torr for two different laser powers, 440 and 220 megawatts. The temperature in each case rises very rapidly and remains fairly constant until about 6 nanoseconds after the peak and then falls very rapidly. Due to the overlap in error bars, these results show that at these power levels there is little or no detectable dependence of temperature upon laser power. Figure 21 at 300 torr, shows a similar temporal behavior. The laser powers in each case shown were within 3 percent of each other yet scatter in the data is evident especially during the 4 nanosecond initial rise. Large shot-to-shot variations in power density may manifest themselves as variations in initial temperature rise. Figure 22 is a graph of maximum electron temperature as a function of deuterium pressure. All shots at a given pressure were averaged regardless of laser power. A general trend toward higher temperature as the pressure is increased is shown, however, the dependence is slight at these pressures. No trend or conclusive dependence of maximum temperature on laser power could be found.

LATE-TIME AIR BREAKDOWN CHARACTERISTICS

The general nature of this paper prompts the inclusion of this section even though it does not conform to the experimental conditions employed so far. It does, however, illustrate the nature of experimentation and results of gas breakdown characteristics during the final stage, i.e., shock wave propagation. The plasmas in this case were produced in air with a high-power($\sim$ 700MW) neodymium laser. Figure 23 is a schematic of the experimental apparatus. Breakdown was initiated in air at pressure of 753 torr and 300 torr at the focus of a 100 mm focal length lens. The initiating laser was a Q-switched, neodymium doped glass, oscillator-amplifier of wavelength 1.06 µm. Typical outputs were 28 joules in a pulse 40 nanoseconds in duration (FWHM). The breakdown region was investigated employing a pulsed ruby laser to obtain interferometric holograms. The single-mode ruby laser had a pulse duration of about 100 ns and an energy per pulse of .5 joules. The measured coherence length was greater than 4 meters.

Experimental Data & Results

Figure 24 is a photographic reproduction of a typical reconstructed hologram obtained in this experiment are similar to those obtained during the development stage to measure electron density. These, however, do not contain artificial fringes and are made with a diffuse scene beam rather than a colimated beam as before. This results in a three-dimensional reconstruction and is made possible by the excellent spatial and temporal coherence of the holocamera laser. Such a hologram is termed infinite fringe because in unchanged regions of space the fringes are infinitely far apart[29]. Figure 24, then, is a record of the change in index of refraction of the breakdown region 16 microseconds after initiation of the air spark. At times greater than 10 microseconds the gas is no longer ionized or excited, thus, from a series of holograms one can conclude that the state of the breakdown is represented by a spherical shock wave. The existence of this shockwave is evident from the audible sound accompanying breakdown.

From a series of holograms taken at various times after

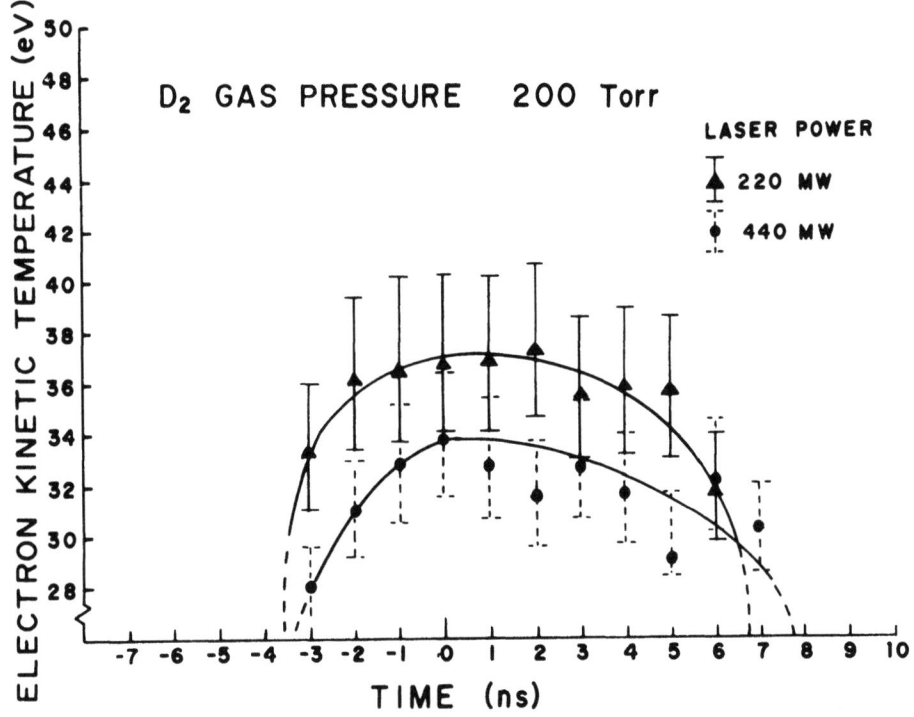
Figure 20. Temporal Electron Temperature at 200 Torr

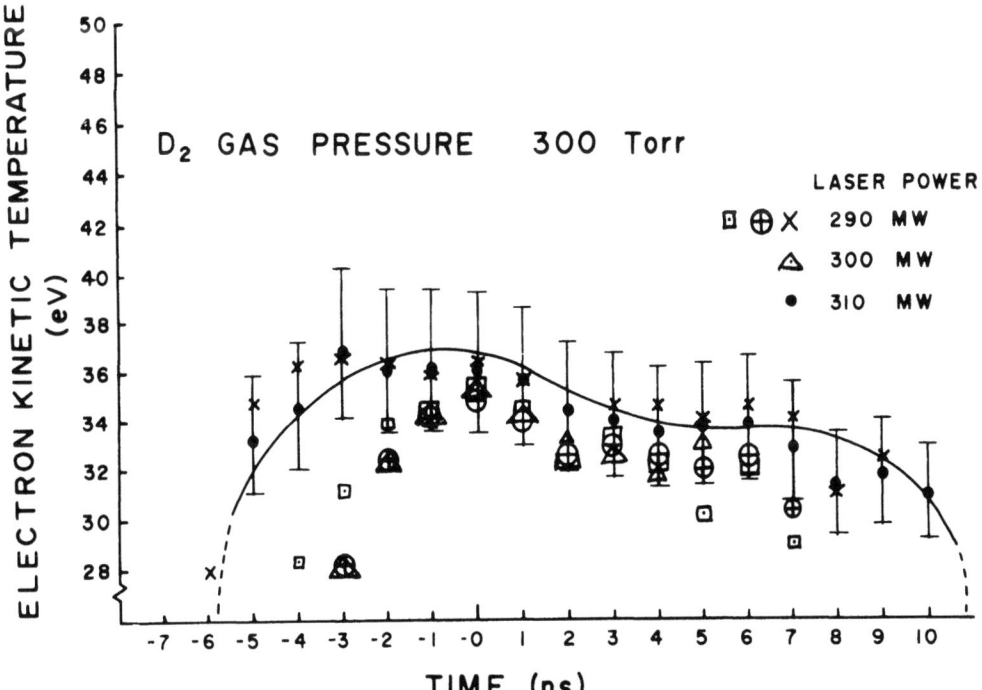
Figure 21. Temporal Electron Temperature at 300 Torr

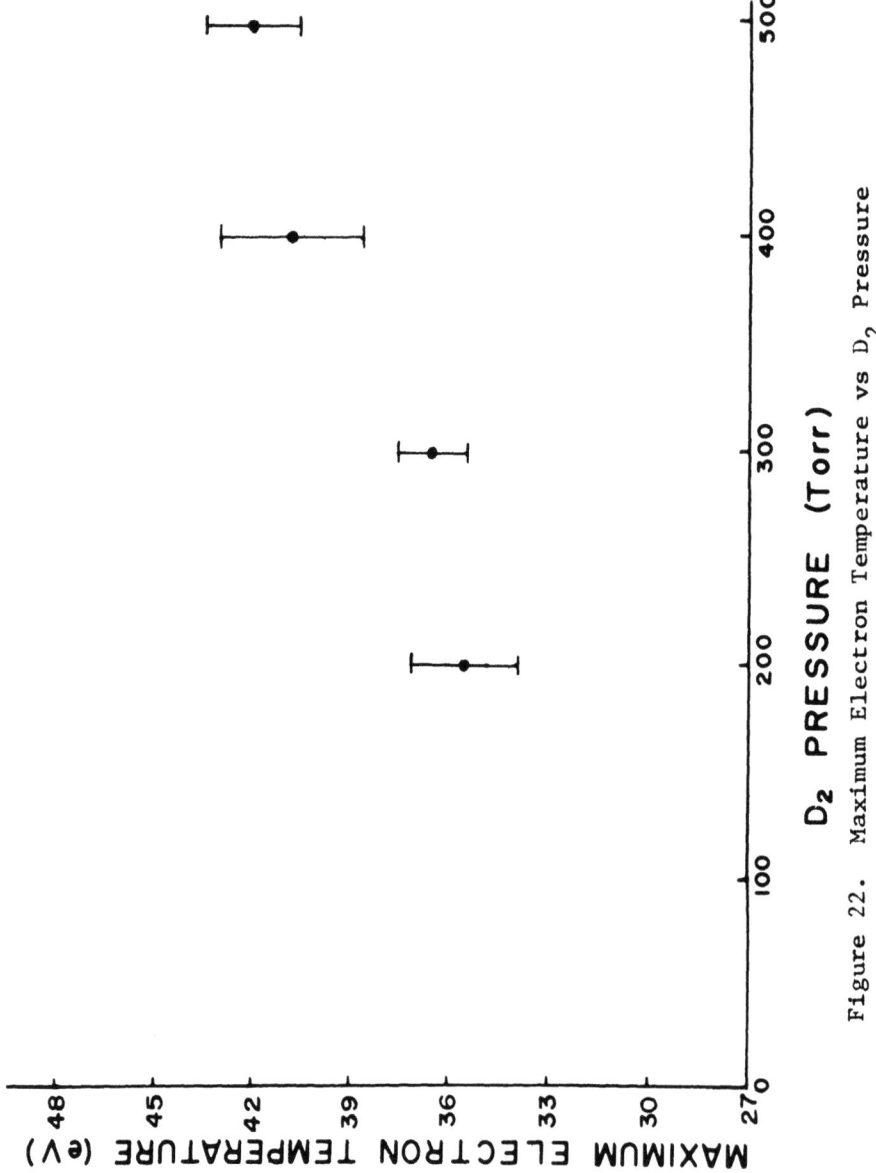

Figure 22. Maximum Electron Temperature vs D_2 Pressure

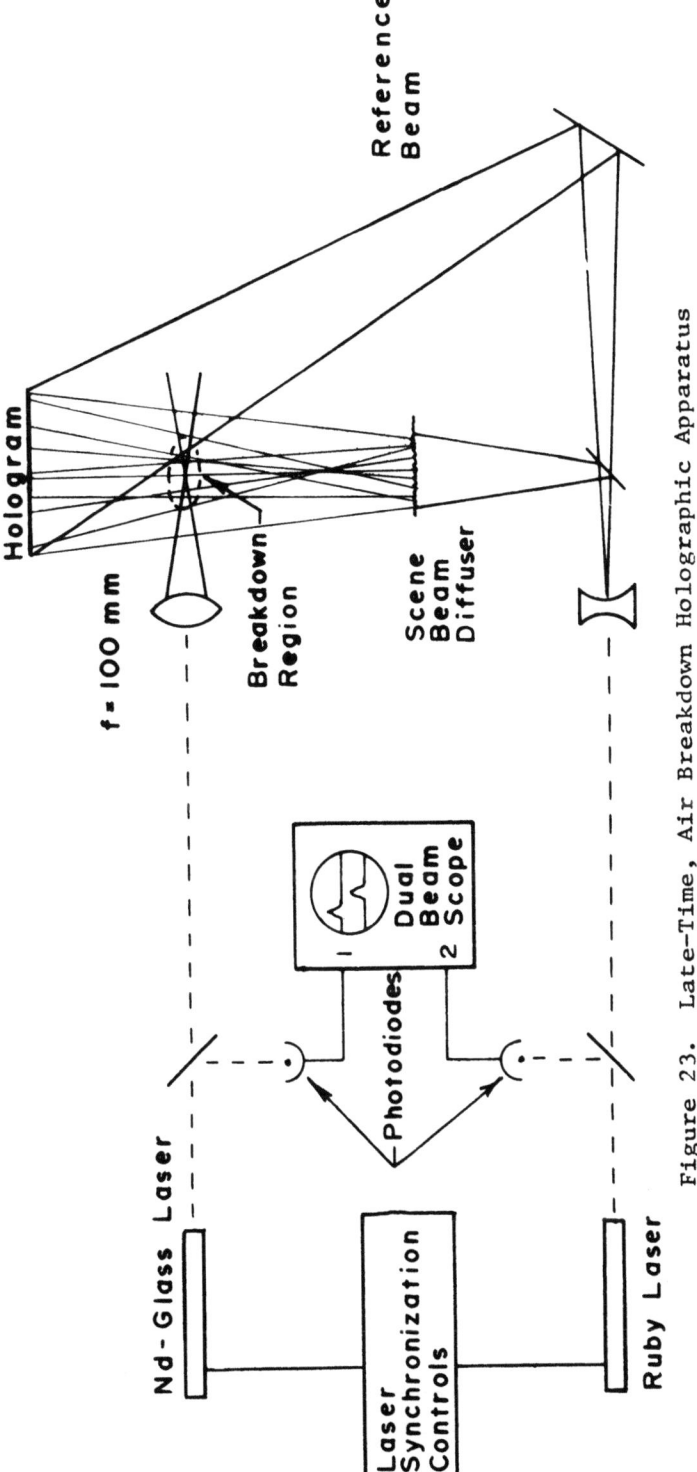

Figure 23. Late-Time, Air Breakdown Holographic Apparatus

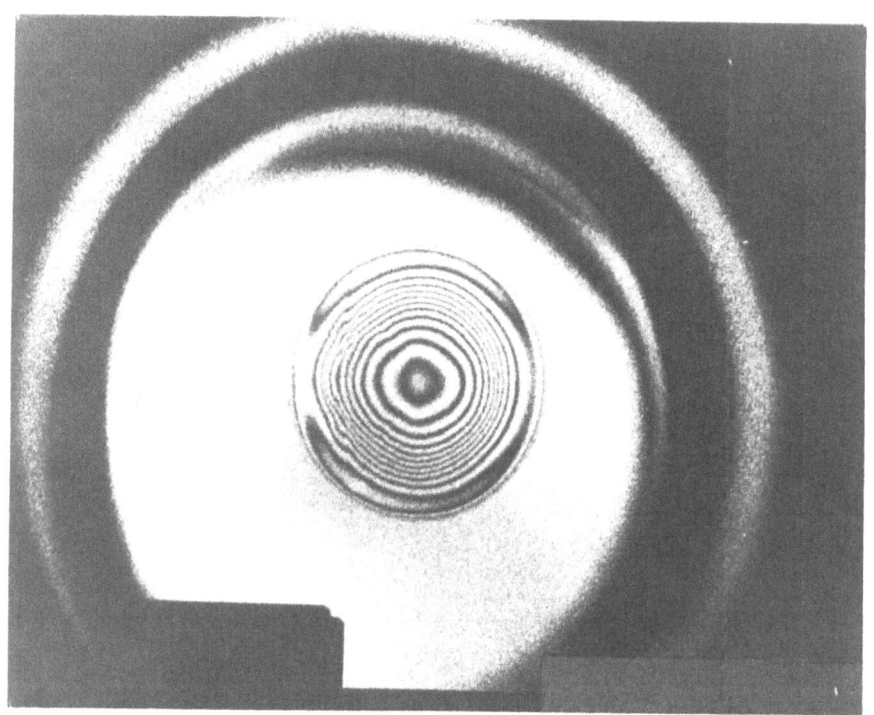

Figure 24. Interferogram of Late-Time, Air Breakdown Blast Wave

initiation (different breakdowns) a shock radius versus time (R - t) diagram was determined, Figure 25. The dimensionless coordinates have been normalized with the parameter

$$R_o = \left(\frac{E/E_o}{4\pi p_o}\right)^{1/3} E_o^{1/3}$$

where E is the initial shock wave energy E_o is the laser energy and p_o is the ambient gas density and ct/R_o (Figure 25) is the dimensionless time after breakdown where c is the acoustic velocity. Such normalization enables all gases with same $\gamma = \frac{C_p}{C_v}$ to be plotted on a common R - t curve regardless of pressure. The data are compared using three blast wave theories. The best fit is that predicted by Sakurai [19]. This theory treats the hydrodynamic mass, energy, and momentum conservation equations and finds solutions in terms of a second order power series in reciprocal Mach number. This theory is especially applicable for low Mach number shocks (<4) as are typical in these experiments at post laser irradiation times greater than 10 μs. Curve fitting was obtained by finding the best value of R_o for each data point and averaging these values. For both 300 torr and 753 torr the data indicate that ∼15 percent (4 joules) of the incident laser energy is contained in the resulting blast wave. This is in good agreement with the work of Buchl, et al. [35]. For comparison, Figure 25 shows the R - t curve predicted by Taylor-Sedov blast wave theory. This is a first order theory, good when the shock wave velocity is at least Mach 4. The third comparison is made with a numerical, one-dimensional, Lagrangian, hydrodynamic computer code, SAP[36] used at the Air Force Weapons Laboratory to predict spherical nuclear detonation wave propagation in the atmosphere. The agreement with the data and the Sakurai model is good.

The interferometric content of the hologram shown in Figure 24 was analyzed to obtain a radial distribution of gas density in the breakdown region at 16 microseconds after initiation. The following assumptions were made:

(1) The breakdown region is spherically symmetric
(2) The index of refraction change due to breakdown at this time results solely from the neutral air density distribution and no contribution comes from ionization or excitation of the air.
(3) The density is less than ambient at the center of the sphere.
(4) The Gladstone-Dale relation holds, i.e., the index of refraction of air is linear with density over the density range within the breakdown region.
(5) Refractive effects at the shock boundary are negligible.

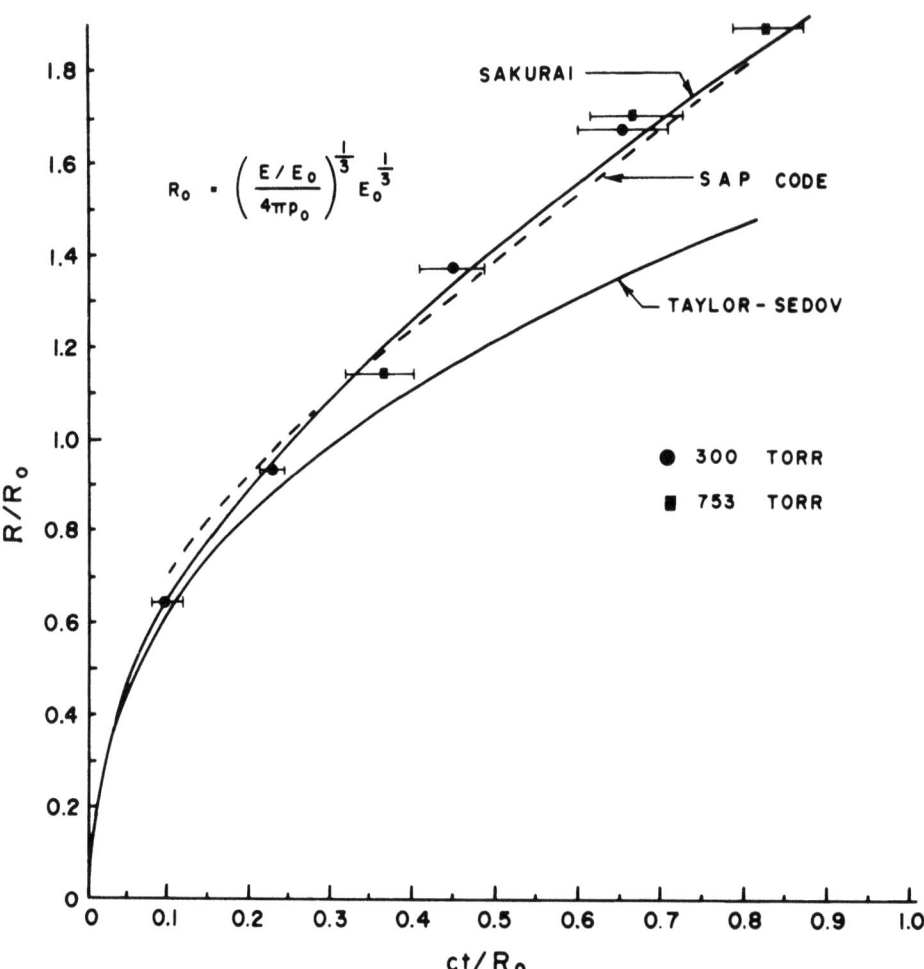

Figure 25. Normalized Spherical Blast Wave R-t Diagram

The geometry of the measurement is the same as described in Figure 10 and the associated integral equation to be solved is

$$\Delta f(y) = \frac{2K}{\lambda} \int_y^{r_o} \frac{\rho(r) - \rho(\infty)}{\sqrt{r^2 - y^2}} \, r \, dr \tag{23}$$

where $\Delta f(y)$ is the integrated phase change across the chord located y centimeters from the center of the sphere, K is the proportionality constant between gas density and index of refraction, and index of refraction, and λ is the holocamera wave-length. The density as a function of radius, $\rho(r)$, is obtained, as before, by Abel inversion of the integral once $\Delta f(y)$ is measured. $\Delta f(y)$ was determined for the vertical plane of the breakdown region orthogonal to the incident laser pulse. Starting from the undisturbed gas, the location of each dark fringe was determined with a traveling microscope. The edge of the disturbance was ignored because grazing incidence reflection would cause a dark boundary not to be considered an interference fringe. The first fringe is very thin and most easily seen on the left-hand side of Figure 23. A phase shift of $-\pi$ was assigned to this fringe as it was assumed that compression occurs near the edge. The next fringe, corresponding to a phase change of -3π is readily evident and is folded. This means that for the plane chosen the second and third fringes each correspond to a phase change of -3π (it is the same fringe). The succeeding fringes must correspond to phase _increases_ of 2π/fringe until the center is reached. If this were not the case the gas density at the center would necessarily be greater than ambient and conservation of mass could not be maintained. The presence of the folded fringe is fortuitous; without it, assigning absolute phase changes to all fringes would be most difficult. Any slight asymmetry in the shock wave is enough to produce an easily identified fringe reversal point.

Once $\Delta f(y)$ was determined for each dark fringe a smooth, symmetric, eighth degree, least-squares fit was made to the measured points and the resulting curve was used in the Abel inversion integral. Figure 26 is the radial density distribution resulting from the data analysis of the hologram in Figure 24. The profile has been normalized with respect to ambient pressure and shock radius. As a check on the validity of this interpretation, the number of gas molecules in the total shocked region is within 3 percent of the original number in the same volume before being shocked. For comparison, the radial distribution was calculated with the above mentioned SAP code and is also shown in Figure 26. The general shape of the calculated shock profile is similar to the experimental result however the peak overdensity and density gradient are somewhat different. These difference may be due to several limitations of the code. First of all, the equation of state used by SAP is restricted to the range 0 - 1.5 eV, far below

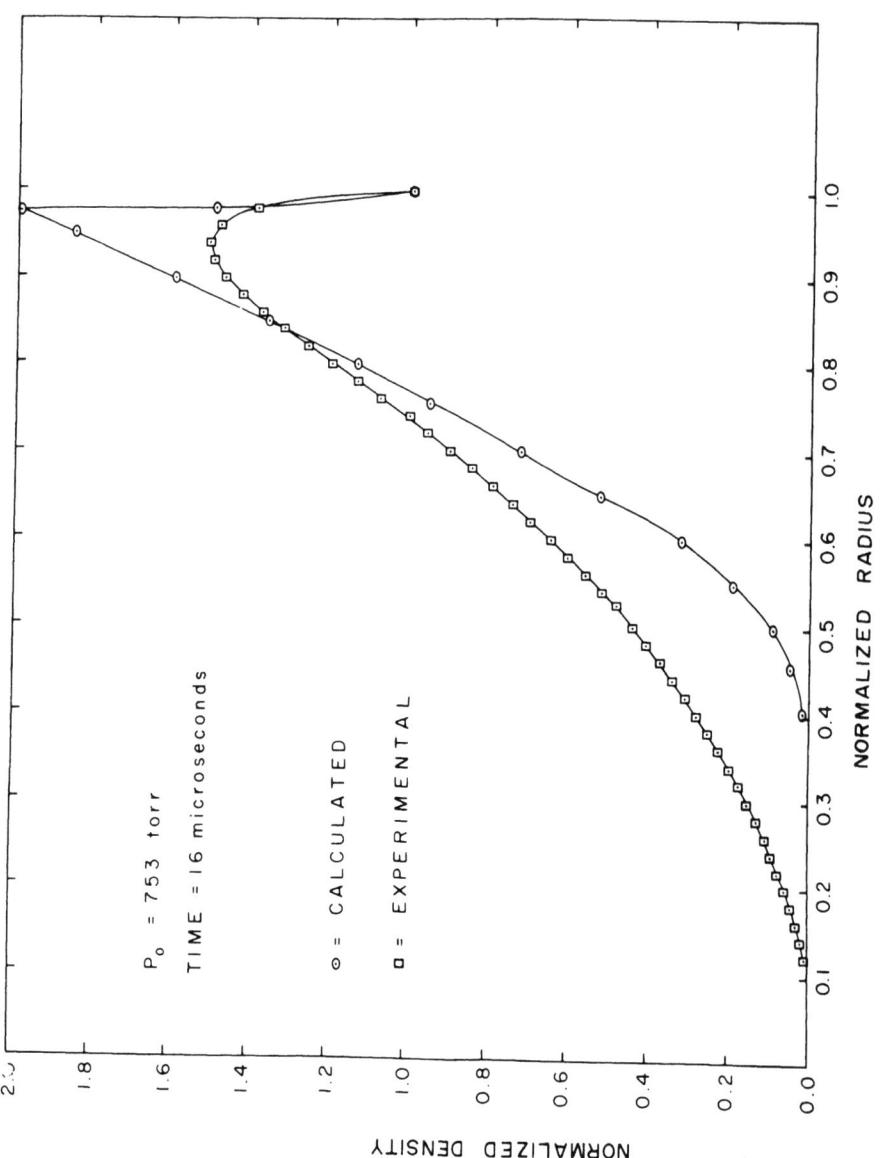

Figure 26. Normalized Radial Shock Profile

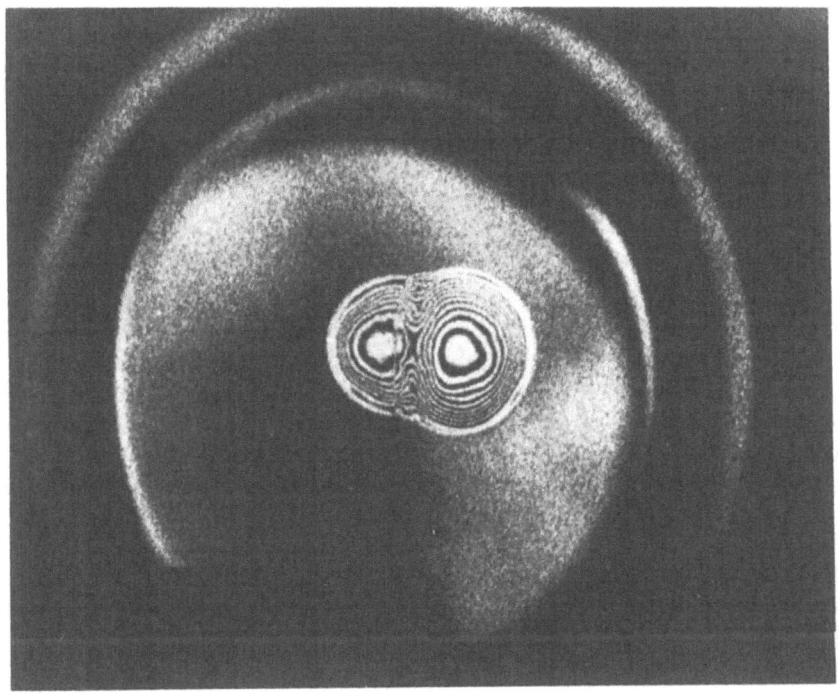

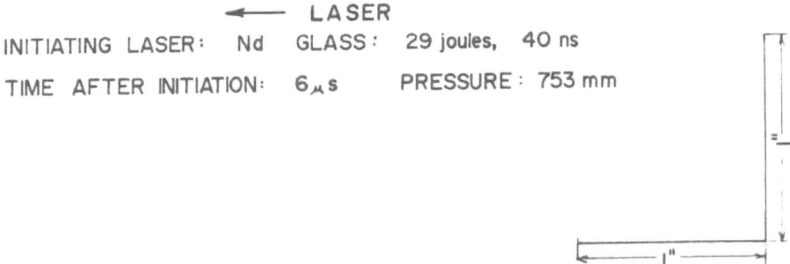

Figure 27. Interferogram of Late-Time Air Double Breakdown Blast Wave

the tens of eV temperatures known to exist during the early history of the plasma. Secondly, there is no provision within SAP to account for the early-time energy dissipation processes such as heat conduction, viscosity, or radiation loss. The third limitation is the fact that the initial focal volume during plasma formation is cylindrical, instead of spherical. Under the circumstances the disagreement between experiment and the SAP code prediction is not considered too severe.

Figure 27 is an example of the late-time shock profile following a multiple breakdown in air. The colliding shock fronts form a nearly plane interface.

INTERPRETATION

The previous four sections have described the techniques and results of a range of laser-produced deuteruim plasma characteristics. Most notably are the determination of laser breakdown threshold dependence on pressure, plasma growth rate, boundary interactions, electron density & temperature and late-time shock characteristics of air plasmas. All of these characteristics have either been observed or calculated before in various gases but until now the entire set of experiments has not been conducted under fixed experimental conditions such that correlation of all measurements could be made. In addition, the direct time-resolved measurement of the electron density distribution during the plasma development phase becomes a unique contribution. Each of the above characteristics will now be discussed briefly.

Thresholds

Figure 3 shows the log of the relative breakdown threshold laser intensity versus the log of the pressure from 150 to 600 torr. A linear least squares fit having a slope of -0.87 ± 0.08 compares with the analytic results of reference 9 showing that a slope of -1.0 corresponds to the pressure region where laser pulse duration is the donimant parameter governing attainment of breakdown. Thus, in the pressure range 150 to 600 torr electron losses due to diffusion and recombination are not as important as the 4 nanosecond laser pulse duration.

Plasma absorption

Figure 4 shows the energy absorbed by the plasma from the laser as a function of pressure. Spatial determination of the electron density such as plotted in Figures 12 and 13 shows that the electron density in approximately constant throughout the plasma. Assuming that the temperature is likewise uniform throughout the plasma allows one to calculate the free-free absorption coefficient given by

$$K_\nu = 3.69 \times 10^8 \; \frac{N_e^2}{T_e^{1/2} \nu^3} \; cm^{-1}$$

$\nu = 4.3 \times 10^{14}$ hertz for 0.69 μm radiation

Te is the electron temperature (°K) and Ne is the electron density (cm^{-3}). At a pressure of 600 torr, the electron density is about 1 x 10^{19}/cm^3 (Fig. 15), the measured temperature is about 4.7 x 10^5 °K, and the axial plasma dimension at the end of the laser pulse is about 0.3 cm (Fig. 5). Equation (23) gives an absorption coefficient for this case of 0.6 cm^{-1} resulting in a value of transmission of 84 percent. This compares well with a measured transmission of 82 ± 6 percent.

The plasma absorption data can also be cross checked against the temperature measurements by means of an energy balance. At 600 torr the plasma volume at the end of the laser pulse is about 3.8 x 10^{-4} cm^3. At an electron density of 1 x 10^{19} there are a total of 3.8 x 10^{15} electrons in the plasma. The average kinetic energy per particle (3/2 Kt$_e$) at this time is about 63 eV, the ionization potential of a molecular hydrogen is 15.4 eV, thus the total energy per electron is about 78 eV. These data were taken from a shot where the incident laser energy was 0.4 joule and from Figure 4, 18 percent or 0.072 joules (4.5 x 10^{17} eV) is absorbed by the plasma. This results in an energy per electron of 118 eV. Considering that the plasma is radiating and that some of the absorbed energy heats the ions and neutrals, the 33 percent difference in particle energy is not unreasonable.

Plasma Growth Rate

The rate of plasma development was determined from high-speed streak photographs such as shown in Figure 5. The main features of these streak records are the rate of growth and the spatial asymmetry. At 600 torr the plasma grows toward the laser at a rate of about 4x10^7 cm/sec and away from the laser at about 1x10^6 cm/sec. The theory section contained a discussion of the three plasma development models: radiation-driven shock wave, radiative transport wave and threshold breakdown wave. The radiation-driven shock wave model proposes to explain asymmetric growth toward the laser by postulating the existence of a strong absorbing shock wave at the upstream side of the plasma. The observations of the present experiments do not show sufficiently strong absorption to justify using this model. Spatial, time resolved shadowgrams and holograms do not show any strongly asymmetric property of the plasma, either. It is clear that the conditions required in the radiation-driven shock wave model are not met. For the remaining two models to explain asymmetric growth there must be a great variation in laser intensity between the entrance and exit sides of the plasma. In terms of average intensity this is clearly not the case for an integrated absorption of only 18 percent. The failure of the three models under the conditions of this experiment cast doubt of their applicability to previous experimental observations. Detailed data about the spatial and temporal variation of laser intensity at the focus seem to be the most likely source of knowledge with which to resolve the asymmetric growth anomaly under condition of low absorption.

Boundary Interactions

The boundary between the plasma and neighboring gas is well-defined, in Fig. 6. In addition, the electron density distribution shown in Figures 12 and 13 has a steep slope at the radial and axial boundaries. From these observations it is possible to use rather simple analysis (in first approximation)to characterize the laser-plasma boundary interactions which occur.

These includes: reflection, refraction, and scattering. The reflection coefficient for normal incidence on an interface between dielectrics is given by the relationship

$$R = \left(\frac{n_p - n_g}{n_p + n_g}\right)^2 \qquad (24)$$

where n_p and n_g are the phase indices of refraction for the plasma and neutral gas. At a pressure of 200 torr and an electron density of $4 \times 10^{18}/cm^3$ equation (24) yields a value for reflection coefficient of about 2×10^{-7}. For a laser pulse of 100 megawatts peak power the instantaneous light reflection would be on the order of 20 watts. Such a reflected power is negligible in terms of the incident laser radiation but can be easily detected photographically.

The refractive effects can also be determined to a first approximation from the data on shape and index of refraction from Fig. 9. The plasma can be considered a nearly symmetrical convex-convex thick lens. Using straightforward ray tracing, matrix optics, or the thick lens formula it is easily shown that the plasma shown in Figure 9 is equivalent to a lens with focal length equal minus 60 mm. Because of the long effective focal length of the plasma it seems certain that none of the 90° ruby light emitted by the plasma is due to refraction.

Scattering laser radiation within the plasma would be dominated by free electron-photon scattering (Thomson scattering). Such scattering would occur throughout the plasma not just at the boundaries as in observed. In addition. The scattering cross section has a value of 6.65×10^{-25} cm^2 [26], which for laser power of 100 megawatts and plasma electron denisty of 4×10^{18}/cm^3 would result in a Thomson scattered intensity of about 2 watt, an order of magnitude lower than that calculated earlier for reflected power from the 100 μm long plasma-gas interface under these conditions.

CONCLUSION

A number of further experiments are suggested by the above results. First, as a check on the $\frac{1}{\tau p}$ dependence of breakdown threshold, the pulse length of the laser should be varied, holding all other variables constant. The time-variable reflectivity technique affords such a capability by changing the oscillator cavity dimensions. A second area for experimental investigation is suggested by the failure of existing plasma growth models to explain the asymmetric growth at low plasma absorption conditions. Existing models have not included detailed laser intensity distributions in the focal volume because they are difficult to produce, to calculate, and to measure. The effect on plasma development of gross changes in focal volume laser intensity distribution should be determined experimentally to show its effect or lack thereof. Thirdly, having demonstrated the ability to study dense plasmas with pulsed holography during their most dynamic stage, improvement in hologram quality (fringe contrast and uniformity) should be made. Reducing the pulse duration and reducing the fringe noise (better coherence) would improve the temporal and spatial resolution of refractive index within the plasma. In more complex gases it would be advisable to do interferometry at two or more wavelengths to separate the effects of ions from the electrons more conclusively. A fourth improvement in the experimental determination of plasma properties is the spatial determination of electron temperature. Such an experiment involves taking high speed streak or framing records of the image from an X-ray pinhole camera.

From an engineering standpoints, measurements taken at different wavelengths and pulse durations will enable determination of scaling laws for further applications. The production of plasma in conjunction with other energy sources such as high current electron guns of the dense plasma focus are important experiments which should be pursued. The effect of magnetic confinement on the plasma density and temperature are presently underway at the Air Force Weapons Laboratory.

ACKNOWLEDGMENT

The authors would like to extend their appreciation to those individuals who made a significant contribution to research presented in this report. From the Los Alamos Scientific Laboratory our thanks go to Dr. Richard E. Siemon who suggested the holographic technique employed. From the Air Force Weapons Laboratory in alphabetical order:

        D. L. Bensinger
        L. D. Boehmer
        R. E. Davies
        D. J. Johnson
        R. V. Wick

REFERENCES

1. P. D. Maker, R. W. Terhune, and C. M. Savage, Quantum Electronics III, 1559, (1964).
2. G. A. Askar'yan, M. S. Rabinovich, M. M. Savchenko, and V. K. Stepanov, JETP Letters, $\underline{5}$, 121, (1967).
3. C. DeMichelis, IEEE Journal of Quantum Electronics, $\underline{QE-5}$, No. 4, $\underline{188}$, (1969).
4. I.K. Krasyuk, P.P. Pashinin, and A.M. Prokhorov, JETP, $\underline{31}$, 860, (1970).
5. N.K. Berezhetshaya, G.S. Voronov, G.A. Delone, and G.K. Piskova, JETP, $\underline{31}$, 403, (1970)
6. L.V. Keldysh, JETP, $\underline{20}$, 1307, (1965).
7. V. Chalmeton, J. de Physique, $\underline{30}$,687, (1969).
8. Blanc and Guyot, Ninth International Conference on Ionization Phenomena in Gases, Bucharest, (1969).
9. F. Morgan, L.R. Evans, and C.G. Morgan, private communication September 1970. Address: Dept. of Physics, University College of Swansea, Swansea, Wales.
10. Ya. B. Zel'Dovich, and Yu. P. Raizer, JETP, $\underline{20}$, 772, (1965).
11. P.E. Nielsen, G.H. Canavan, and S.D. Rockwood, Proceedings of the IEEE , 709, (April, 1971).
12. Yu. P. Raizer, Uspekhi, $\underline{8}$,650, (1966)
13. A.J. Alcock, C. DeMichelis, and M.C. Richardson, IEEE Journal of Quantum Electronics, $\underline{QE-6}$, 662, (1970)
14. L.R. Evans, and C.G. Morgan, Physical Review Letters, $\underline{22}$, 1099, (1969).
15. V.V. Korobkin, S.L. Mandel'shtam, P.P. Pashinin, A.V. Prokhindeev, A.M. Prokhorov, N.K. Sukhodrev, M. Ya. Shchelev JETP, $\underline{26}$, (1968),79.
16. Ya. B. Zeldovich, A.S. Kompaneets, Theory of Detonation, (1960).
17. Yu. P. Raizer, JETP Letters $\underline{7}$, 55, (1968)
18. M.M. Litvak, D.F. Edwards, IEEE Journal of Quantum Electronics, QE-2, 486, (1966).
19. A. Sakurai, Journal of the Physical Society of Japan, 8, 9, 662, 256, (1953), (1954).

20. G.A. Hardway, A.H. Guenther, A.K. Graf, Annals of the New York Academy of Sciences, 168, 440, (1970)
21. R.W. Minck, and W.G. Rado, Physics of Quantum Electronics, 527, (1966)
22. J.L. Bobin, C. Canto, G.F. Floux, J. Reuss, and P. Veyrie IEEE International Quantum Electronics Conference, Miami (1968)
23. R.V. Wick, Ph.D. Thesis, Pennsylvania State University, (1966)
24. G.G. Dolgov, and S.L. Mandel'shtam, Zh. Eksperim i. Teor. Fiz, 24, 691, (1953)
25. R.Z. Alpher and D.R. White, Physics of Fluids, 2, 162, (1959)
26. Jr. L. Spitzer, Physics of Fully Ionized Cases, 2nd Ed. Wiley Interscience, (1962)
27. Landolt-Bornstein, Zahlenwerte und Funktionen aus Physik, Chemie, Astronomie, Geophysik und Technik, 6th Ed., Vol. II, No. 8, Optical Constants, Berlin, Springer-Verlag, (1962).
28. W. C. Marlow and D. Bershader, Physical Review, 133, A629, (1964).
29. R. E. Brooks, L. O. Heflinger, and R. F. Wuerker, IEEE Journal of Quantum Electronics, QE-2, 275, (1966).
30. H. M. Smith, Principles of Holography, New York, Wiley-Interscience, (1969).
31. F. B. Hildebrand, Methods of Applied Mathematics, 440, (1952).
32. H. E. Fettis, Mathematics of Computation, XVIII, 491, (1964).
33. E. H. Beckner, Journal of Applied Physics, 37, 4944, (1966).
34. B. L. Henke, R. L. Elgin, R. E. Lent, and R. B. Ledingham, Advances in X-Ray Analysis, 13, (1967).
35. K. Büchl, K. Hohla, R. Wienecke, and S. Witkowski, Physics Letters, 26A, 248, (1968).
36. C. Smith, Air Force Weapons Laboratory Technical Report (to be published).

INFLUENCE OF PARTICLES ON LASER INDUCED AIR BREAKDOWN[*][†]

Robert J. Hull, Donald E. Lencioni, Louis C. Marquet

Massachusetts Institute of Technology, Lincoln Laboratory

Lexington, Massachusetts

We shall describe a portion of our experimental program on laser induced air breakdown. The purpose of our experiments is to measure the effect of particles in the focal volume of the laser beam on the breakdown threshold and to investigate the nature of the initiation process and the growth mechanisms of the plasma. Although we are interested primarily in particle induced breakdown we have done some experiments in clean air and on a carbon surface.

The experimental layout is shown in Fig. 1. We use a TEA CO_2 laser with a helical arrangement of electrodes-resistors and pins for the cathode and a copper bar anode. We obtain approximately 0.2 joule in the pulse with a 300 nsec full width at half maximum and a low intensity shoulder of 2 μsec duration. The beam is focussed to approximately a 100 μm diameter spot with a 2-inch focal length Irtran-4 lens.

Particles are dropped through a hypodermic needle into the focus of the laser beam. A camera microscope system is focussed onto this spot. A He-Ne laser is also focussed at this point. When a particle is exactly at the focus, red light from the He-Ne

---

[*]Presented at the Second Workshop on "Laser Interaction and Related Plasma Phenomena" at Rensselaer Polytechnic Institute, Hartford Graduate Center, August 30 - September 3, 1971.

[†]This work was sponsored by the Advanced Research Projects Agency of the Department of Defense.

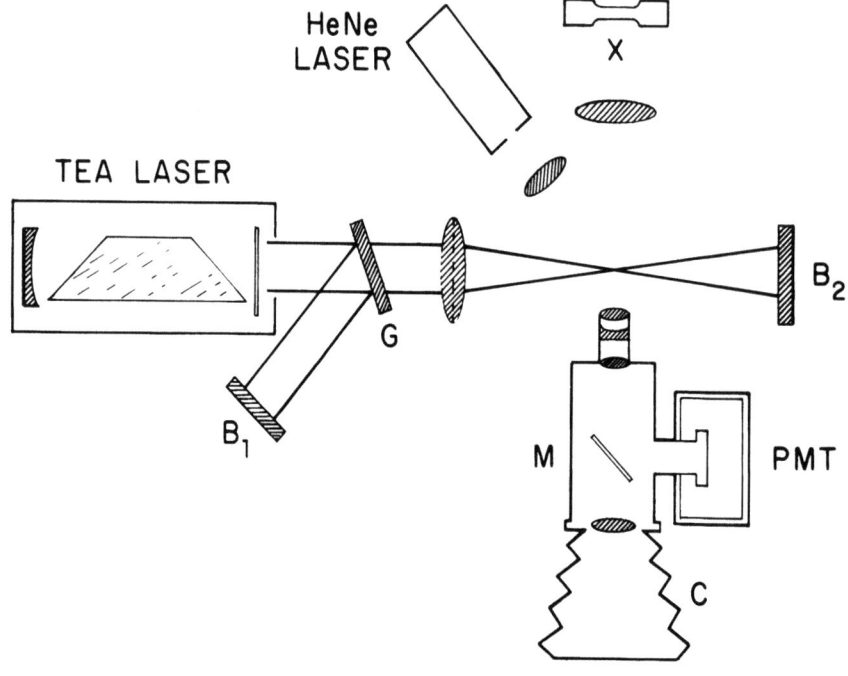

Fig. 1

A schematic diagram of the experimental apparatus. B_1 and B_2 are barium titanate energy detectors. M is a microscope providing 20 to 100 times magnification. PMT is a photomultiplier tube used to detect the presence of particles. G is a germanium beam splitter. C is a polaroid camera. X is xenon flash lamp.

laser is scattered into the microscope. This triggers a xenon flash lamp and the CO_2 laser. The lamp is fired a few microseconds before the laser to provide a photograph of the particle from which the size of the particle is determined. A beam splitter provides a fraction of the beam for diagnostics: total energy and pulse shape. A second energy transducer B_2, shown in Fig. 1, measured the energy transmitted through the focal spot. An electrostatic probe of collecting area 7 mm^2 was used for measuring the ionization of the air in the region surrounding the breakdown. A streak camera having a rotating mirror assembly was used to obtain high speed streak photographs of the motion of the luminosity of the plasma.

Measurements were made of the breakdown threshold in air at one atmosphere pressure with optically thin particles (sodium chloride and powdered germanium) and optically thick particles (glass beads). Sodium chloride was used extensively since it is a known atmospheric aerosol. Particles in the range of 40 to 100 μm diameter were observed to trigger breakdown at power densities as low as 10^8 watt/cm^2, i.e., a value about a factor 20 lower than the so-called clean air threshold. The total absorption of salt crystals of this size is less than 10^{-5}, a value obtained from the bulk absorption coefficient of crystalline NaCl. The energy absorbed by these particles is too low to account for sufficient heating to liberate enough electrons to lower the threshold significantly. An equal lowering of the threshold was found for germanium particles and optically thick glass beads.

The clean air threshold was checked by doing experiments in a chamber which contained filtered air at 1 atmosphere pressure. A TEA laser of approximately 5 joules/pulse energy was used for these experiments. With ordinary air in the chamber breakdown occurs every laser pulse at the expected intensity of 2×10^9 watts/cm^2. However, when the chamber was evacuated and then filled with air filtered through a 250 Å Millipore filter, no breakdowns were observed at intensities up to 10^{10} watts/cm^2 for the first few laser shots. Then breakdown would recur, apparently due to particles knocked from the walls of the chamber.

This hypothesis was confirmed by sampling the air within the chamber with a particle distribution analyzer. The analyzer uses forward light scattering to detect particles down to approximately 1 μm diameter. The particle distribution was measured for four different conditions in the chamber: a) ambient lab air; b) after pumping and filtration through a 250 Å filter; c) after one hundred pulses of the laser beam had passed through chamber and out through an AR/AR coated Ge window; and d) after one hundred pulses had been fired into the chamber with the exit window blocked by a barium titanate energy detector. The results, shown in Fig. 2, indicate a substantial reduction in particle density for case b and an increase in density for case d. Case c shows only a slight increase in particle density. These results show a clear correlation between breakdown threshold and particle density.

Streak camera photographs were also taken of the breakdown region. A typical streak record of a breakdown in laboratory air is shown in Fig. 3. The velocity of the luminous front is initially 10^6 cm/sec and decreases to approximately 2×10^5 cm/sec at the end of the initial (intense) portion of the laser pulse. A one dimensional detonation wave[1] would travel at approximately this velocity for the intensity of our focussed laser beam. For our 2-inch focal length lens we estimate that a breakdown wave[2] would travel at 5×10^5 cm/sec. The luminous front shows a 15 nsec modulation in intensity. This time corresponds to the round trip time of a photon in our 2.2 m laser cavity. Our laser pulse shows a 15 nsec modulation caused by various longitudinal modes present

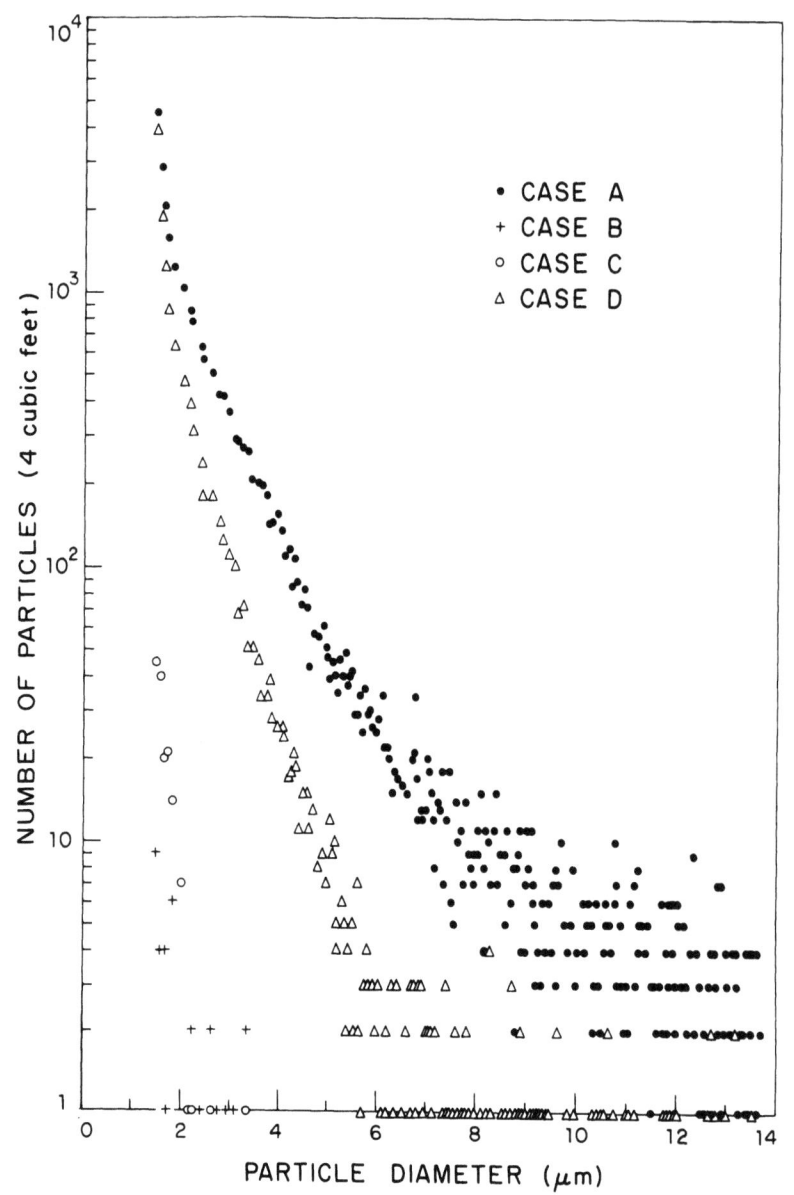

Fig. 2

Particle distribution measured for a) laboratory air; b) filtered air; c) filtered air after 100 laser pulses transmitted through the chamber; and d) after 100 laser pulses fired onto a barium titanate detector inside the chamber.

INFLUENCE OF PARTICLES ON LASER INDUCED AIR BREAKDOWN

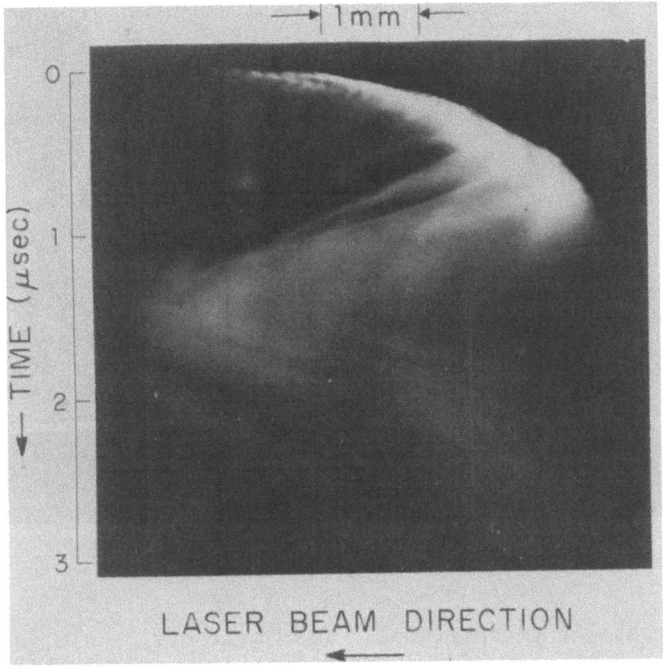

Fig. 3

Streak photograph of a laboratory air breakdown.

within the laser cavity (self mode locking). Hence, we conclude that the luminosity of the breakdown region is driven by the laser power.

The ionization produced by the laser induced breakdown was studied using electrostatic probes. The probe was placed well outside the focal volume of the laser and was biased to collect either positive or negative charges. A typical probe signal is shown in Fig. 4. The time of appearance of the probe signal did not depend on the distance away from the breakdown region. This indicates that the ionization did not diffuse out to the probe but rather was due to photoionization from the primary breakdown. The dip in the signal at 10 μsec results when the sound wave coming from the breakdown produces a local perturbation in the ion density. The time of arrival of this signal was obtained as a function of distance from the breakdown. The first 0.5 cm is described well by a shock wave model of a point explosion with 0.2 joules of energy input. From that point on the velocity is

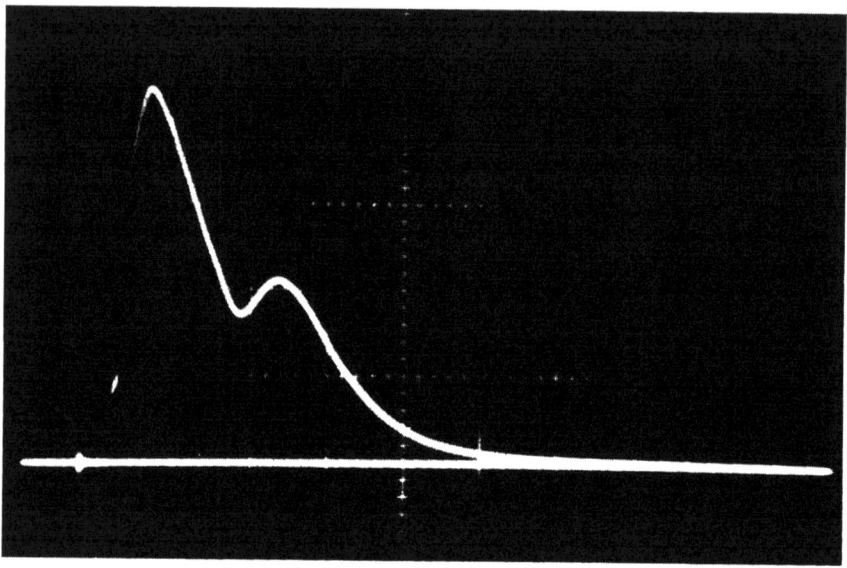

Fig. 4

An electrostatic probe signal. The horizontal scale is 5 μsec/division.

constant at 5×10^4 cm/sec which is slightly faster than the speed of sound at room temperature and may signify a local heating of the air. The strength of the ionization was obtained as a function of distance from the breakdown. The ionization decreases away from the breakdown with an e-folding distance of 1 mm. This corresponds to the mean free path of 200 eV photons.

REFERENCES

1. Yu. P. Raizer, Soviet Physics, Uspekhi **8**, 650 (1966).

2. The breakdown wave front is defined by the points at which the breakdown threshold has been exceeded. The velocity of the resulting wave toward the laser depends upon the geometrical convergence of the laser beam produced by the focussing lens. A derivation of this velocity is given in Ref. 1.

LASER PRODUCED PLASMA - STREAMER INTERACTION*+

Laird P. Bradley

Sandia Laboratories, Albuquerque, N. M.

One aspect of laser-plasma interaction which has received little attention is the effect of laser-produced ionization on streamer propagation. Vail, et al.,[1] have shown spatial channeling of a streamer along a laser ionized path in air. Recently, using a pulsed nitrogen laser to produce ionization, we have demonstrated control of streamer velocity over orders of magnitude for N_2 and SF_6. Experimentally, a pulsed voltage was applied across a nonuniform field gap to launch a streamer. The laser-induced ionization was produced spatially ahead of the already propagating streamer. The streamer velocity was thereby controlled. This technique is currently being applied to provide subnanosecond jitter in megavolt switches; the process is different from the triggering method of Guenther, et al.[2]

1. J. R. Vail, D. A. Tidman, T. D. Wilkerson, and D. W. Koopman, Appl. Phys. L. <u>17</u>, 20(1970).

2. A. H. Guenther and J. R. Bettis, IEEE J. Q. Elect. <u>QE3</u>, 581(1967).

*Work supported by the U.S. Atomic Energy Commission. Published in detail: L. P. Bradley and T. J. Davies, IEEE J. Q. Elect., QE-7, 464 (1971).

+ Presented at the Second Workshop on "Laser Interaction and Related Plasma Phenomena" at Rensselaer Polytechnic Institute, Hartford Graduate Center, August 30 - September 3, 1971.

EXPERIMENTS ON SELF-FOCUSING IN LASER-PRODUCED PLASMAS [*]

A.J. Alcock

Division of Physics, National Research Council of

Canada, Ottawa, Ontario, Canada

ABSTRACT

Recent investigations of laser induced gas breakdown have revealed a number of phenomena which are compatible with the occurrence of self-focusing at the time of breakdown. Some examples of the effects observed are: 90° scattering of laser light from regions having transverse dimensions as much as an order of magnitude less than the diameter of the focal volume, intense forward scattered radiation at a wavelength close to that of the laser, and plasma filaments of $\sim 10\mu$ diameter which have been detected by means of high spatial resolution, sub-nanosecond, Schlieren photography and interferometry. In this talk the large amount of experimental evidence for self-focusing is reviewed and a number of possible mechanisms discussed.

INTRODUCTION

The first indication that self-focusing might be associated with laser-produced plasmas was reported by Basov et al.[1] who observed the creation of long sparks, extending over a distance of ~ 2 meters, in air by means of weakly focused neodymium:glass laser radiation. Although this observation was explained in terms of the temporal variation of the laser beam divergence, the possibility of self-focusing leading to the observed effects was also introduced. Within a short time a similar suggestion was made by Korobkin et al.[2] who carried out a detailed study of the spark in air produced

---

[*] Talk presented at the 2nd Workshop on "Laser Interaction and Related Plasma Phenomena", Rensselaer Polytechnic Institute, Hartford Graduate Center, Aug. 30 - Sept. 3, 1971.

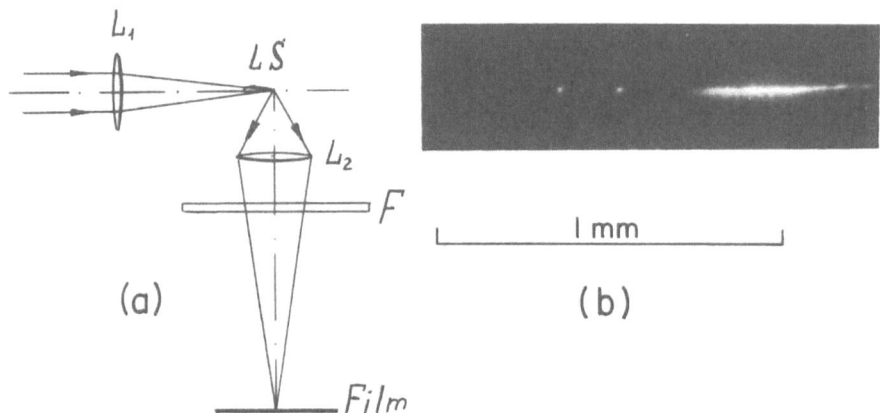

Fig. 1 (a) Experimental arrangement used to obtain high magnification time integrated photographs of air sparks by means of laser radiation scattered at 90°: L_1-focusing lens; L_2-imaging optics; LS-spark; F-narrow band filter. (b) Spark in air showing filamentary and discrete scattering regions. The direction of the laser beam is from right to left.

by means of a 100 MW, Q-switched ruby laser. From photographs of the spark, obtained with the scattered laser radiation emerging at 90° to the beam axis, and laser-illuminated shadowgraphs, the authors concluded that a number of discrete breakdown regions were formed during the development of the plasma. Although this phenomenon was discussed in terms of a radiation wave mechanism the authors also referred to earlier theoretical work by Askaryan[3] and Litvak[4] who discussed the possibility of intense electromagnetic radiation being trapped within a plasma.

EXPERIMENTAL STUDIES IN THE NANOSECOND REGION

Shortly after the possibility of a connection between self-focusing and plasma production with lasers had been pointed out, additional studies of spark structure, carried out by means of scattered light photography, were reported.[5,6] In the case of (5) a multimode ruby laser, Q-switched by means of a rotating prism, was used to investigate sparks produced in hydrogen, nitrogen, oxygen, carbon dioxide, chlorine, methane and inert gases. The results obtained indicated a bead-like structure in molecular gases, when the laser power was slightly above threshold, and the formation of fork-like scattering regions as the laser intensity was increased. Further studies of the polarization of the scattered light were carried out and from these, and photographs taken simultaneously in two directions, it was concluded that a "surface" scattering process was responsible for the observed structure.

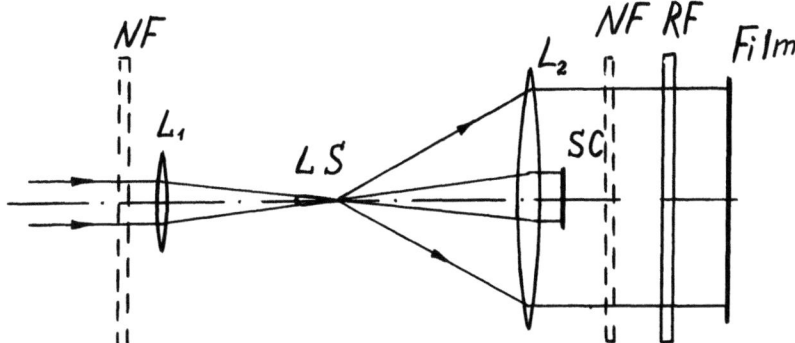

Fig. 2 Experimental arrangement used to detect forward scattered radiation emerging from a laser spark: L_1, L_2-lenses; LS-spark; NF-neutral density filters; RF-narrowband filter; SC-obstacle.

Some evidence for much weaker scattering was also obtained in the case of noble gases, and in particular, the authors noted the appearance of 3 to 5 widely separated scattering points in the case of argon.

Somewhat different results, obtained with a passively Q-switched ruby laser, operating in a single axial and transverse mode, were reported by Korobkin and Alcock[6] who investigated sparks in air by means of the scattered laser radiation emerging in the forward direction and at approximately 90° to the laser beam axis. Photographs of the sparks were obtained with a spatial resolution of ∼5μ using the type of experimental arrangement shown in Fig. 1(a). The diffraction limited laser beam which had a divergence of 0.5×10^{-3} radians was focused by means of a 10cm focal length lens and the highly magnified image of the breakdown region was recorded on infrared film using a narrow band filter to select only the scattered laser radiation. A typical result, obtained with the aid of this technique, is presented in Fig. 1(b) where it can be seen that the scattered radiation originated in filamentary regions having a diameter which did not exceed the resolution of the optical system. Thus the transverse dimensions of the scattering region were at least an order of magnitude smaller than the estimated focal spot diameter of 50μ.

The strong resemblance of the filamentary scattering region to the self-focusing of intense laser radiation in liquids prompted an investigation of the forward scattered radiation and, using the arrangement shown in Fig. 2, forward scattered light emerging from the focal region was detected both photographically and photo-

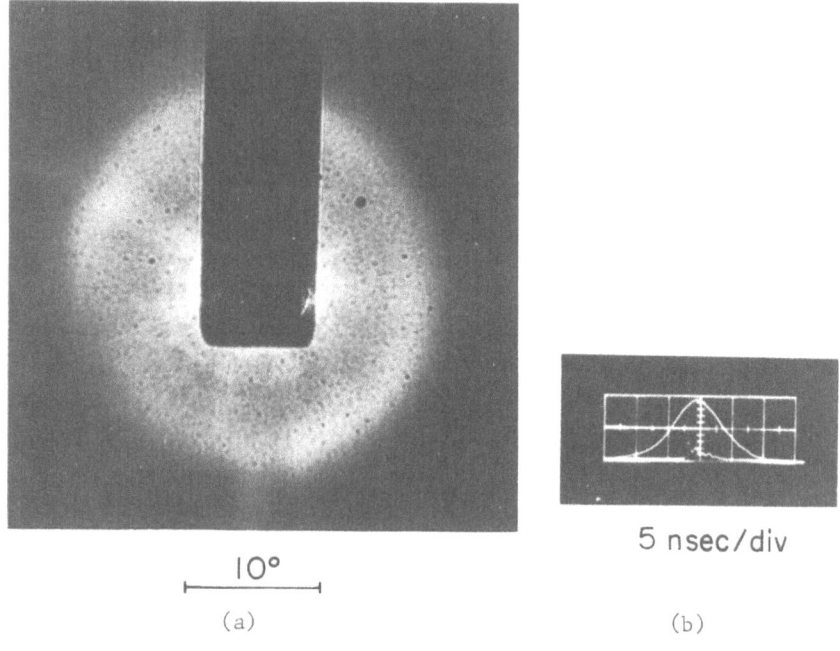

Fig. 3 (a) Photograph of forward scattered light. (b) Oscilloscope trace showing both the incident laser pulse and the forward-scattered pulse.

electrically. Photographs of the forward scattered radiation, such as that shown in Fig. 3(a), revealed that the scattered radiation from an air spark emerged in a well-defined cone having an included angle of $\sim 30°$. When the camera was replaced by a planar photodiode coupled to a Tektronix 519 Oscilloscope it was found that, within the 0.5 nanosecond resolution of the detection system, the scattered light was produced at the instant of breakdown* and had a risetime which could not be resolved. As shown in the oscilloscope trace of Fig. 3(b), the peak power of the forward scattered light corresponded to approximately 30% of the incident 3 MW pulse.

Having confirmed that the polarization of the forward scattered radiation was always the same as that of the incident laser light, its coherence was also investigated by allowing two beams emerging at opposite sides of the cone to interfere with one another. The results of this test clearly demonstrated that the forward scattered light was coherent and on the basis of this observation, the spatial distribution of the scattered light, and its intensity relative to that of the incident beam, it was

*ie., the time at which the transmission of the non-scattered laser radiation decreased sharply.

considered unlikely that scattering from the plasma alone could account for the observed effects.

Although the above results, obtained in air, were initially interpreted in terms of self-focusing in the neutral gas it was clear that much additional evidence was needed to confirm the presence of self-focusing effects and to determine the processes involved. Valuable experimental data has been provided by more detailed studies of scattered laser radiation which were carried out by a number of authors.[7,8,9] Both Ahmad et al.[7] and Tomlinson[8] carried out quantitative investigations of the fraction of incident laser radiation scattered at 90°. On the basis of results obtained in helium[7] and a number of gases[8] it was concluded that the scattering was not Thomson scattering but was probably due to reflection from regions of high electron density. An important, additional, feature of both of these investigations was the emphasis given to the transverse dimensions of the source of scattered laser radiation. In the model proposed by Ahmad et al., the geometry of the optical system, used to collect the scattered light, was taken into account, and, on the assumption of reflection from a spherical shock front, it was predicted that the image of the reflecting region would have transverse dimensions corresponding to the diffraction limit. Although it was suggested that anomalous features of previously reported streak photographs[10] might be due to this 'apparent' reduction in the dimensions of the reflecting surface, no direct comparison was made with time integrated photographs obtained by means of scattered radiation. Such photographs were presented by Tomlinson[8] who applied both time-integrated and high speed streak photography to sparks produced in atmospheric pressure air and argon by the 5 MW, 30 nanosecond pulse from a rotating-mirror Q-switched ruby laser. When obtained with scattered laser radiation, both photographic techniques revealed the existence of discrete scattering centres having dimensions apparently less than the 14μ resolution limit of the recording optics. In the case of air these scattering centres were sufficiently close to one another, along the axis of the focal region, to form an almost continuous scattering 'filament', while it was noted that the centres in argon were clearly separated. The possibility that self-focusing might be occurring in the plasma sometime after the beginning of the ionization process was proposed as an explanation for the observed structure.

A similar suggestion was put forward by the author and co-workers[9] following a detailed study of sparks produced in a number of gases by means of the 4 MW peak power, 10 nanosecond, pulse from a passively Q-switched single-mode ruby laser having a measured linewidth of 0.003 cm^{-1} and a full angle beam divergence of 0.63 milliradians. The gases investigated were nitrogen, freon, methane, helium, argon, neon, krypton and xenon, and the pressure was varied in the range from 760 to 9000 Torr. Time-integrated

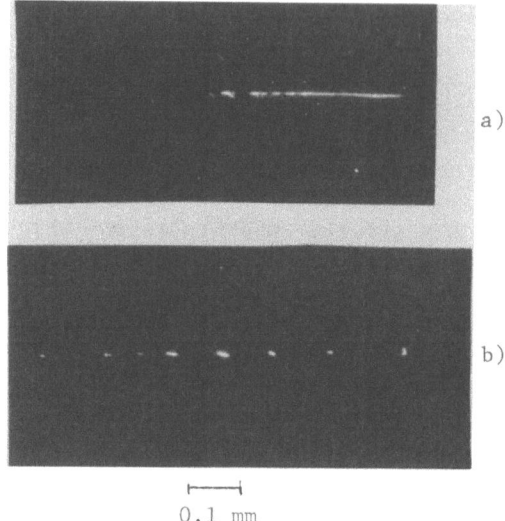

Fig. 4   Photographs of ruby laser light scattered at 90° from sparks in (a) nitrogen at 3800 torr and (b) argon at 1800 torr. The direction of the laser beam is from left to right.

photographs having a spatial resolution of ∿5μ were obtained in nitrogen and argon and typical results are presented in Fig. 4. The two photographs in this figure demonstrate both the small transverse dimensions of the scattering regions and the sharp contrast between the spatial structure observed in the molecular and noble gases. In all cases the transverse dimensions of the scattering regions did not exceed the resolution of the optical system even though the Gaussian beam from the laser was focused by several different lenses to give estimated focal spot diameters as high as 80 microns. In addition it was found that the spacing of the scattering centres in noble gases was not only a function of the type of gas but decreased from a value of several hundred microns as the pressure was increased from 760 torr. This result showed very clearly that the observed structure did not result from intensity variations in the focal region, produced by spherical aberration of the focusing optics.[11]

Image-converter streak photographs of the scattering regions were also obtained and typical results, such as those presented in Fig. 5, show both the smoothly developing scattering filament characteristic of molecular gases (Fig. 5a) and the smaller discrete centres which develop at intervals of a few nanoseconds in a gas such as argon (Fig. 5b). Similar photographs (Fig. 5c), obtained with the visible radiation emitted by the argon plasma itself, confirmed that these small scattering points did in fact

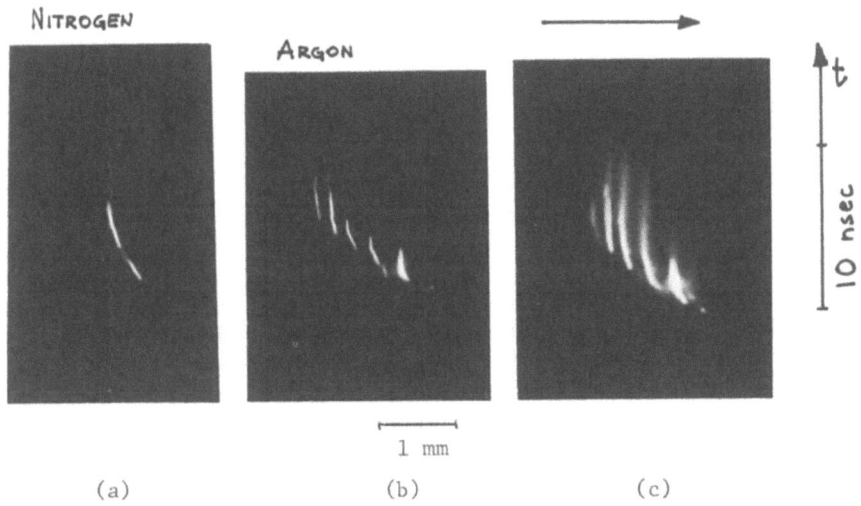

Fig. 5 90° streak photographs showing longitudinal development of sparks (a) scattered laser light from nitrogen at 3800 torr (b) scattered laser light from argon at 1800 torr (c) visible radiation emitted by spark in argon at 1800 torr. The arrow indicates the direction of the laser beam.

correspond to isolated plasma blobs.

An examination of the forward scattered radiation emerging from the sparks produced in noble gases yielded a number of unexpected results. The latter were obtained by detecting light emerging in the forward direction both photoelectrically and photographically, and oscilloscope traces showing the radiation emerging from a spark in argon at 8300 torr are presented in Fig. 6. On each oscilloscope trace the incident laser pulse is displayed as the first half while the second, delayed, photodiode signal shows the forward scattered light transmitted by a narrow band interference filter centred at the laser wavelength. In the absence of a spark equal signals were obtained from both detectors. Fig. 6(a) shows the effect observed when only radiation travelling in the same direction as the incident beam was permitted to reach the second photodiode. In this case the occurrence of breakdown was accompanied by the apparent absorption of approximately 80 percent of the remainder of the pulse. However, when all scattered light, emerging at angles up to 30° from the forward direction was collected, the output of the second detector corresponded to the trace displayed in Fig. 6 (b). This observation revealed that more than 80 percent of the incident pulse emerged from the focal region and that the true

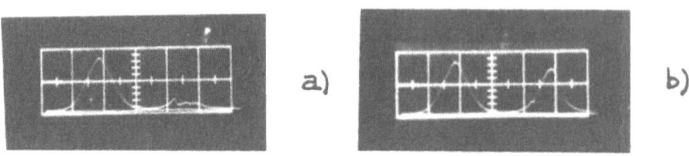

$\sim$2.5MW/div^n - vertical
10 nsec/div.^{n} - horizontal

Fig. 6 Oscilloscope traces showing forward scattered ruby laser light in argon at 8300 torr. The first signal on each trace corresponds to the incident laser pulse, while the second shows the forward scattered pulse. (a) signal obtained when only non-scattered laser radiation is permitted to reach the second photodiode. (b) signal obtained by collecting all light, at the laser wavelength, emerging from the focal region at angles up to 30° from the forward direction. The two detectors were normalized to give equal signals when no spark was produced.

absorption was in fact quite small. Although the largest signals were observed in the case of argon, similar results were obtained with all the gases investigated, and in all cases the polarization of the scattered radiation was found to be the same as that of the incident light. Both the incident power and the gas pressure determined the time at which breakdown occurred, however it was found that the ratio of the scattered radiation's maximum intensity to that of the incident light did not vary by more than $\sim$10%.

An additional property of the forward scattered radiation, revealed by recording it photographically, was the dependence of the angle of emission on gas pressure. This effect is illustrated by Fig. 7 which shows the scattered laser light emerging from argon sparks at pressures of 760 and 9000 torr. These photographs were obtained by using an obstacle to block out the direct laser radiation transmitted through the focal region and permitting only light that passed the obstacle to fall on the photographic film after passing through a narrow band filter. Fig. 7(a) shows scattered light emerging at an angle of $\sim$5° from the forward direction when a spark was produced in high pressure argon. At lower pressures the higher power required for breakdown resulted in a noticeable leakage around the obstacle, however, in the case of atmospheric pressure argon the radiation emerging at an angle of $\sim$9° was clearly visible, (Fig. 7b).

An investigation of the spectral characteristics of the scattered radiation was also carried out and this revealed a slight broadening of the spectrum towards longer wavelengths in the case of molecular gases, while both anti-stokes shifts and a

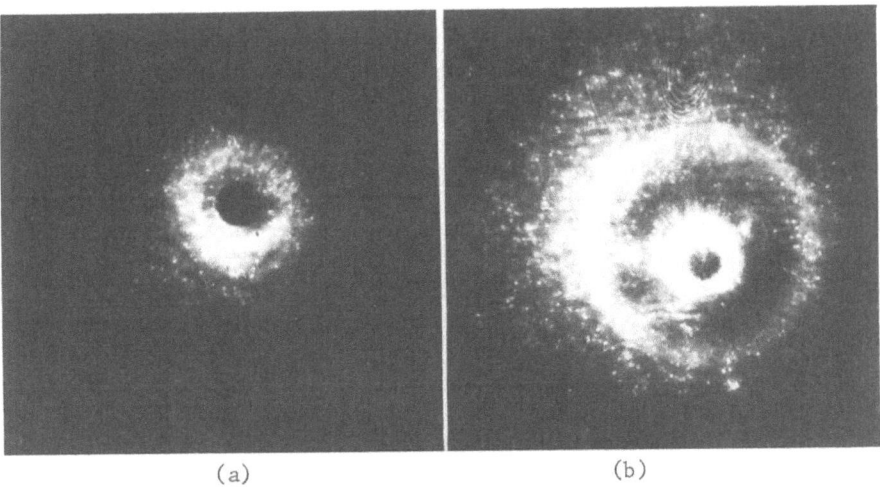

Fig. 7 Forward scattered laser light in argon. a) Pressure of 9000 torr. b) Pressure of 760 torr.

spectral width of ~ 0.1 cm^{-1} were observed in noble gases. In the latter case the intensity of scattered radiation permitted the pressure dependence of the shift to be measured, and it was found that it decreased smoothly from an initial value of 0.2 cm^{-1}, at a pressure of one atmosphere, and could no longer be resolved at a pressure of ~ 7000 torr. Although it was expected that the non-linear effects associated with a self-focusing process might influence the spectral properties of the scattered radiation the shift observed in the case of noble gases strongly resembled a Doppler shift due to rapid motion of the 'scattering' source. However, it is by no means clear why a similar effect was not observed in molecular gases, and this discrepancy still lacks a satisfactory explanation.

Although, at the present time, the processes responsible for the spatial and spectral characteristics are not fully understood the high intensity of radiation emerging in the forward direction does provide a strong indication of self-focusing, particularly if it is assumed that the sources for 90° scattering and forward scattering are one and the same. However, the need for more direct evidence has prompted recent investigations where either high resolution Schlieren photography[12] or interferometry[13] have been applied to study the initial development of the spark plasma.

Key et al.[12] used a Nd:glass laser, operating in the pulse transmission mode to produce breakdown in atmospheric pressure argon. By passing part of the same laser beam through a gas breakdown cell, the 1 nanosecond rise time, 6 nanosecond duration, pulse was cut off to yield a 1 nanosecond pulse, which after frequency doubling and an appropriate optical delay, illuminated a Schlieren system. Photographs with 5µ resolution revealed filamentary regions, of high refractive index gradient, having a diameter less than 10µ and running between blobs of plasma separated by several hundred microns. The use of interferometric techniques by Richardson and Alcock[13] has permitted the identification of the observed filaments as extremely narrow channels of high density plasma. The same single-mode ruby laser, used in previous studies, was employed in the experimental arrangement shown in Fig. 8 and interferograms of sparks were obtained with a temporal resolution of ∼700 picoseconds and a spatial resolution of ∼10µ. As can be seen from the schematic diagram of the experiment, the oscillator beam was split into two parts, one of which passed through a subnanosecond electro-optical light gate, while the other passed twice through an amplifier rod to yield peak powers of ∼20 MW. The 1 MW peak power, 700 picosecond long, pulse transmitted by the polarizer-Pockels cell combination illuminated a conventional Mach-Zehnder interferometer while the amplified output from the oscillator was focused, with a 15 cm focal length lens, into a pressure cell situated in one arm of the interferometer. By means of appropriate optical delays the interferometer could be illuminated at various times within the first few nanoseconds after the initiation of breakdown, and, since the subnanosecond probe pulse was polarized orthogonally to the main laser beam, the orientation of a polarizer permitted either the interferogram or scattered light to be recorded.

The intensity distribution of the high power radiation within the focal region was investigated, by recording the light scattered from a dilute solution of milk, and revealed that the full width of the focal spot was ∼190µ at the half power points (Fig. 9a). An independent confirmation of this value was obtained by comparing the minimum power required for breakdown in argon with previously reported measurements of the breakdown threshold.[9]

Interferograms were obtained in a number of gases at pressures in the range of 0.5 to 8.0 atmospheres and examples of the results obtained are presented in Figs. 9 and 10. In Fig. 9 a series of interferograms, obtained at various times after breakdown in atmospheric pressure argon, provides a clear indication of how the plasma develops. As can be seen from the figure a filamentary region, with a fringe shift corresponding to a negative-going refractive index change, is created first at the time of breakdown (Fig. 9b). This filament, which initially has transverse dimensions of no more than 13µ, rapidly develops to a length of ∼300µ at which time a larger region of plasma is formed at the end of the filament farthest from the laser (Fig. 9c). As the filament continues to move rapidly towards

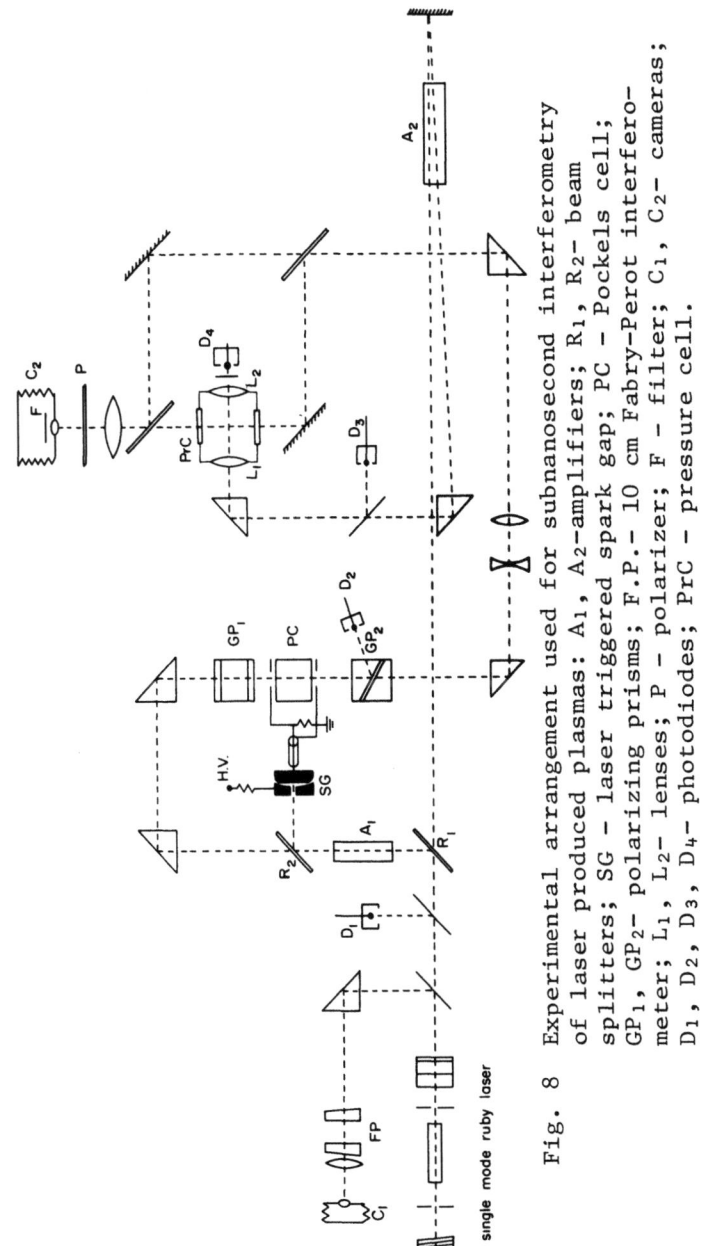

Fig. 8 Experimental arrangement used for subnanosecond interferometry of laser produced plasmas: A_1, A_2-amplifiers; R_1, R_2- beam splitters; SG - laser triggered spark gap; PC - Pockels cell; GP_1, GP_2- polarizing prisms; F.P.- 10 cm Fabry-Perot interferometer; L_1, L_2- lenses; P - polarizer; F - filter; C_1, C_2- cameras; D_1, D_2, D_3, D_4- photodiodes; PrC - pressure cell.

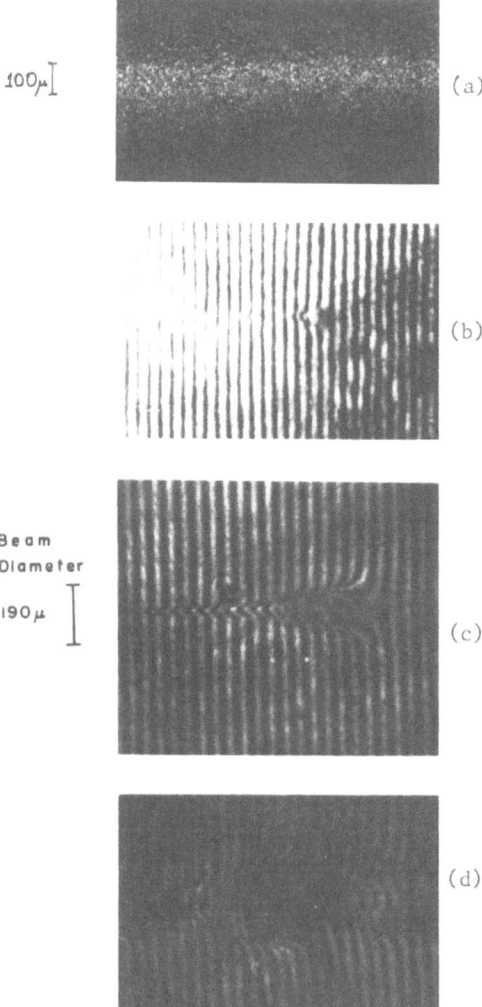

Fig. 9 Subnanosecond interferograms of breakdown in argon at 760 Torr; (a) laser light scattered by dilute solution of milk in focal region; (b) interferogram of spark obtained at time of breakdown; (c) and (d) show interferograms obtained 4 nsec and 10 nsec after breakdown respectively. The laser beam is incident from the left.

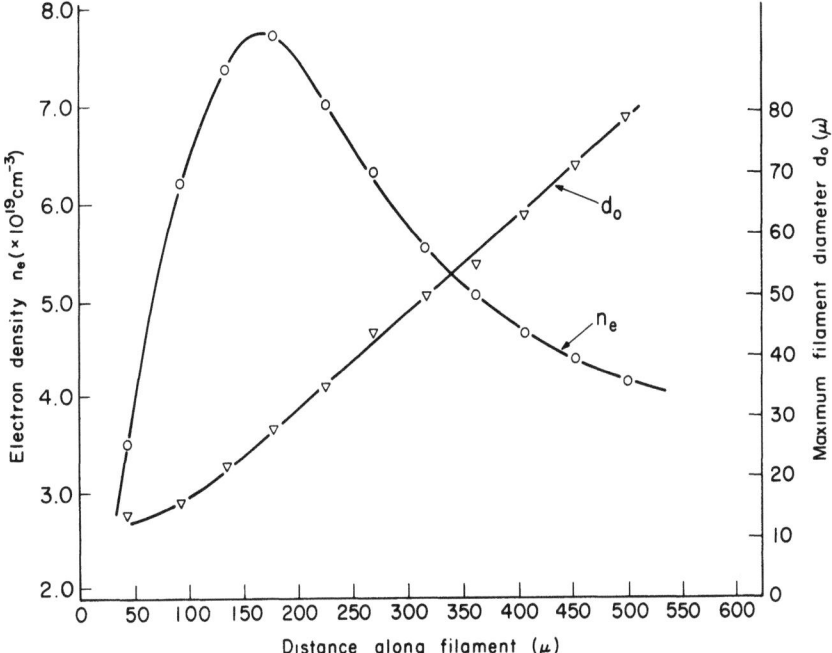

Fig. 10 Variation of electron density and maximum filament diameter along filament length, for the interferogram shown in Fig. 9c.

the laser additional blobs of plasma are generated at intervals of 200-400μ and following their formation the filament between them gradually decays (Fig. 9d).

From such interferograms an estimate of the average electron density in the direction of the illuminating beam can be obtained by assuming that the contribution of ions and neutral atoms to the refractive index change is negligible. Thus from the width, d, (cm) of the region where fringes are displaced, and the fringe shift, S, the average electron density, $\bar{n}_e$, is given by

$$\bar{n}_e = 3.25 \times 10^{17} \, Sd^{-1} \, cm^{-3}$$

For the interferogram of Fig. 9c the width of the plasma filament and the average electron density are plotted in Fig. 10. As can be seen from this figure the electron density has a maximum value of $\sim 8 \times 10^{19}$ cm^{-3} near the end of the filament closest to the laser.

In the case of the larger regions of plasma which develop along the length of the filament, regions of high electron density are observed within a few nanoseconds after the blob's formation. Such a high density core is clearly visible in the interogram

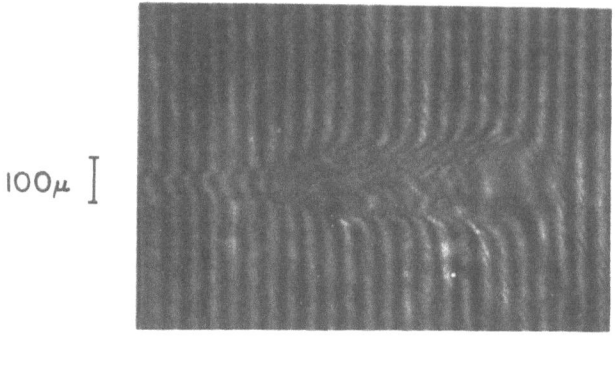

Fig. 11 Subnanosecond interferogram of a spark in argon, at a pressure of 760 Torr, obtained 5.5 nsec after breakdown and showing the internal structure of the plasma blob. The laser beam direction is indicated by the arrow.

of Fig. 11 and it appears likely that the discrete centres observed previously by means of 90° scattered light[5,8,9] have their origin in these regions.

In addition to the interferograms obtained in noble gases the development of sparks in molecular gases, such as hydrogen, has also been investigated. Although no extended filamentary regions of plasma were detected, and the plasma expanded smoothly during the laser pulse, the results indicated that the plasma developed initially in the form of small filament and then expanded rapidly behind the filament as it progressed towards the laser.

The direct observation of filamentary plasma regions, having transverse dimensions more than an order of magnitude smaller than those of the focal region, has provided the most significant evidence for self-focusing in laser-produced plasmas reported so far.

EXPERIMENTS INVOLVING SUBNANOSECOND LASER PULSES

Although all of the results described above were obtained with lasers generating pulses longer than a nanosecond, a number of similar observations were made on sparks produced by means of sub-nanosecond pulses. Both 90° scattering and forward-scattering from sparks produced by mode-locked ruby and neodymium:glass lasers have yielded preliminary evidence for self-focusing[14] as has an examination of the spectral structure of the forward

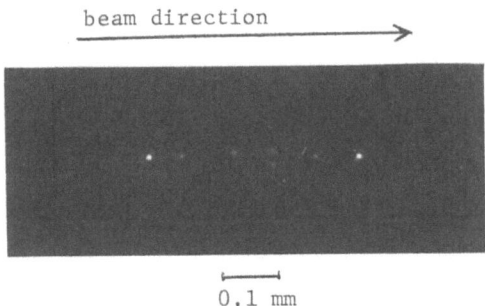

Fig. 12 Time-integrated photograph of spark produced in air by mode-locked Nd:glass laser. The photograph shows second harmonic laser radiation scattered at 90° to the incident beam. A 2 cm focal length lens was used.

scattered light.[15] These experiments were carried out with a mode-locked neodymium:glass laser and a mode-locked ruby laser generating pulses of 5 and 30 psecs respectively. Peak powers of ∼4 GW were obtained from the glass laser, in a beam which had a divergence of ∼2 milliradians, while the ruby system generated pulses of 100 MW peak power and operated in a single transverse mode to yield a diffraction limited beam divergence of 0.5 milliradians. Light scattered at 90° from sparks in air was detected photographically be means of the scattered ruby laser light or a second harmonic beam which was generated co-linearly with the neodymium laser radiation. In both cases it was found that the scattered light emanated from a number of discrete points distributed along the axis of the focal region (Fig. 12). The scattering centres bore a strong resemblance to those observed previously with nanosecond pulses and noble gas breakdown, however, it should be noted that in this case the sources of scattered light correspond to different breakdown regions formed by successive pulses in the mode-locked train. A more significant feature of the scattering points was their small diameter ($\lesssim 5\mu$) which, as in the nanosecond case, was found to be independent of the focusing lens used. Since the measured beam divergences indicated focal spot diameters varying from 25 to 200μ, and time-integrated measurements of the near and far field patterns provided no evidence for filamentary structure, it again appeared probable that some type of self-focusing mechanism was involved. Additional evidence was provided by observations of the forward scattered light and as can be seen from a typical result, obtained with the mode-locked ruby system, (Fig. 13), the scattered radiation emerged from the focal region within a cone of 15°-20° included angle. As described already in the preceding section an obstacle was used to block the laser light transmitted directly through

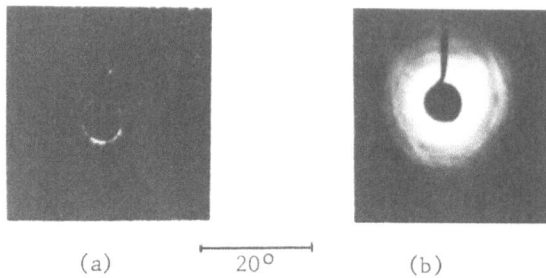

(a)　　20°　　(b)

Fig. 13 Photographs of forward scattered radiation when the mode-locked ruby laser beam was focused by means of a 5 cm focal length lens. In (a) no spark was produced.

the focal region. Similar measurements made with either laser and a photodiode detection system revealed that approximately 10% of the incident power was transmitted beyond the obstacle when breakdown occurred. A study of the spectral width of this transmitted light was carried out with the neodymium:glass laser and indicated a broadening from the initial laser linewidth of ∼75 Å to a value of 200 Å. Unlike the results obtained with nanosecond pulses of single-mode laser radiation the broadening occurred towards both longer and shorter wavelengths with the shift to longer wavelengths being slightly greater. So far these observations have not been augmented by more detailed studies, however, the spectral broadening does indicate the presence of a non-linearity compatible with self-focusing effects.

Although similar observations of micron scale scattering centres have since been reported[16] only one experiment has indicated the occurrence of self-focusing when breakdown is produced by means of a single picosecond pulse. This investigation was carried out by Bunkin et al.[17] Who used a 2 GW ruby laser pulse, with a duration in the range of 20-100 psec., to produce breakdown in air, nitrogen and argon. With the laser intensity reduced well below the value required for gas breakdown, both the diameter of the beam and the intensity distribution in the focal region were measured by means of the fluorescence from a solution of Rhodamine 6G in ethanol. In addition to revealing a Gaussian intensity distribution in each section of the beam measurements performed with a 15 cm focal length lens yielded a minimum cross section of ∼200μ.

When sparks were produced in air and nitrogen, time-integrated and high speed streak photographs showed that a number of discrete breakdown regions separated by distances of 1.5-2 mm were formed in the vicinity of the focus. In some instances the individual breakdown points had dimensions less than 15μ and, from streak photographs, appeared to develop in the direction of propagation of the laser radiation with a velocity comparable to that of the actual pulse. Since the laser intensity decreased by a factor of 2 only 0.2 mm from the focus, neither of these observations could be accounted for without invoking a self-focusing process. Further evidence was provided by the observation of scattered laser radiation at power levels slightly below those at which the breakdown points appeared, and by measurements of the actual threshold intensity. The value of 3.5×10^{12} W/cm^2 was considerably smaller than that required to create a spark when the beam was focused to a 17μ diameter spot by a short focal length lens. In view of the fact that diffusion effects are negligible during a picosecond pulse such a dependence on the dimensions of the focal region would not be expected under linear focusing conditions. Although further experiments are required these results appear to confirm the earlier suggestion that self-focusing effects are associated with picosecond breakdown.

SELF-FOCUSING MECHANISMS

Although many of the experimental results described in the preceding sections indicate the presence of self-focusing effects, they do little to reveal the actual physical processes involved. However, during the last few years a number of possible mechanisms have been discussed and a brief review of these gives some insight into the more probable explanations.

Since all of the experimental evidence obtained so far applies to laser induced gas breakdown, self-focusing could occur during one or more of the following stages of spark formation:-
a) immediately prior to breakdown in the neutral gas
b) during the cascade ionization process
c) after a high density plasma has been produced.

During each of these stages it is possible to identify non-linear processes capable of producing refractive index changes of appropriate sign. In the case of (a), either electrostriction or the optical Kerr effect could provide the required nonlinearity and by using the expression for critical power derived by Kelley[18] threshold powers have been estimated.[9] The values obtained for molecular and noble gases, at atmospheric pressure, are of the order of tens of megawatts for electrostriction and gigawatts for the Kerr effect. Thus in the case of nanosecond pulses, at power levels of a few megawatts, a neutral gas process appears highly unlikely since the Kerr effect threshold is much too high and electrostriction requires both higher powers and longer pulse durations. In addition self-focusing in the neutral gas would inevitably lead to breakdown and a reasonable argument[8,9] can be based on the fact that there is little evidence to support the hypothesis that self-focusing rather than cascade ionization determines the breakdown threshold. However, processes occurring in the neutral gas cannot be ruled out very easily in the case of picosecond pulses where power levels may be well in excess of the Kerr effect self-focusing threshold.

Once the cascade ionization process, stage (b), has begun, the neutral gas in the focal region is replaced by a mixture of electrons, ions and neutral atoms with various degrees of excitation. The presence of free electrons results in a reduction of the refractive index which, in the absence of any compensating effect, will eventually lead to a defocusing effect. However, in many cases it appears likely that ionization occurs by photo-ionization of excited atoms and that the populations of ions, excited atoms and overexcited atoms (ie., atoms excited to within one or two photon energies of ionization) are approximately equal.[19] Since one would expect the absolute polarizabilities of the excited atoms to exceed those of unexcited or ionized atoms it is possible that the effect of the electrons might be cancelled and a net self-focusing effect result. Although there appears to be little experimental data on excited atom polarizabilities the possibility of such an effect can be illustrated in the case of argon where the observed polarizability of the $^3P_2$ state is $\sim 10^{-22}$ and exceeds by approximately two orders of magnitude that of the ground state.[20] The resulting refractivity at a wavelength of 7000Å is $\sim 6 \times 10^{-22} N_{AI}$ while the electronic contribution is $-2.3 \times 10^{-22} N_e$ (where N_e and N_{AI} are the densities of electrons and excited argon atoms respectively). Although from the above it is obvious that extremely high excited atom densities are required to produce a significant change in the refractive index, ie., of the order of 1%, it has been suggested that the effect of excited atoms could be considerably enhanced due to the resonant character of the nonlinear polarizability.[12] Furthermore the presence of excited molecules, at an early stage of the breakdown process might well initiate the self-focusing effect and, although this explanation has been proposed in the case of picosecond pulses,[17] it might also explain some of the observed discrepancies between molecular and noble gases.

Finally, the possibility of self-focusing occurring in the resulting plasma, stage (c), must be considered since the existence of numerous nonlinear plasma effects is well known. However, it appears that only two of these will be significant at the relatively low power levels where most of the experimental evidence for self-focusing has been obtained. Both of these processes involve localized changes in plasma density which arise either from thermal energy deposition or ponderomotive forces governed by the intensity gradient across the focal region. These mechanisms have been treated theoretically by Shimoda[21] and Hora[22] respectively and the minimum powers required to sustain a self-trapped filament have been estimated. In each case the power required is comparable to or less than the breakdown threshold and thus it appears possible that such a filament could exist within a laser produced plasma. However, in neither theory is any attempt made to determine the self-focusing length and from an order of magnitude estimate[12] it is difficult to see how the ponderomotive mechanism could account for the plasma filaments observed during the very early stage of spark development. In a more recent consideration of such effects Palmer[23] has applied the stimulated scattering theory of Herman and Gray[24] to a dense plasma and estimated the power densities required for the onset of several stimulated scattering processes. Although the values obtained indicate an extremely high threshold for thermal self-focusing effects the results for ponderomotive self-focusing in a high density plasma ($N_e \approx 10^{19}$) are compatible with observed breakdown thresholds.

CONCLUSION

Although investigations of laser-induced gas breakdown, carried out over a wide range of experimental conditions, have yielded a substantial amount of evidence for self-focusing, a satisfactory explanation for all of the observed effects is still lacking. Several mechanisms which could have an important role in such phenomena have been proposed and it appears that the effect of excited atoms and molecules during the breakdown process must be taken into account. However, before this can be done, additional experimental data obtained from more carefully controlled experiments, is essential.

ACKNOWLEDGEMENT

It is a pleasure to acknowledge the significant contributions of C. DeMichelis, V.V. Korobkin and M.C. Richardson to the investigations of self-focusing in laser-produced plasmas carried out at the National Research Council of Canada.

REFERENCES

1. N.G. Basov, V.A. Boiko, O.N. Krokhin and G. Sklizkov, Sov. Phys.-Doklady, 12, 248 (1967).

2. V.V. Korobkin, S.L. Mandel'shtam, P.P. Pashinin, A.V. Prokhindeev, A.M. Prokhorov, N.K. Sukhodrev, and M. Ya Shchelev, Sov. Phys. JETP, 26, 79 (1968).

3. G.A. Askaryan, Sov. Phys.-JETP, 15, 1088 (1962).

4. A.G. Litvak, Izv. Vuz. Radiofizika, 9, 675 (1966).

5. M.M. Savchenko and V.K. Stepanov, JETP Letters, 8, 281 (1968).

6. V.V. Korobkin and A.J. Alcock, Phys. Rev. Letters, 21, 1433 (1968).

7. N. Ahmad, B.C. Gale and M.H. Key, J. Phys. B. (Atom. Molec. Phys.) Ser. 2, Vol. 2, p. 403 (1969).

8. R.G. Tomlinson, Bull. American Phys. Soc. 14, 1021 (1969); IEEE J. Quant. Elect., QE-5, 591 (1969).

9. A.J. Alcock, C. DeMichelis and M.C. Richardson, IEEE J. Quant. Elect., QE-6, 622 (1970).

10. N. Ahmad, B.C. Gale and M.H. Key, in "Advances in Electronics and Electron Physics" edited by J.D. McGee, D. McMullen, E. Kahan and B.L. Morgan, vol. 28B, p. 999 (1969).

11. L.R. Evans and C. Grey Morgan, Phys. Rev. Letters 22, 1099 (1969).

12. M.H. Key, D.A. Preston and T.P. Donaldson, J. Phys. B. (Atom. Molec. Phys.), 3, L88 (1970).

13. M.C. Richardson and A.J. Alcock, Appl. Phys. Letters, 18, 357 (1971); also Kvantoviya Electronica 1, no.5, p.37 (1971).

14. A.J. Alcock, C. DeMichelis, V.V. Korobkin and M.C. Richardson, Appl. Phys. Letters 14, 145 (1969).

15. A.J. Alcock, C. DeMichelis, V.V. Korobkin and M.C. Richardson, Phys. Letters 29A, 475 (1969).

16. Charles C. Wang and L.I. Davis Jr., Phys. Rev. Letters 26, 822 (1971).

17. F.V. Bunkin, I.K. Krasyuk, V.M. Marchenko, P.P. Pashinin and A.M. Prokhorov, ZhETF 60, 1326 (1971).

18. P.L. Kelley, Phys. Rev. Letters 15, 1005 (1965).

19. G.A. Askaryan and M.S. Rabinovich, Sov. Phys. JETP 21, 190 (1965).

20. R.H. Huddleston and S.L. Leonard, Eds., "Plasma Diagnostic Techniques". New York: Academic Press, 1965, p. 440.

21. K. Shimoda, J. Phys. Soc. (Japan), 24, 1380 (1968).

22. H. Hora, Z. Phys. 226, 156 (1969).

23. A.J. Palmer, these Proceedings - see page 367.

24. R.M. Herman and M.A. Gray, Phys. Rev. Letters 19, 824 (1967); R.M. Herman and M.A. Gray, Phys. Rev. 181, 374 (1969).

SCATTERING AND BEAM TRAPPING IN LASER-PRODUCED PLASMAS IN GASES*

R. G. Tomlinson

United Aircraft Research Laboratories

East Hartford, Connecticut 06108

ABSTRACT

Experimental studies of laser light scattered from laser-induced breakdown plasmas in gases have shown that this scattering is dominated by reflections from the interfaces of minute localized regions of high electron density. These observations clarify some properties of laser-induced plasmas which appeared to be anomalous when the scattering was attributed to Thomson scattering from free electrons. The geometry of the scattering regions indicates that self-focusing may occur in the plasmas generated in some gases.

*Presented at the Second Workshop on "Laser Interaction and Related Plasma Phenomena" at Rensselaer Polytechnic Institute, Hartford Graduate Center, August 30 - September 3, 1971.

SUMMARY OF DISCUSSION

(II. Laser Induced Gas Breakdown)

Much discussion was evoked by the surprising results of R. Papoular who measured a high number of free electrons and a great amount of emission of radiation originating from a laser irradiated gas at much lower intensities than necessary for gas breakdown. It was concluded that the laser intensities creating these photons are several orders of magnitude lower than is needed for the multiphoton ionization of single atoms or molecules. Questions still to be answered are where the "first electrons" come from, and how the avalanche multiplication for breakdown occurs. Possible solutions are the quantum kinetic model of Zeldovich and Raizer[1], or the concept of inverse bremstrahlung absorption according to R. G. Tomlinson[2].

The problem of calculating radial density and temperature profiles from side-on measurements of cylindrical plasmas was brought up by A. J. Alcock, A. H. Guenther and W. K. Pendleton. It was pointed out that a very extensive mathematical analysis of this inaccuracy problem and its optimized numerical solution has been carried out by R. Gorenflo[3].

In regard to Tomlinson's experiments on scattering diagnostics of laser produced plasmas, B. Kronast pointed out that one should not expect any Thompson scattering from plasmas with densities larger than $10^{19} cm^{-3}$ and temperatures lower than $10^2 eV$.

References
1. Y. B. Zeldovich and Y. P. Raizer, Sov. Phys. JETP <u>20</u>, 772(1965).
2. A. MacDonald, <u>Microwave Breakdown in Gases</u>, Wiley, New York,1966.
3. R. Gorenflo and Y. Kovetz, Institut für Plasmaphysik, Garching, Report IPP6/29 (Nov. 1964).

RECENT DEVELOPMENTS IN LIGHT SCATTERING EXPERIMENTS ON LABORATORY

PLASMAS*

B. Kronast

Division of Physics, National Research Council of Canada

Ottawa, Ontario, Canada

INTRODUCTION

In the past, light scattering techniques were mainly aimed at measuring plasma parameters such as the electron density and the temperatures of electrons and ions. These efforts were most successful and, thus, have demonstrated that such techniques represent a superior tool of plasma diagnostics also for laboratory plasmas. A review of this earlier development of the field can be found for example in the articles by Kunze[1] and by Evans and Katzenstein[2]. With the further refinement of the employed techniques a wider range of plasmas as well as the finer details of spectra became accessible. As a result deviations from the light scattering spectra calculated under the assumption of Maxwellian velocity distributions for electrons and ions became apparent. In addition, the measurement of magnetic fields by light scattering was shown to be feasible as was the detection of enhanced plasma waves in turbulent plasmas. These more recent developments in the field of light scattering by plasmas form the subject of this article which is not claimed to be fully comprehensive but is intended to demonstrate these developments by typical cases.

The various light scattering experiments are grouped according to the influence which the magnetic field has on the spectra. The first large group of measurements in which the magnetic field does not reveal itself or where its influence is not recognized as yet,

---

*Presented at the Second Workshop on "Laser Interaction and Related Plasma Phenomena" at Rensselaer Polytechnic Institute, Hartford Graduate Center, August 30 - September 3, 1971.

is classified due to the familiar parameter $\alpha_e = (k \cdot \lambda_D)^{-1}$, (k absolute value of the scattering wavevector; λ_D electron Debye length). Its value reflects the extent to which correlations between electrons become apparent in the light scattering spectra in the absence of a magnetic field. Thus, $\alpha_e \ll 1$ corresponds to Thomson scattering, i.e., the scattering by statistically independent electrons, whereas $\alpha_e \gtrsim 1$ represents scattering from more or less Landau damped waves and is often termed collective, coherent or cooperative scattering.

THOMSON SCATTERING ($\alpha_e \ll 1$)

When Gerry and Rose[3] had demonstrated the feasibility of rather accurate measurements of the electron velocity distribution by means of light scattering in a low density plasma, they felt that they had demonstrated "that highly ionized plasmas with density $n_e > 10^{12}$ cm^{-3} and temperature $T_e < 10^3$ eV are now amenable to study, with careful application of techniques known today". This feeling was confirmed when a Culham group, M.J. Forrest et al.,[4] investigated the T3A-Tokamak by means of light scattering. In measuring the space and time distribution of the temperature and density of electrons at conditions of $n_e > 10^{13}$ cm^{-3} and several hundred eV temperature, it was intended to operate the ruby laser in a relaxation mode and thus deliver about 100 Joules of light energy through the investigated plasma volume of about 0.1 cm^3. However, the background plasma light exceeded by a factor of 200 the expected bremsstrahlung and - in order to overcome it - the laser had to be operated in the Q-switched high power mode, delivering only 6 Joule in a 25 nsec pulse. Although the entrance slit of the 10 channel 1 m Ebert polychromator was illuminated directly by the scattered light, and not via a brightness deteriorating glass fibre bundle, the number of photoelectrons per pulse in a single wavelength channel was only of the order of 10-100. In order to attain reasonable statistical errors of 10-15% the spectra were evaluated from ten shots. The results are shown in Fig. 1.

The fact that a Gaussian line shape fitted the spectra very well is evidence for a Maxwellian velocity distribution of electrons transverse to the main magnetic field. The temperatures derived from these best fitting curves are accurate to within 5 to 15% and also the absolute values of n_e as obtained after calibration with a tungsten ribbon lamp agreed within 10% with measurements of the microwave interferometer.

In an experiment on a θ - pinch plasma of a much higher density of $6 \cdot 10^{16}$ cm^{-3} and a temperature of 100 eV Gondhalekar and Kronast[5] investigated possible deviations from the often assumed Maxwellian velocity distribution of electrons. The problems arising from the need for a highly accurate measurement of the spectrum are not only associated with an accurate measurement of the scattering intensity

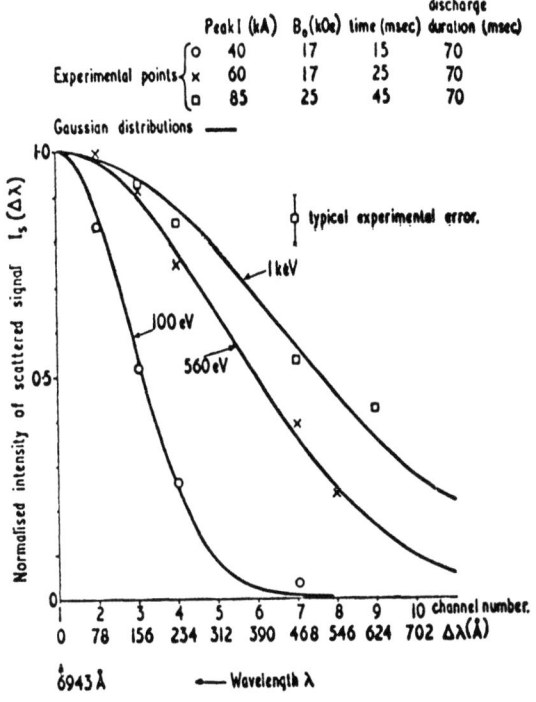

Fig. 1 The scattered light spectrum for various discharge conditions of the TA-3 device (after Forrest et al.[4])

in a single spectral point but also with a precise determination of the relative calibration of the wavelength channels. The first problem was solved in a satisfactory way by feeding a 5 Joule ruby laser pulse through the scattering volume and calculating the single spectral points from a number of measurements using an eight channel wavelength analyser after reasonable reproducibility of the discharge had been achieved. In order to solve the problem of relative calibration, extensive studies were carried out to determine the spectral sensitivities of the eight channels, accounting also for risetime differences of the photomultipliers and the associated electronics, as well as for their non-linearities. In order to reduce further the systematic error in the relative calibration the scattering intensities at particular wavelengths were measured with various photomultiplier channels. The scattering wave number $\underline{k}$ determines the direction onto which the electron velocities are projected. This wave vector was at an angle of 131.5° to the axial magnetic field of the θ - pinch. Light scattering spectra were determined from a center position in the plasma column and from its sheath where gradients of magnetic field and density are highest. The results are given in Fig. 2. There was an indication of a dip in the red wing of the center measurements which became somewhat more pronounced in the off centre spectrum. Whether this deviation was from an excess of scattering intensity in the last three points or a deficiency of scattering intensity at the particular wavelength of this dip could not be determined from these spectra. With more refinement of the above procedures and an increase in spectral resolution the red wing of this spectrum was then investigated in more detail in a separate experiment. The results showed that the scattering intensities in the last three points did exceed the values expected for a Maxwellian velocity distribution of electrons.

Another feature of these light scattering spectra was a blue-shift of about 12Å common to both locations of the scattering volume and, therefore, not attributable to Doppler-shifts due to azimuthal electron drift velocities which should exist only in the region of density and magnetic field gradients but not in the center of the plasma. Further efforts to explain this blueshift were successful in that a relativistically correct treatment of the scattering process[6] can at least account for a major part of this blueshift.

An experiment aimed at demonstrating another aspect of such relativistic corrections to the light scattering process was

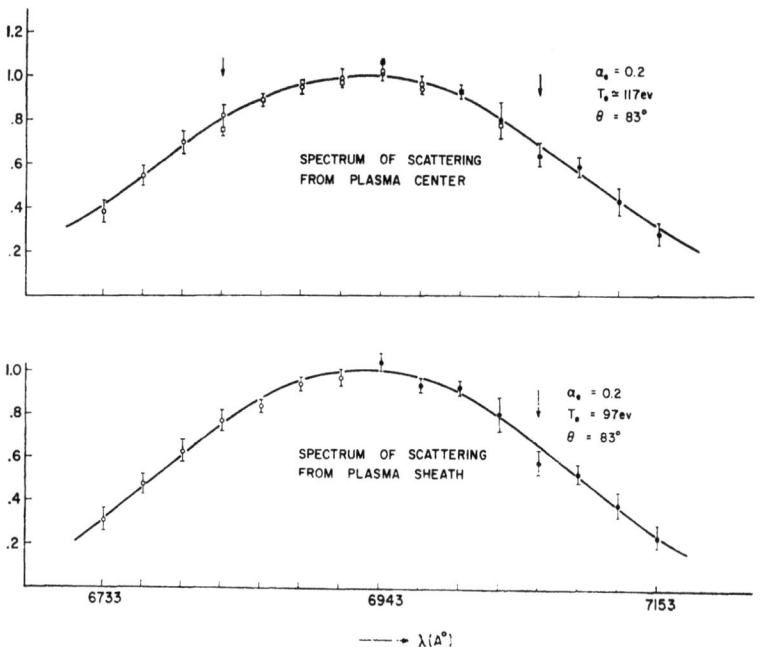

Fig. 2 Light scattering spectra from two locations in a θ - pinch plasma with best fitting curves based on the theory of Fejer[8] (after Gondhalekar and Kronast[5]).

performed by Ward, Pechacek and Trivelpiece.[7] For relativistic plasmas there should be a difference in the wavelength integrated scattering crossection depending on whether electrons stay in the scattering volume during the irradiation or traverse it in a short time. This wavelength integrated scattering crossection was investigated on a 50-kV confined flow electron beam. Because of the very low electron densities of about 7×10^9 cm^{-3} in the beam, photon saving was mandatory. So a conical mirror was used with its axis coinciding with the direction of the electron beam to accept all the light scattered into the cone about the electron beam. With the help of a plane mirror this light was reflected out of the vacuum system where it was focused onto the photomultiplier.

This total scattered light was determined for three different cone angles as shown in Fig. 3. The full line representing the theoretical predictions indicates satisfactory agreement between experiment and the theory which accounts for a size of scattering volume which is small in comparison to the electron path during the time of laser irradiation.

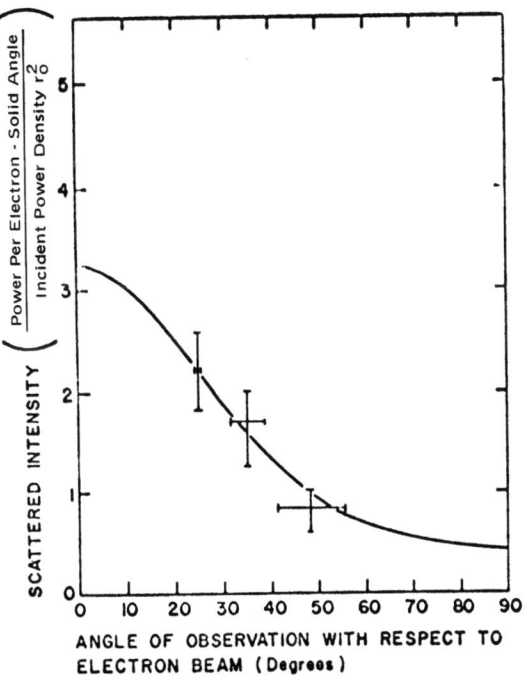

Fig. 3 Measured and theoretical intensity (in watts per electron-solid angle/incident power density γ_0^2) of radiation scattered from a laser beam by a beam of 50-kV electrons as a function of scattering angle. The 0.6943-μ laser photons are Doppler-shifted to 0.4350 μ at 25° scattering angle (after Ward et al.[7])

COLLECTIVE LIGHT SCATTERING

1) The Electron Feature

Fig. 4 shows the theoretical spectra as predicted in the work of Fejer[8] and Salpeter[9] for a Maxwellian velocity distribution of electrons.

a) $(k\lambda_D)^{-1} \approx 1$.

As pointed out by Kunze[10] this kind of light scattering spectrum lends itself with great advantage to the simultaneous measurements of electron temperature and density since its shape depends on the ratio $\frac{T_e}{n_e}$ whereas its width is proportional to temperature. Thus, from the early beginning of light scattering investigations experiments were aimed at these scatter-in conditions. As a result there are available a number of measurements of such spectra.

However, whilst the early measurements (see e.g. Kunze[10], Anderson,[11] Izawa et al.[12]) showed satisfactory agreement with theoretical predictions for thermal plasmas more recent measurements of e.g. Evans, Forrest and Katzenstein[13] (Fig. 5) and of Siemon and Benford[14] (Fig. 6) show deviations from these beyond the given error brackets. The main difference between these two groups of experiments seems to be the state of the plasma. The

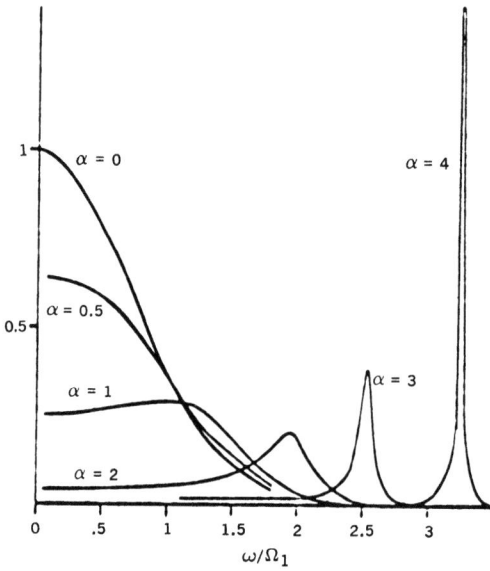

Fig. 4 The frequency spectrum of the electron feature of light scattering for different values of the correlation parameter $\alpha=(k\cdot D)^{-1}$. The frequency ω is normalized to the thermal Doppler broadening of electrons Ω_1 (after Fejer[3]).

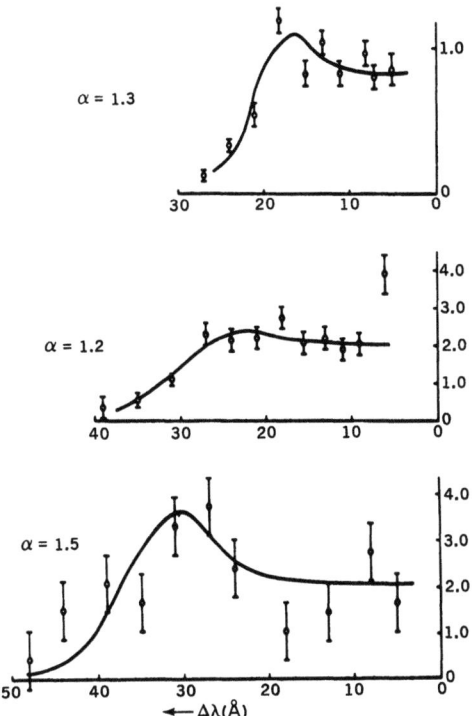

Fig. 5 The electron feature of light scattering for three different operating conditions of a θ - pinch. Curves represent best fit theoretical spectra (after Evans, Forrest and Katzenstein[13]).

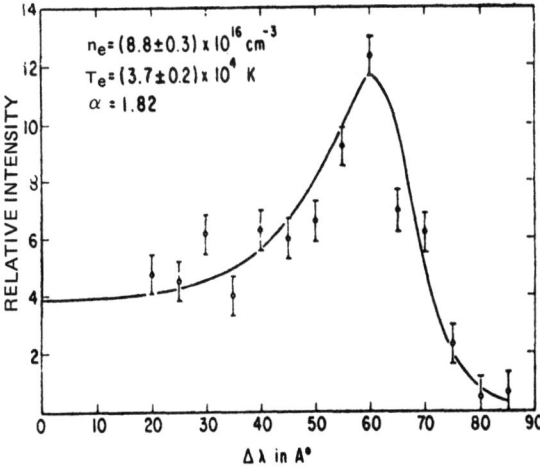

Fig. 6 The electron feature of light scattering as measured in the imploding current sheath of a small θ - pinch device (after Siemon and Benford[14]).

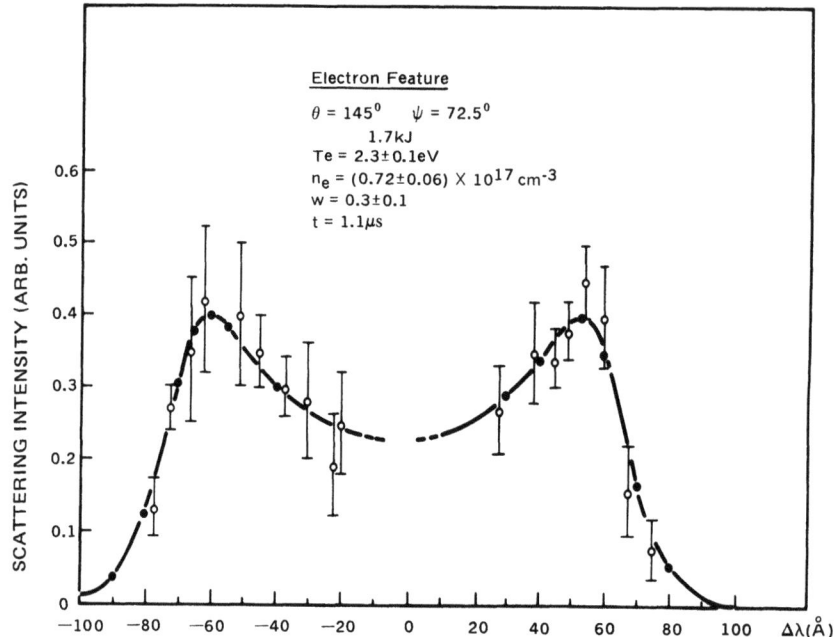

Fig. 7 The electron feature in a z-pinch plasma with the best fitting theoretical curve for a thermal plasma and the corresponding plasma parameters (after Kronast and Pietrzyk[15]).

first group of investigations dealt mainly with rather cold and dense plasmas after cessation of their dynamic phase. In addition, the plasma centre, where gradients in density and magnetic field are minimal, was the subject of investigations. These plasmas are most likely near to thermal equilibrium and it is, therefore, to be expected that their behaviour can be described by the assumption of a Maxwellian for the electron velocities. Also the inclusion of a moderate electron drift velocity does not change this situation drastically in a cold and dense z-pinch plasma as it was demonstrated by Kronast and Pietrzyk[15] (Fig. 7).

However, the situation seems to be different in case of a hot, less dense θ - pinch plasma (Evans et al.[13]) and also when the imploding current sheet of a small θ - pinch device is investigated rather than the late phase of such a plasma long after cessation of the dynamic period.

The results of Evans et al.[13] prompted W.H. Kegel[16] to try an interpretation of such deviations on the basis of non-Maxwellian velocity distributions of electrons. Indeed, assuming a velocity

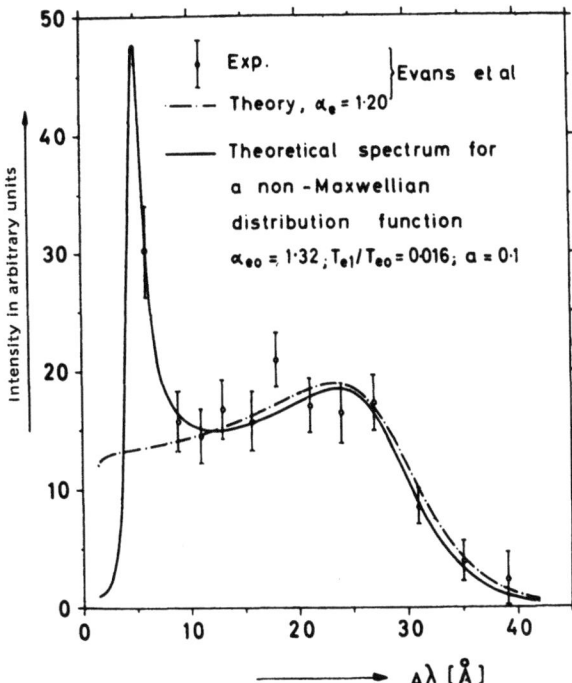

Fig. 8 Comparison of experimental data given by Evans et al. with two theoretical spectra; one calculated for a plasma in thermodynamic equilibrium and the other one obtained for a plasma with a non-Maxwellian distribution function (after Kegel[16]).

distribution made up from a superposition of two Maxwellians with different temperatures and varying relative contributions, he succeeded in explaining a widely deviating point near the centre of the spectrum as shown in Fig. 8. However, this treatment cannot account for the deviating point at 18Å and it cannot account for many of the deviations in the first and third spectrum of Fig. 5. It might well be that different phenomena contribute to these deviations.

For instance, another anomaly in such light scattering spectra was observed by Ringler and Nodwell[17,18]. They investigated the electron and ion feature of light scattering in a magnetically stabilized, hydrogen arc plasma and detected light scattering peaks which were superimposed on the expected spectrum of decreasing amplitude at half the plasma frequency, at the plasma frequency

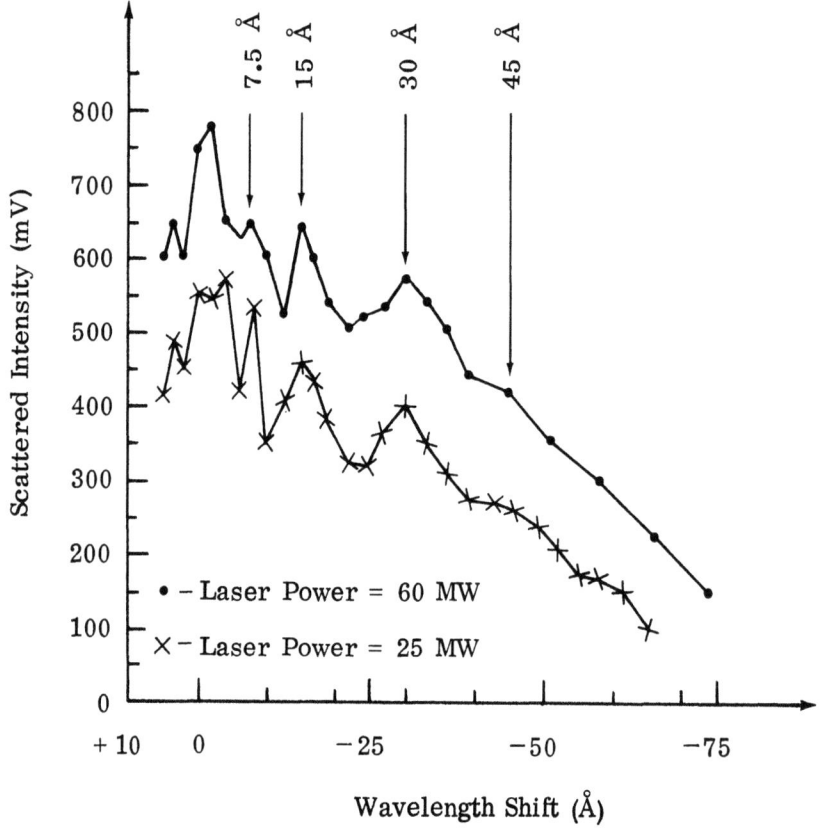

Fig. 9 The electron and ion feature of light scattering as obtained on a magnetized arc plasma (after Ringler and Nodwell[18]).

and at harmonics of it, (Fig. 9). When the electron density was changed the peaks shifted in wavelength accordingly. Since the scattering wavenumber was at right angles to the stabilizing magnetic field one could think of Bernstein waves being associated with such peaks. This aspect has been under investigation in a continuation of such light scattering experiments conducted by Ludwig and Mahn[19] on the same plasma. Provisions were made for measuring the spectra both with the scattering wave vector perpendicular and parallel to the magnetic field. As can be seen from

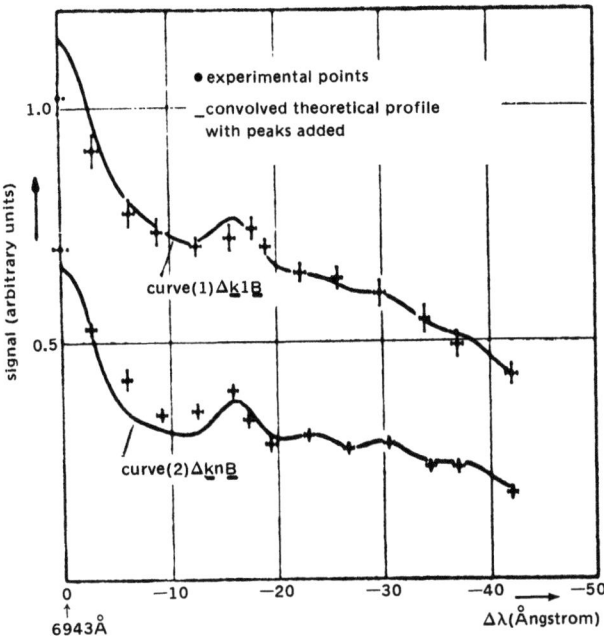

Fig. 10 The electron feature for parallel and perpendicular orientation of scattering wave vector $\Delta \underline{k}$ and magnetic field $\underline{B}$ (after Ludwig and Mahn[19]).

Fig. 10 there was not much of a difference. In addition, the indicated peak was resolved with the higher spectral resolution provided by a Fabry-Perot interferometer. The result was that the indicated peaks had a width of only about 1.5Å and their amplitude exceeded the thermal background by 100 to 200%.

So far all the deviations mentioned in this chapter are lacking explanations. It might be that they represent another revelation of effects which make themselves felt also in the next chapters.

b) $(k\lambda_D)^{-1} \gg 1$.

In this case Landau damping of electron plasma waves is greatly reduced and such waves are well established. Thus, the electron feature of light scattering exhibits satellites well defined in frequency as shown for larger values of the correlation parameter $\alpha_e = (k \cdot \lambda_D)^{-1}$ in Fig. 4. Whereas the existence of such satellites was demonstrated in a number of experiments[20,21,22], Röhr[23] tried to resolve this satellite in the light scattering spectrum obtained

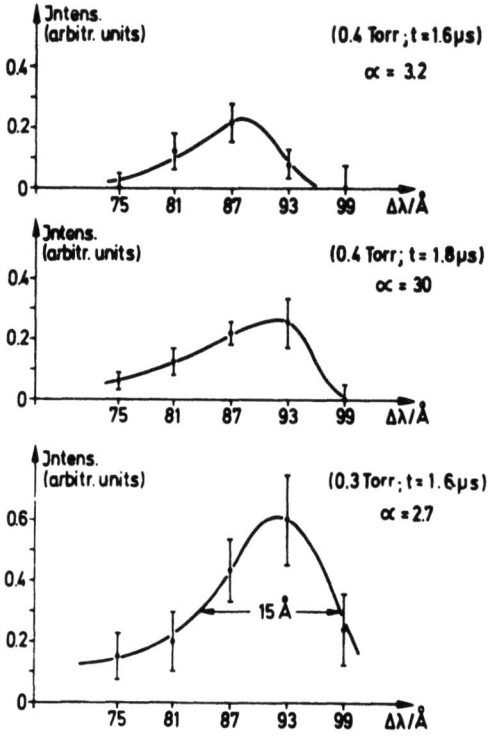

Fig. 11 Satellite spectra obtained in a cold but dense θ - pinch plasma for various values of the correlation parameter α (after Röhr[23]).

from a cold, but dense θ - pinch plasma. Its width as given in Fig. 11 was found to be four times that predicted both by collision dominated as well as collisionless theoretical treatments of thermal plasmas. This broadening turned out to be attributable to the density gradients in the scattering volume.

Rather than avoiding this influence Kronast and Benesch[24] made deliberate use of the effect which the density distribution in the scattering volume has on the satellite spectrum. They showed that the density distribution in a plasma is mapped as a wavelength distribution of such satellites as long as the α_e-values of these are larger than two and the laser beam is homogeneous over the plasma regions where this applies.

It was in this context that spectra of such satellite distributions revealed a line structure. A more detailed investigation of this phenomenon was performed and helped to shed some light on it. It will be described in the last Chapter on magnetic field effects.

A considerable enhancement of such satellites, well beyond

thermal levels, which is due to microinstabilities or other causes inherent in the plasma has not been observed in laboratory plasmas. However, the wave amplitudes were driven to enhanced levels with the help of two dye lasers in an experiment conducted by Stansfield, Nodwell and Meyer[25]. Two dye lasers of 2 MW power each illuminated a plasma jet at an angle such that the difference wave vector of the two lasers assumes a value for which an electron plasma wave can be expected to exist at a higher value of the correlation parameter α_e. The light scattering set up of this experiment was arranged such that scattering was observed from the same wavenumber. When the dye lasers were not in operation or when their frequency difference did not coincide with that of the electron plasma wave the scattering intensity was that expected from a thermal plasma. However, when the difference frequency of the dye lasers coincided with it an enhancement in scattering intensity was observed as shown in Fig. 12, indicating that the density fluctuation associated with this plasma wave were excited beyond thermal levels.

2. The Ion Feature

In contrast to the electron feature the ion feature displays an even greater susceptibility to deviations from a thermal plasma. For the purpose of comparison, spectra of a thermal plasma are

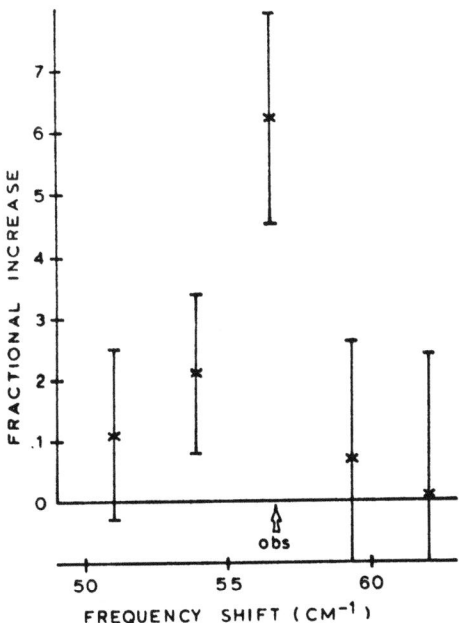

Fig. 12 The fractional increase of the satellite intensity as a function of the frequency difference between the dye lasers. The shift indicated by the arrow is that of the normally observed satellite (after Stansfield et al.[25]).

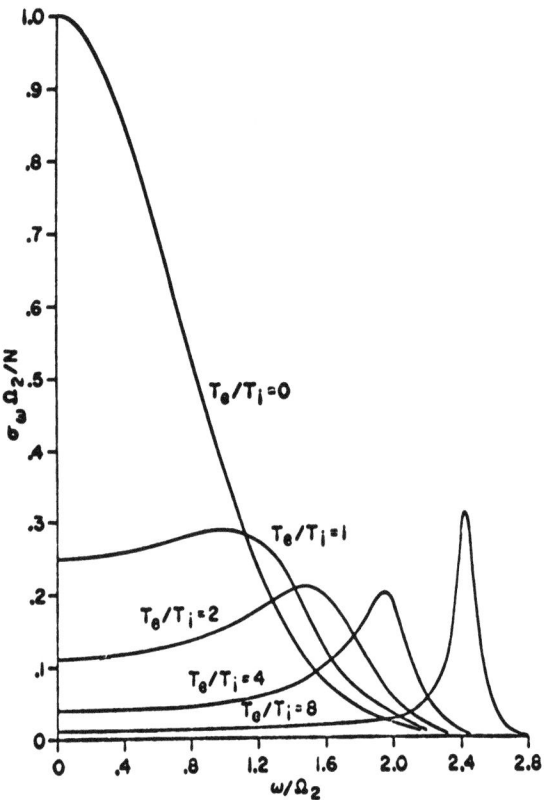

Fig. 13   The ion feature of a thermal two temperature plasma for large values of the correlation parameter α_e (after Fejer[8]).

shown in Fig. 13. In these cases the plasma is described by Maxwellians for electrons and ions though their temperatures may differ. Only one half of these symmetric features is plotted for large values of $\alpha_e = (k \cdot \lambda_D)^{-1}$ against the frequency ω which is normalized to the thermal Dopplershift of ions Ω_2.

The symmetry of these spectra is destroyed when the Maxwellian velocity distribution of electrons is changed to a shifted one. How this influence changes the character of the spectra is pictured in Fig. 14, for various values of the electron drift velocity w which is normalized to the thermal electron velocity. The abscissa Y gives the frequency in units of the thermal Doppler broadening of ions as above. This showed up already in the earlier measurements as for instance in[27,28,29].

Amongst the more recent observations only those of Röhr and Decker[30] (Fig. 15) are compatible with spectra of thermal plasmas

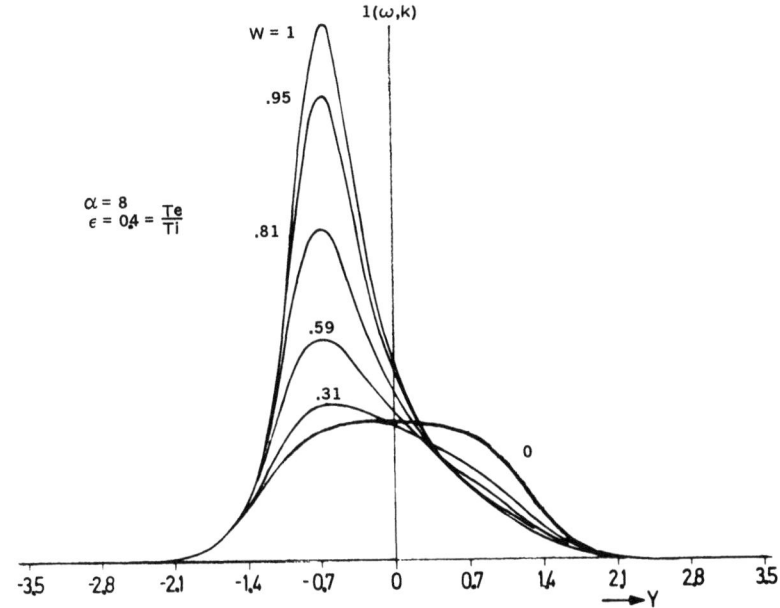

Fig. 14 The distortion of the symmetric ion feature with the electron drift velocity w which is measured in units of the r.m.s. - velocity of electrons. Y is the frequency of density fluctuations in units of the r.m.s. - Doppler shift of ions (after Theimer[26]).

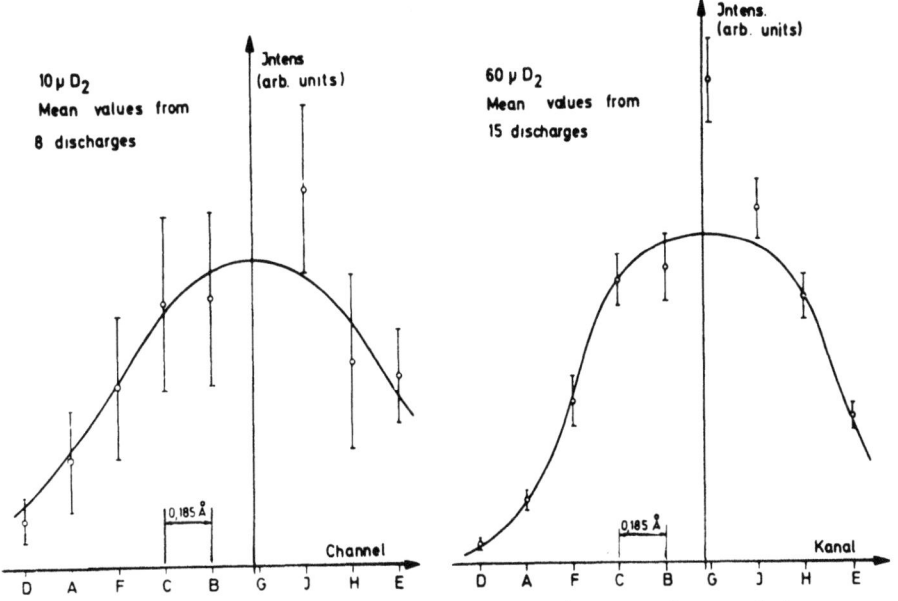

Fig. 15 The ion feature of light scattering as observed in a θ - pinch plasma for two different filling pressures of deuterium (after Röhr and Decker[30]).

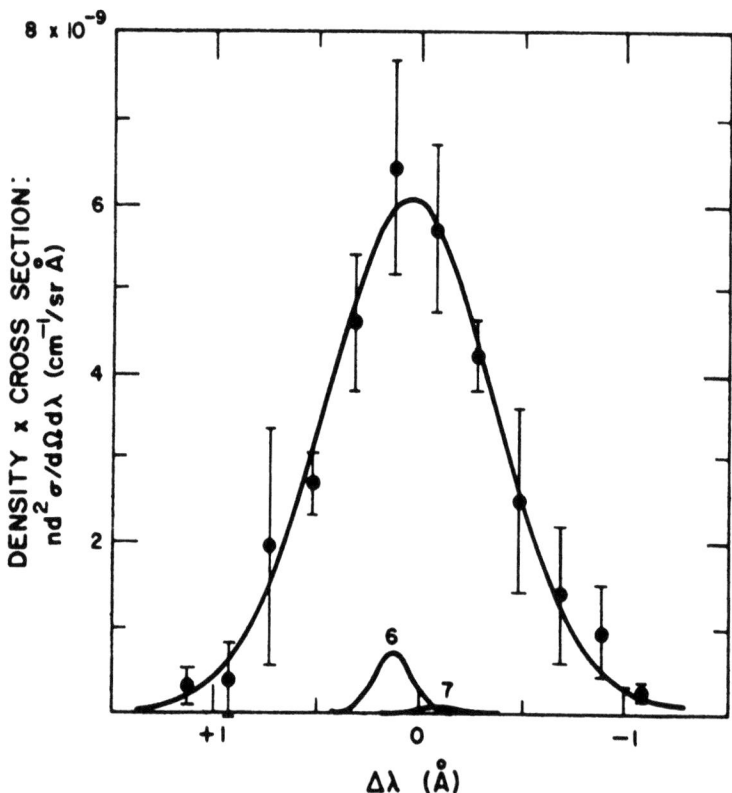

Fig. 16 A symmetrically enhanced ion feature as obtained on a high temperature θ - pinch plasma (after Daehler et al.[32]).

although the plasma was produced in a medium sized θ - pinch device at relatively low densities but temperatures high enough to produce neutrons. However, also here, guided by discrepancies in the interpreted ion temperature and observed neutron yield, the authors were led to conjecture the presence of non-thermal density fluctuations. In a θ - pinch plasma of similar parameters enhanced light scattering in the ion feature was observed by Daehler, Sawyer and Thomas[31] who confirmed earlier measurements[32] by means of a careful comparison between light scattering intensities and values for the plasma density obtained from a combination of bremsstrahlung measurements, interferometry and streak photographs of the plasma column. Whereas this central feature (Fig. 16) was by a factor of two too narrow to account for the neutron production by ions of the corresponding temperature its amplitude was increased by a factor of six in comparison to the predictions of a two temperature thermal plasma. This central peak has a unique feature in that it is symmetrical with respect to $\Delta\lambda=0$. All measurements of enhanced scattering show an asymmetry (see e.g., [33,34]) as do all the spectra indicating deviations from thermal plasma.

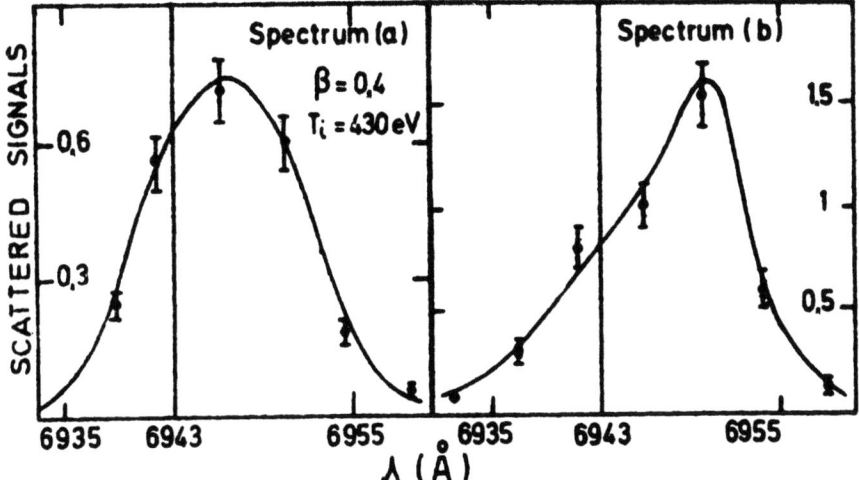

Fig. 17 Two kinds of ion features which were observed in a plasma focus (after Baconnet et al.[36]).

For instance, also the observations of Ringler and Nodwell[17,18] of an enhanced ion feature in the scattering from an hydrogen arc plasma show such an asymmetry at closer inspection (Fig. 9). However, it lacks an explanation as does the above enhancement in a θ - pinch plasma.

An asymmetry and possible enhancement of the ion feature was also observed by Baconnet et al.[36] who performed light scattering measurements on a plasma focus. With the scattering wave vector at an angle of 45° to the axial drift of the electrons they detected two kinds of ion features. One was symmetrical though shifted to the red side as could be accounted for by an ion drift, and the second kind exhibiting a pronounced asymmetry, which seems to indicate an enhancement in the peak by comparison to the symmetric spectrum (a) [note the different scales in Figs. 17a) and b)]. Whether this asymmetric feature represents the spectrum of a homogeneous plasma or is composed of the symmetric feature of Fig. 17a) and a superimposed contribution from another region which adds a broad line at about the ion plasma frequency cannot be decided without further information. The latter possibility does not seem unlikely in a plasma focus which is suspected of being turbulent, and which exhibits great inhomogeneities within the crossection of the probing laser beam. Thus, it is conceivable that the enhanced line, located somewhat below the ion plasma frequency, indicates the presence of turbulent regions similar to the observations made on turbulent shock plasmas (see below), whereas the underlying symmetric contribution to the spectrum represents scattering from the main part of the plasma which moves away from the inner electrode. The possibility that an

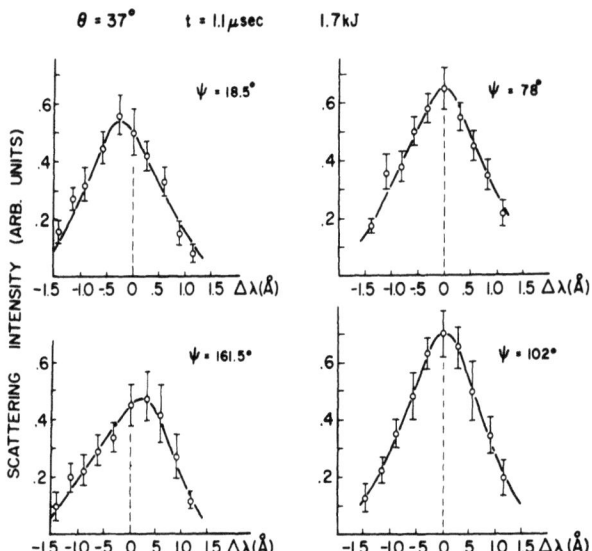

Fig. 18 The ion feature of light scattering spectra in a 1.7 K J z-pinch plasma with best fitting spectra, for various angles ψ between scattering wave vector and the axial drift velocity of electrons (after Kronast and Pietrzyk[15]).

axial electron drift velocity distorts the ion feature of Fig. 17b) can be excluded since it would enhance the blue side of the spectrum in the chosen scattering geometry and with the electrons drifting towards the inner electrode. The excitation of ion waves by drifting ions, however, cannot be discounted on these grounds.

The influence an electron drift velocity has on the ion feature was investigated in a small z-pinch plasma by Kronast and Pietrzyk[15]. The observed asymmetry of the spectra (Fig. 18) varied with the angle ψ between the scattering wave vector and the axial drift velocity of electrons in a way which was in qualitative accord with linear light scattering theory. The spectra were, therefore, interpreted in terms of the latter and values for the electron drift velocity as well as temperatures of electrons and ions were derived from the computer aided bestfitting procedure. On the other hand, the electron feature of light scattering was observed from the same scattering volume (Fig. 7) at an angle of 145° providing independent values for electron drift velocity, and temperature and also density. Whereas the electron temperature agreed well within the error brackets, the normalised drift velocity derived from the Doppler shift of the electron feature of $w=0.3\pm0.1$ was greatly at variance with the value of $w=1.4\pm0.3$ which was required to explain the asymmetry of the ion feature. Also here, a satisfactory explanation is still lacking.

Genuinely asymmetric and enhanced ion features can be observed

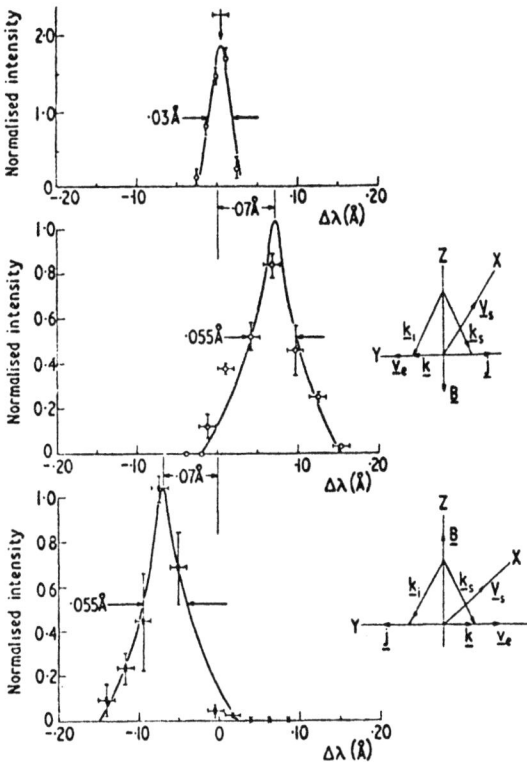

Fig. 19 Frequency spectrum as obtained from a collisionless shock plasma. The spectral profiles show a) incident laser light; b) and c) largely enhanced scattered light for parallel and antiparallel direction of scattering wave vector k and electron drift velocity v_e as shown (after Daughney et al.[37]).

in collisionless shock plasmas. This was demonstrated by Daughney, Holmes and Paul[37] who performed light scattering measurements in the turbulent shock front of a collisionless plasma. The shock was produced by the radial compression of a highly preionised and magnetic plasma in a z-pinch device. The magnetic field gradient in the shock front is associated with an azimuthal current density which is high enough to give rise to a corresponding electron drift velocity, perpendicular to the magnetic field, which is sufficient to drive ion acoustic waves unstable. The spectra of Fig. 19 show broad lines which are enhanced beyond thermal levels by more than two orders of magnitude and are located near the ion plasma frequency. Depending on whether the scattering wave vector is selected to be parallel or antiparallel to the azimuthal drift velocity the line appears on the blue or red side of the spectrum as required for a drift induced instability. By varying the density at constant magnetic field the nature of these scattering waves could be shown to correspond to that of ion waves and not Bernstein

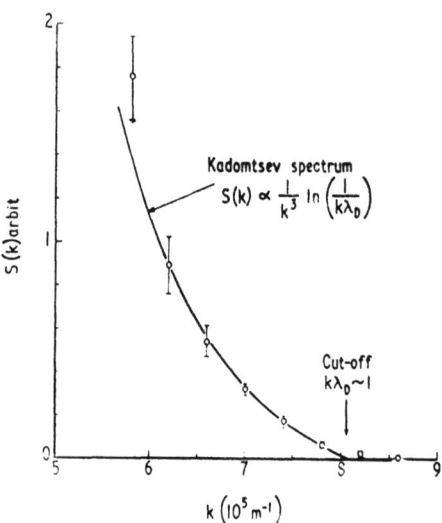

Fig. 20 The wave number spectrum of a collisionless shock plasma with the theoretical predictions for current driven ion wave turbulence due to Kadomtsev[38] (after Daughney et al[37]).

waves of similar frequency. Furthermore, in addition to these frequency spectra, a wave number spectrum was determined by measuring the frequency integrated scattering intensity for various scattering angles. The agreement with Kadomtsev's[38] predictions (Fig. 20) for a current driven ion wave turbulence in the absence of a magnetic field is excellent.

In a similar light scattering experiment performed on a θ-pinch produced, turbulent shock plasma of high β, Keilbacker and Steuer[34] were able to confirm various features such as wave enhancement by two orders of magnitude, the dependence of this enhancement on the direction of the drift velocity, and that the frequency of these waves scaled with about one half the ion plasma frequency. Also the wavenumber spectrum was shown to comply with at least the general features though not the exact shape of Kadomtsev's prediction[38]. In addition, the authors were successful in measuring the frequency spectrum of such enhanced waves from three spatially resolved positions across the shock. The spectra of Fig. 21 show an increase in their width which reflects the increasing perturbation and associated lifetime reduction of these waves across the shock. The authors also succeeded in determining the parameters governing the electrostatic ion wave turbulence, i.e., temperatures of electron and ions and the electron drift velocity in this "collisionless" shock plasma. Since it turns out that ion acoustic waves cannot become unstable under these condition, the excitation of electron cyclotron drift

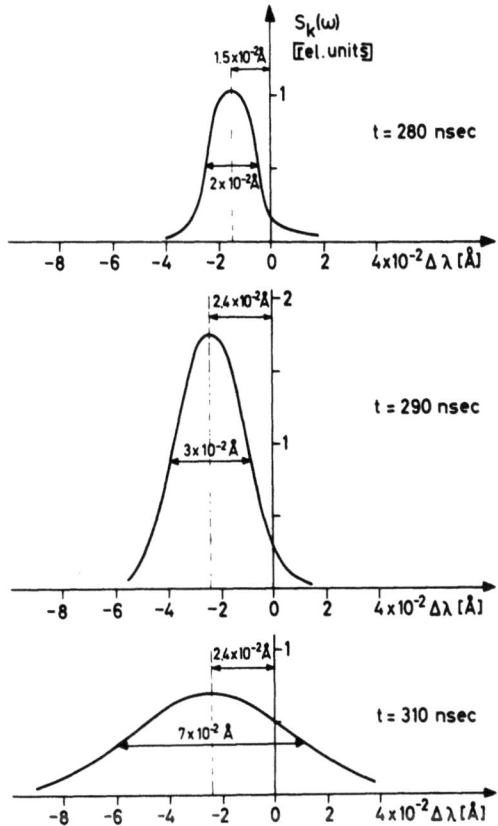

Fig. 21 Frequency spectra of enhanced fluctuations for three positions across a collisionless shock plasma (after Keilhacker and Steuer[34]).

waves[39,40] was preferred as an explanation for the observed large wave amplitudes.

LIGHT SCATTERING MEASUREMENTS OF THE MAGNETIC FIELD IN THE PLASMA

The magnetic field in the plasma represents a parameter of basic importance. Nevertheless, only with the refinement of light scattering techniques has it become promising to approach its local measurement. Even then, it required the creation of a special plasma to observe the effect of a magnetic field in the Thomson scattering spectrum. The plasma device chosen by Kellerer[41,42] was a combination of a z- and θ - pinch apparatus in which the θ - pinch coils produced a field of about 125 K Gauss and in which the z- discharge created a plasma of $1.2 \times 10^{16} \text{cm}^{-3}$ electron density at 3.2 eV electron temperature. Both the incident

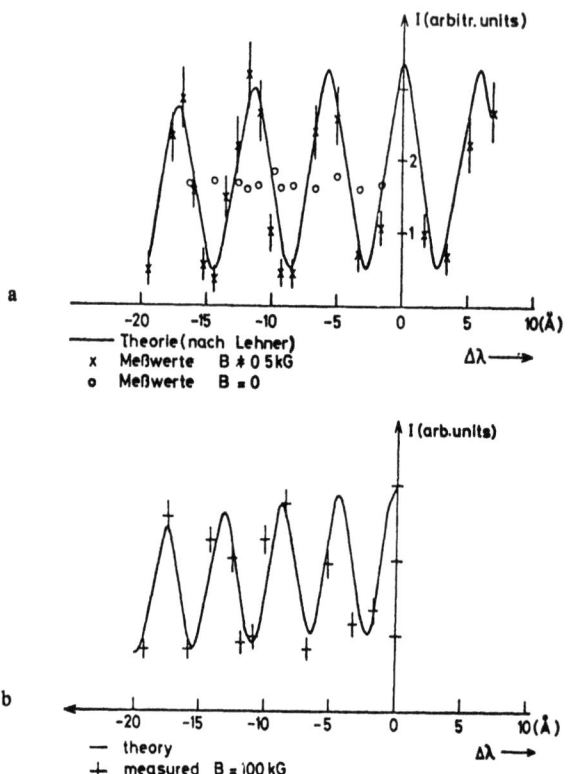

Fig. 22. Electron feature of light scattering in a magneto plasma for perpendicular orientation of scattering wave vector and magnetic field; case a) B = 125 K Gauss (crosses) and B=0 (open circles); case b) B=100 K Gauss (after Kellerer[42]).

laser beam and the scattered light direction were carefully aligned to guarantee perpendicular orientation of magnetic field and scattering wave vector within a divergence angle of about 40 mrad. This is required in order to avoid smearing of the light scattering spectrum which is modulated at the electron cyclotron frequency, as was discussed and shown in computations in a proposal by Lehner and Pohl[43]. The spectra obtained for a scattering angle of 90° and for a value of the correlation parameter of 0.6 is given in Fig. 22. The modulation of the grossly incoherent scattering spectrum agrees well with the cyclotron frequency due to the applied magnetic field of 125 K Gauss. The modulation disappears when this magnetic field is not applied (open circles in Fig. 22a and it changes its frequency when the magnetic field is varied as in Fig. 22b).

Unfortunately, the difficulties associated with such measure-

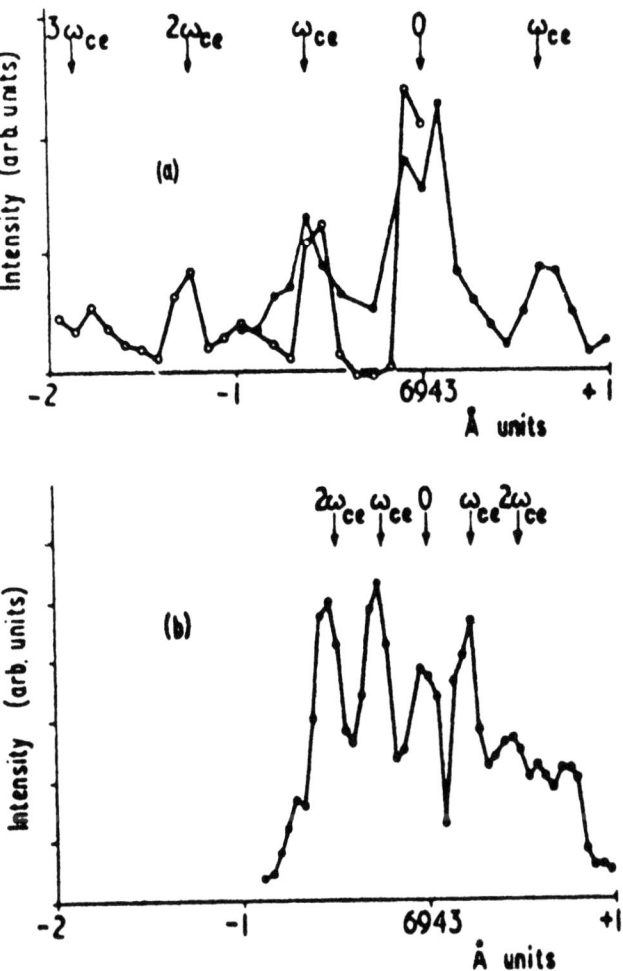

Fig. 23 (a) The ion feature of the scattered-light spectrum for perpendicular orientation of scattering wave vector and magnetic field for B=14 kG (16 kG by Faraday rotation). Two experimental runs are shown: closed circles, premonochromator centered at 6943 Å; open circles, premonochromator centered at 6941 Å. (b) Scattered-light spectrum for B= 5.5 kG (8 kG by Faraday rotation),(after Evans and Carolan[44]).

ments become greater as the electron temperature increases in comparison to the magnetic field since the scattering intensity per analyser channel is then low and the available total scattering intensity is decreased because of the simultaneously required reduction in the acceptance solid angle.

In this situation it might often be preferable to measure the ion feature which exhibits a similar modulation at harmonics of the electron cyclotron frequency provided the total width is wider than this modulation frequency. Thus, at least the spectral brightness of the scattered light is increased in the ion feature. Evans and Carolan[44] demonstrated in an experiment on a θ - pinch plasma that this technique allows the measurement of magnetic fields as low as 5.5 K Gauss. Fig. 23a) and b) show two such spectra taken at peak field a) and half the applied magnetic field b) in a θ - pinch plasma of about 20 eV electron temperature at an electron density of a few times $10^{15} cm^{-3}$. With a scattering angle of 30° the correlation parameter α_e was between 0.2 and 0.3 in case a) and somewhat higher in case b). The accepted set of wavevectors occupied a cone of half-angle 0.85^o about the normal to the magnetic field. The β-value of the chosen plasma was of the order of a few percent. Thus, a measurement of the magnetic vacuum field by means of Faraday rotation in a cylinder of dense glass yielded an independent check of the interpretation of the spectrum modulation. It confirmed the evaluated fields reasonably well.

Recently, Ludwig and Mahn[19] found evidence for such a modulation of the ion feature also in the magnetically stabilized arc plasma when they investigated the spectrum shown in Fig. 10 with higher spectral resolution and for perpendicular orientation of the scattering wave vector to the magnetic field.

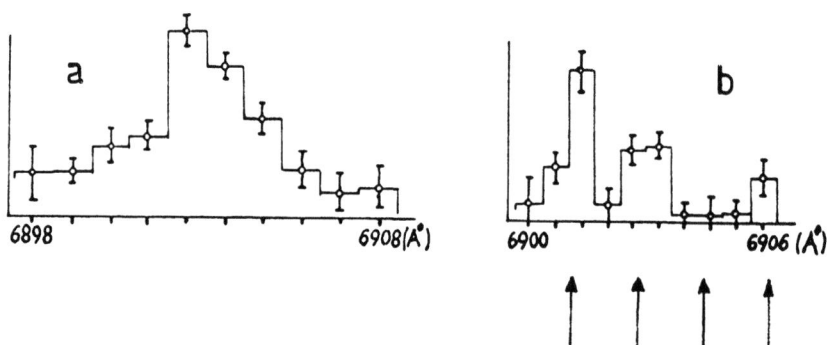

Fig. 24 The electron feature of light scattering for a value of the correlation parameter $\alpha_e > 3$ and for near perpendicular orientation of scattering wave vector and magnetic field; a) for a position in the centre of the θ - pinch plasma; b) for an off-centre position with the arrows indicating harmonics of the electron cyclotron frequency (after Kronast and Benesch[45]).

An observation, which belongs to this group of magnetic field effects in light scattering spectra and which may also lend itself to a determination of local magnetic fields in a plasma was made by Kronast and Benesch[45] when they investigated in more detail an apparent structure in the satellite spectrum of the collective electron feature as mentioned already in the corresponding chapter. When they chose a scattering volume in the centre of the plasma column, where density gradients are minimal, they observed a satellite line (Fig. 24a) the width of which was in accord with the broadening due to the axial density gradient. However, when the scattering volume was moved radially the feature (Fig. 24b) did not only broaden, as expected from the additional radial density gradients, but showed a splitting into lines the separation of which corresponded closely to that of electron cyclotron harmonics though the amplitude ratios were not particularly reproducible.

REFERENCES

1. H.J. Kunze in "Plasma Diagnostics", edited by W. Lochte-Holtgreven, p. 550 (1968).

2. D.E. Evans and J. Katzenstein, Rep. Prog. Phys. 32, 207 (1969).

3. E.T. Gerry and D.J. Rose, J. Appl. Phys. 37, 2715 (1966).

4. M.J. Forrest, N.J. Peacock, D.C. Robinson, V.V. Sannikov, and P.D. Wilcock, Culham Laboratory Report CLM-R107 (1970).

5. A.M. Gondhalekar, B. Kronast, and R. Benesch, Phys. Fluids 13, 2623 (1970).

6. C.H. Papas and K.S.H. Lee, in Proc. Fifth Int. Conf. on Ionization Phenomena in Gases, Amsterdam 1962, Vol. II, p.1204.

   R.A. Pappert, Phys. Fluids 6, 1452 (1963).

   R.E. Pechacek and A.W. Trivelpiece, Phys. Fluids 10, 1688 (1967).

   O. Theimer and J.E. Sollid, Phys. Rev. 176, 198 (1968).

7. Gray Ward, R.E. Pechacek, and A.W. Trivelpiece, Phys. Rev. A 3, 1721 (1971).

8. J.A. Fejer, Can. J. Phys. 39, 716 (1961).

9. E.E. Salpeter, Phys. Rev. 120, 1528 (1960).

10. H.J. Kunze, Z. Naturforschg. 20a, 801 (1965).

11. O.A. Anderson, Phys. Rev. L 16, 978 (1966).

12. Y. Izawa, Y. Nakanishi, M. Yokoyama, and C. Yamanaka, J. Phys. Soc. Japan 23, 1185 (1967).

13. D.E. Evans, M.J. Forrest and J. Katzenstein, Nature 211, 24 (1966).

14. R.E. Siemon and J. Benford, Phys. Fluids 12, 249 (1970).

15. B. Kronast and Z.A. Pietrzyk, Phys. Rev. L 26, 67 (1971).

16. W.H. Kegel, Plasma Physics 12, 295 (1970).

17. H. Ringler and R.A. Nodwell, Physics L 29A, 151 (1969).

18. H. Ringler and R.A. Nodwell, Proc. Third Europ. Conf. on Controlled Fusion and Plasma Physics, Utrecht, p. 111 (1971).

19. D. Ludwig and C. Mahn, Physics L 35A, 191 (1971).

20. P.W. Chan and R.A. Nodwell, Phys. Rev. L 16, 22 (1966).

21. S.A. Ramsden and W.E.R. Davies, Phys. Rev. L 16, 303 (1966).

22. C. Yamanaka, Y. Izawa and M. Yokoyama, Conf. on Plasma Diagnostics, Culham, 1968, paper A-5.

23. H. Röhr, Institut für Plasmaphysik Report IPP 1/58 (1967); H. Röhr, Physics L 25A, 167 (1967).

24. B. Kronast and R. Benesch, Proc. Ninth Int. Conf. on Phenomena in Ionized Gases, Bucharest, Romania, Sept. 1-6, 1969, p. 651.

25. B.L. Stansfield, R.A. Nodwell, and J. Meyer, Phys. Rev. L 26, 1219 (1971).

26. O. Theimer, Institut für Plasmaphysik Report IPP 1/48 (1966).

27. U. Ascoli-Bartoli, J. Katzenstein, and L. Lovisetto, Nature 207, 63 (1965).

28. B. Kronast, H. Röhr, E. Glock, H. Zwicker, and E. Fünfer, Phys. Rev. L 16, 1082 (1966).

29. D.E. Evans, M.J. Forrest, and J. Katzenstein, Nature 212, 21 (1966).

30. H. Röhr and G. Decker, Z. Physik 214, 157 (1968).

31. M. Daehler, G.A. Sawyer, and K.S. Thomas, Phys. Fluids 12, 225 (1969).

32. M. Daehler and F.L. Ribe, Phys. Rev. 161, 117 (1967).

33. J.W.M. Paul, C.C. Daughney, and L.S. Holmes, Nature 223, 822 (1969).

34. M. Keilhacker and K.H. Steuer, Phys. Rev. L 26, 694 (1971).

35. D.E. Evans and M.J. Forrest, Proc. Ninth Int. Conf. on Phenomena in Ionized Gases, Bucharest, Romania, Sept. 1-6, 1969, p. 646.

36. J.P. Baconnet, G. Cesari, A. Coudeville, and J.P. Watteau, Proc. Ninth Int. Conf. on Phenomena in Ionized Gases, Bucharest, Romania, Sept. 1-6, 1969, p. 643.

37. C.C. Daughney, L.S. Holmes, and J.W.M. Paul, Phys. Rev. L 25, 497 (1970).

38. B.B. Kadomtsev, Plasma Turbulence (Academic Press, New York, 1965).

39. C.N. Lashmore-Davies, J. Phys. A: Proc. Phys. Soc., London 3, L40 (1970).

40. D.W. Forslund, R.L. Morse, and C.W. Nielson, Phys. Rev. L 25, 1266 (1970).

41. L. Kellerer, Z. Physik 232, 415 (1970).

42. L. Kellerer, Z. Physik 239, 147 (1970).

43. G. Lehner and F. Pohl, Z. Physik 232, 405 (1970).

44. D.E. Evans and P.G. Carolan, Phys. Rev. L 25, 1605 (1970).

45. B. Kronast and R. Benesch, Proc. Tenth Int. Conf. on Phenomena in Ionized Gases, Oxford, Sept. 13-18, 1971, p. 415.

ELECTRON-PHOTON INTERACTION*

Helmut Schwarz

Rensselaer Polytechnic Institute-Hartford Graduate Ctr.

275 Windsor St., Hartford, Connecticut 06120

ABSTRACT

A short survey of recent experimental and theoretical results on the Kapitza-Dirac Effect (the reflection of electrons from a standing light wave) will be given. Higher order calculations show that the effect under certain conditions may disappear while increasing the intensity. -- In the second half, results on the optical modulation of an electron beam will be presented. The modulation experiment is performed by an electron beam passing through a thin crystal modulated by laser light which shines through it. The light is then reproduced when the electron beam strikes a nonluminescent target. Further experimental and theoretical results, especially on the monochromaticity requirements for the electron beam, and the quantum mechanical nature of the effect, are reported.

INTRODUCTION

With the advent of lasers and their special properties in monochromaticity, coherence, and high intensities, interaction experiments with electrons that were predicted a long time ago and some of which seemed to be utopian have become possible.

I. KAPITZA-DIRAC EFFECT

The Kapitza-Dirac effect belong to this category; an electron beam was predicted to reflect from a standing light wave. The theory

---
*Presented at the Second Workshop on "Laser Interaction and Related Plasma Phenomena" at Rensselaer Polytechnic Institute, Hartford Graduate Center, August 30-September 3, 1971.

of this effect was first described by Kapitza and Dirac[1] in 1933, but it was not until 1965 that it could be observed in a laboratory[2,3]. Kapitza and Dirac considered the effect as the inverse of the diffraction of photons (x-rays) by standing electron waves in solid crystals. The intensity maxima in the standing light wave acts like a crystal pattern; the "lattice" spacing is one-half of the light wave length λ_p.

Angle of Deflection

From this model, Kapitza and Dirac arrived at a Bragg relationship for the angle of incidence θ being equal to the angle θ' of reflection. Taking into account the elastic nature of the effect and the conservation of momentum, one arrives at a relationship

$$2\hbar\omega/c = 2p_e \cos\theta \qquad (1)$$

in which $\hbar$ is Planck's constant divided by 2π, ω the angular frequency of the light, $\omega = 2\pi c/\lambda_p$ (c velocity of light), and p_e momentum of electron. Entering with the de Broglie relationship for the electron momentum $p_e = h/\lambda_e$ (λ_e being the wave length of the electron) into Eq. (1), one arrives at the simple formula:

$$\cos\theta = \lambda_e/\lambda_p \qquad (1a)$$

For slow electrons the electron wave length is given by:

$$\lambda_e = 12.3/U^{1/2}$$

λ_e comes out in Ångstrom if the electron beam acceleration voltage U is entered in volts.

For the normal experimental conditions $\lambda_p \gg \lambda_e$, the total angle θ_o under which the electron is deflected by the standing light wave will be $\theta_o = 2(\frac{\pi}{2} - \theta)$ which leads to

$$\theta_o = 2\frac{\lambda_e}{\lambda_p} = \frac{24.6}{\lambda_p U^{1/2}} \qquad (2)$$

θ_o in radian, λ_p in Ångstrom and U in volts

Applying a neodymium doped glass laser, $\lambda_p = 10,600$ Å, and an electron beam slowed down in the interaction zone to $U = 10$ volts (as this was done in the experiment[2,3]), results in an angle $\Theta_o \simeq 7 \times 10^{-4}$ radian or approximately 2.5 minutes.

Reflection Probability

Although Kapitza and Dirac in their original paper discussed the effect from a wave mechanical model, they calculated, however, the probability of reflection from the consideration of a particle model. Hereby, they assumed that a photon travelling towards the reflecting mirror would be absorbed by a traversing electron, which would be a Thomson scattering effect. Since the electrons are slow, another photon coming in the opposite direction from the mirror would still encounter this particular electron and stimulate the re-emission of the previously absorbed photon which would be ruled by Einstein's coefficient of stimulated emission. This then leads to a formula for the ratio of the number N_r of the electrons being reflected and to the number N_o of incident electrons:

$$N_r/N_o = B_o \ell \lambda_p^6 I^2 / (\Delta\lambda_p U^{1/2}) \qquad (3)$$

Herein the constant B_o has the approximate value $B_o \simeq 10^{-38} \text{ cm}^3 \text{V}^{1/2} \text{W}^{-2} \text{Å}^{-5}$, expressing the path length ℓ of the electron beam inside the light beam in centimeters, λ_p and $\Delta\lambda_p$ (the wave length spread of the light) in Ångstrom, I the intensity of the laser light in the interaction zone in Watts per square centimeter, and the electron acceleration voltage U in volts.

Kapitza and Dirac calculated an example using data which they felt might be experimentally possible at that time. The electrons were assumed to have an energy corresponding to $U = 25$ V and a path length inside the light of $\ell = 10$ cm. The green light ($\lambda_p = 5460$ Å) of a mercury discharge was thought to have a wave length spread of $\Delta\lambda_p = 0.1$ Å and an output of 1 Wcm^{-2}. This would result in the reflection of one electron out of 10^{14} and a total angle of deflection $\Theta_o \simeq 0.05^0$. Kapitza and Dirac said in 1933 "...the experiment could scarcely be made with ordinary continuous sources of light...." But today the laser light can produce such high intensities, for which Kapitza-Dirac's formula (see Eq. (3)) is not valid any more, since intensities higher than 10^7 Wcm^{-2} would lead the formula to absurdity, viz. yielding more reflected electrons than there are available. Therefore, it is obvious that their theory needs extension to higher order approximations. A first order quantum mechanical treatment leads[3] practically to the same formula (see Eq. (3)) as derived by Kapitza and Dirac in 1933.

We have observed more than 50% reflected electrons, whereas others[4-6] -- at higher intensities -- have not observed any reflections and therefore, believe that the Kapitza-Dirac effect does not exist[5,7]. The seeming discrepancy can be cleared up, if higher order quantum mechanical calculations are performed. Gush and Gush[8] have shown that the relative amount of reflected electrons decreases if at a given interaction time the light intensity is increased. They found[8] that for intensities up to approximately $10^7 Wcm^{-2}$, the probability is a periodic function of interaction time t and intensity I, viz.:

$$N_r/N_o = \sin^2 a_o t \qquad (4)$$

whereby the parameter a_o is given by:

$$a_o = \frac{r_o \vec{A}_o^2}{16 \hbar} \qquad (5)$$

$r_o = e^2/(mc^2)$ being the classical electron radius, and $\vec{A}_o$ being the peak value of the magnetic vector potential of the light.

$$\vec{A}_o^2 = 16 \pi c I/\omega^2 \qquad (6)$$

Entering with Eq. (6) and the expression for r_o into Eq. (5) leads to:

$$a_o = \frac{\pi e^2 I}{mc\omega^2 \hbar} \qquad (7)$$

Using practical units I in Wcm^{-2} and t in sec, Eq. (4) can be written:

$$N_r/N_o = \sin^2(2.52 \times 10^{32} It/\omega^2) \qquad (8)$$

In all our experiments, we kept the interaction time t constant, since the geometrical light optics as well as the electron optics remained unchanged; the light beam diameter ℓ was always 0.3 cm and the electron energy inside the light beam was 10 eV, which

ELECTRON-PHOTON INTERACTION

results in a constant interaction time of $t = 1.6 \times 10^{-9}$ sec. The light originated from a neodymium doped glass laser of $\lambda_p = 10,600$ Å, so that $\omega = 1.78 \times 10^{15}$ sec^{-1}. We arrived, therefore, at the simple relationship

$$N_r/N_o = \sin^2(1.275 \times 10^{-7} I) \qquad (8a)$$

This function is plotted on Fig. 1 for intensities I between 1 and 40 MWcm^{-2} applied for the wave length $\lambda_p = 10,600$ Å and an interaction time $t = 1.6$ nsec. A second abscissa is drawn indicating the argument $a_o t$ of Eq. (4) for values $a_o t = 0...5.5$. Even though a correct determination of the laser intensities in our experiments was not possible, a relatively good measurement for N_r/N_o could be performed. And since ω is very accurately known and t could also be measured quite precisely, one medium value of N_r/N_o - i.e. 60% - was fitted to the curve. This yielded an intensity $I = 7.0 \times 10^6$ W cm^{-2}. The intensities were varied by inserting filters in the light path before the laser beam entered the vacuum system. The filter factor allowed actual values in W cm^{-2} for the other five settings of I normalizing them to the 60% reflectivity conditions. These five different intensities were then entered in Fig. 1; they seem to fit the theoretical curve. Intensities higher than 9.2×10^6 Wcm^{-2} could not be obtained.

In Table I we have summarized the results of all experimental work on the Kapitza-Dirac effect up-to-date.

TABLE I

	designation in Fig. 1	U [v]	ℓ [cm]	t [sec]	ω [sec^{-1}]	$a_o t$	N_r/N_o
Bartell et al[4]	▽	1,640	1.2	4.7x10^{-10}	2.72x10^{15}	3.2	0.004
Pfeiffer[5]	◩	36	1.1	3.1x10^{-9}	2.72x10^{15}	3.16	0.0004
Schwarz[3]	⊗ X	10	0.3	1.6x10^{-9}	1.78x10^{15}	0.89 1.173 0.52 0.32 0.165	0.60 ⊗ 0.80 0.22 0.08 0.06
Takeda et al[6]	⛊	300	0.5	4.85x10^{-10}	2.72x10^{15}	0.165	0.027

In the first column of Table I are listed the authors with their respective references; in the second, the designation of the points as marked on Fig. 1 as calculated from the respective experimental data; the third column, the acceleration voltage of electrons within the laser light beam; the fourth, the path length ℓ of the electrons in the interaction zone; the fifth, the interaction time t; the sixth, the angular frequency of the laser light; the seventh, the argument of the sinus function in Eq. (4); and in the last column, the probability of the electron reflection.

In our experiments, we applied five different intensities and these are also evaluated in the seventh and eighth columns. Only the values in the seventh column are changed since all other parameters (electron velocity, interaction time, electron path and light sources) remained unchanged. With the normalization of the 60% reflectivity point fitted to the curve of Fig. 1, the other four points follow quite closely the calculated curve.

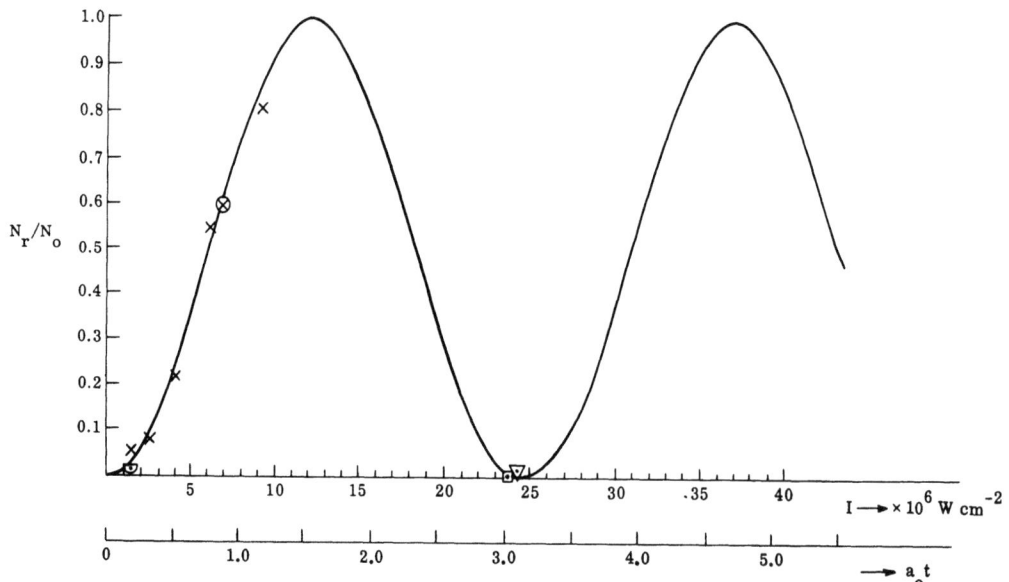

FIG. 1 - Reflection Probability as a Function of Light Intensity for Interaction Time t = 1.6 nsec, using a neodymium laser.

II. OPTICAL MODULATION OF ELECTRON BEAMS*

Description of Effect

In earlier experimental work, we observed[10,11] modulation of an electron beam at optical frequencies. An electron beam of 50 keV energy, 0.4 microamps current and several micrometers diameter was directed perpendicularly through a freely supported dielectric thin film of approximately 1000 Å thickness. The films used so far were SiO_2, A_2O_3 and SrF_2. These materials were chosen since they are optically transparent. Therefore, it was possible to shine a laser beam through these films. If the electrical vector was perpendicular to the thin film and parallel to the electron beam, the electrons could be made visible on a nonfluorescent screen at a distance varying between 10 and 35 cm. In order to observe the light on the nonfluorescent screen, the dielectric transparent thin film was absolutely necessary. A momentum transfer between photons and electrons had to take place within the dielectric.

Mostly the electron beam passed through areas of the thin film which were single crystals so that the light spots on the screen corresponded to the Laue pattern of electron diffraction. The light spots appeared in the same color as the laser beam. However, only two lines of an argon ion laser were used, namely the lines 4880Å and 5145Å.

Modulation and Demodulation

Several theoretical papers have been published explaining the modulation of the electron beam[12-28,30-38,44-45].

We believe that we have established the quantum mechanical nature of the modulation effect, since we have observed[29] a "spatial beating wave length" (characteristic length Λ). Independently from other theoretical papers which predicted such a spatial beating wave length, we found Λ also theoretically and reported the theory and experiments in Reference 29.

E. T. Jaynes suggested that the effect would give means to "observe for the first time experimentally the initial size and shape of electron wave packets"[22].

The effect was not just discovered as a serendipity, but we were looking for scattering effects involving electrons and photons in direct interactions. In a discussion with Professor Dirac on the reflection of electrons from a standing light wave[1-8], we raised the

---

*This section represents excerpts of a paper "Quantum Mechanical Bunching of Electron Beams Scattered by a Laser Light Within a Solid Dielectric"[9]

question whether or not it might be possible to modulate a particle wave by an electromagnetic wave, as it is well known in communication technology (AM radio), viz. that one electromagnetic wave can modulate another one. Professor Dirac pointed out after some calculations that this would not be possible with visible photons in a vacuum since the photon momentum would not be sufficient. However, we came to the conclusion that if the interaction did take place within a dielectric, the effect might be possible, since the dielectric material would provide for the necessary recoil of the electron momentum.

A simple explanation at first sight seemed to be a classical bunching process as it occurs for example in traveling wave tubes. This possibility was mentioned in our paper in Applied Physics Letters[11] and also was suggested by Harris and Smith[23]. Hereby a simple velocity modulation of the electrons is assumed as this was worked out later on more extensively by B. M. Oliver and L. S. Cutler[24] and at the beginning of a paper by L. L. Van Zandt and James W. Meyer[15]. However, in our case the classical theory cannot explain the effect since the "frequencies" of the two waves (light wave and electron wave) are too far apart. Also the peak velocity is much too small in relation to the dc velocity of the original electron beam.

In Van Zandt and Meyer's[15] paper, besides quantum mechanical treatment, an expression for the electron density n, as it fluctuates along the electron beam path, was developed classically:

$$n = n_0 \left\{ 1 - \frac{v_1}{v} (1 + \frac{\omega^2 r^2}{v^2})^{1/2} \sin\left[\omega(t - \frac{r}{v}) + \cos^{-1}(1 + \frac{\omega^2 r^2}{v^2})^{1/2}\right] \right\} \quad (9)$$

n_0 is the linear density of the electrons in an unmodulated beam, v_1 the peak velocity due to the electric field of the light, v the velocity of the primary beam, ω the frequency of the laser light, r the distance measured along the electron beam. In order that the Equation (9) yield a positive value (n is the number of electrons), the expression

$$\frac{v_1}{v} (1 + \frac{\omega^2 r^2}{v^2})^{1/2}$$

from Eq. (9) has to be smaller than unity, which means that it can only be valid for distances r obeying the inequality:

$$r < \frac{v}{\omega}\frac{v}{v_1} \qquad (10)$$

$\frac{v}{v_1}$ being much larger than unity

or in practical units this inequality can be written approximately as

$$r < 2.3 \times 10^{-2} \frac{v}{\omega}\sqrt{\frac{U}{s\sqrt{I}}} \qquad (10a)$$

r in meter, v in msec^{-1}, ω in Hz, U the acceleration voltage of the electron beam in volt, s the thickness of the dielectric thin film in m, I the intensity of the laser in Wcm^{-2}. Entering with the experimental values, we arrive at the conclusion that the classical formula can only be valid for distances below 10 μm. Therefore, a classical theory cannot be applied. This will also be shown later from a quantum mechanical criterion.

QUANTUM MECHANICAL BUNCHING

One of the first quantum mechanical theories was presented by A. Salat[12]. Salat considered the effect as a quantum mechanical scattering process due to the laser induced periodical distortion of the scattering potential within the crystal. The demodulation of the electron wave at the nonfluorescent target was considered as a high frequency induced charge modulation at the screen by which the lattice electrons of the screen caused radiation at the frequency of the laser light.

The interesting result of Salat's[12] investigation is that he also found an electron probability density from his quantum mechanical scattering theory which gives a periodic modulation of the intensity with increasing distance between screen and interaction zone. This characteristic distance is exactly the same as others[13,17,21,25-29, 31-38,44-45] have calculated. Salat[12] arrived at his electron probability density through a quantum mechanical scattering theory applying a first order solution of the Schroedinger equation introducing in the Hamiltonian a perturbation term containing the periodic variation of the magnetic vector potential

Most others have taken the direct approach but assume the superposition of three waves. These three waves should be the ground wave with an eigenvalue equal to the energy of the entering electron beam, and two secondary waves - one with an energy increased by one photon energy and the other being decreased by one photon energy. At the same time a momentum transfer between photons and electrons should take place within the dielectric. In our paper[29], we do not go into the mechanism of such a momentum transfer and we just assumed the existence of the three electron waves being superimposed after leaving the dielectric. The momentum does not have to be conserved in the interaction, since the film surface provides the necessary recoil momentum.

Electron Wave Functions

In our earlier treatment[29] with which we arrived essentially at the same results, we assume the three electron waves with eigenvalues, E, $E + \hbar\omega$, and $E - \hbar\omega$:

$$\Psi_0(E) = \exp\left[\frac{i}{\hbar}(p \cdot r - Et)\right] \quad (11)$$

$$\Psi_+ = e^{\frac{i}{\hbar}\left[(p + \Delta p_+) \cdot r - (E + \omega\hbar)t\right]} \quad (12)$$

$$\Psi_- = e^{\frac{i}{\hbar}\left[(p - \Delta p_-) \cdot r - (E - \omega\hbar)t\right]} \quad (13)$$

The superposition wave function then reads

$$\Psi = a_-\Psi_-(E - \omega\hbar) + a_0\Psi_0(E) + a_+\Psi_+(E + \omega\hbar) \quad (14)$$

The wave function represented in Eq. (12) has an eigenvalue which represents an electron that absorbs one photon and the wave function in Eq. (13) represents the electron wave that had given up a photon by stimulated emission. In order to obey the energy conservation law, the electron must have gained in the first case the momentum Δp_+ and in the second case lost the momentum Δp_-. This then leads to the equation

$$\frac{(p \pm \Delta p_\pm)^2}{2m} = E \pm \omega\hbar \quad (15)$$

ELECTRON-PHOTON INTERACTION

Equation (15) allows the calculation of the increase and decrease respectively of the momentum, namely,

$$\Delta p_{\pm} = \tfrac{1}{2} p(\varepsilon \pm \tfrac{1}{4} \varepsilon^2 + \text{-----}) \qquad (16)$$

Hereby ε means the ratio of the two energies involved, namely the photon energy and the electron energy, which is, of course, very small - in our case approximately 5×10^{-5}. The momentum transfer to the electron can only come from the photon, i.e. $\Delta p_{\pm} \lesssim \hbar\omega/c$ which combined with Eq. (16) leads then to the condition

$$\frac{v}{c} \cos \theta = 1 \pm \tfrac{1}{4} \varepsilon \qquad (17)$$

where v means the velocity of the electron and c the velocity of light, and θ the angle between electron beam and laser beam as scattered within the crystal. The experiments gave a value for $v/c = 0.41$ and for $\tfrac{1}{4} \varepsilon \simeq 10^{-5}$. Eq. (17) can, therefore, only hold if the value of c is not taken in vacuum.

Electron Probability Density

The wave function Eq. (14) allows the calculation of the electron probability density to be

$$\psi^*\psi = A_0 + A_1 \cos(\tfrac{2\pi}{\Lambda} r) \cos\left[\omega(\tfrac{r}{v} - t)\right] \qquad (18)$$

where $A_0 = a_0^2 + a_+^2 + a_-^2$. The reasonable assumption was made that the probability for absorption of a photon as well as for desorption of a photon is relatively small and that the two probabilities are almost equal. A higher order term with the double frequency of the laser is left out in Eq. (18) since again its amplitude would be much smaller than $A_1 \simeq 4a_0 a_+ \simeq 4a_0 a_-$. One can see that the electron probability density is a periodic function of time and space. The spatial beating wave length is given by

$$\Lambda = \frac{16\pi\hbar}{p\varepsilon^2} = \frac{16\pi E^2}{p\omega^2\hbar} = 2(\tfrac{2}{\varepsilon})^2 \lambda_e \qquad (19)$$

in which λ_e is the wave length of the electron. Using the experimental values, one arrives for Λ at the value of 1.65 cm. A relativistic calculation of Λ will not change it by more than 10%. This spatial beating wave length or characteristic wave length Λ was observed to be 2 x 0.85 cm for the blue light of λ = 4880Å. See also Figure 4 of Reference 29. A similar curve was also observed for the green color of the laser light, λ = 5145Å. The blue and green light followed to a precision of about 10% of Eq. (19). The green light showed experimentally a distance between the intensity peaks of about 1.1 cm.

The strongest theoretical argument for a basic quantum mechanical effect and against a classical bunching of electrons is based on the fact that the balance of energy and momentum in the electron-photon interaction process can only be preserved if allowance is made for the existence of a traveling wave mode inside the dielectric sheet, as this is described by R. E. Collin[39]. An electron crossing the film can then absorb or emit the photon with the surfaces of the dielectric providing the necessary recoil momentum. The expectation value of the energy of the modulated electron is essentially unchanged by the interaction, whereas in the classical bunching theory as proposed by Harris and Smith[23] as well as Oliver and Cutler[24] it would require a change of velocity.

LIGHT INTENSITY

A classical treatment would also require that the light intensity on the nonfluorescent screen should be proportional to the square of the electron current. Experimental results seem to indicate that there is no evidence for a purely quadratic dependence of the intensity on the electron current.

L. L. Van Zandt[18] has derived an expression for the light intensity based on quantum mechanical considerations. His formula contains a linear and a quadratic term; however, the linear term predominates.

On the other hand, C. Becchi and G. Morpurgo[17] claim that even if the demodulation on the screen were to be explained quantum mechanically, the dependence should be quadratic. They stated that a one electron wave function could not result in radiation from the screen and they were, of course, right if the necessity existed to have two electron interactions. But the modulation effect is not like the Kapitza-Dirac effect[1-3,8] where no energy is transferred and, therefore, two processes of electron-photon interactions have to be involved. In our case of modulation, we have only one interaction process occurring within the dielectric film.

COHERENCE LENGTH AND REQUIRED MONOCHROMATICITY OF ELECTRON WAVE

The basic difference between classical and quantum mechanical bunching can also be expressed in the relationship between coherence length of the electron wave and the light pulse as has been pointed out by Varshalovich and D'Yakonov[16]. In order that the effect be quantum mechanical, the coherence length Δs_c of the electron wave should be larger than the light pulse $\frac{v}{\omega}$. This leads for quantum mechanical bunching to the criterion $\Delta s_c \gg \frac{v}{\omega}$, and for classical bunching to $\Delta s_c \ll \frac{v}{\omega}$.

The coherence length of the electron wave Δs_c is, of course, a function of the energy spread ΔU of the electron beam and can roughly be expressed as

$$\Delta s_c \simeq \frac{\lambda_e^2}{\Delta \lambda_e} \tag{20}$$

whereby λ_e the wave length of the electron

$$\lambda_e = \frac{h}{p} \text{ and } \Delta\lambda_e = \frac{h \Delta p}{p^2}$$

p being the momentum of the electron.

$$\Delta s_c \simeq \frac{h}{\Delta p} = \frac{h}{e} \frac{v}{\Delta U} \tag{20a}$$

e = elementary charge.

We have shown theoretically and experimentally that the modulation effect "washes" out if the variation of

$$\frac{\omega r}{v} \tag{21}$$

as taken from the second part of Eq. (18) does not remain below π over the distance r, which leads to the inequality for the energy spread[40-42]

$$\Delta U < \frac{v}{c} \frac{\lambda}{r} U \tag{22}$$

For the original experiment, the cutoff energy spread equals 30 mv. Entering with $\Delta U = 0.03$ into Eq. (20a) leads to a minimum coherence length of $\Delta s_c = 2$ μm, which is much larger than the light pulse $\frac{v}{\omega} \simeq 0.03$ μm.

A similar inequality as (22) can be derived from Reference 26, except that it is by a factor of $2/\pi \simeq 0.6$ smaller.

Experiments were performed whereby the 5 μm diameter holes of the smallest diaphragm through which the slowed down electron beam passes before it is reaccelerated to 50 keV and before it penetrates the thin film were increased in order to change the energy spread of the electrons.

The electron current was maintained at its low level of approximately 0.5 μa. Naturally it is not possible to achieve such a relatively high current of 10 eV electrons with the energy spread required within a single beam. The beam had to be given a "filamentary" structure[42], after it passed through the 90° magnetic electron monochromator. This was achieved by letting the electron beam shine under a relatively large aperture angle on a thin tantalum disc (diameter 0.5 - 0.8 cm) with precisely perforated holes (5, 10, 20, and 100 μm) being separated at well defined distances and having a hole density of 10,000 to 100,000 per cm². The light intensity observed at the screen was always adjusted to optimum by properly changing the electrical and magnetic characteristics of the velocity analyzer so that there was no phase difference between the single beams other than integral multiples of 2π. Each single beam filament fulfilled the required energy spread[40] whereby the energy differences between the beam filaments yielded a change of the argument of the last cosine term in Eq. (18) being equal to $2n\pi$ (n integral number) or:

$$\omega r e \Delta U_o / (mv^3) = 2n\pi \qquad (23)$$

$e\Delta U_o$ is the energy difference between the single beam filaments. For each diaphragm a series of measurements of light intensity were performed, changing continuously the distance between interaction zone (thin dielectric film) and nonfluorescent observation screen. Through a mirror arrangement the light entered a photomultiplier whose output was traced on a strip chart recorder driven with the same motor as the moving mechanism for the screen. On Fig. 2 points are plotted which indicate light intensity maxima during the distance r (see Fig. 4 of Ref. 29). The intensity at distance r = 10.2 cm, using the smallest diaphragm $\Delta s = 5$ μm, was taken as absolute maximum intensity I_o. For the diaphragm of diameter 100 μm no effect could be observed. The points were taken from the recorder chart. The curves on Figure 2 are not an indication of the <u>continuous</u> change of

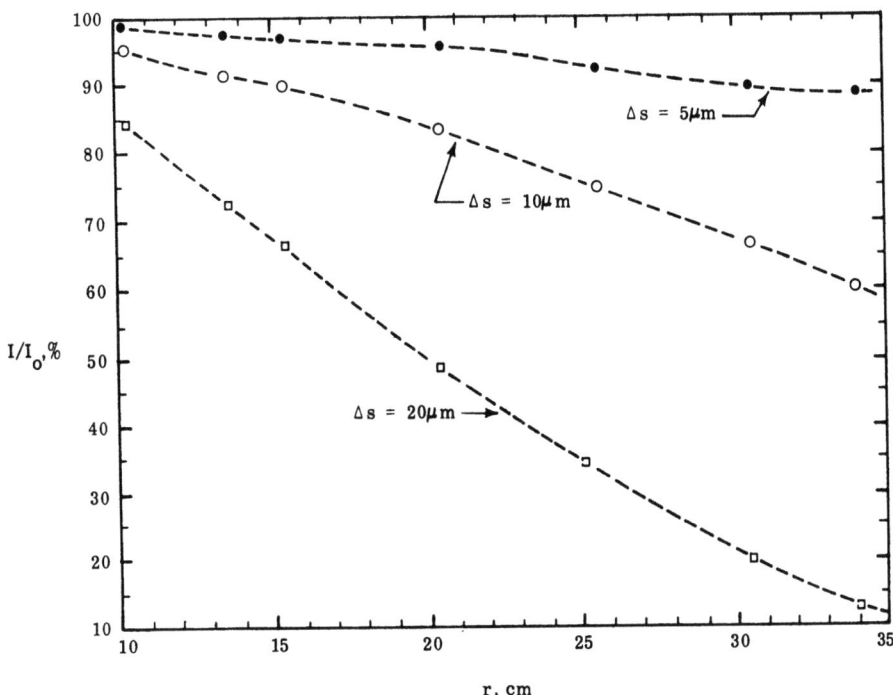

FIG. 2 - Relative Light Intensity for Different Degrees of Electron Monochromaticity as a Function of Distance r between Interaction Zone and Non-fluorescent Observation Screen

light intensity with distance. Their real shapes are like the representation of a damped oscillation, only the upper envelope curves are shown here.

A recent attempt[43] to reproduce the modulation effect shows clearly that good monochromaticity of the electron beam is an indispensable prerequisite. As calculated by L. D. Favro et al[44], the damping factor of these attempts[43] is equal to $\exp(-90)$ for a distance of 30 cm, whereas in our case it is only $\exp(-0.13) \simeq 90\%$.

O. Scherzer[45] in an interesting theoretical approach could derive a very general formulation of the modulation effect which would cover the classical as well as the quantum mechanical bunching. He set up a wave function which contains many eigenvalues of which each one consisted of the base electron wave with eigenvalue E and modulating waves with eigenvalues $E \pm n\hbar\omega$ (n being integral number). He could show that if n goes to very large numbers and $\hbar\omega$ to very small values, one arrives at classical bunching formula for electrons. In quantum mechanical bunching, the energy of the modulating wave is transferred to the wave to be modulated in quanta, whereas in classical bunching the energy is transferred continuously.

REFERENCES

1. P. L. Kapitza and P. A. M. Dirac, Proc. Cambridge Phil. Soc. $\underline{29}$, 297 (1933).
2. H. Schwarz, H. Tourtellotte and W. W. Gaertner, Phys. Letters $\underline{19}$, 202 (1965).
3. H. Schwarz, Zeitschrift für Physik $\underline{204}$, 276 (1967).
4. L. S. Bartell, H. B. Thompson and R. R. Roskos, Phys. Rev. Letters $\underline{21}$, 581 (1965); L. S. Bartell, R. R. Roskos and H. B. Thompson, Phys. Rev. $\underline{166}$, 1494 (1968).
5. H. Chr. Pfeiffer, Phys. Letters $\underline{26A}$, 362 (1968).
6. Y. Takeda and I. Matsui, J. Phys. Soc. Japan $\underline{25}$, 1202 (1968).
7. H. Schoenebeck, Phys. Letters $\underline{27A}$, 286 (1968).
8. R. Gush and H. P. Gush, Phys. Rev. $\underline{D3}$, 1712 (1971).
9. H. Schwarz, *Proceedings of the Internl. Conf. on Light Scattering in Solids*, Paris, 1971, ed. by Balkanski, Flammerion, Paris, 1971, pp. 123-127.
10. H. Schwarz, Bull. Am. Phys. Soc. $\underline{13}$, 897 (1968).
11. H. Schwarz and H. Hora, Appl. Phys. Letters $\underline{15}$, 349 (1969).
12. A. Salat, J. Physik C., $\underline{3}$, 2509 (1970).
13. A. R. Hutson, Appl. Phys. Letters $\underline{17}$, 343 (1970). Errata: Appl. Phys. Letters $\underline{18}$, 208 (1971).
14. A. D. Varshalovich and M. I. D'Yakonov, JETP Letters $\underline{11}$, 411 (1970).
15. L. L. Van Zandt and James W. Meyer, J. Appl. Phys. $\underline{41}$, 4470 (1970).
16. A. D. Varshalovich and M. I. D'Yakonov, "Quantum Theory of Modulation of an Electron Beam at Optical Frequencies", JETP; Russian Reference: Vol. $\underline{60}$, 90 (1971) (English Translation to be published by AIP).
17. C. Becchi and G. Morpurgo, Phys. Rev. $\underline{D4}$, 288 (1971).
18. L. L. Van Zandt, Appl. Phys. Letters $\underline{17}$, 345 (1970).
19. P. L. Rubin, JETP Letters $\underline{11}$, 239 (1970).
20. C. S. Owen, Bull. Am. Phys. Soc. $\underline{16}$, 118 (1971).
21. L. D. Favro, D. M. Fradkin, P. K. Kuo and W. B. Rolnick, Bull. Am. Phys. Soc. $\underline{16}$, 25 (1971).
22. E. T. Jaynes, Abstract of an "Electrophysics Colloquium", Polytechnic Institute of Brooklyn, entitled "A Theory of the Schwarz Effect - Optical Modulation of Electron Beams", May 21, 1970.
23. R. L. Harris and R. F. Smith, Nature $\underline{225}$, 502 (1970).
24. B. M. Oliver and L. S. Cutler, Phys. Rev. Letters $\underline{25}$, 273 (1970).
25. L. D. Favro, D. N. Fradkin and P. K. Kuo, Phys. Rev. Letters $\underline{25}$, 202 (1970).
26. L. D. Favro, D. M. Fradkin and P. K. Kuo, Phys. Rev. $\underline{D3}$, 2934 (1971).
27. L. D. Favro, D. M. Fradkin, P. K. Kuo and W. B. Rolnick, Appl. Phys. Letters $\underline{18}$, 352 (1971).
28. L. D. Favro, D. M. Fradkin and P. K. Kuo, Lettere al Nuovo Cimento $\underline{4}$, 1147 (1970).
29. H. Schwarz, Transactions N. Y. Academy of Sciences $\underline{33}$, 150 (1971).

30. B. Ya. Zeldovich, JETP, Russian Issue, Vol. 61, 135 (1971).
31. D. Marcuse, J. Appl. Phys. 42, 2255 and 2259 (1971).
32. J. Kundo, J. Appl. Phys. 42, 4458 (1971). G. T. diFrancia, Nuovo Cimento 37, 1553 (1965).
33. H. J. Lipkin and M. Peshkin, Appl. Phys. Letters 19, 313 (1971).
34. C. R. Hadley, E. J. Stanek, and R. H. Good, Jr., J. Appl. Phys. 43, 144 (1972). Part II. Relativistic Treatment (to be published)
35. H. J. Lipkin and M. Peshkin, "Interference Effects Produced by Modulated Particle Beams", Appl. Phys. Letters (to be published)
36. P. L. LaFleur, "Dynamical Theory of Electron Diffraction by a Laser Illuminated Crystal", Physica (to be published).
37. C. S. Chang and P. Stehle, "Interference and Correlations in Photon and Electron Optics", Phys. Rev. A (to be published).
38. C. S. Chang and P. Stehle, "Quantum Effects in Accelerating Gaps", Phys. Rev. A (to be published).
39. R. E. Collin, *Field Theory of Guided Waves*, (McGraw-Hill, New York, 1960), Sec. 11.5.
40. H. Schwarz, Appl. Phys. Letters 19, 148 (1971).
41. The calculations were done neglecting relativistic effects.
42. H. Schwarz, "Energy-Spread Structure of Electron Beams to be 'Optically' Modulated", Appl. Phys. Letters (to be published 15 Febr. 1972).
43. R. Hadley, D. W. Lynch, E. Stanek and E. A. Rosauer, Appl. Phys. Letters 19, 145 (1971).
44. L. D. Favro, D. M. Fradkin, P. K. Kuo, and W. B. Rolnick, Appl. Phys. Letters 19, 378 (1971).
45. O. Scherzer, Technische Hochschule Darmstadt, Private Communication, July 1971.

INTERPRETATION OF CYLINDRICAL LANGMUIR PROBE SIGNALS FROM STREAMING LASER-PRODUCED PLASMAS*

Stephen B. Segall

Institute for Fluid Dynamics and Applied Mathematics

University of Maryland, College Park, Maryland 20742

ABSTRACT

Experimental procedures for the use of cylindrical Langmuir probes to diagnose streaming laser-produced plasmas are described, and the steps involved in the development of a theory for the probe in a high velocity flowing plasma are summarized. The probe-plasma interaction for the front half of the probe facing the plasma flow is simulated using a stationary plasma model. Qualitative conclusions obtained from this analysis are then used to derive expressions for current collected in the flowing plasma case. Theoretical electron current characteristics are calculated and fitted to the experimental curves. Comparison is made with other works using particle collection techniques to diagnose laser-produced plasmas.

---

*Presented at the Second Workshop on "Laser Interaction and Related Plasma Phenomena" at Rensselaer Polytechnic Institute, Hartford Graduate Center, August 30-September 3, 1971.
Work supported by the U.S. Atomic Energy Commission.

INTRODUCTION

Electrostatic probes may be used to diagnose the properties of streaming, laser-produced plasmas in the region from about a centimeter to several tens of centimeters from the surface of the target in which the plasma is produced. At these distances most of the energy of the plasma is in the directed kinetic energy of the ions. The ion thermal velocity is much less than the ion flow velocity and the energy of the electrons is much less than the ion energy. The flowing plasma exists for several microseconds and its properties change rapidly as functions of space and time.

Because they are simple to construct and operate and can give localized, time resolved information on plasma properties, Langmuir probes have become an important means of plasma diagnostics. Interpretation of the probe signals, however, has often turned out to be a very complicated problem. The electrostatic probe in a stationary plasma has been studied extensively and is fairly well understood, but the application of stationary probe theory to the case of a probe in a flowing plasma may lead to incorrect interpretation of the probe signal. Theories have been developed for spherical and cylindrical Langmuir probes in flowing plasmas [1,2,6], but in these theories the central field of the probe is assumed undisturbed by the plasma flow. This is certainly not the case for a probe in a laser-produced streaming plasma where sharply defined wakes behind the probe greatly distort the fields so that they are no longer central. The use of cylindrical electrostatic probes to diagnose streaming laser-produced plasmas has been investigated by the author at the Institute for Fluid Dynamics and Applied Mathematics of the University of Maryland. Others associated with the laser-produced plasma experiment are David Koopman and Augustine Cheung.

EXPERIMENTAL APPARATUS AND PROCEDURES

A diagram of the experimental apparatus is shown in Fig. 1. The plasma is produced by focusing a pulsed laser beam on a solid target in the vacuum chamber. Pressures in the chamber were typically of the order of 10^{-6} Torr. In most cases the target material has been either copper or aluminum. The laser used in this experiment is a Korad ruby laser with a six joule 30 ns pulse. The laser pulse can be monitored by a photodiode located in back of the laser. Plasma streams down the vacuum chamber and is sampled by diagnostic instruments placed in the chamber ports. Both microwave and probe diagnostics have been used. The microwave diagnostics are described in a report by A. Cheung.[3] The oscilloscopes are triggered by the voltage pulse which activates the pockels cell. Time delays on the scopes can be adjusted to begin the sweep just before the plasma arrives at the probe, permitting the use of faster sweep speeds and giving better time resolution.

CYLINDRICAL LANGMUIR PROBE SIGNALS FROM STREAMING PLASMAS

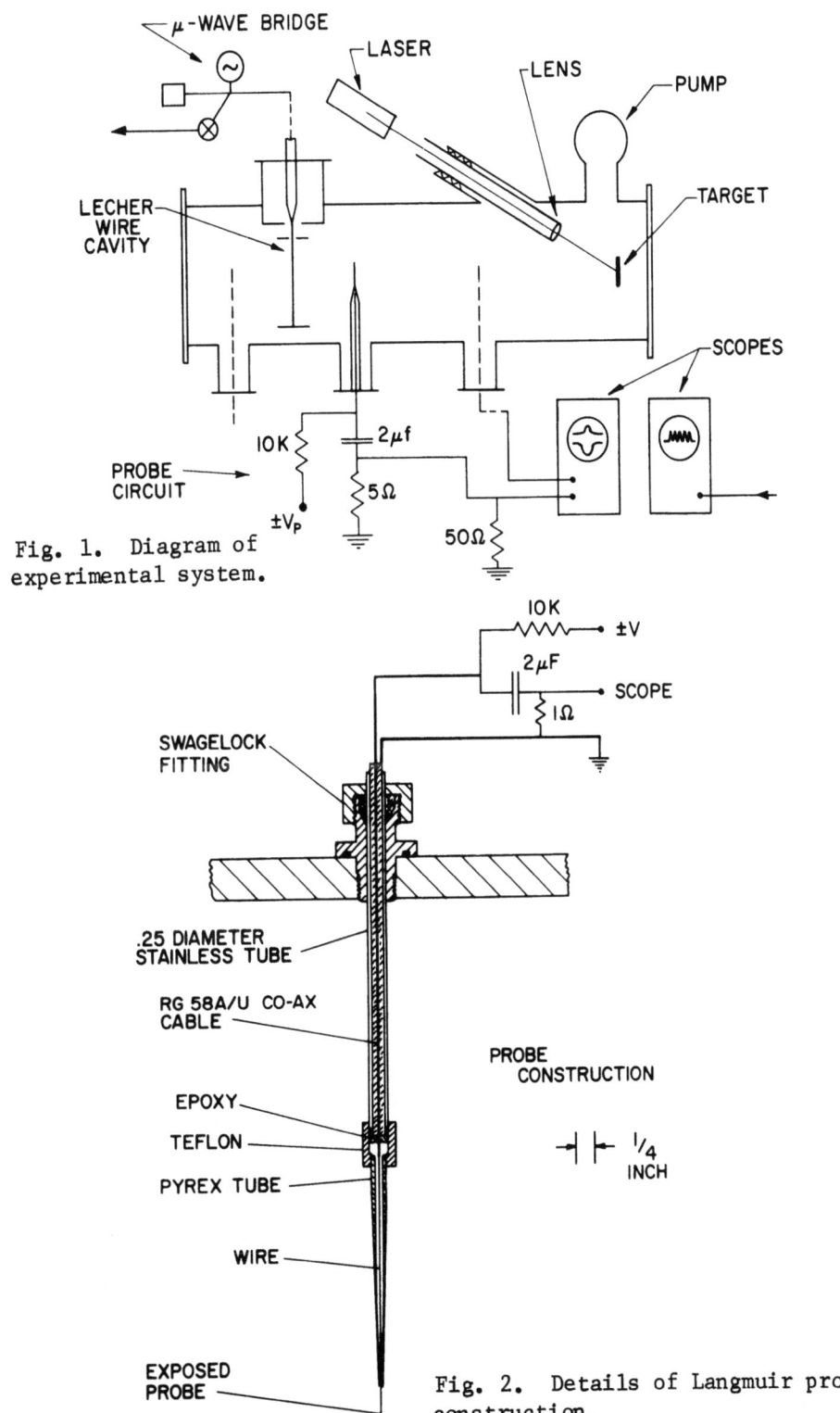

Fig. 1. Diagram of experimental system.

Fig. 2. Details of Langmuir probe construction.

A diagram of the probe construction is shown in Fig. 2. The load resistance is made small to insure a fast time response. For a load resistance of 1Ω and cable and stray capacitance of ~200 μμf we obtain an RC time constant of ~2 × 10^{-10} sec, which for flow velocities of the order of 10^7 cm/sec means that resolution is limited by the scope rise time (<50 ns) and probe sampling dimensions (probe diameter = .025 cm). The probe axis is perpendicular to the direction of plasma flow. Probe signals are photographed by an oscilloscope camera. Current-voltage characteristic curves are obtained by taking measurements for a series of probe voltages. Then, for a given time after the firing of the laser, currents can be determined from the photographs for the different probe potentials.

Since many laser firings must be made to obtain a single current characteristic curve it is important that plasma conditions be repeatable from shot to shot. Shots are taken at intervals of a few minutes to prevent thermal distortions and insure uniform laser pulse shapes. Plasma parameters are also sensitive to the position of the laser focus. The position of the focusing lens can be adjusted using a collimator and is fixed for a given series of shots.

Small shot to shot variations can be compensated for by introducing a normalization procedure. Two current collecting probes were used for each measurement and the signals appeared on a double beam oscilloscope. One probe was held at a constant negative voltage while the potential of the second probe was varied over a series of voltages. The current values for the probe with the varying potential were later adjusted in proportion to the deviations of the signals from their average value for the probe held at constant potential. Signals with large deviations from average behavior were rejected.

Noise from the laser electronics and other possible sources was minimized by putting two signals on each oscilloscope beam, the probe signal from the current collecting probe plus the negative of the signal from a dummy probe circuit not in contact with the plasma but otherwise identical to the active probe circuit. Larger load resistances can be used to reduce the signal to noise ratio provided they are not so large that they affect the time response of the probe. Larger probes also reduce the signal to noise ratio.

After all precautions had been taken to obtain repeatable signals with very low noise levels, characteristic curves could be obtained with a random scatter of data points about the line of best fit of not more than 10%. Typical smoothed experimental curves for current as a function of voltage for copper are shown in Fig. 3. In this figure the current values of each curve have been multiplied by an appropriate factor to make the currents equal in the ion saturation region. The most easily identifiable feature of the characteristic curves is that for sufficiently large negative

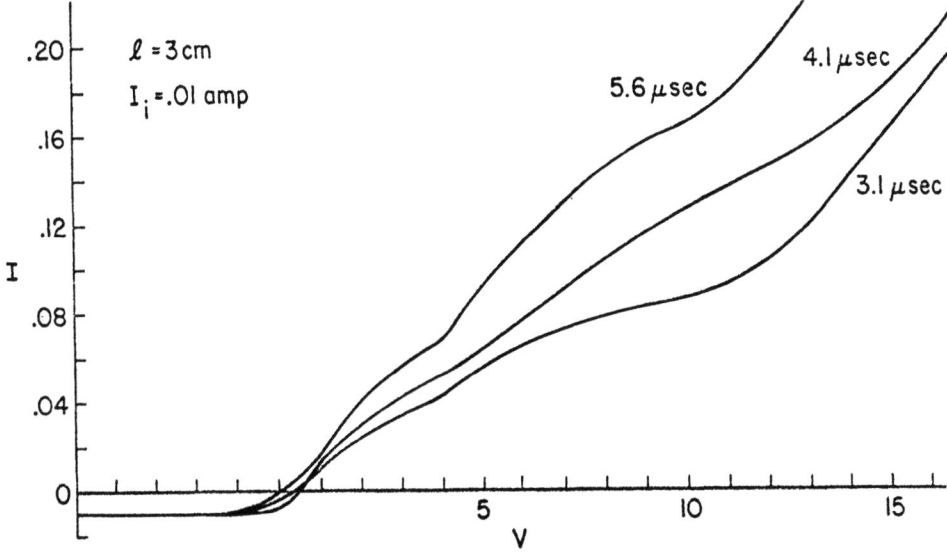

Fig. 3. Current-voltage characteristic curves for probes in a Cu plasma. Current values have been adjusted so that the currents in the ion saturation region are equal.

potentials on the probe, the current to the probe is virtually constant. If we assume that the ions are so energetic that they are not affected by probe potentials of up to a few tens of volts positive or negative, the ion density can be calculated from the ion current. Using the simple model that the current to the probe is just the flux of ions moving with the flow velocity times the cross sectional area of the probe, we obtain for the ion density

$$n_i = \frac{I_i}{AeU} , \qquad (1)$$

Where I_i is the ion saturation current, A the cross sectional area of the probe, e the magnitude of the charge on the electron, and U the flow velocity. Flow velocity is obtained from time of flight measurements assuming that, except for a very small fraction of their transit time, the ions move at constant velocity. For probes located 30 cm from the target densities obtained using equation (1) were of the order of 10^{12} charges/cm^3 during the first few microseconds after the arrival of the plasma. The density values calculated from equation (1) will tend to be too large, because I_i includes both the ion current to the probe and the current of secondary electrons away from the probe.

Subtracting the saturation current I_i from the total current, the electron current to the probe could be determined. When the electron current is plotted on semilog paper there is a linear region a few volts wide from which preliminary values for electron temperature could be calculated using the expression from stationary Langmuir probe theory

$$T_e (ev) = \frac{\Delta V \text{ (volts)}}{\Delta \log I_e \text{(amperes)}}, \qquad (2)$$

where T_e is the electron temperature and I_e the electron current. Typical data points for electron current as a function of potential are shown in Fig. 4. Electron temperature measured in this way varied from about .5 ev to several electron volts with electron temperature generally increasing as a function of time.

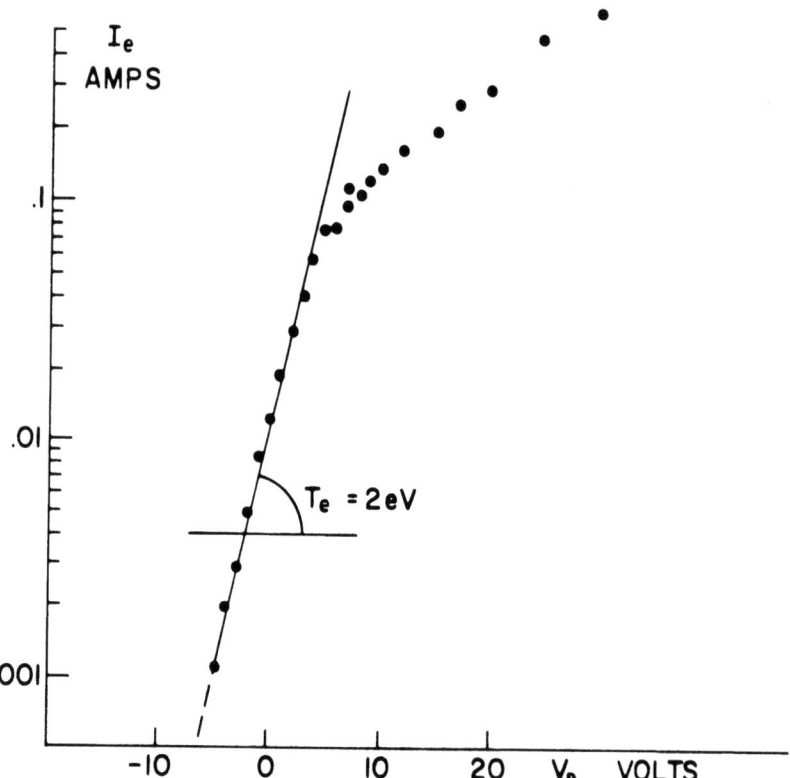

Fig. 4. Semi-log plot of electron current as a function of probe voltage.

The behavior of the current characteristic in the electron repelling region seemed, therefore, to be fairly well understood both with regard to the behavior of the electrons and the ions. The current characteristic in the electron attracting region, however, could not be understood in terms of any existing theory. No collisionless theory for probes in stationary or flowing plasmas predicts the observed behavior (Fig. 3). Collision lengths in the plasma for the densities and temperatures measured were of the order of a few centimeters. The probes used for the measurements had a diameter of .025 cm and the Debye length for the plasma was of the order of 10^{-4} cm, so collisions seemed unlikely as a source for the anomalously large currents collected in the electron attracting region. Without some explanation for the behavior of the probe in this region, the validity of the interpretation of the probe characteristic in the electron repelling region also becomes subject to question.

STATIONARY PLASMA MODEL FOR THE FRONT HALF OF THE PROBE

In order to gain an understanding of the phenomena which occur when a probe interacts with a flowing plasma, the case of a probe in a stationary plasma was first investigated to determine if some change of parameters for the stationary case could produce effects similar to those for a flowing plasma. Since the wake in back of the probe greatly disturbs the central field in this region, only conditions in front of the probe can be simulated using a stationary model. However, since only a small fraction of the probe current is collected by the back half of the probe, this model should explain the principle features of the current characteristic. Because the field of the probe has little effect on the flowing ions at the potentials being considered, the ion density in front of the probe remains practically constant up to the probe surface, and an approximation to conditions in front of the probe can be obtained by considering the stationary case in which ion temperature is much greater than electron temperature and a significant fraction of the ions hitting the probe are reflected.

A study of the status of stationary Langmuir probe theory revealed that the more general the theory being considered, the more difficult it becomes to extract useful information. However, in the process of examining these theories this author has developed a modified theory for the cylindrical electrostatic probe in a stationary collisionless plasma which can yield detailed qualitative results for arbitrary variations of probe and plasma parameters without resorting to computer calculations. The theory is based on the work of Bernstein and Rabinowitz[4] and may be considered a simplification and extension of the work done by Laframboise[5]. A computer program has been written to obtain quantitative information from the theory. A report including this work will be published shortly.

Using the computer program mentioned above, calculations were carried out for the stationary case which simulates the probe-plasma interaction for the front half of a probe in a flowing plasma. Results for potential as a function of distance with a positive potential on the probe surface are shown in Fig. 5.

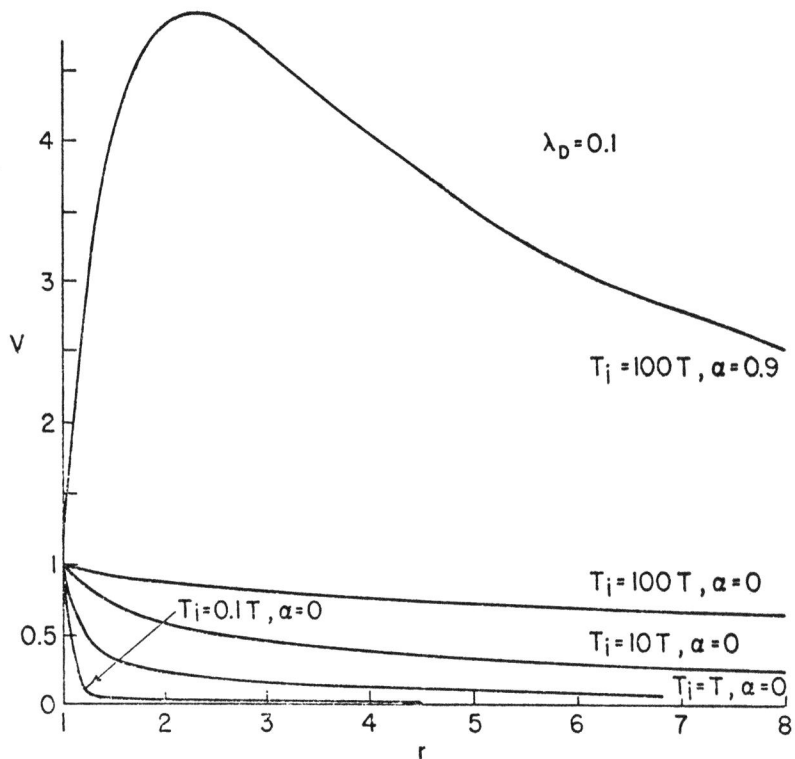

Fig. 5. Computed results for potential as a function of distance for various ratios of ion to electron temperature and ion reflectivity, α. The probe potential is normalized with respect to electron temperature.

Potential is normalized with respect to electron temperature; $V = e\phi/kT_e$ where ϕ is the probe potential and k the Boltzmann constant. As the ratio of ion temperature to electron temperature increases, the ions become less affected by the probe potential. Ion density in the vicinity of the probe increases, shielding of the electric field of the probe by the plasma sheath decreases, and the potential decreases more slowly as a function of distance from the probe surface. If the coefficient of reflectivity for the ions, α, is increased while the electrons continue to be totally absorbed, the ion density may become greater than the electron density in the immediate vicinity of the probe, and the plasma sheath will no longer shield the electric field of the probe from the plasma and may even amplify the probe potential. The plasma sheath, the region of significant difference between the charge densities of the ions and electrons will still only extend out several Debye lengths from the probe surface, but the value of the potential at the sheath edge will be radically different than in the usual stationary plasma case. Outside the sheath region the potential will fall off only slowly with distance, the rate of falloff being determined by the condition that at every point the density of ions be equal to the density of electrons.

The basic qualitative conclusions that can be drawn from stationary probe theory for the front half of a Langmuir probe in a flowing plasma are that when the probe is attracting electrons the potential can be expected to have significant values for distances large compared to the Debye length and probe radius, but when the probe is repelling electrons the potential can be expected to fall-off even more rapidly than for the case of a stationary plasma with the same density and temperature.

ELECTRON CURRENT COLLECTED BY A LANGMUIR PROBE IN A FLOWING PLASMA

Using these conclusions, the electron current characteristic for the flowing plasma case can be derived by a computational procedure similar to that used by Langmuir[6]. The assumptions made in the derivation are: (1) particles are collected only on the front half of the probe; (2) for the electron attracting region the field of the probe penetrates deeply into the plasma; and (3) for a negative voltage on the probe the distance the field penetrates into the plasma is much smaller than the probe radius. The electron current for the electron attracting region obtained from this derivation is given by

$$i = \left(\frac{2m}{\pi kT}\right)^{3/2} \int_0^\infty du\, u(u^2 + 2\frac{e}{m}V)^{1/2} e^{\frac{-m(u^2+U^2)}{2kT}} \quad (3)$$

$$\left[\sum_{n=0}^\infty \left(\frac{muU}{kT}\right)^{2n} \left(\frac{\pi}{2^{2n+1}(n!)^2} + \frac{\left(\frac{muU}{kT}\right)}{\left((2n+1)!!\right)^2}\right)\right]$$

where u is the radial component of the electron velocity, m the electron mass and $(2n+1)!! = (2n+1)(2n-1)(2n-3)\ldots$. The current is normalized with respect to the thermal flux which would be collected by the front half of the probe for zero probe voltage if the plasma were stationary. The sum is carried out for as many terms as necessary until the desired convergence is obtained. The repelled particle current is given by

$$i = \left(\frac{2m}{\pi kT}\right)^{3/2} \int_{\sqrt{\frac{-2e}{m}V}}^{\infty} du\, u^2\, e^{\frac{-m(u^2+U^2)}{2kT}} \left[\sum_{n=0}^{\infty} \left(\frac{muU}{kT}\right)^{2n}\left(\frac{\pi}{2^{2n+1}(n!)^2} + \frac{\left(\frac{muU}{kT}\right)}{\left((2n+1)!!\right)^2}\right)\right]. \quad (4)$$

The integrals for attracted and repelled electron currents have been calculated by computer. Several computed current characteristics for various ratios of flow to thermal velocity are given in the accompanying table. Current characteristics normalized to unit current at plasma potential are shown in Fig. 6.

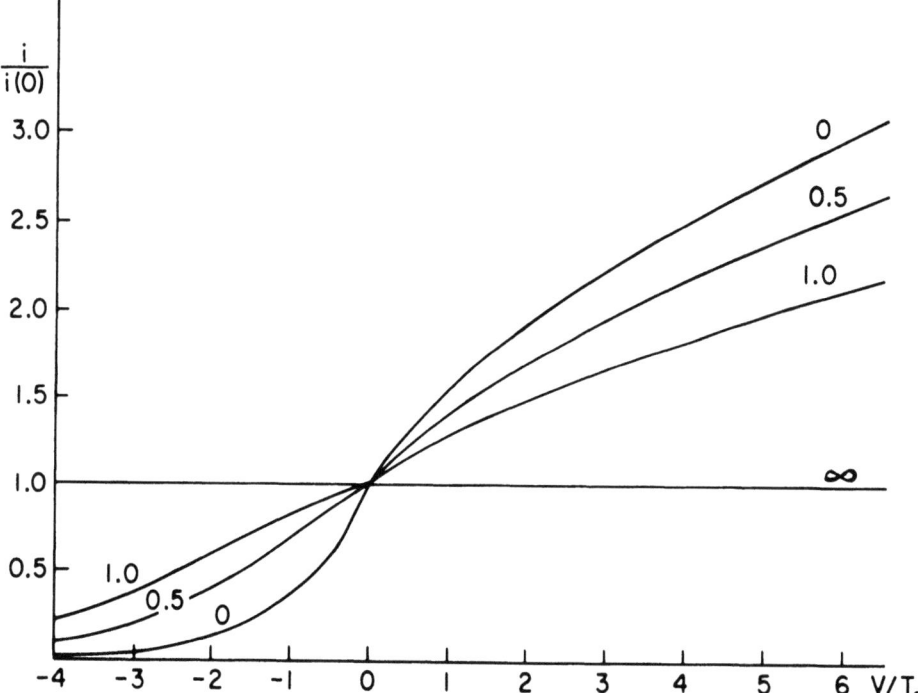

Fig. 6. Theoretical electron current characteristics for a cylindrical probe in a flowing plasma, normalized with respect to current at plasma potential. Numbers on curves indicate ratio of flow velocity to thermal velocity.

ELECTRON CURRENT CHARACTERISTICS FOR FLOWING PLASMAS
VALID FOR SUPERSONIC FLOW
UFLOW/UTHERMAL

V/TE	.05 I	.10 I	.20 I	.40 I	.60 I	.80 I	1.00 Y
-10.00	.0002	.0003	.0004	.0010	.0027	.0066	.0158
-9.00	.0005	.0007	.0010	.0025	.0059	.0138	.0310
-8.00	.0014	.0017	.0025	.0057	.0129	.0283	.0597
-7.00	.0035	.0042	.0062	.0131	.0276	.0569	.1130
-6.00	.0089	.0105	.0148	.0295	.0585	.1127	.2087
-5.00	.0217	.0254	.0349	.0655	.1212	.2173	.3743
-4.00	.0530	.0612	.0814	.1425	.2444	.4059	.6475
-3.00	.1267	.1438	.1848	.3005	.4752	.7266	1.0688
-2.00	.2921	.3257	.4030	.5817	.8746	1.2251	1.6548
-1.00	.6271	.6853	.8127	1.1124	1.4701	1.8813	2.3788
0.00	1.0730	1.1485	1.3058	1.6434	2.0038	2.3797	2.7664
1.00	1.6549	1.7552	1.9590	2.3712	2.7771	3.1681	3.5428
2.00	2.0502	2.1694	2.4093	2.8838	3.3351	3.7522	4.1344
3.00	2.3762	2.5115	2.7823	3.3114	3.8043	4.2478	4.6416
4.00	2.6611	2.8106	3.1089	3.6873	4.2185	4.6878	5.0945
5.00	2.9175	3.0800	3.4035	4.0271	4.5940	5.0880	5.5081
6.00	3.1528	3.3272	3.6739	4.3396	4.9402	5.4578	5.8915
7.00	3.3714	3.5570	3.9255	4.6307	5.2631	5.8034	6.2506
8.00	3.5765	3.7727	4.1616	4.9042	5.5670	6.1291	6.5897
9.00	3.7704	3.9765	4.3850	5.1631	5.8549	6.4379	6.9117
10.00	3.9547	4.1704	4.5974	5.4095	6.1291	6.7324	7.2192
11.00	4.1307	4.3555	4.8003	5.6450	6.3915	7.0145	7.5138
12.00	4.2995	4.5331	4.9950	5.8711	6.6434	7.2855	7.7972
13.00	4.4619	4.7039	5.1823	6.0886	6.8859	7.5467	8.0705
14.00	4.6185	4.8687	5.3630	6.2987	7.1202	7.7991	8.3348
15.00	4.7700	5.0280	5.5378	6.5019	7.3470	8.0434	8.5908
16.00	4.9168	5.1825	5.7073	6.6989	7.5669	8.2806	8.8394
17.00	5.0593	5.3325	5.8718	6.8903	7.7806	8.5111	9.0811
18.00	5.1980	5.4783	6.0318	7.0765	7.9885	8.7354	9.3166
19.00	5.3330	5.6204	6.1877	7.2578	8.1912	8.9542	9.5461
20.00	5.4646	5.7590	6.3398	7.4348	8.3889	9.1676	9.7703
21.00	5.5932	5.8943	6.4882	7.6076	8.5821	9.3762	9.9894
22.00	5.7188	6.0265	6.6334	7.7766	8.7710	9.5803	10.2037
23.00	5.8418	6.1559	6.7754	7.9420	8.9559	9.7801	10.4137
24.00	5.9622	6.2826	6.9145	8.1039	9.1370	9.9758	10.6194
25.00	6.0802	6.4069	7.0509	8.2628	9.3146	10.1678	10.8213

Theoretical curves are fitted to the data by assuming a value for the ratio of flow to thermal velocity and trying to fit the theoretical curve for this ratio to the smoothed experimental electron current characteristic. An initial value for the ratio of flow to thermal velocity can be calculated from time of flight measurements together with the value for electron temperature obtained from equation (2). A value for the electron current at plasma potential, I_o, is estimated and the potential values for the experimental curves are then recorded for a series of current values taken from the theoretical curve at intervals $\Delta V = T_e(ev)$. This is done for a number of different values of I_o. For the correct value of I_o the potential intervals in the attracted and repelled region should be equal and the size of the interval will give the electron temperature in electron volts. This method will give approximately the correct value for T_e even if the assumed ratio of flow to thermal velocity is incorrect. This new value for T_e will usually be lower than that obtained from equation (2). The new value for electron temperature can then be used to calculate a new value of flow to thermal velocity which can then be used to obtain a better fit to the data and a more accurate value for the electron temperature.

The method for obtaining the correct value of I_o can also be applied to the stationary plasma case and has the advantage that it provides a much more accurate determination of the plasma potential than the traditional method of intersecting tangents.

Theoretical current characteristics have been fitted to experimentally derived curves with excellent results. It is possible to fit the experimental curves to within the accuracy to which the curves can be determined. Values of T_e obtained using this method are consistent with the ratios of flow to thermal velocity for the theoretical current characteristics used to obtain them. An example of agreement between theory and experiment for a Cu plasma is shown in Fig. 7. Agreement exists, however, only for a limited range of potentials from several volts below plasma potential to a few volts above plasma potential. This range is sufficient for the determination of plasma properties.

Electron density can be calculated using the values for electron temperature and electron current at plasma potential obtained from the experimental characteristic curve. Making use of the fact that the theoretical current characteristic is normalized with respect to electron thermal flux to the probe at zero flow velocity, the following expression for electron density can be derived:

$$n_e = \frac{(I_o/I_{th})}{\pi R \ell e} \sqrt{\frac{2\pi m}{kT_e}}, \qquad (5)$$

where I_o is the measured electron current at plasma potential, I_{th} is the normalized theoretical value for current at plasma potential, R is the probe radius, ℓ the probe length, and kT_e the electron temperature in energy units. Electron densities obtained in this way are generally lower than the ion densities obtained from equation (1). For copper the calculated ion density may be almost twice the electron density at early times, but the difference between the ion and electron densities decreases with time. These results may be completely explained in terms of secondary emission from the probe surface due to ion bombardment. For ion energies in the kev range and average ionization levels of two or greater, emission of one or more secondary electrons per ion is to be expected.[7]

A few volts above plasma potential the electron current starts to deviate from its theoretical values. This may be explained in terms of the deep penetration of the electric field into the plasma for positive probe voltages. Although the Debye length for the plasma is much smaller than any collision length, the distance over which the potential may have values which are a significant fraction of the potential at the probe surface is no longer limited to a few Debye lengths. The increase in collisionless current to the probe due to end effects and collection of particles from the region in back of the probe as a result of the slow potential falloff is, however, insufficient to account for more than a part of the total deviation of the current from its theoretical values at higher potentials. The rest of the increase must be due to the collection

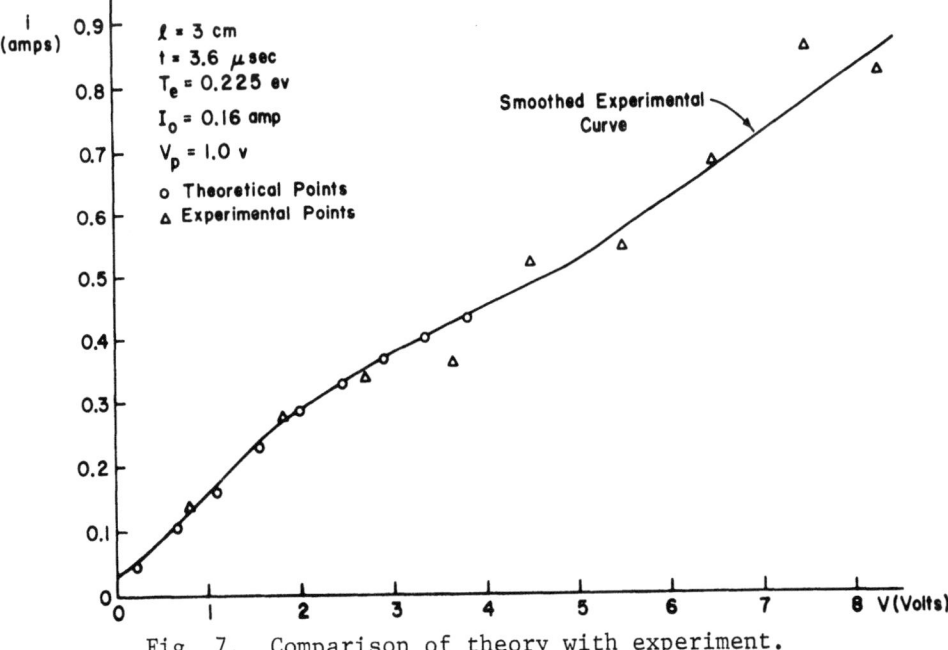

Fig. 7. Comparison of theory with experiment.

of particles which have become trapped as a result of collision processes. Yamanaka[11] has shown that when a plasma is produced from a planar solid target, a high density flow of neutral atoms appears immediately following the main body of laser-produced plasma. The density in this neutral flow is more than an order of magnitude greater than peak plasma density. Collisional ionization of this background gas may contribute to the electron current at higher potentials and later times. Collision lengths for ion-electron collisions for laser-produced plasmas may be smaller than for other types of plasmas with the same temperatures and density, because many ions are highly charged and the ion-electron collision length for large angle scattering of electrons varies inversely with the square of the ionic charge. In addition to collisions between particles, most of the electrons in the immediate vicinity of a positively charged probe which do not hit the probe will be scattered by the wake. As the probe potential increases, an increasing percentage of these electrons will be scattered into orbits which intersect the probe. None of these effects are significant when the probe is repelling electrons.

The development of the present theory provides a basis for the interpretation of Langmuir probe signals from a streaming laser-produced plasma. Proper application of the theory may change the status of the Langmuir probe from an unreliable device to an accurate instrument for measuring plasma properties in a high velocity flowing plasma.

COMPARISON OF RESULTS WITH EARLIER WORKS

Several attempts have been made to diagnose streaming laser-produced plasma by means of particle collecting devices. Often the results of these investigations have been inconclusive or in error because insufficient attention was given to proper experimental procedure, or because the experimental results were incorrectly interpreted. Hirono and Iwamoto[8] tried to interpret their electron characteristic curves using simple stationary Langmuir probe theory. Because the Debye length was much less than any collision length, they ruled out the possibility of collisional processes producing anomalously high currents in the electron attracting region. Since the currents collected by their probe started to level off above 95 volts, they concluded that plasma potential wa 95 volts above the ground potential on the target and chamber walls. Analyzing the current characteristic for potentials below 95v, they concluded that there were two species of electrons, one with an electron temperature of 12ev and a second with an electron temperature of 50-60 ev. The leveling off of electron current which was observed may have been due to depletion of plasma electrons by the probe. The conclusions with regard to electron temperature are probably also incorrect.

Namba, Kim et al.[9,10] attempted to analyze the ion signal for plasmas produced from a metallic target using a grounded circular collector located from 7 to 32mm from the target and connected to the target through a 50 Ω resistor. The signal observed on the oscilloscope was the voltage across the 50 Ω resistor. They claimed to have observed two groups of ions, a fast group and a slow group, and observed the effect of applying a magnetic field to the plasma on the shape and position of the ion signals. Because the collector was grounded, the signals observed on the scope were the result of the plasma-target interaction and were not necessarily directly related to the current drawn by the collector.

It is possible that under certain conditions more than one peak may appear in the ion signals obtained from laser-produced plasmas. Koopman[12] has shown that two peaks may develop when a laser produced plasma expands into a background plasma created by photoionization of a neutral background gas at chamber pressures of several times 10^{-3} torr. Magnetic fields also have a pronounced effect on the ion signals when a background plasma exists.[13] The evidence which has been presented that such phenomena may occur at background pressures of the order of 10^{-6} torr, however, is inconclusive.

REFERENCES

[1] W.A. Clayden, Proc. 3rd Symp. on Rarefied Gas Dynamics, J. Laurman, ed., Academic Press, Vol. 2, p. 435 (1963).
[2] F.O. Smetna, Proc. 3rd Symp. on Rarefied Gas Dynamics, J. Laurman, ed., Academic Press, Vol. 2, p. 65 (1963).
[3] A. Cheung, Univ. of Md. Tech. Note BN-696 (1971).
[4] I.B. Bernstein and I.N. Rabinowitz, Physics of Fluids 2, 112 (1959).
[5] J.G. Laframboise, Univ. of Toronto Inst. for Aerospace Studies Report No. 100 (1966).
[6] I. Langmuir and H.M. Mott-Smith, Collected Works of Irving Langmuir, Vol. 4, Macmillan (1961). Originally presented in Phys. Rev. Vol. 28, No. 4 (1926).
[7] M. Kaminsky, Atomic and Ionic Impact Phenomena on Metal Surfaces, Chapters 12 and 14, Academic Press (1965).
[8] M. Hirono and I. Iwamoto, Journal of the Radio Research Laboratories, Tokyo, Vol. 14, No. 72, p. 79 (1967).
[9] S. Namba, P.H. Kim, T. Itoh, T. Arai, and H. Schwarz, Scientific Papers of the Inst. of Phys. and Chem. Res., Vol. 60, No. 4, p. 101 (1966).
[10] H. Schwarz, Proc. of Second Workshop on Laser Interaction and Related Plasma Phenomena, (1971), H. Schwarz and H. Hora, eds., Plenum Press, p. 207.

[11] C. Yamanaka, <u>Proc. of the Second Workshop on Laser Interaction and Related Plasma Phenomena</u> (1971), H. Schwarz and H. Hora, eds., Plenum Press, p. 481.
[12] D.W. Koopman, Univ. of Md. Tech. Note BN-698 (1971).
[13] D.W. Koopman, <u>Modern Optical Methods in Gas Dynamics Research</u>, Dosanjh, ed., p. 177, Plenum Press (1971).

THE DIELECTRIC STRENGTH OF ALKALI-HALIDE CRYSTALS AT OPTICAL FREQUENCIES[+]

Eli Yablonovitch

Gordon McKay Laboratory, Harvard University

Cambridge, Massachusetts 02138

CO_2 laser induced breakdown was studied in ten of the alkali-halides. The bulk intrinsic breakdown thresholds are intimately related to the corresponding D.C. dielectric strengths. It is therefore concluded that the same mechanism is responsible in both types of experiment. A method is proposed for designing more damage-resistant materials. In addition, the question of inclusions is dealt with.

+Supported under ARPA Contract No. DAHC-15-67-C-0219

*Presented at the Second Workshop on "Laser Interaction and Related Plasma Phenomena" at Rensselaer Polytechnic Institute, Hartford Graduate Center, August 30-September 3, 1971 and published in Appl. Phys. Letters, 19, 495 (Dec. 1971).

PLASMA PRODUCTION BY CO_2 LASER[+]

Klaus P. Büchl

Max-Planck-Institut für Plasmaphysik

Euratom Association, Garching, Germany

Plasma production from solid hydrogen targets with laser radiation at o.69 μ and 1.o6 μ has been well investigated in past years and good agreement has been found between a model based on hydrodynamic calculations and experiments. The model was developed by Afanasyev, Krokhin and Sklizkov[1], Caruso and Gratton[2], Mulser[3] and others. It assumes that the absorption of laser radiation is due to inverse bremsstrahlung at the surface of the solid target. A hot plasma is produced and expands towards the laser into the vacuum. Then the radiation will be absorbed in the plasma near the critical density n_e, which is given by

$$n_c = \frac{\omega_L^2 \, m_e}{4 \pi e^2} \tag{1}$$

where ω_L is the laser frequency.

As a consequence of momentum conservation a compression or shock wave moves into the solid.

[+] Presented at the Second Workshop on "Laser Interaction and Related Plasma Phenomena" at Rensselaer Polytechnic Institute, Hartford Graduate Center, August 3o - September 3, 1971.

The properties of the plasma created and of the shock wave are determined by the power density ϕ in the focus and the wavelength λ of the laser radiation. Caruso and Gratton[2] give formulas developed with dimensional analysis for a finite target. These calculations are valid for 10^9 W/cm^2 < ϕ < 10^{14} W/cm^2. The mean velocity v_p of the total number N of particles, and the mean temperature T_p of the expanding plasma can be expressed by

$$v_p = \text{const} \cdot r^{1/9} \cdot \lambda^{2/9} \cdot \phi^{2/9} \qquad (2)$$

$$N = \text{const} \cdot r^{16/9} \cdot \lambda^{-4/9} \cdot \phi^{5/9} \cdot \Delta\tau \qquad (3)$$

$$T_p = \text{const} \cdot r^{2/9} \cdot \lambda^{4/9} \cdot \phi^{4/9} \qquad (4)$$

with

r = radius of focal spot
$\Delta\tau$ = time length of laser pulse

The velocity v_s of the shock wave in ionized hydrogen of solid density is given by

$$v_s = \text{const} \cdot r^{-1/18} \cdot \lambda^{-1/9} \cdot \phi^{7/18} \qquad (5)$$

One can take the results of experiments using, for example, a ruby laser[4] to determine the constants in the above equations. Then the values were calculated for a plasma produced by a CO_2 laser. The laser parameters used correspond to a TEA-CO_2 laser described in the following section. The theoretical values for the mean expansion velocity v_p, total number of particles N, mean temperature T_p of the plasma, and shock wave velocity v_s in the dense matter are tabulated in Table 1.

For the experimentes a transverse electrically pulsed CO_2 laser at atmospheric pressure[5] was used. The laser is described in detail in[6]. The resistively loaded pin electrodes are arranged in a helix along the laser tube (fig. 1). Thus, a nearly diffraction limited beam was produced. Its divergence was measured to be 1.1 mrad (fig. 2). The energy was determined to be 0.2 - 0.3 joule. A typical laser pulse is shown in fig. 4 (incident laser light). The sudden rise after 4.5 usec is due to signal mixing with the image converter camera. The step in the signal represents the closing of the shutter of the streak camera. The laser beam is focused onto a target of solid hydrogen in vacuum by a rock salt lens with a focal length of 5 cm. The diameter of the

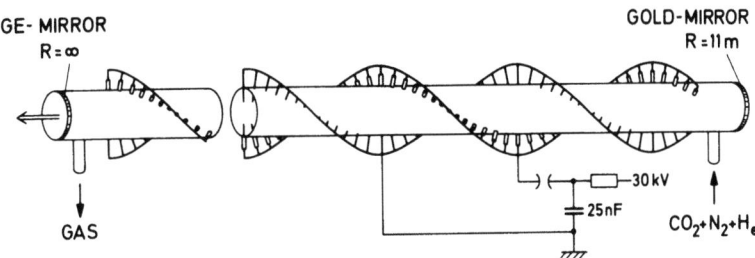

Fig. 1

Principle of TEA laser

focal spot is calculated to be 110 µ. For a mean laser power of 0.5 MW the radiation flux at the survace of the target is $5 \cdot 10^9$ W/cm^2. This is just above the lower limit of validity of the described theory.

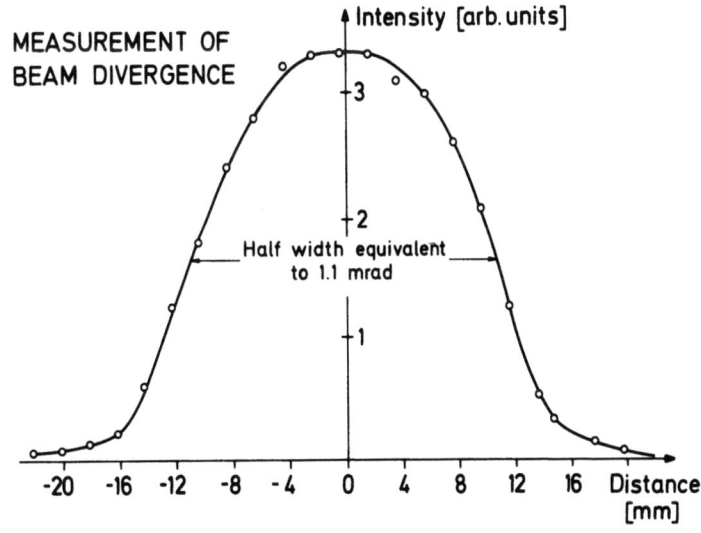

Fig. 2

Beam divergence of TEA laser

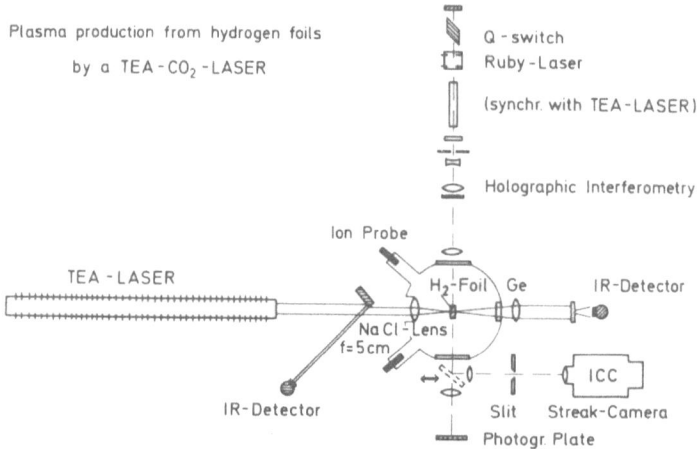

Fig. 3
Experimental set-up

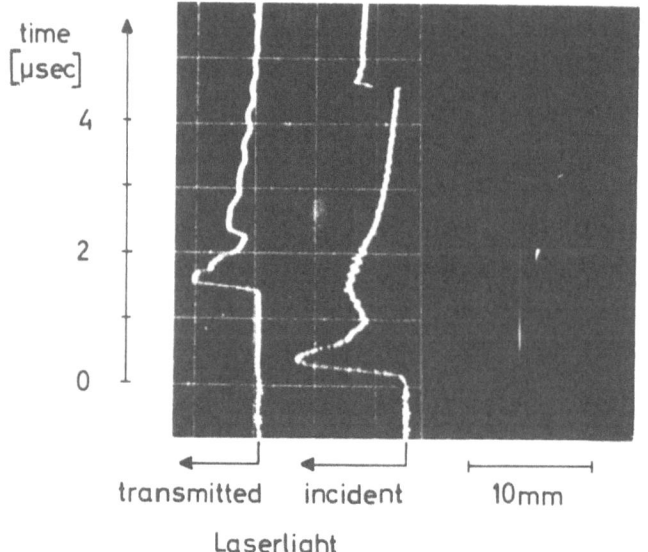

Fig. 4
Streak photography and incident and transmitted laser radiation

The experimental set-up is shown in fig. 3. Several methods of plasma diagnostics have been used for investigating the plasma. By infrared detectors transmitted and incident 1o.6 µ radiation is measured. A streak camera records the expanding plasma (fig. 4). When irradiation of the hydrogen disk (thickness about 1 mm) with the CO_2 laser is started, a plasma is created near the surface and expands towards the laser. Inside the target a shock wave moves to the rear. The dense shocked matter is pushed through the disk. When it appears at the opposite surface of the target, light is emitted from that side and at about the same time the target becomes transparent to the 1o µ radiation. A time integrated photograph of the plasma is given in fig. 5. The laser beam hits the target from the left. The bounding lines of the target are drawn in this figure. The smear slit of the streak camera picture (fig. 4) is directed parallel to the laser beam through the center of the plasma. The shock wave velocity can be estimated from the streak photograph and the time of opacity to the laser light for known thickness of the target. These measurements give a velocity of $1.1 \cdot 10^5$ cm/sec. This velocity includes the transverse expansion of the material which is pushed through the hydrogen disk.

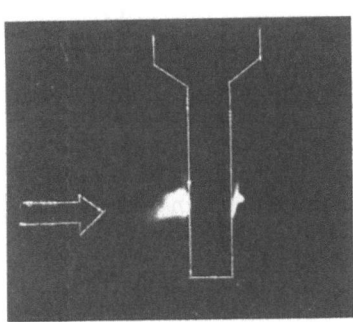

Fig. 5

Time integrated photography of plasma production from a H_2 target

The ion collecting probes measure the arrival of ions at the surface of the probes. The probes have different distances to the target. From the time of flight measurements the expansion velocity of the plasma can be determined. In eq.(2) the expansion velocity is proportional to $\phi^{2/9}$. This relation has been examined in experiments where the intensity of the laser was varied. The intensity was changed by thin absorbing plastic foils in the laser beam. The result is shown in fig. 6.

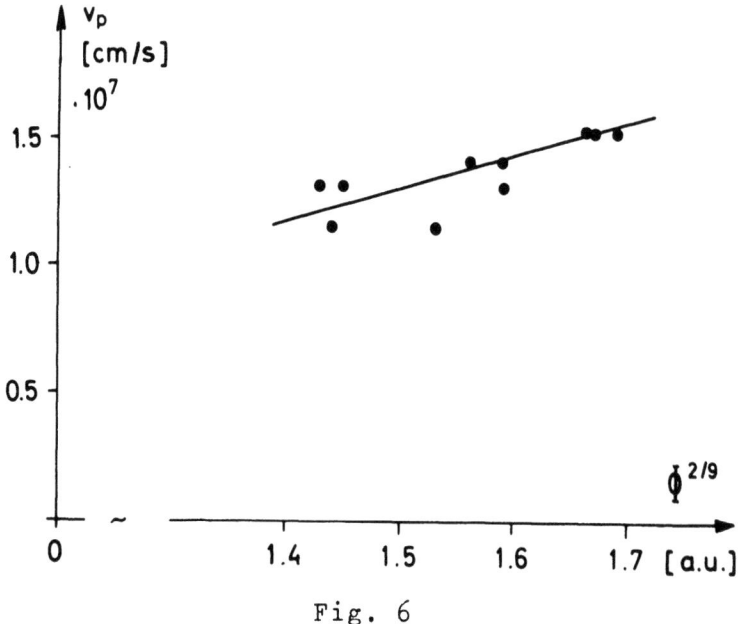

Fig. 6

Expansion velocity versus laser intensity

Another test of the validity of the above mentioned theory is the number of ions emitted from the plasma produced. The ions are emitted primarily into a cone with an apex angle of about 35°. The axis of the cone is the axis of the incident laser beam. It was shown that the shape of the emission distribution was independent of the incident laser energy. The total number of ions was calculated by integrating over the spatial distribution of the emission of ions measured by charge collecting probes. The total number of ions is a function of the power density ϕ in the focus. From eq.(3) it can be seen that the number of ions is proportional to $\phi^{5/9}$. Instead of the total number of ions we use the signal of one of the charge collecting probes, because the spatial distributions of expanding ions at different laser energies are similar to each other. Figure 7 shows the number of ions versus $\phi^{5/9}$. The experimental points are arranged along a straight line as predicted by the theory. Deviations can be observed at small incident laser energy. At low energy the assumptions of the theory do not hold. Probably the ionization energy cannot be neglected as was done in the calculatiogns of Caruso and Gratton[2].

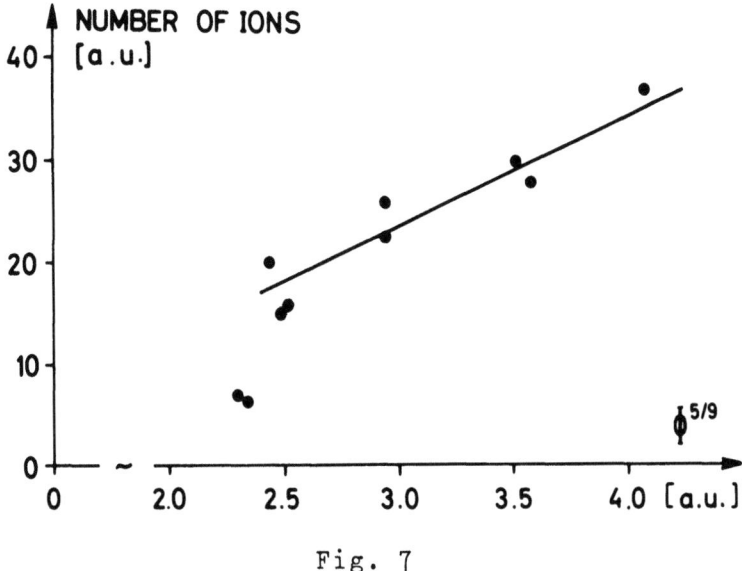

Fig. 7

Number of ions measured by one charge collecting probe versus laser intensity.

The total number of ions in the plasma was determined by three methods. Charge collecting probes are used to measure the ion current into a certain solid angle. Integrating over the half-space we calculate the total number of ions to be $0.2 \cdot 10^6$. This is checked by interferometric measurements with a ruby laser. From these interferograms the number of electrons near the target is estimated to be less than $4 \cdot 10^{16}$. However, we have so far not considered that recombination takes place during expansion. To estimate the number of recombined atoms we set up the energy balance. The absorbed laser energy is consumed in vaporizing the hydrogen in kinetic energy of the cold matter and of the plasma produced. The total energy in the cold vaporized matter can be calculated from observations by shadowgrams. The energy of the ions is also measured. Thus, we can calculate the energy of the neutrals in the plasma. If we assume that they are created by recombination and therefore their mean kinetic energy equals that of the ions, then the number of neutrals is $N_o \approx 0.8 \cdot 10^{16}$. The sum of the number of ions and atoms is in agreement with the value obtained from the hydrodynamic calculations. The result is given in Table 1.

	calculated	measured
v_p	1.0×10^7 cm/sec	1.3×10^7 cm/sec ion probes
$N = N_i + N_o$	0.74×10^{16}	$N_e < 4 \times 10^{16}$ interferometric $N_i = 0.2 \times 10^{16}$ ion probes $N_o \approx 0.8 \times 10^{16}$ energy balance
KT_p	5...15 eV	(10...25 eV·)
v_s	3.1×10^5 cm/sec	1.1×10^5 cm/sec transm. laser

Table 1

The production of a hydrogen plasma by a CO_2 laser can be described by the hydrodynamic model. In the described experiment the recombination has to be taken into account. These experiments are of great interest, because plasma production by powerful CO_2 lasers will be an important method in plasma physics in the near future.

REFERENCES

1. Afanasyev, Krokhin and Sklizkov, IEEE J. Quant. El. QE-2, 483 (1966).
2. A. Caruso and R. Gratton, LGI 68/2 (1968).
3. P. Mulser, Z. Naturforsch. 25a, 282 (1970).
4. R. Sigel, Z. Naturforsch. 25a, 488 (1970).
5. A.J. Beaulieu, Appl. Phys. Lett. 16, 504 (1970).
6. K. Büchl, Inst. f. Plasmaphysik Garching, IPP IV/16, (1971).

ACKNOWLEDGMENT

The author wishes to thank P. Mulser for presenting this paper at the Second Workshop.

INTENSE ELECTRON EMISSION FROM LASER PRODUCED PLASMAS[+]

G. Siller, K. Büchl and H. Hora[++]

Max-Planck-Institut für Plasmaphysik/Euratom
Association, Garching, Germany

ABSTRACT

Measurements are reported in which laser pulses are used to produce plasmas at tantalum targets biased as cathode against an anode by a voltage U_A = 3o to 2oo kV. The electron emission currents during 1oo nsec are a few hundred amperes up to 1 kA. The brightness of the emitted electron beams is more than a hundred times better than in beams generated from comparable field emission sources. The energy spread at 1oo kV anode voltage is less than 8oo eV. The measured high electron currents cannot be explained by classical thermionic emission because space charge effects permit currents a factor of 10^{-4} less than observed. We suggest that self-focusing filaments are created by the laser at the ends of which the nonlinear force preaccelerates the electron in the space-charge free surface region of the plasma up to some keV energy. With such initial velocities the electrons in vacuum are not restricted by space-charge effects. Thus model also explains the independence of the emission current J of the focusing, the linear continuation of the measured dependence of J on U_A and the slope of J on the laser power P by a $P^{3/4}$ law.

[+]Presented at the Second Workshop on "Laser-Interaction and Related Plasma Phenomena" at Rensselaer Polytechnic Institute, Hartford Graduate Center, August 3o - September 3, 1971.
[++] also from Rensselaer Polytechnic Institute.

INTRODUCTION

One of the early applications of a laser was to heat solids in vacuum and measure the properties of the electron emission. While a completely regular behavior of the thermionic emission was established for laser powers up to a few megawatts[1-3] with emission currents of a few milliamperes, very unexpectedly high currents of up to 100 amperes were observed with laser powers exceeding 10 MW[4,5,6]. Meanwhile emission currents of 100 amperes and more have been measured[7,8].

Electron currents of such magnitude are technologically important, e.g. for relativistic electron ring accelerators (smokatron)[9], where about 10^{13} and more electrons of a beam of 10^{-8} sec duration are accelerated beyond a few MV and formed into a ring by a nearly homogeneous magnetic field. The compressed ring consists of a relativistically stabilized cloud of electrons which can finally be accelerated. High current electron sources are also very important for the techniques of nuclear fusion with relativistic megaampere electron beams[10]. This paper reports measurements of laser produced electron currents and their properties of beam divergence (brightness) and energy width, which are much better than the properties of field emission cathodes, used in smokatrons at present.

The problems entailed in explaining the anomaleously high electron currents are discussed more qualitatively. The classical description is not sufficient by orders of magnitudes. We suggest a mode where a laser induced high initial velocity of the electrons is concluded heuristically on the basis of a nonlinear acceleration. This mechanism should also explain why the field emission cathode cannot, in principle, achieve the properties of the laser produced electron emission principally.

EXPERIMENTAL SETUP

Two sets of apparatus were used, one for voltages between the cathode and anode up to 60 kV and another for voltages up to 200 kV. Figure 1 describes the first apparatus. A Q-switched, two-stage ruby laser with an energy output up to 8 joule, a pulse half-width of 17 nsec, and a beam divergence of 5 mrad was focused on the cathode of a thick tantalum target. The focal lengths were 5 and 15 cm with focal radii of 2.5 and 7.5×10^{-2} cm

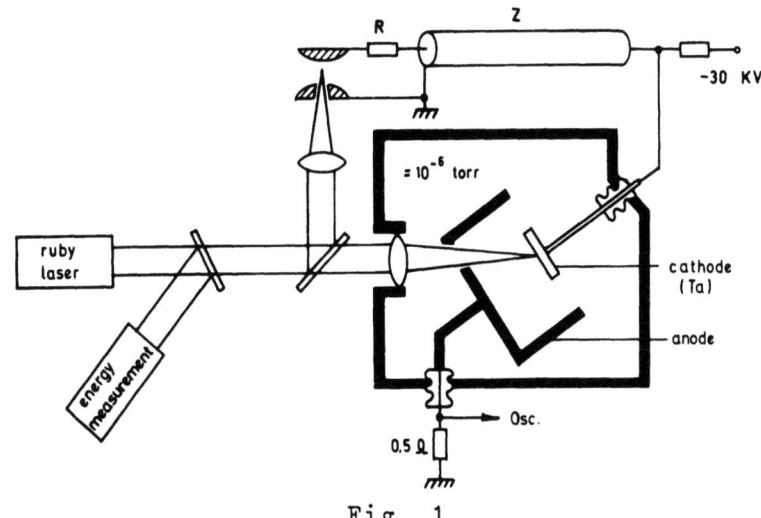

Fig. 1

Experimental setup for laser irradiation of a target biased by -30 to -50 kV against a collector anode during the time of electron emission. The switch-off after this time is performed by a laser switched line.

respectively, resulting in 10^{10} W/cm^2 intensities. The target was within a grounded vessel of 10^{-6} torr pressure. The total emission current was measured with an anode as described in Fig. 1, while the cathode was initially at a voltage of U_c = -30 to -60 kV. In order to measure at the anode only the primary electrons and not the following ions and to avoid a plasma discharge, the voltage U_c was switched off at the time of the arrival of the fastest ions at the anode, corresponding to the well known velocities[11] of 10^7 cm/sec, by means of a coaxial cable with a characteristic impedance Z of, for example, 30 Ω. This line was terminated at one end by a series connection of an ohmic load R equal to Z and a laser triggered spark gap (Fig. 1), the light of which was taken from the primary laser pulse.

The second apparatus (Fig. 2) used a Blümlein cable to keep the cathode at a voltage of -200 kV for the times before arrival of the ions at the anode. The spark gap was triggered by a separate laser, while the main laser was obliquely incident on a cathode designed with a Pierce profile. The grounded anode had holes and only the electron current passing through the central hole was measured.

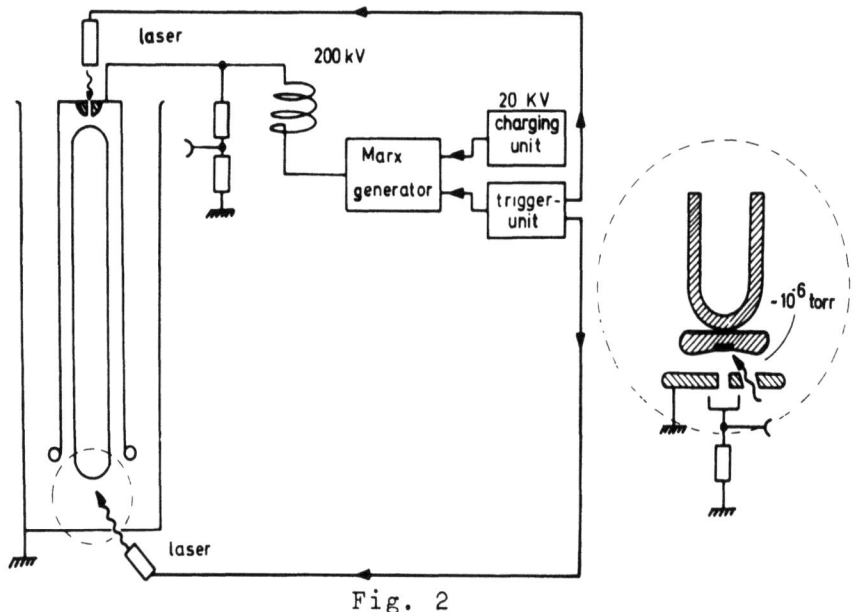

Fig. 2

Experimental setup similar to that in Fig. 1 with a Blümlein cable biasing the cathode up to -200 kV. The wavy arrow indicates the incident laser pulse.

RESULTS OF THE ELECTRON EMISSION CURRENT

Identification of the Current Carriers

It is not evident from the very beginning that the measured negative currents are due to a free electron beam from the cathode to the anode. Therefore, it is necessary to identify the current carriers. The first indirect argument that the current carriers are electrons is as follows: It was found that the rise of the current pulse coincided with the laser pulse within the resolving time of the Tektronix 519. The cathode-to-anode distance is 4 cm. For electrons of 30 keV this means a transit time of about 0.8 nsec. The expansion velocity of the plasma cloud is at least two orders of magnitude smaller than the velocity of the electrons. One method of direct identification of the current carriers is the deflection in electric fields we used.

Figure 3 shows a sketch of the experimental arrangement. The charge carriers fly through a hole (3 mm diameter) in the anode into the electric deflection field. Preference was given to identification of the deflection on a photographic film. The film was wrapped in a

20 µ aluminum foil. The range of 30 keV electrons in an aluminum foil is about 5 µ. The X-rays produced by the decelerated electrons in the foil will expose the film.

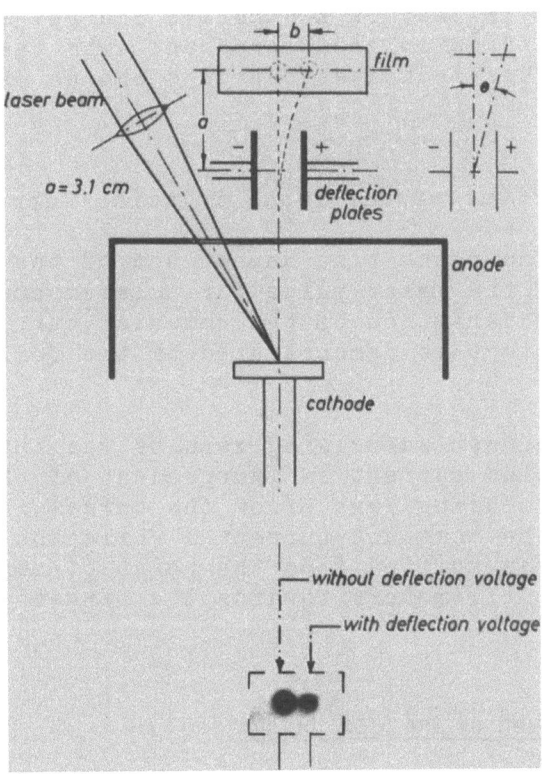

Fig. 3

Reflection of the emission current in an electric field for giving evidence of electrons. Scheme of the apparatus (upper part) and photographic film exposed by the electrons (lower part).

The deflection angle θ (see Fig. 3) is given by

$$\operatorname{tg} \theta = \frac{\ell\, E}{2\, U_B} \tag{1}$$

where ℓ is the length of the deflection plates (2.5 cm), E the field strength (6×10^3 Vcm^{-1}), and U_B the acceleration voltage of the electrons. In the case of Fig. 3 we used a coaxial line with a characteristic impedance of 5 Ω, and the amplitudes of the current pulses were of the order of 500 amperes. So the voltage drop across the inner impedance was about 2.5 kV. The accelerating voltage U_B of the electrons between the cathode and the anode was therefore about 2.75×10^4 V. This yields tg $\Theta = 0.273 = b/a$ (see Fig. 3) and $b \approx 8.5$ mm. From the exposed film in Fig. 3 (lower part) it can be seen that the deflection is that of a negative charge. The measured distance b is in good agreement with the computed value. This indicates that the charge carriers are electrons.

Properties of the Total Emission Current

Figure 4 shows the time dependence of the emission current and of the laser pulse for a laser energy of 1.3 joule. The variation of the emission current with various parameters is demonstrated by the following figures.

One of the most surprising results was that the electron emission current is independent of the distance of the focusing lens from the target. Figure 5 demonstrates the constant current J while the distance of the lens was shifted from the point of complete focusing five millimeters towards the target and ten

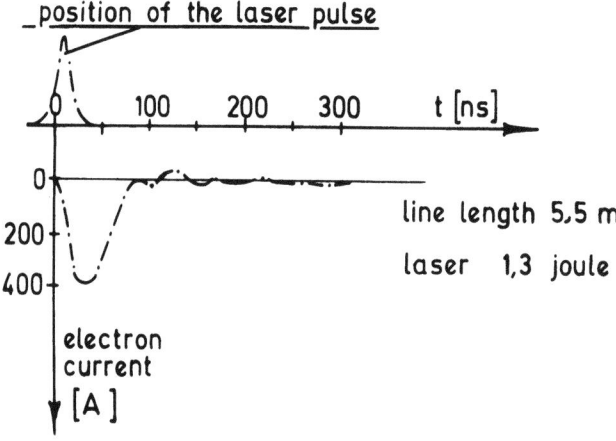

Fig. 4

Time dependence of the laser pulse and the electron emission current for a line length of 5.5 m and a laser pulse of 1.3 joule. The cathode voltage U_c was -60 kV.

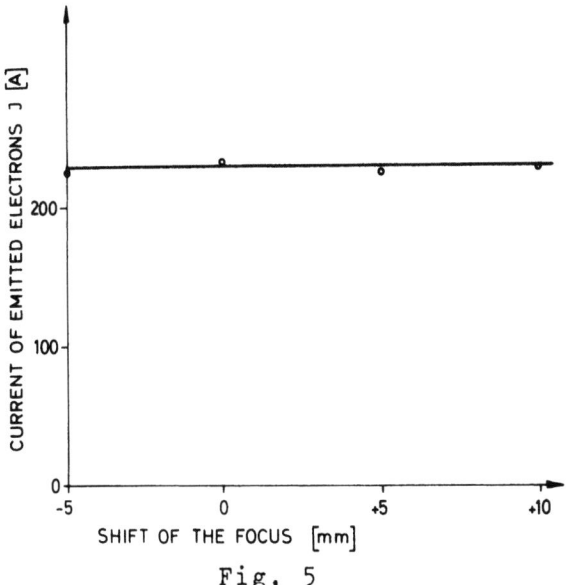

Fig. 5

Variation of the focus distance for a 15 cm lens shows no variation of the emission current.

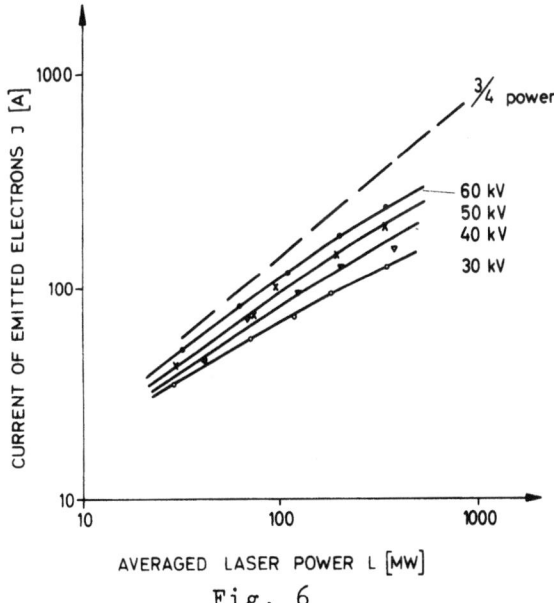

Fig. 6

The current of the emitted electrons as a function of the laser power for various voltages (30 to 60 kV) between anode and cathode.

millimeters in the reverse distance, where a lens of 15 cm focal length was used. The laser pulse length was 13 nsec and the power was 350 MW $\pm$ .15 % for these measurements.

Figure 6 demonstrates the measurements of the emission current J at varying laser power P when the cathode voltage U_c had values between 30 and 60 kV. A theoretical line of the power $\alpha = 3/4$

$$J \sim P^\alpha \qquad (2)$$

was drawn for the following discussion.

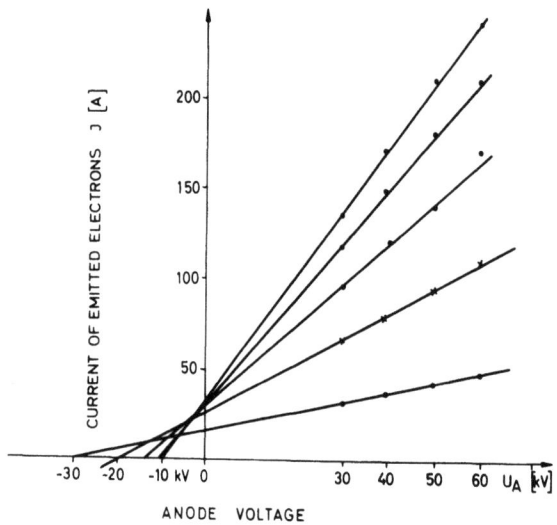

Fig. 7

Reversed diagram of Fig. 6 where the interpolated currents are used for laser powers of 400, 300, 200, 100, and 30 MW for lines from up to down, indicating a nearly linear increase of the emission current on the anode voltage.

Figure 7 is a reversed diagram of Fig. 6, where the dependence of J on $U_A = -U_c$ is demonstrated with the laser power as a parameter.

Emittance Measurements

The most interesting quantity of the electron source is the current that can be obtained in a certain emittance. The experimental setup used can be seen in Fig. 8.

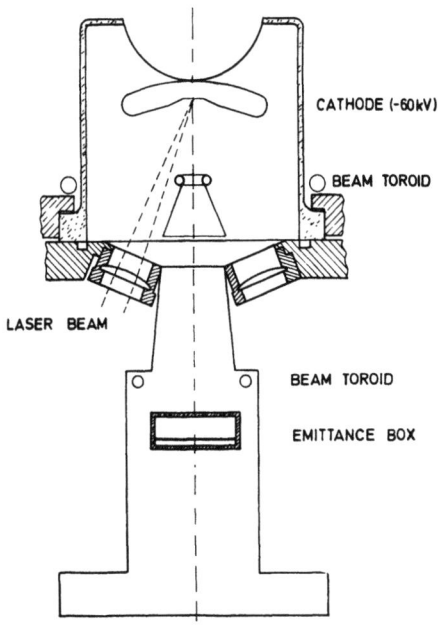

The tantalum plate of the cathode was surrounded by an electrode to shape the electric field lines (Pierce optical system). The anode was a grounded ring with an inner diameter of 1 cm. This ring was continued on one side by a cone so that only electrons passing through the ring can reach the emittance box. The cathode-to-ring-distance was 4 cm. The cathode was charged to -60 kV (see Fig. 1). The total current emitted by the cathode was measured by the upper beam toroid (Rogowski coil) and the current to the emittance box was measured by the lower one.

Fig. 8

Experimental setup for measuring the emittance

We determined the emittance in the central part of the electron beam. The emittance ε is defined as an integral value of partial divergence of the electron beam (measured by diaphragms) over the radial coordinate of the cylindrical electron beam. This can be explained by the product of the averaged beam divergence of all parts of the electron

Fig. 9

Array of the holes in the emittance box (upper part) and photograph of the irradiated film (lower part).

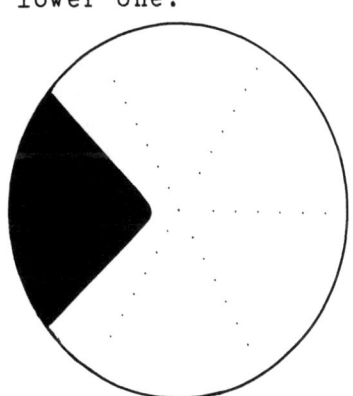

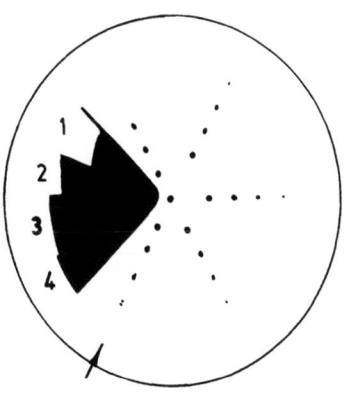

beam times the cross section of the beam. A good electron source should have a small value ε, i.e. a small beam divergence at a small beam diameter. The divergences were measured with an emittance box consisting of a 1.5 mm thick metal plate with an array of apertures 0.3 mm in diameter. The array of the holes in the box is shown in Fig. 9 (upper part). A film (Agfa Gevaert N 33) wrapped in 10 μ aluminum foil was mounted at a distance of 2 cm behind the plate. The range of 60 kV electrons in aluminum is 20 μ. A photograph of the irradiated film can be seen in Fig. 9 (lower part). The distance of the holes in the emittance box must be so large that the density curves of the holes on the film do not overlap. The black sectors in Fig. 9 are for calibrating the density of the film. In this area there is a piece cut out of the metal plate of the emittance box and the electrons hit the wrapped film directly. The film is exposed to four electron pulses. In Fig. 9 (lower part) the black sector is divided into four subsectors. The subsector 1 is hit by one electron pulse, the next by two, the next by three and the darkest one by four equal electron pulses. The difference in the density of the four subsectors cannot be seen very well in this reproduction, but is clear on the exposed film. The photometric curves along these four subsectors in the radial direction are evaluated and calibrated, as described in detail in a separate report[12], where the emittance ε is evaluated by integration over the measured two-dimensional phase-space diagram (Fig. 10).

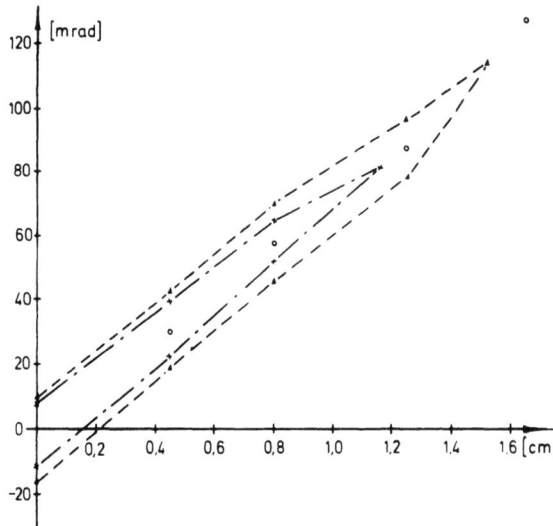

Fig. 10

Measurement of the angle (mrad) at which the electron density of a particular beam passing through the holes of 0.3 mm at a distance from the center of the beam (cm) has decreased to 1/2 (×) or 1/4 (▲) of the central intensity. The half-value of emittance ε is given by the area surrounded by curves.

The total current emitted by the tantalum plate was of the order of several hundred amperes. The current within the area where the emittance was measured was J = 17 A. We measured a current of J = 9.12 A in an emittance of ε = 31.4 mrad.cm or a current of J = 12.3 A in an emittance of ε = 63.6 mrad.cm. With these values of the emittance and the current one can compute the normalized brightness. This quantity is given by the formula

$$B = \frac{J}{\varepsilon^2 \, eU \, (1+eU/2m_o c^2)}$$

where eU is the kinetic energy and $m_o c^2$ the rest energy of the electrons. With the evaluated values one gets

$$B_{1/2} \approx 1.5 \; 10^{-1} \; \frac{A}{sterad.cm^2 \, eV}$$

and

$$B_{1/4} \approx 4.8 \; 10^{-2} \; \frac{A}{sterad.cm^2 \, eV}$$

We can compare these values with the electron emission system, where a brightness[8] of the order of 10^{-3} A/sterad cm^2 eV is measured, and with hot cathodes (Ardenne), where a brightness of 10^{-4} A/sterad.cm^2.eV is measured.

In our opinion the values evaluated for the emittance are only an upper limit and the real values are smaller. Therefore, the values of the real brightness are larger than those given here, because of the photometric method and with respect to our time integrated measurement[12].

Measurements of the Energy Spread

The next quantity in which we were interested was the energy spread of the electrons in the pulse. The energy of the electrons was measured by deflecting them in a static magnetic field. The chosen radius of curvature of the electrons in the magnetic field was about 10 cm. This corresponds to a field strength of about 110 gauss for 100 keV electrons. The measuring device for the electrons was an arrangement of 14 collectors, each in the form of a strip with a width of 1 mm, a height of 14 mm, and a distance of 1 mm from the next strip. A change of 1 % in the energy of the particles resulted in a deflection of about 1 mm at the place where the 14 collectors were mounted. For these measurements the voltage U_c at the cathode was of the order of -100 kV. Evaluating the time dependence of the currents from the strips we immediately found[12] an energy spread of less than 800 V.

THE DIFFICULTIES OF A CLASSICAL INTERPRETATION

Two mechanisms have to be considered for a theoretical interpretation of the electron emission current on the basis of classical theories: the thermionic electron emission laws and space charge effects. We present here a generalized equation of electron emission from a body (plasma) of temperature T including the Richardson-Dushman equation and the revised Richardson equation of the photoelectric emission[13] which becomes effective for plasmas at high temperature[14] for thermodynamic equilibrium[15] with respect to the blackbody radiation.

In the case of no degeneration
$$n_e h^3 \ll (2m_e kT)^{2/3} \quad , \qquad (3)$$
the emission current density is

$$j = \begin{cases} A_{1/2} \cdot T^{1/2} \cdot \exp(-h\nu_0/kT) & \text{if } (kT)^{5/2} < \frac{c^2 h^3}{2\sqrt{2\pi m_e}} n_e & (4a) \\ M \cdot T^3 \cdot \exp(-h\nu_0/kT) & \text{if } (kT)^{5/2} > \frac{c^2 h^3}{2\sqrt{2\pi m_e}} n_e & (4b) \end{cases}$$

where[16]
$$A_{1/2} = n_e e \sqrt{k/2\pi m_e} \qquad (4c)$$
and the photoelectric efficiency constant[14]
$$M = \frac{2}{c^2}\left(\frac{k}{h}\right)^3 . \qquad (4d)$$

h is Planck's constant, k is Boltzmann's constant, c is the velocity of light, e is the charge, m_e the mass, and n_e the density of electrons, T is the temperature and $h\nu_0$ is the work function for electrons. In the case of degeneration
$$n_e h^3 \gg (2m_e kT)^{2/3}$$
the emission current density is

$$j = \begin{cases} A_2 \cdot T^2 \cdot \exp(-h\nu_0/kT) & \text{if } kT < 2\pi m_e c^2 & (4e) \\ M \cdot T^3 \cdot \exp(-h\nu_0/kT) & \text{if } kT > 2\pi m_e c^2 & (4f) \end{cases}$$

where
$$A_2 = 4\pi mk^2 e/h^3 \tag{4g}$$
is Sommerfeld's "universal constant".

The work function $h\nu_o$ is a well defined value for the electron emission from solids or liquid metals. The problems of semiconductors and insulators can be neglected in this discussion because of the considered high temperatures. In the case of a plasma located in vacuum and subjected to inertial or magnetic or radiation pressure confinement[17], the Debye sheath at the surface produces a work function $h\nu_o$ which has the value kT. Therefore, when the electrons of the Debye sheath are separated, a thermionic emission of the maximum value of equations (4) with an exponential factor exp(-1) can be expected.

The current densities (4) are the saturation values if no space charge processes reduce the emission. These saturation currents can indeed account for the measured high electron emission currents from laser produced plasmas even at very unfavorable conditions. If a plasma surface of an electron density of $n_e = 10^{19}$ cm^{-3} is assumed, and if the electron temperature T is 10^4 °K - these values being lower bounds of those of plasmas produced by lasers from solids -, the electron emission current from a surface of 10^{-3} cm^2 cross section is found from Eq.(4a) to be
$$J = 9.2 \times 10^3 \text{ amp} \tag{5}$$
where $h\nu_o = kT$ was used.

Such high saturation currents are usually prevented by the space charge laws[18]. The electron emission current J_s from a surface F(cm^2) of a parallel plate cathode to an anode at a distance d_1(cm) biased by a voltage U_L (volts) is
$$J_s = 2.33 \times 10^{-6} \frac{F}{d_1^2} U_L^{2/3} \text{ amp} \tag{6}$$
which gives for the case F = 10^{-3} cm^2; d_1 = 4 cm; U_2 = 60 kV
$$J_s = 2.14 \times 10^{-3} \text{ amp} \tag{7}$$

A case more comparable with our experiment is a cylindrical geometry with a cathode wire of length l, radius r_1 of a temperature T within a coaxial anode of radius r_2. The maximum current J_M in the case of minimumfree potential is for r_2 = 6 cm, r_1 = 2 x 10^{-2} cm, l = 4x10^{-2}cm, and the extreme temperature $T = 10^7$ °K

$$J_M = 2.39 \times 10^{-3} \text{ amp} \qquad (8)$$

which is much too small to explain the measured currents of kA.

The electron emission currents at the moderate laser intensities of Ready[3] of about mA are somehow within the values given by space charge effects and are describable in terms of the usual classical theory of electron emission. In the case of the experiment of Siller et al.[7], the measured 10^3 amp was five magnitudes as large as the space charge limitation (eq.(7)), though the pure thermionic emission from plasmas (eq.(5)) would permit the measured values. Another necessary condition for the high electron currents is a sufficient electric conductivity of the plasma or of the metal below the emitting surface. The aim is to reach a current density of 10^6 amp/cm^2. With an electric conductivity of almost 10^5 Ω^{-1}cm^{-1} for tantalum, we find a necessary field strength of $E = 10$ V/cm, which is low enough - also if it were increased by orders of magnitudes owing to the heating of the material - to explain the observed[5-7] currents up to 1 kA at the applied voltages.

NONLINEAR THEORY

The impossibility of explaining the high emission currents because of space charge effects is a serious difficulty. Other results for the space charge properties can only be expected by postulating boundary conditions for the electrons other than those used[18], e.g. by assuming a certain initial energy of the electrons of some keV which they must have received within the space charge free interior of the laser produced plasma. The fact that such accelerated electrons enter the vacuum within the cross section of the laser irradiated spot of the cathode can indeed result in other space charge limited currents, as could be shown from the change of the boundary conditions of the differential equations involved.

One mechanism of the preacceleration of the electrons within the space charge free interior of the inhomogeneous surface region of the laser created plasma is well known from the nonlinear force of the collisionless interaction of the laser radiation with the plasma due to the spatial change of the complex refractive index η [20]. For a reflectionless penetration of the light into the plasma with negligible collision induced absorption, the nonlinear force density in the plasma for perpendicular incidence of the light along a coordinate x

$$f_{NL} = - \frac{E_v^2}{8\pi} \frac{\partial}{\partial x} \left(\frac{1}{|\eta|} + |\eta| \right) \qquad (9)$$

where E_v is the amplitude of the electrical vector of the light in vacuum. It was proved[21] that this nonlinear force exceeds the gasdynamic forces, if for ruby laser radiation the intensity I exceeds a value I^*

$$I > I^* = 2.08 \times 10^{14} \, T^{1/4} \, W/cm^2 \quad (T > 10^2 eV) \qquad (10)$$

or if for lower intensities[22] T exceeds 10^4 eV. These intensities do not occur in our experiments, where I was near 10^{10} W/cm^2. However, it is well known from the application of the nonlinear theory[20] to self focusing[21] in plasma that laser powers P exceeding one megawatt can create filaments the cross section F_{SF} of which is simply given by

$$F_{SF} = \frac{P}{I^*} \qquad (11)$$

because then the gasdynamic pressure of the filaments is equal to the thermokinetic pressure. The diameter measured of these filaments on gas breakdown[23] agree very well with the calculated values of a few μm. At the beginning of the filaments at the plasma surface the nonlinear acceleration can be effective because the high intensity I reaches I^* there.

This model gives the following explanation of the measurements:

(1) The high electron emission currents are explainable, in principle, in terms of the nonlinear space-charge free preacceleration in the plasma surface to avoid the well known space-charge limitations.

(2) The linear continuation of the measurement of Fig. 7 to vanishing emission currents J indicate an initial energy of the electrons of a few keV. If a J were to be measured between 0 and +30 volts of the anode voltage $U_A = -U_c$ to show a deviation from the linear relation, this may not upset our conclusion because at low values of U_A one can expect a mixing up of several mechanisms which overlap the very linear behavior measured for U_A exceeding 30 kV.

(3) The fact that the emission current J is independent of the laser focusing (Fig. 5) indicates once more the self-focusing mechanism, because the diameters of the resulting filaments are independent of the incident laser intensity.

(4) The nonlinear theory also provides a fair explanation of the slope of the curves of Fig. 6. It can be assumed that the emission current J is proportional to the cross section F_{SF} of the self-focusing areas.

$$J \sim F = \frac{P}{I^*} \qquad (12)$$

If the temperature T in the filaments is assumed to be proportional to the laser power P^α by an exponent α, we find from eq.(10) and (12)

$$J \sim P^{1-\frac{\alpha}{4}} \qquad (13)$$

The slope drawn in Fig. 5 is a $P^{3/4}$-line according to $\alpha=1$, as can be expected from heating by thermokinetic processes.

The authors gratefully acknowledge the encouragement and stimulating remarks of Prof. A. Schlüter on the work presented.

REFERENCES

1. D. Lichtman and J.F. Ready, Phys.Rev.Lett. 10, 342 (1963).
2. C.M. Verber and A.M. Adelman, Battelle Techn. Rev. 14, (7), 3 (1965)
3. J.F. Ready, J. Appl. Phys. 36, 462 (1965); Phys. Rev. 137A, 620 (1965); A.J. Alcock, H. Motz and D. Walsh, Quantum Electronics, 3rd Int. Congr. Paris 1963 (P. Grivet and N. Bloembergen Eds.) Dunod Paris 1964, Vol. II, p. 1687.
4. R.E. Honig, Appl. Phys. Lett. 3, 8 (1963).
5. R.E. Honig, Laser Interaction and Related Plasma Phenomena, (H. Schwarz and H. Hora Eds.) Plenum Press New York 1971, p. 85.
6. G. Bourrabier, T. Consoli and L. Slama, Phys. Lett. 23, 236 (1966).
7. G. Siller, K. Büchl and E. Buchelt, Max-Planck-Institut für Plasmaphysik, Rpt. o12, 3/1oo (1969).
8. C. Andelfinger, K. Büchl, E. Buchelt, W. Ott, G. Siller, Proceedings Electron, Ion and Laser Beam

Technology, 4th Int. Conf. Los Angeles, May 1970 (R. Bakish, Ed.) The Electrochem. Soc., New York 1970, p. 3.
9. V. I. Veksler, G.I. Budker and Ya.B. Fainberg, CERN Symp. on Accelerators, 1956; see Symp. Electron Ring Accelerators, Febr. 1968, UCLR-18103.
10. F. Winterberg, Phys. Rev. $\underline{174}$, 212 (1968); M.V. Babykin, E.K. Zavoiskii, A.A. Ivanov, L.I. Rudakov, Proc. 4th Int. Conf. Controlled Thermonuclear Fusion, Madison, June 1971, IAEA, Vienna 1971.
11. W.I. Linlor, Appl. Phys. Lett. $\underline{3}$, 210 (1963); Energetic Ions Produced by Laser Pulses, Laser Interaction and Related Plasma Phenomena, (H. Schwarz and H. Hora Eds.) Plenum 1971, p. 173.
12. G. Siller, E. Buchelt and H.B. Schilling, Max-Planck-Institut für Plasmaphysik, Report 0/7 (1971).
13. P. Görlich, H. Hora and W. Macke, Exp. Tech. Phys. $\underline{5}$, 217 (1957); Jenaer Jahrb. 1957, p. 91; P. Görlich, and H. Hora, Optik $\underline{15}$, 116 (1958).
14. H. Hora and H. Müller, Z. Physik $\underline{164}$, 359 (1964).
15. W. Klose (private communication) see ref. 14.
16. A. Sommerfeld and H. Bethe, Handbuch der Physik, (A. Scheel Ed.) Springer Berlin 1953, Vol. 24/2; C. Herring and M.H. Nicols, Rev. Mod. Phys. $\underline{21}$, 266 (1949).
17. H. Hora, Application of Laser Produced Plasmas for Controlled Thermonuclear Fusion, in Laser Interaction and Related Plasma Phenomena (H. Schwarz and H. Hora Eds.) Plenum New York 1971, p. 437.
18. H. Rothe and W. Kleen, Grundlagen und Kennlinien der Elektronenröhren, Akad. Verlagsges. Leipzig 1953, p. 21; Henry F. Ivey, Advances in Electronics and Electron Physics (L. Marton, Ed.) Acad. Press New York 1954, Vol. 6, p. 137.
19. H.G. Möller and F. Detlefs, Jb. drahtl. Telegraphie und Telephonie $\underline{27}$, 74 (1926).
20. H. Hora, D. Pfirsch and A. Schlüter, Z. Naturforschg. $\underline{22a}$, 278 (1967); A. Schlüter, Plasma Physics $\underline{10}$, 471 (1968); H. Hora, Phys. Fluids $\underline{12}$, 182 (1969).
21. H. Hora, Nonlinear Effect of Expansion of Laser Produced Plasmas, Laser Interaction and Related Plasma Phenomena, (H. Schwarz and H. Hora, Eds.) Plenum 1971, p. 383; Opto-Electronics $\underline{2}$, 201 (1970).
22. L.C. Steinhauer and H.G. Ahlstrom, Phys. Fluids $\underline{13}$, 1103 (1970).
23. V.V. Korobkin and A.J. Alcock, Phys. Rev. Lett. $\underline{21}$, 1433 (1968).

SUMMARY OF DISCUSSION

(III. Plasma Diagnostics and
Special Interaction Processes)

B. Kronast pointed out that for the diagnosis of laser plasma, the classical electron scattering process of radiation from plasmas could be combined with the scattering experiments of H. Schwarz, whereby electrons are scattered by a standing light wave. If for very high density plasmas the classical laser scattering diagnosis does not work out, then it may be possible that the recently discovered quantum mechanical modulation[1] of electrons due to photons or to plasmons[2] could be applied.

The interpretation of probe signals from time-of-flight measurements of the laser produced plasma expanding into vacuum is still in some cases ambiguous. Segall criticized the interpretation of signals observed from time-of-flight measurements as reported in the paper of H. Schwarz. However, Hirono and Iwamoto[3] have confirmed these measurements by applying other methods, for example microwave absorption techniques, as pointed out by C. Yamanaka. Segall pointed out that in his paper that with his treatment Langmuir probes could be better applied to such problems.

While discussing the high electron emission current from laser produced plasmas due to space charges, Yonas commented that from cold cathodes very high electron densities of equal magnitude could be produced.

References
1. H. Schwarz, Bull. Am. Phys. Soc. 13, 897 (1968); Trans. N.Y. Acad. Sci. 33, 150 (1971); Proceedings Int. Conf. Light Scattering in Solids, Paris, July 1971 (M. Balkanski, ed. Flamarion, Paris, 1971), pp. 123-127; H. Schwarz and H. Hora, Appl. Phys. Letters 15, 349 (1969).
2. H. Hora, Phys. Stat. Solidi 42, 131 (1970); Proc. Int. Conf. Light Scattering in Solids, Paris, July 1971 (M. Balkanski, ed., Flamarion, Paris 1971), pp. 128-132.
3. M. Hirono and I. Iwamoto, Japan J. Appl. Phys. 6, 1006 (1967).

INTERACTIONS OF LASER-PRODUCED PLASMAS WITH BACKGROUND GASES*

John A. Stamper, Stephen O. Dean and Edgar A. McLean

Naval Research Laboratory, Washington, D. C.

INTRODUCTION

A laser-produced plasma (referred to hereafter as a laser-plasma) has such a high initial pressure that its expansion into a background gas produces a variety of interesting interactions. There has been particular interest recently in collision-free ion-ion interactions (e.g., heating and momentum transfer) between interstreaming plasmas. Such interactions, occurring between the expanding laser-plasma and a background plasma depend on the state of the initial laser-plasma so that a study of these interactions must begin with the laser-plasma. For example, the laser-plasma emits energetic photons which ionize the background gas. We found also, surprisingly, that large magnetic fields can be generated in the laser-plasma.[1] These spontaneous magnetic fields are convected outward where they affect interactions with the background plasma. The spontaneous magnetic fields are considered in some detail after discussing interactions between the laser-plasma and background. The discussion is motivated by experimental work being carried on by the authors at the Naval Research Laboratory (NRL).

Experimental

Our laser-plasma was produced by a Nd-doped glass laser system having an oscillator and two amplifiers with a rotating prism Q-

---

*Presented at the Second Workshop on "Laser Interactions and Related Plasma Phenomena", at Rensselaer Polytechnic Institute, Hartford Graduate Center, August 30 - September 3, 1971.

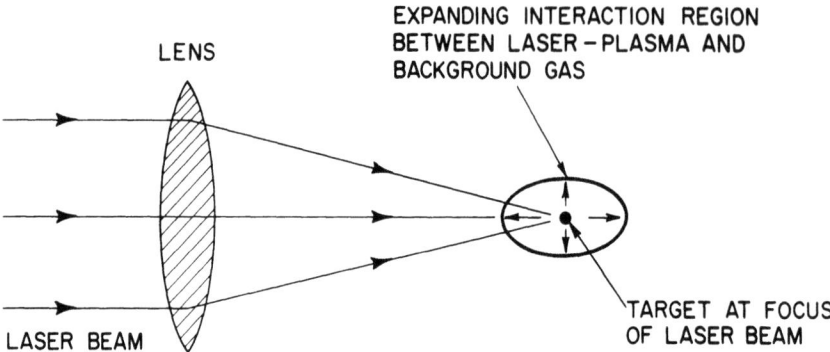

Fig. 1. Schematic of Experiment

switch. The output was 60 Joules in 30 nsec with a 32 mm diameter beam having a full angle, half-power beam divergence of 300 μrad. A focal spot size of about 250 μm was thus obtained for the 74 cm focal length lens used in the experiments. The target, in most of our studies, was the tip of a 250 μm diameter fiber of lucite ($C_5H_8O_2$). The target and a low pressure background gas were contained in a vacuum-tight glass chamber. Nitrogen or helium background gases were used, typically in the pressure range of 10 to 200 mTorr. The incident and transmitted laser pulses were monitored with fast photodiodes. Other diagnostics are mentioned in the text.

Figure 1 illustrates some major features of our experiment. The laser beam is focused, by means of a lens, onto a small solid target located in a low pressure background gas. Absorption of the laser energy in the target produces a hot laser-plasma which radiates energetic photons, ionizing the background gas. The laser-plasma then expands into and interacts with the background plasma, producing a shell-like interaction region or front.

BACKGROUND

Interactions

Interactions between the laser-plasma and background start when energetic photons from the laser-plasma ionize the background gas. Heat transfer in the hot, initial laser-plasma is due to electrons and optically dense radiation. Because of non-linear effects, the heat propagates in a sharply defined front called a heat wave.[2] Due to spherical divergence effects, these heat waves don't affect the background appreciably beyond a small radius (~ 1 mm).

Expansion of the laser-plasma begins when the ions receive energy from the electrons (heated by laser radiation). The expanding laser-plasma then heats and transfers momentum to the background plasma. These interactions can be either collisional or collision-free (collective) depending on densities and velocities. The background densities in the NRL experiment have typically been such that electron processes are collisional while the ion processes (momentum transfer, ion heating) are collision-free. Ordinary collisions have a negligible effect at high energy due to the decrease in the Coulomb cross-section.

The various collisional processes which have been discussed may have their counterparts under collision-free conditions. Plasma collective effects (wave-particle and wave-wave scattering and wave damping) can be very effective in randomizing the particle velocities. A turbulent spectrum of waves can be generated by various plasma instabilities such as streaming instabilities. When the collective effects occur on a sufficiently small spatial scale they can be included in the fluid description by defining effective or "anomalous" transport coefficients such as resistivity or viscosity.

Since a typical laser-plasma expands with a velocity greater than the local sound speed in the background plasma, a shock may develop in the background. However, this supposes (1) a (collision-free) mechanism for transferring momentum to the background and (2) there is sufficient time for the development of a shock. Our studies show a sharply defined momentum coupling interaction region between the laser-plasma and background but have not shown the existence of a shock.

Spherical Divergence Effects

The background is changed by fluxes of photons, electrons and ions from the laser-plasma. We can treat these fluxes as resulting from a spherically diverging source since the laser-plasma is initially very small. A simplified discussion of spherical divergence, valid when time integrated fluxes are meaningful, is given here. If ℓ and b refer to laser-plasma and background, respectively, the binary reaction rate per unit volume is $z = n_\ell n_b V \sigma = f/r^2 \lambda$, where n is density, V is the relative velocity, σ is the cross-section and $\lambda = V_\ell / V n_b \sigma$ is the mean free path of a laser-plasma particle having a velocity V_ℓ in the background. When the interacting background particle is an ion or atom (e.g., as in photoionization and charge exchange) then $V_\ell \cong V$ or $\lambda \cong 1/n_b\sigma$. However, if the interacting background particle is an energetic electron (e.g., as in electron impact ionization), heated by photoionization, we can have $\lambda \ll 1/n_b\sigma$. The flux per unit solid angle

$f = f_0 \exp(-r/\lambda)$ is used since, for $r \ll \lambda$, it is independent of r. Due to spherical divergence, all of the background particles out to a critical radius r^*, defined by $n_b = \int z(r^*)dt$, will interact. If, as is often the case, $r^* \ll \lambda$, we have

$$r^* = \sqrt{\sigma F_0} \qquad (1)$$

where F_0 is the time integrated flux/steradian. For example, photoionization is complete out to a radius $\sqrt{\sigma F_0}$ where σ is the photoionization cross-section and F_0 is the total number of photons per steradian.

We have experimentally determined[3] the electron density in a photoionized background of nitrogen to which a small amount of hydrogen was added for diagnostics. Measurements of the Stark profile of the Hβ line showed, in a 200 mTorr background at r = 2 cm, that the degree of ionization was about 5%. This implies (assuming the degree of reaction varies as r^{-2} for $r^* < r < \lambda$) that photoionization was complete out to a radius of about $\tfrac{1}{2}$ cm.

Absorption of Laser Radiation

A simple form of the absorption coefficient is given here and is used later in discussing magnetic sources. The laser radiation is absorbed primarily via inverse bremstrahlung.[4] In the following discussion, n (cm^{-3}), T (without subscripts) refer to the electrons, z is the ion charge state, n^* ($\sim 10^{21}$ cm^{-3} at 1.06 μm) is the critical density and $\bar{n} = (1 - n/n^*)^{\frac{1}{2}}$ is the plasma index of refraction in the underdense region. It has been pointed out by Hora[5] that an accurate expression for the inverse-bremstrahlung absorption ceofficient can be obtained from the 2-fluid description. The values which this expression assumes in the underdense ($n < n^*$) and critical ($n \cong n^*$) regions are

$$K = \frac{n}{n^*}\frac{\nu}{\bar{n}c} \text{ if } n \ll n^*; \quad K = \frac{\sqrt{\omega_p \nu}}{c} \text{ if } n = n^* \qquad (2)$$

where[6] ν (sec^{-1}) = 3×10^{-6} zℓnΛ n(cm^{-3})/T$^{\frac{3}{2}}$ (eV) is the electron-ion collision frequency. If the laser intensity is sufficiently high (I > 10^{14} W/cm^2) the absorption will be enhanced by collective effects.[7] However, Eq. (2) should be valid for our experiment where the intensity was 10^{12} W/cm^2.

HEATING AND MOMENTUM TRANSFER

Heating

Radiation from the laser-plasma photoionizes the background gas and produces an initial heating of the background plasma. The electrons are heated directly and the ions are heated by collisions with the electrons. However, in our experiment, the interaction region between the expanding laser-plasma and background plasma arrives at small radii ($r \leqslant 2$ cm) before the electrons and ions have equilibrated in the background. The initial ion temperature should then be very low. We have used spectroscopy to estimate the electron temperature in a photoionized background of 200 mTorr of nitrogen to which a small amount (10%) of hydrogen was added for diagnostics.[3] From the Stark profile of the Hβ line and the absolute intensities of the N_2 (3371 Å) and N_2^+ (3914 Å) lines, we estimate an electron temperature of 3.3 eV ± 1 eV at $r = 2$ cm.

Heating of both electrons and ions occurs in the interaction region between the laser-plasma and background. Our initial studies show that both the electrons and ions are heated in this region but the individual temperatures have not been measured. We give a brief discussion of heating due to elastic collisions in a fully ionized plasma. As noted earlier, the collisional processes may have their counterparts under collision-free conditions. The total heating rate per unit volume for electrons or ions is described by a heat equation.[8]

$$\dot{W}_H = \frac{kn}{\gamma-1} \frac{dT}{dt} - kT \frac{dn}{dt} = -\nabla \cdot \underline{q} - \underline{\underline{\Pi}} : \nabla \underline{V} + Q \qquad (3)$$

Heating due to collisions between particles of the same species can be represented by a divergence of heat flow. For electrons, the primary contribution is from conducted heat ($-\nabla \cdot \underline{q}$) while the ion contribution is usually due to convected heat (i.e., viscous effects: $-\underline{\underline{\Pi}} : \nabla \underline{V}$). Q represents heating due to collisions between particles of different species and charge. This includes Joule heating (ρJ^2) due to electron-ion collisions in an electromagnetic field and heat transfer between the electrons and ions when they have different temperatures.

When considering the particle heating rate $d(kT)/dt$, one must include compressional heating $(\gamma-1)(kT/n)dn/dt$ with the total heating rate. Under adiabatic conditions, the total rate vanishes so that the particle heating rate is just that due to compression. The particle heating rates are of interest since kT can be measured experimentally by light scattering and spectroscopy.

Momentum Transfer

In addition to heating, the laser-plasma can transfer momentum to the background plasma producing a mass motion of the background. Momentum transfer cross-sections are usually calculated in the center of mass (primed) frame but it is the mean free path in the laboratory frame which is of interest experimentally. We thus note the relation between the two mean free paths. The force on the moving (#1) particle $p_1 = vp_1/\lambda$ is taken to be the same in both frames and $p_1/p_1' = m_1/\mu$ where $\mu = m_1 m_2/(m_1 + m_2)$. Thus the momentum transfer mean free path is always larger in the laboratory frame.

$$\lambda = (1 + A_1/A_2) \lambda' \qquad (4)$$

where A is the ionic (or atomic) mass number.

Most of the momentum of the expanding laser-plasma is carried by the ions. In the velocity range of interest to laser-plasmas ($\sim 10^7$ cm/sec) the appropriate ion-ion momentum transfer cross-section is obtained from a screened Coulomb potential.[9] The mean free paths λ are typically 10 or more centimeters for background filling pressures less than a few hundred mTorr. We are thus in a collision-free regime with respect to momentum transfer since the total streaming distances, in the region studied, are of the order of a centimeter.

Our diagnostics have shown a momentum coupling between the laser-plasma and background. Figure 2 shows the luminous fronts recorded with an image converter camera for a lucite target with 3 different background pressures. The laser pulse is incident from the right. There was no applied magnetic field in any of the data to be discussed. The photographs are framing pictures with three 5 nsec exposure frames taken at each pressure. Time zero is when the leading edge of the laser pulse first arrives at the target. In each case, there is a fast expansion followed by a slower expansion. The fast front velocities V_o vary with background density in such a way that $V_o \rho_o^{\frac{1}{3}}$ is constant. This dependence is expected for a detonation wave driven by the laser radiation.[11] The duration (~ 100 nsec) also agrees with a phenomenon driven by the incident laser pulse. The fronts continue to slow down, with the radii at later times ($t \gtrsim 100$ nsec) having the approximate time dependence ($r \propto t^{.4}$) of a spherical blast wave. Velocities are so low (few $\times 10^6$ cm/sec) in the blast wave region that the momentum transfer there may be primarily collisional. This deceleration of the fronts, both in the detonation wave and blast wave regions, implies momentum transfer to the background.

The fronts were also studied using shadowgraphy.[10] Figure 3 is a shadowgraph made in a 200 mTorr background of nitrogen, 53 nsec

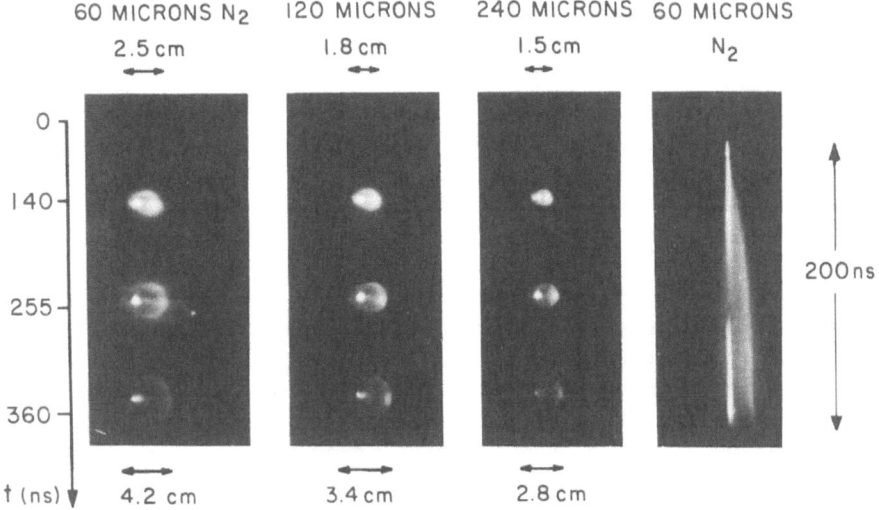

Fig. 2. Luminous Fronts in a Nitrogen Background

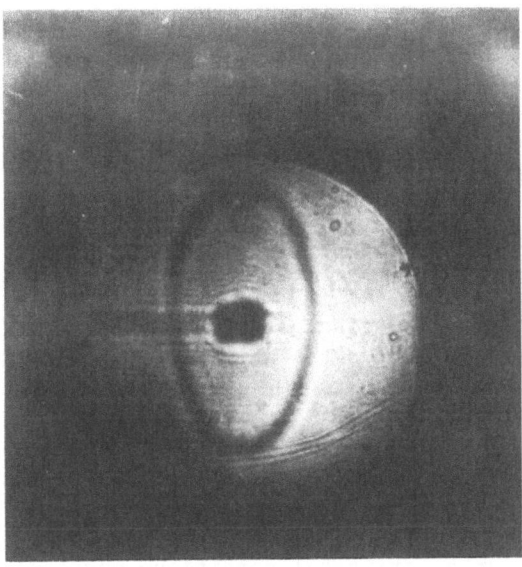

Fig. 3. Shadowgraph Showing the Interaction Region with a Nitrogen Background

after the leading edge of the laser pulse is incident on the lucite target. The bright-dark-bright pattern in the front indicates a shell of enhanced density since, in a plasma, light rays are refracted towards regions of lower density. Part of the main laser beam was split off and frequency doubled into the visible to provide a light source. Shadowgraphs taken at different times and with different background pressures, showed that the density front dynamics agreed with the luminous front dynamics (detonation wave and blast wave). Shadowgraphy further revealed that the front is a shell of enhanced density.

Spectroscopy provided direct evidence for momentum coupling to the background nitrogen ions.[10] Spectroscopic observations of Doppler-shifted nitrogen ion lines from the luminous front were made along the line of sight perpendicular to the laser beam and 8 mm to the laser side of the target. These data, taken in a shot-to-shot scan, showed weak satellite lines on the wings of strong NII lines. Due to other strong lines, we could not observe both wings of any one NII line. However, the red wing of NII 4621 and the blue wing of NII 4631 were scanned and are shown in Fig. 4. A velocity scale is given for each line with the luminous front velocities toward (V-) or away (V+) from the observer indicated by tic marks. The peak of the NII 4631 satellite line is shifted 2.65 Å toward the blue from line center. This corresponds to a velocity of 1.75×10^7 cm/sec toward the observer compared to the luminous velocity V_- of 2.3×10^7 cm/sec. Similarly, the red shift of the NII 4621 satellite line gives a velocity of 1.25×10^7 cm/sec compared to the luminous front velocity of V_+ of 1.4×10^7 cm/sec away from the observer. The Doppler data agrees, within experimental error, with the luminous front data including an asymmetry in the expansion. Spectroscopic data thus shows that the background ions are swept up in the expanding front and are given the approximate velocity of the front.

As mentioned earlier, we are in a collision-free regime with respect to momentum transfer. The required force must then come from electric fields produced in the plasma. Electric fields were, in fact, observed[10] by floating potential electric poles. The potentials observed ($\sim$ 100 Volts) were large enough to account for the kinetic energy which a nitrogen ion had at the end of the collision-free interaction.

We discuss now the theoretical explanation of this collision-free momentum transfer. Such a coupling could result from a strong ion-ion streaming instability. According to zero magnetic field linear theory[12], our experimental conditions correspond to a linearly stable regime: $C_s < V < V_t$ where C_s is the ion sound speed, V is the speed of plasma interstreaming, and V_t is electron thermal speed. However, recent theoretical work[13] has shown that this is an

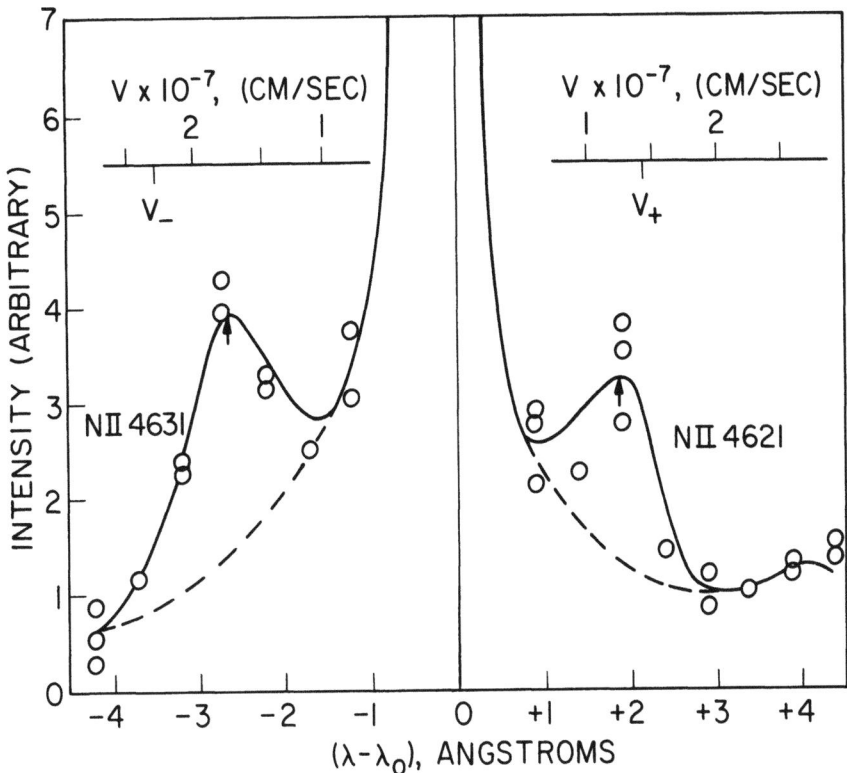

Fig. 4. Doppler-Shifted Satellite Spectral Lines on the Wings of Two Nitrogen Lines

unstable regime in the presence of a small magnetic field (electrons only tied to field lines). Magnetic fields (spontaneous) have, in fact, been found to exist in the experiment. The front thickness and its dependence on background density and (spontaneous) magnetic field are in agreement[10] with this theory.

SPONTANEOUS MAGNETIC FIELDS

Observation of the Fields

Magnetic probes were made from small (∼ 1 mm dia.) coils of wire coated with glass or epoxy for protection from the plasma. The probes were connected, via 50 ohm coaxial signal cables, to an oscilloscope which recorded the time derivative or Ḃ signal. Large signals were recorded in the absence of any applied field and after careful studies, we concluded that spontaneous or self-generated magnetic fields were responsible for the signals.[1]

The spontaneous magnetic fields, as studied by the probes, were pulses, which moved with the velocity of the front or interaction region between the laser plasma and background plasma. No signal was seen until the front reached the probe position. The pulse duration at the smallest radii (~ 5 mm) was comparable to the laser pulse width. At larger radii (r > 1 cm), the fields were primarily in an azimuthal direction about the laser beam but showed a more complicated behavior at smaller radii. Figure 5 shows the radial variation of the maximum azimuthal field observed in the mid-plane (z = 0) for a lucite fiber target in a 200 mTorr background of nitrogen. The fields at radii of a centimeter or less were in the kilogauss range and fell-off with increasing radii roughly as 1/r. In this range, the front passes the probe while the laser pulse is still incident on the target. The fields at radii greater than a centimeter decreased rapidly with increasing radius -- approximately as the inverse fourth power. The laser pulse is over and the front has slowed down at these larger radii, before the front reaches the probe. Some magnetic probe data were also taken for aluminum and silver surface targets. The fields

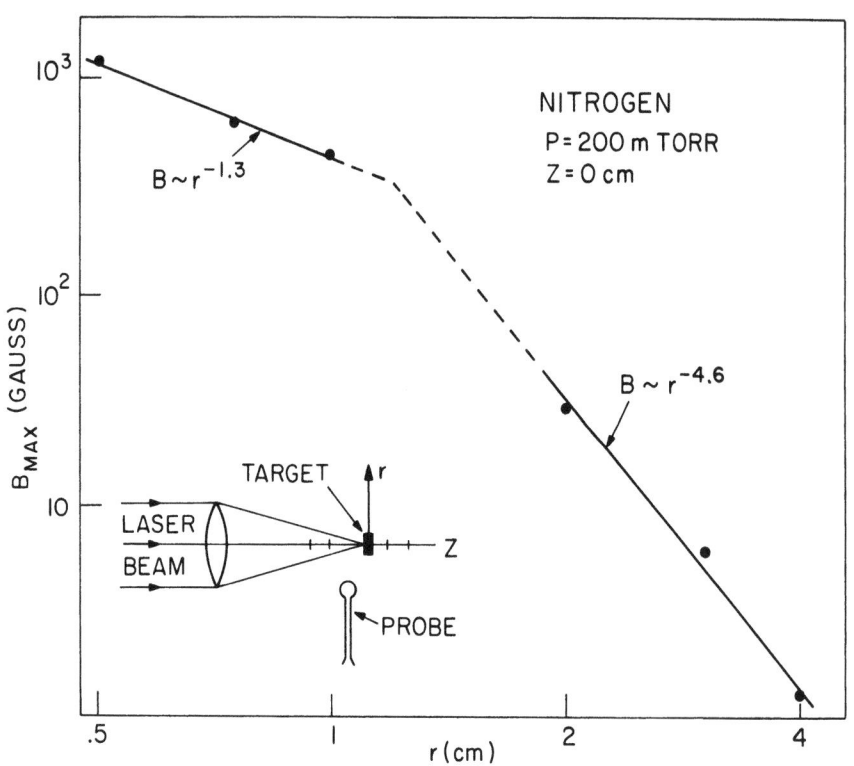

Fig. 5. Log-Log Plot Showing Radial Variation of Spontaneous Magnetic Fields

for a silver target were considerably smaller than for an aluminum target. This could be due to a larger radiative energy loss from the higher z target. The fields, in either case, were primarily in an azimuthal direction about the incident laser beam. Other diagnostics were not used in these studies with surface targets.

Theory of Field Development

Due to the approximately spherical expansion of many laser-plasmas, the generation of magnetic fields may be surprising. The only magnetic field which can be generated with perfect spherical symmetry is a radial field and this is ruled out by the absence of magnetic charge. However, the preferred direction of laser propagation provides, in itself, a deviation from spherical symmetry.

The origin of these fields is connected with phenomena occurring in the sub-millimeter size region where the laser energy is absorbed. The 100 eV electron temperature (NRL) and the large pressure and temperature gradients are equivalent to a thermoelectric "battery" or "emf" which drives currents associated with the fields. A precise description of the phenomena involved must be based on local or field quantities. An emf, in this field description, corresponds to a solenoidal electric field, i.e., one having a curl. Faraday's law says that the production ($\partial B/\partial t$) of a magnetic field is associated with an initial solenoidal electric field. We must find an expression for the electric field.

Consider the initial evolution of the laser-plasma. The density is sufficiently high that we can use the 2-fluid (electron, ion) model of a collision dominated plasma. The electron and ion fluids, together, constitute a neutral plasma. Nevertheless, we can have an electric current, i.e., a flow of one fluid with respect to the other. The equation of motion of an electron fluid element is

$$nm \frac{dV_{-e}}{dt} = -ne(E + \frac{1}{c} V_{-e} \times B) - \nabla \cdot P + R \qquad (5)$$

The force density ($nm dV_e/dt$) on an electron fluid element is due to averaged electromagnetic fields (E, B), to the pressure (P) of adjacent electrons and to collisions (R) with the ions.

$$R = ne (J/\sigma + \alpha \cdot \nabla T); \quad J = ne (V_i - V_e) \qquad (6)$$

The collisional force R contains the Joule or resistive drag term, proportional to the electric current density and a cross term, the thermoelectric force, proportional to the temperature gradient.

We will be interested in changes which are sufficiently slow that the electron inertia term ($nmdV_e/dt$) can be neglected. We thus obtain from (5) the needed expression for the electric field.

$$\underline{E} = \underline{J}/\sigma - \frac{1}{c} \underline{V}_e \times \underline{B} + (\underline{\underline{\alpha}} \cdot \nabla T - \frac{1}{ne} \nabla \cdot \underline{\underline{P}}) \qquad (7)$$

One can obtain an equation describing the development of the magnetic field by taking the curl of Equation (7) and combining the results with Maxwell's equations.

$$\frac{\partial \underline{B}}{\partial t} = \nabla \times (\underline{V}_e \times \underline{B}) + \frac{c^2}{4\pi\sigma} \nabla^2 \underline{B} + \underline{S} \qquad (8)$$

with

$$\underline{S} = -c\nabla \times (\underline{\underline{\alpha}} \cdot \nabla T - \frac{1}{ne} \nabla \cdot \underline{\underline{P}}) \qquad (9)$$

This (8) is the basic equation which will be used in the description of the generation and development of the magnetic field. It will be referred to as the field development equation. The equation shows how the magnetic field changes due to convection, diffusion or wave propagation and field generation. If only the first term were present on the right-hand side, the magnetic field would be convected or carried with the electron fluid. The second term describes diffusion of the magnetic field and represents the eventual conversion of field energy to internal energy through Joule heating. Displacement current could be ignored in the NRL results[1] and was omitted in deriving (8). In general, however, one must include displacement current and thus allow (as $\sigma \to 0$) for wave propagation by replacing ∇^2 with $\nabla^2 - c^{-2} \partial^2/\partial t^2$. The flow velocity ($V_e$) and the electrical conductivity (σ) are sufficiently large at early times in our experiment that the diffusion term can be neglected compared to the convection term. This statement can be made quantitative by defining the magnetic Reynold's number $R = VL/D$ where $D = c^2/4\pi\sigma$ is the magnetic diffusivity and V and L are average velocities and lengths characteristic of the flow. The magnetic field changes due to diffusion are small compared to those resulting from convection when $R \gg 1$ - as is true at early times.

The primary interest here is in the last term S which represents sources for the generation of a magnetic field. Solenoidal electric fields associated with either a pressure gradient ($-(1/ne)\nabla \cdot \underline{\underline{P}}$) or the thermoelectric effect ($\underline{\underline{\alpha}} \cdot \nabla T$) are responsible for these sources. There are other ways to see that these electric fields can act as a source for magnetic fields. The rate per unit volume at which particle energy is converted to electromagnetic field energy is given, in the electron fluid frame, by

$$-\underline{J}\cdot\underline{E} = \underline{J}\cdot(\underline{\underline{\alpha}}\cdot\nabla T - \frac{1}{ne}\nabla\cdot\underline{\underline{P}}) - J^2/\sigma; \quad \underline{V}_e \equiv \underline{0}. \qquad (10)$$

The thermoelectric and pressure fields can thus act as batteries to pump energy into the magnetic field while the field energy is dissipated through Joule heating. (10) can be derived using Eq. (7) for the electric field.

In reference (1), we were able to account, at early times, for many of the observed features of the spontaneous fields by assuming a simple, heuristically constructed source function, ignoring diffusion and assuming a spherically symmetric expansion of the electron fluid. The magnetic source S was assumed to exist only at the origin ($S \propto \delta(\rho)$) and to have the time dependence (pulsed) of the laser radiation. In this model, the magnetic fields are generated at the focus of the laser beam and are convected with the expanding laser-plasma (interaction region) to the observation point. The $1/r$ dependence of the convected fields is a spherical divergence effect. At later times ($t \gtrsim 100$ nsec) the rapid decay of the magnetic field ($B \sim r^{-4}$) indicates that magnetic field diffusion is important. This is expected since the magnetic Reynolds number is not so large as at early times. Thus, the development of the magnetic field (once generated) appears well understood at early times. However, the detailed mechanisms responsible for the generation of the magnetic fields are not well understood. A study of magnetic sources will lead to a better understanding of how laser radiation interacts with a plasma, and an experimental study of the resulting fields can help in the diagnosis of these interactions.

Magnetic Sources

Sources for the generation of a magnetic field can be attributed to two effects of laser radiation on the electrons: non-adiabatic effects and the development of an anisotropic velocity distribution. The laser radiation can be sufficiently intense in a focused laser beam that both of these effects must be considered. We consider first only the non-adiabatic effects and assume the pressure and thermoelectric power are scalars.

The electron heat equation (3) states that the rate per unit volume $\dot{W}_H$ at which heat is supplied to the electrons equals the rate of change of internal energy $\dot{W}_I = [nk/(\gamma-1)] \, dT/dt$ plus the rate $\dot{W}_W = -(kT/n) \, dn/dt$ at which work is done by the electrons. Adiabatic conditions exist when energy is supplied at such a slow rate ($\dot{W}_H \ll \dot{W}_I, \dot{W}_W$) that there is an approximate balance ($\dot{W}_I = \dot{W}_W$) between the internal energy and work rates. We then have

$T n^{1-\gamma}$ = const. (implying n, P are functions of T) so that $\nabla n \times \nabla T = \underline{0}$ and $\nabla P \times \nabla T = \underline{0}$. Magnetic sources vanish under these conditions.

If the laser energy is supplied to the electrons at a sufficiently high rate then magnetic sources can exist in the resulting non-adiabatic state. $\dot{W}_H/\dot{W}_I$ can serve as a measure of non-adiabatic conditions. $\dot{W}_H$ is equal to KI where K (cm^{-1}) is the absorption coefficient and I (W/cm^2) is the radiation intensity. $\dot{W}_I$ is represented as nkT/τ_h where τ_h is an average electron particle heating time. For the experimental results discussed here, the laser pulse width is large compared to the electron-ion equilibration time[14] τ_{eq}. Heating of the electrons should continue until the energy is shared with the ions. Thus τ_h should be of the order of τ_{eq}. Using Eq. (2) for K gives

$$\frac{\dot{W}_H}{\dot{W}_I} = \frac{\tau_h}{\tau_c} \frac{I/\bar{n}c}{n^* kT} \qquad (11)$$

In our experiment, $I = 10^{12}$ W/cm^2 and $kT = 10^{-17}$ J so that $(I/c)/n^* kT = .003$. Thus, if τ_h is of the order of τ_{eq}, $\dot{W}_H/\dot{W}_I \sim 1$ and the electrons are highly non-adiabatic. $\tau_c = \nu^{-1}$ is collision time.

This non-adiabatic state is a necessary condition for magnetic sources. The nature of the resulting source depends on whether or not the laser radiation is sufficiently intense to produce an anisotropy in the electron velocity distribution. Without this anisotropy, the only direction characterizing the effect of laser radiation is its direction of propagation. We then have, for a spherically symmetric expansion, azimuthal symmetry about the laser pulse. However, if anisotropy is important, then the polarization of the laser radiation also enters.

Linearly polarized electromagnetic radiation can produce an anisotropic electron velocity distribution so that both the electron pressure and thermoelectric power are anisotropic tensors. Magnetic sources will then exist even when all quantities vary exactly in a radial direction. Electrons receive directed energy in the field of the focused, linearly polarized laser beam. The electron velocities become randomized through binary collisions and collective effects. If the radiation is sufficiently intense, a steady-state can be maintained where the temperature is relatively larger along the direction of the laser electric field. Numerical studies at Princeton[7] have shown a relatively higher heating rate along the electric field. The required intensities may involve a radiation pressure I/c comparable to the electron pressure n kT. Such conditions can be produced with existing lasers but may not exist in the NRL experiments to date.

CONCLUSION

We have discussed how a laser-plasma interacts with a background gas. The enormous concentration of power in a focused laser beam presents unique opportunities for studying interactions of radiation with plasmas as well as interactions between interstreaming plasmas. A collision-free momentum coupling was described which may depend on magnetic fields generated in the initial laser-plasma. The focused laser radiation produces conditions in the laser-plasma which lead to the generation of these magnetic fields. This intimate relation between laser radiation, laser-plasma and electromagnetic fields makes the problem complicated but also opens the door for many important studies.

REFERENCES

1. J. A. Stamper, K. Papadopoulos, R. N. Sudan, S. O. Dean and E. A. McLean and J. M. Dawson, Phys. Rev. Lett. $\underline{26}$, 1012 (1971).

2. Ya. B. Zeldovich and Yu. P. Raizer, Physics of Shock Waves and High-Temperature Hydrodynamic Phenomena, Vol. II, p. 652, Academic Press, N.Y. (1967).

3. E. A. McLean, A. W. Ali, J. A. Stamper and S. O. Dean, Bull. Am. Phys. Soc. $\underline{15}$, 1411 (1970).

4. J. M. Dawson and C. R. Oberman, Phys. Fluids $\underline{5}$, 517 (1962).

5. H. Hora, Garching Report IPP-6127 (2nd Print.), Sept. 1964.

6. L. Spitzer, Jr., Physics of Fully Ionized Gases, 2nd Ed., Interscience, N.Y. (1964), p. 133.

7. P. Kaw, J. Dawson, W. Kruer, C. Oberman and E. Valeo, Princeton University PPL Report MATT-819, (1970).

8. S. I. Braginskii, Chapter in Review of Plasma Physics, Vol. I, M. A. Leontovich, Ed. (Consultants Bureau, New York, 1965), p. 211.

9. E. Everhart, R. J. Carbone and G. Stone, Phys. Rev. $\underline{98}$. 1045 (1955).

10. S. O. Dean, E. A. McLean, J. A. Stamper and H. R. Griem, Phys. Rev. Lett. $\underline{27}$, 487 (1971).

11. Yu. P. Raizer, Sov. Phys. Uspekhi $\underline{8}$, 650 (1965).

12. T. E. Stringer, Plasma Physics (Journal of Nuclear Energy, Part C) 6, 267 (1964).

13. K. Papadopoulos, R. C. Davidson, J. M. Dawson, I. Haber, D. Hammer, N. A. Krall and R. Shanny, Phys. Fluids 14, 849 (1971).

14. Reference 6, pg. 135.

MAGNETIC FIELD CONFINEMENT OF LASER IRRADIATED SOLID

PARTICLE PLASMAS*

A. F. Haught

United Aircraft Research Laboratories

East Hartford, Connecticut 06108

ABSTRACT

High energy, spherically symmetric, free plasmas are produced by electrically suspending a small, solid, lithium hydride particle in vacuum at the focus of a lens where the particle is vaporized, ionized, and the resulting plasma heated by the focused beam of a Q-spoiled laser. Experimental studies with such plasmas formed within mirror and minimum-B magnetic fields up to 8kG show that the expanding plasma can be captured by the magnetic field and the expansion kinetic energy thermalized, in agreement with a simple magnetohydrodynamic model of the plasma-magnetic field interaction. Both the magnitude and the temperature dependence of the plasma decay from a minimum-B containment field are consistent with plasma loss by Coulomb collisional scattering into the magnetic field loss cones.

---

*Presented at the Second Workshop on "Laser Interaction and Related Plasma Phenomena" at Rensselaer Polytechnic Institute, Hartford Graduate Center, August 30-September 3, 1971. See also A. F. Haught, D. H. Polk and W. J. Fader, Phys. Fluids 13, 2842 (1970)

DYNAMICS OF A RESISTIVE PLASMOID IN A MAGNETIC FIELD*

Dilip K. Bhadra

Gulf General Atomic Company

San Diego, California 92112

I. INTRODUCTION

The behavior of a plasma produced by irradiation of a solid target by a laser in the absence of a magnetic field has been discussed by Basov and Krokhin,[1] Engelhardt,[2] Dawson,[3] Hora,[4] Ascoli-Bartoli, DeMichelis, and Mazzucato,[5] and Haught and Polk.[6] The general behavior of the plasma produced by Haught and Polk appears to be describable in terms of Dawson's simple hydrodynamic theory. Such a simple hydrodynamic model has been used by Bhadra[7] to estimate the effect of a magnetic field on the expansion of a plasmoid, including the effects of finite resistivity. A moderate amount of resistivity superposes upon the collisionless solution a slow diffusion of plasma across the magnetic field. Without resistivity, the expanding plasma bounces repeatedly off the magnetic field in periodic fashion. With some resistivity, at each bounce the plasma penetrates a little deeper into the field because of the diffusion.

In a recent comprehensive article, Haught, Polk, and Fader[8] have discussed both the experimental and theoretical aspects of magnetic field confinement of laser irradiated solid particle plasmas. Using numerical integration of the magnetohydrodynamic equations for a one-dimensional spherical model of a plasma, the details of the internal motion of the plasma and the plasma-magnetic field interaction have been studied.[9] The numerical solutions yield radial distributions of the plasma temperature and density characterized by an inward-facing shock between a cool, expanding interior

---
*Presented at the Second Workshop on "Laser Interaction and Related Plasma Phenomena" at Rensselaer Polytechnic Institute, Hartford Graduate Center, August 30-September 3, 1971.

region and a dense, high temperature region adjacent to the plasma boundary. It has been found that the magnitude and effect of the plasma-magnetic field interaction depend primarily on the properties of the shocked high temperature outer region which effectively shields the interior of the plasma from the magnetic field.

In the following, we first give a very brief description of the theory and results obtained from employing a spherical model of the expansion of a laser-produced spherical plasma in order to investigate the radial motion of the plasma boundary. It is assumed that the plasma temperature and density distributions are uniform in space at any given time and a skin-depth parameter is used to account for the penetration of the field into the resistive plasma. We then give a brief theoretical description of this plasma-magnetic field interaction phenomenon with the aid of the theory of strong shock waves. We assume a point-like explosion and consider the bouncing off of the generated strong shock waves from the magnetic field. Very briefly, we next consider the effect of any rotation of the plasmoid on its dynamic behavior and suggest the possibility that a vortex-like configuration may be established through nonradial motions during the collapse phase.

II. EXPANSION IN A MAGNETIC FIELD

In the following, we write complete equations describing both the heating and the expansion phases of the plasmoid, but only selected terms would be considered during each portion of plasma development.

The basic equations are Ohm's law and the momentum and energy equations, the latter two being modified to include current driving and heating terms. Taking the magnetic field to decay exponentially into the plasma with an e-folding length λ, we find for the energy, momentum, and integrated Ohm's law equations

$$\frac{3}{2} k(N_e+N_i) \frac{dT}{dt} + \frac{3}{r} \frac{dr}{dt} kT(N_e+N_i) - W$$
$$= \frac{\eta B^2}{8\pi\lambda} \left[r^2 - r\lambda + \frac{\lambda^2}{2} (1-e^{-2r/\lambda}) \right], \quad (1)$$

$$k(N_e+N_i) \frac{T}{r} \frac{dr}{dt} - \frac{1}{2} M \frac{d}{dt} \left(\frac{dr}{dt} \right)^2 = B^2 \frac{dr}{dt} [r^2 - 2r\lambda + 2\lambda^2(1-e^{-r/\lambda})], \quad (2)$$

$$\frac{d}{dt} \lambda[r - \lambda(1-e^{-r/\lambda})] = \frac{\eta}{4\pi} \frac{r}{\lambda}, \quad (3)$$

where W is the absorbed laser power, $M = (3/5)N_i m_i$, and m_i is the mass, N_e and N_i are the total number of electrons and ions in the plasma, respectively. The resistivity given by Spitzer is

$$\eta = \frac{5.22 \times 10^6}{T^{3/2}} \ln \Lambda , \qquad (4)$$

where T is expressed in electron volts, and Λ is the ratio of the maximum to minimum impact parameter. We have assumed $T_i = T_e$ and temperature uniform throughout the plasma volume.

The right-hand sides of Eqs. (1) and (2) are ηj^2 and $j \times B$ terms integrated over the plasma volume, respectively.

To simplify the problem, we assume spherical symmetry, B^2 in our formulas being the actual field near the plasma surface averaged over angles. Also, we assume that

$$B \sim B e^{-\rho/\lambda}, \quad 0 < \rho < r .$$

We are interested in the regime of parameters such that the resistive diffusion time τ_r (at time t_1 when the plasma becomes transparent to the laser) is much longer than t_1. In that case, we may regard the flux as being approximately conserved during the time $W \neq 0$. One can then easily estimate the skin depth λ, at the time t_1 from the relation

$$r(t_1)\lambda(t_1) = \tfrac{1}{2} r_o^2, \qquad (6)$$

where r_o is the initial pellet radius, and $r(t)$ is given by Eq. (7b) later.

We shall then be interested in the behavior of the expanding plasma at times $t > t_1$, i.e., after the laser source is shut off, taking the skin depth at t_1 to be given by Eq. (6) and taking the temperatures and radii at t_1 given by (see Ref. 3):

$$kT = \frac{Wt_1}{3(N_e + N_i)} \left[\frac{2r_o^2 + 5Wt_1^3/9 m_i N_i}{r_o^2 + 10 Wt_1^3/9 m_i N_i} \right] , \qquad (7a)$$

$$r(t_1) = \left(\frac{10}{9} \frac{Wt_1^3}{N_i m_i} + r_o^2 \right)^{\frac{1}{2}} . \qquad (7b)$$

One then numerically solves Eq. (1) to (3).

Figures 1 and 2 illustrate the periodic bouncing of the plasma against the magnetic field. The oscillations have an amplitude $R \sim (NKT_o/2B^2)^{\frac{1}{2}}$ and a period $t \sim R/V_o$. Because of resistivity, the envelope of the oscillations depends on time, the inner and outer bounce radii increasing with the skin depth, λ. Note also that R and T oscillate out of phase since almost free expansion occurs except at the bounce points so that $TV^{2/3} = TR^2 =$ const over most of the cycle. Also R and λ are out of phase, but this relation involves a more complicated competition between the non-resistive phase, when flux trapped in the cloud is preserved and $R\lambda =$ constant, and the resistive phase when $\lambda \propto \eta^{\frac{1}{2}} \propto T^{-3/4}$.

It is obvious from these results that there exist certain periodic solutions of the problem. With finite resistivity, the plasma bounces back after a certain time and, it is found that the

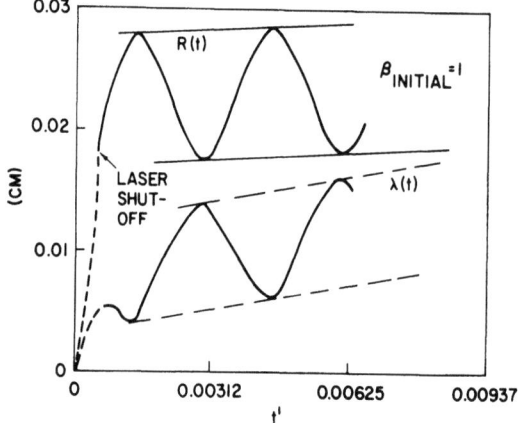

Fig. 1. Oscillatory behavior of the expanding plasmoid

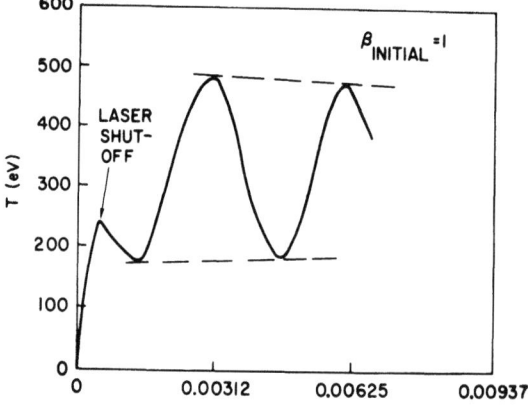

Fig. 2. Time variations in temperature of the expanding plasma

radius at which this happens is larger than that for the nonresistive case, the latter being the β ≃ 1 radius. Also, the periodic solution in the resistive case has a slanted envelope, representing a slow diffusion of the pulsating resistive plasma across the magnetic field. A comparison with the nonresistive case also shows that over and above the periodicity in time, the radius changes with time due to finite skin-depth and finite resistivity. It is also noticed from these results that the amplitude of oscillation increases with increasing values for initial β. The results obtained show the expansion of a hydrogen plasma against a magnetic field of strength 3000 G. The plasma has attained a size $r = 2.43 \times 10^{-2}$ cm at the time when it has become transparent to the laser (total energy supplied is ~ 53 J at the rate of 10^{10} W to a plasma with $r = 0.015$ cm). The plasma has a total number of 2.2×10^{16} H atoms and it reaches a temperature of ~ 240 eV at that time.

III. STRONG SHOCK WAVE IN A MAGNETIC FIELD

In the following, we consider the interaction of the magnetic field with a strong shock wave generated because of sudden release of large amount of energy inside a plasma by means of laser beam.

At the beginning, the temperature and the conductivity on the front of the shock wave are sufficiently large so that the radius of the region from which the field becomes forced out is very close to the radius of the shock wave. This diamagnetic behavior is connected with the appearance of eddy currents in the conducting layers of the medium that move behind the shock wave in the external magnetic field. For a spherical shock wave, a good approximation for the radius of the shock is given by[10]

$$r_{sh} \simeq [75(\gamma-1)(\gamma+1)^2/16\pi(3\gamma-1)]^{1/5} (Et^2/\rho)^{1/5} \qquad (8)$$

where γ is the ratio of the specific heats, E is the energy absorbed from the laser and ρ is the initial density of the plasma. With decreasing temperature on the shock front as it expands, the region of eddy currents penetrates deeper into the shock wave. It is well known that the temperature of the shock is minimum on its front while it increases rapidly as one moves from the front to the center where the energy is released. The temperature on the shock front is given by

$$T_{sh} \simeq \{8m_i/[25(Z+1) k(\gamma+1)^2]\} (E/\rho r_{sh}^3) \qquad (9)$$

where Z is the ion charge number and k is the Boltzmann constant. The ratio of the temperature at each point to the temperature T_{sh} on the shock front is given by

$$\Theta = T/T_{sh} = F^{6/5} A_1^{-\xi_0} [2\{(\gamma-1)/2 - F\}/(\gamma-1)] A_0^{-2\xi_1} \qquad (10)$$

where $A_0 = 2(3Z+5) \{5(\gamma+1)/2(3\gamma-1) - F\}/(7-\gamma)$

$A_1 = 2\gamma \{F - (\gamma+1)/2\gamma\}/(\gamma-1)$

$\xi_0 = 3/(2\gamma+1)$, $\xi_1 = (13\gamma^2 - 7\gamma + 12)/5(6\gamma^2 + \gamma - 1)$

and the parameter F is connected to the quantity r/r_{sh} as follows

$$r/r_{sh} = F^{-2/5} A_1^{\xi_2} A_0^{-\xi_1} \qquad (11)$$

with $\xi_2 = (\gamma-1)/(2\gamma+1)$.

The parameter F lies in the range $1 \geq F \geq \gamma+1/2\gamma$. Thus, $\Theta \gg 1$ for $F \approx \gamma+1/2\gamma$. In the interior region where $T > T_{sh}$, we have

$$\Theta = (r/r_{sh})^{-3/(\gamma-1)} \qquad (12)$$

where $G(\gamma)$ is some specified function of γ.

As the temperature grows rapidly away from the shock wave towards the interior and since the layer of shock wave is thin, viz., $\delta_{sh} \approx r_{sh}(\gamma-1)/3(\gamma+1)$, there develops a high rate of magnetic field diffusion through this layer. The radius of the magnetic field penetration region may be roughly estimated by equating the radial velocity v_r of the shock to the rate of penetration of the magnetic field, v_m. Now, $v_r = \dot{r}_{sh} r/r_{sh}$ and $v_m = c^2/4\pi\sigma\delta_{eff}$ where $\sigma \approx AT^{3/2}/Z$, $A = K^{3/2}/(\pi e^2 m_e^{2} \ln \Lambda)$, and $\delta_{eff} \approx (T^{-1} dT/dr)^{-1} \sim Cr$, C being same constant; this yields the magnetic field penetration radius

$$r_m \sim G_0(\gamma) [m_i^3 (E/\rho)^4 / Z^2 (Z+1)^3]^{\xi_3} r_{sh}^{-\xi_4} , \qquad (13)$$

where $\xi_3 = (\gamma-1)/(13-4\gamma)$, $\xi_4 = (14\gamma-23)/(13-4\gamma)$ and $G_0(\gamma)$ being some function of γ.

It also follows that at some instant of time, the pressure of the plasma in the region where the field is excluded becomes equal to the magnetic field pressure $B^2/8\pi$. The critical radius and the critical time is obtained from the equation $p \sim E/(4/3)\pi r_{sh}^3 \sim B^2/8\pi$, which yields

$$r_{sh}^{crit} \approx (E/B^2)^{1/3} \quad \text{and} \quad t^{crit} \approx E^{1/3}\rho^{1/2}/B^{5/3} \quad . \qquad (14)$$

After this time, the magnetic field exerts a strong pressure on the transverse motion of the central part of the plasma with $r < r_m$. One may point out that the pressure of the thrust of the plasma motion is much smaller than the pressure inside the plasma. As a matter of fact, $\rho u^2 < \rho_{sh} \dot{r}_{sh}^2 \approx p_{sh} \approx p$. This slowing down of the expansion can accelerate the penetration of the external magnetic field and the shock wave may also cease to be strong, i.e., the condition $p_{sh} \gg p_0(\gamma+1)/(\gamma-1)$, p_0 being the surrounding pressure, is not satisfied. But, it is possible to use high enough magnetic field so that the magnetic pressure stops the transverse motion of the internal plasma before the wave ceases to be strong and then all the estimates obtained by using the theory of strong shock theory are valid. Under such conditions, there would result an inward facing shock similar to the one observed in the numerical calculations mentioned in Ref. 8.

In the above calculations, one has assumed that the thermal conductivity plays a minor role, thereby assuming that the change in temperature in the medium behind the shock wave is determined mainly by the adiabatic expansion and not by the conduction of heat from the hotter layers. However, at the initial stage when there is a large energy input and the dimension of the plasma is small, the characteristic time of heat conduction may be comparable with the time of the process (remembering that the electronic heat conductivity $\kappa \sim T_e^{5/2}/Z$ which can be sufficiently large for high temperatures even at large densities. If $t \sim r^2/\kappa$, one may use the shock wave solutions with zero temperature gradient.[11] These solutions are characterized by a region from the center up to r_0 where the gas is at rest ($r_0 \approx \sqrt{\alpha r_{sh}^2}$, $\alpha \approx .244$ for a spherical shock), and for $r > r_0$, the gas velocity changes almost linearly with distance:

$$v(r) \approx v_{sh} (r-r_0)/(r_{sh}-r_0) \quad .$$

The conductivity now depends only on time but not on space and the temperature is roughly given by:

$$T(t) \approx 4\alpha m_i (E/\rho)^{2/5} t^{-6/5}/25 \text{ k} \quad . \qquad (15)$$

The radius of the shock is still given by Eq. (8) during the early stages but when the magnetic field has time to penetrate, the characteristic radius of penetration is now given by

$$r_m \sim (E/\rho)^{8/15} t^{-4/15} \quad . \qquad (16)$$

IV. EFFECT OF ROTATION

Plasma rotation has often been observed in experiments where the plasma is permeated by time - and/or space-varying magnetic fields. In particular, much work has been done to studies of the rapid plasma rotation occurring in magnetic compression arrangements known as theta-pinch experiments. In a comprehensive work, Haines[12] has analyzed eight different mechanisms proposed by himself and by others. The problem is sufficiently complex as is shown by the fact that only two of these mechanisms can be ruled out to be unlikely.

Using a particle orbit approach, one can show that the total angular motion of a plasma in a magnetic field is the sum of the guiding center motion and the diamagnetic circulating motion. The angular frequency may be written as

$$\omega = - CF_r/erB_z + \partial/\partial r (P_{i_\perp}/\Omega_i)/\rho r \qquad (17)$$

where the magnetic field is in the z-direction with respect to a cylindrical plasma. F_r is the radial component of the force exerted on the plasma, Ω_i is the ion gyrofrequency and $P_{i_\perp}$ is the ion pressure perpendicular to B. For example, the exerted force could be due to a radial magnetic field gradient $\partial B_z/\partial r$.

For an isolated azimuthally symmetrical plasma, the canonical angular momentum of the whole plasma is conserved. However, if the plasma and magnetic field are not azimuthally symmetrical then this conservation does not apply even when the plasma is isolated. Haines[13] suggested that if there are perturbing radial components of the magnetic field which change in magnitude or sign around the azimuthal direction it is possible to produce a torque on the isolated plasma, with an equal and opposite reaction on the external conductors responsible for the perturbing field. This reaction is transferred through a magnetic stress tensor.

If we now consider the case of an expanding plasmoid in a magnetic field, then it is possible that some angular momentum may be exerted to it as it bounces off the magnetic field, assuming the magnetic field has appropriate gradients. It is possible that some slight disturbance would now distribute this angular momentum uniformly over the mass of the plasmoid thereby creating a vortex-like configuration. In the following, we present some arguments borrowed from studies done in fluid dynamics and astrophysics in support of this possible configurations.

In an attempt to explain the formation of hurricanes, Scorer[14] considered the effect of turbulence on a rotating mass of air. He proposed that angular momentum is transported radially by large eddies and then mixed into the surroundings by smaller scale turbulence. In this way a uniform distribution of angular momentum is set up in a well-stirred fluid and a strong vortex is concentrated at the center of activity.

We now consider a simplified model of a rotating gaseous medium having only one component $\bar{v}$ of velocity. We also consider an idealized mixing process in which particles are repeatedly released with randomly distributed velocities, travel with those velocities for a short time τ, and are then mixed again with their new surroundings. This is similar to mixing length theories of turbulence. We now make the statistical assumption that the probability distribution of a particle at the time of release is an even function of u', v', and ω' which are the various components of velocity perturbation. Assuming no radial flux of mass, one can show that there is an average radial flux of angular momentum per particle given by

$$L = -\tau/2 \left[\overline{u'^2} \, \partial/\partial r \, (r\bar{v}) - \overline{2v'^2} \, \bar{v} \right] + O(\tau^2) \quad . \quad (18)$$

If $\overline{u'^2} = \overline{v'^2}$, then $L = 0$ if $\partial/\partial r(\bar{v}/r) = 0$, i.e., when there is solid body rotation. On the other hand, if $\overline{u'^2} \neq \overline{v'^2}$ (nonsymmetric mixing), then there is a possiblity that the mean motion would differ from solid body rotation, but given suitable anisotropy (e.g., $\overline{v'^2} = 0$, $\overline{u'^2} \neq 0$) a vortex can indeed be set up.

A numerical calculation[15] has recently been done in order to determine the effect of rotation and magnetic field on the collapse of a star. This study was made in connection with a possible supernova model. Starting out with a very small amount of kinetic energy of rotation of about 0.25 percent of the energy of collapse (corresponding a given solid-body angular rotation), it was found that the effect of rotation is enhanced by nonradial motions which carry angular momentum from regions where it is high to regions where it is low and the angular velocity distribution gradually approaches a vortex-like configuration. This state is arrived at within a length of time not greater than the time necessary to halt the collapse. It is also interesting to note that, in the final state, the kinetic energy of roation is higher than that arising from a simple radial contraction of a star. The vortex motion produced in such a rotating collapse is the most important feature of the effect of rotation.

When a plasmoid expands against a strong magnetic field with appropriate gradients and derives a certain amount of rotation, it may be possible that during the phase when it bounces off the field, the plasmoid would gradually go into a vortex configuration with smoothed-out angular momentum distribution. In a real three-dimensional world it is not clear how long it would take for an isotropically turbulent condition to set in, but it should take a few sound-transit times. At early times, one expects that the above calculations done for the supernova model would apply, but at later times these calculations probably would overestimate the large-scale order in the spatial distribution of the velocity. If such a vortex configuration is achieved, then a much larger part of the expansion or contraction energy would go into rotational motion than would happen for the case of a constant angular velocity distribution. Possibly, this may give rise to a pseudo-stabilization by rotation of plasmoids in a magnetic field.

REFERENCES

1. N. G. Bosov and O. N. Krokhin, in Proceedings of the Conference on Quantum Electronics (Dunod Cie., Paris, 1963) Vol. 2, p. 1373.
2. A. Engelhardt, Bull. Am. Phys. Soc. $\underline{9}$, 305 (1964).
3. J. Dawson, Phys. Fluids $\underline{7}$, 981 (1964).
4. H. Hora, Institut fur Plasmaphysik, Report No. 6/23 (1964).
5. U. Ascoli-Bartoli, C. deMichelis, and E. Mazzucato, in Plasma Physics and Controlled Nuclear Fusion Research (International Atomic Energy Agency, Vienna, 1966) Vol. II, p. 941.
6. A. F. Haught and D. H. Polk, Phys. Fluids $\underline{9}$, 2047 (1966).
7. D. K. Bhadra, Phys. Fluids $\underline{11}$, 234 (1968).
8. A. F. Haught, D. H. Polk, and W. J. Fader, Phys. Fluids $\underline{13}$, 2842 (1970).
9. W. J. Fader, Phys. Fluids, $\underline{11}$, 2200 (1968); I. B. Bernstein and W. J. Fader, Phys. Fluids $\underline{11}$, 2209 (1968).
10. Ya. B. Zeldovich and Yu. P. Raizer, Physics of Shock Waves and High-Temperature Hydrodynamic Phenomena (Academic Press, New York, 1967).
11. L. I. Sedov, Similarily and Dimensional Methods in Mechanics (Academic Press, New York, 1959); A. S. Kompaneets, Sov. Phys.-Doklady $\underline{5}$, 146 (1960).
12. M. G. Haines, Advances in Physics, Vol. 14, p. 167 (1965).
13. M. G. Haines, Phys. Letters 6, 313 (1963).
14. R. S. Scorer, Sci. J. $\underline{2}$, 46 (1966).
15. J. M. Leblanc and J. R. Wilson, Astrophys. J. $\underline{161}$, 541 (1970).

MAGNETIC FIELD ENHANCED ENERGY INCREASE OF IONS

EMITTED FROM LASER IRRADIATED SOLID TARGETS*+

Helmut Schwarz and Heinrich Hora#

Rensselaer Polytechnic Institute-Hartford Graduate Ctr.

275 Windsor Street, Hartford, Connecticut 06120

ABSTRACT

A ruby laser adjusted to four different power levels, 6, 9, 15, and 30 Megawatt, was focused to a diameter of 0.1 mm on solid targets (W, Ta, Cr, and Al). A magnetic field of flux density B was applied perpendicular to the target. Hereby the ion energy increased; the increase decreased with increasing laser power. Up to a magnetic field of 2000 gauss and within the range of the laser intensities applied the increase $\Delta\varepsilon$ of the kinetic energy ε of the ions followed the relationship: $\varepsilon^3 \Delta\varepsilon = \text{const } B^4$

EXPERIMENTAL SET-UP

Well cleaned solid surfaces of tungsten, tantalum, chromium, and aluminum were irradiated in an ultra high vacuum (background pressure lower than 10^{-10} Torr) by a ruby laser at power levels varying between 6 and 30 megawatt. While varying the laser power, the geometrical optics and the pulse duration of 50 nsec remained unchanged. This was accomplished by passing the laser light through

---

*Presented at the Second Workshop on "Laser Interaction and Related Plasma Phenomena" at Rensselaer Polytechnic Institute, Hartford Graduate Center, August 30-September 3, 1971.
+Preliminary results were already presented at the International Conference on "Laser Plasmas", Moscow, USSR, Nov. 17-20, 1970, in Russian in "Kvantovaya Elektronika" Vol. 1, 102(1972), (English translation to be published by A.I.P. in Soviet Journal of Quantum Electronics).
#Also from Max-Planck-Institut für Plasmaphysik, Garching, Germany

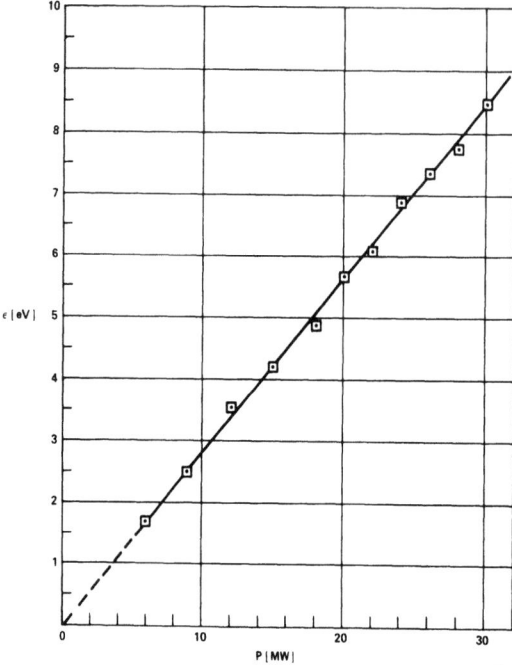

FIG. 1 - Kinetic Energy of "Thermal" Ions $\varepsilon = \frac{1}{2} m_i v_i^2$ in electron volts as a Function of Laser Power P in Megawatt at constant beam diameter of approximately 0.1 mm.

several filters of $CuSO_4$ solutions, before it entered the vacuum system and before the beam was focused to a spot of 0.1 mm diameter on a disc target. This disc target was connected with one plate of an oscilloscope whereas another parallel disc with a center hole positioned at a distance d was connected with the other plate of the oscilloscope; both were not on ground and therefore, were floating. The discs were electrically connected parallel to the oscilloscope by a resistor R. Any net charged particle flow created a voltage drop across R and could be detected by the fast sweeping oscilloscope. In other words, time-of-flight measurements led to the determination of the energy of the ions. More details of the set-up were reported previously[1-3].

As a special feature, we had already observed before[1-3] two discrete energy peaks, one being linear in relation to the laser intensity (see Fig. 1) and the other depending on almost the square of laser intensity (see Fig. 6 of Reference 1 and Fig. 2 of Reference 2).

In this paper we investigate only the slow peak or thermal peak*

---

*Temperature measurements performed with the retarding field method applied to the electrons thermionically emitted confirmed[3] that the slow peak must correspond to the temperature of the target.

MAGNETIC FIELD ENHANCED ENERGY INCREASE OF IONS 303

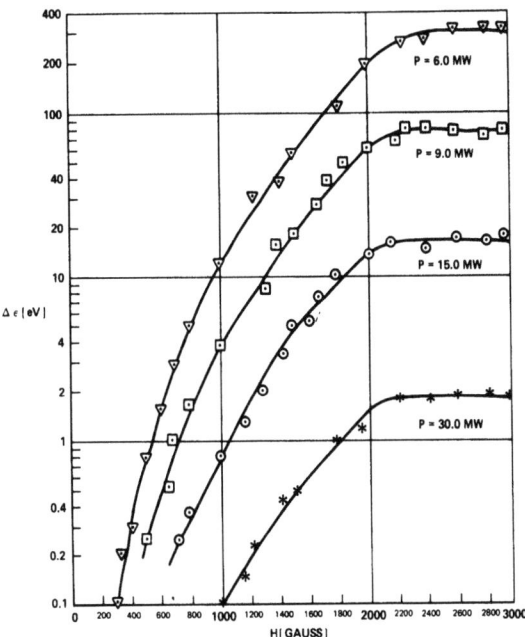

FIG. 2 —Family of Curves depicting measurements of the increase of kinetic ion energy $\Delta\varepsilon$ as a function of magnetic field H at 4 different laser powers (P = 6.0; 9.0; 15.0; and 30.0 MW).

as it changes with a static magnetic field applied perpendicular to the target and parallel to the laser beam. For the intermediate laser power level, we have previously[1-3] found an initial sharp increase of the ion energy with the application of a magnetic field. This sharp increase diminished drastically at magnetic fields of approximately 2000 gauss. We can see this from Figure 2 which contains a series of four curves at different laser power levels: 6 megawatts, 9 megawatts, 15 megawatts, and 30 megawatts. The energy increase of the ions is quite high for magnetic fields below 2000 gauss. Without a magnetic field, the thermal ion peak at these four different laser powers showed energies corresponding to 1.7 eV, 2.5 eV, 4.2 eV, and 8.5 eV (see Fig. 1). At each intensity the increase of ion energy $\Delta\varepsilon$ was measured and is plotted on Fig. 2. From a theory reported in Reference 1, it follows that the increase of the ion velocity can be expressed:

$$\Delta v_i = \frac{H_x^2}{4\pi} \frac{\Delta\mu}{\Delta x} \frac{t}{Nm_i} \qquad (1)$$

H_x is the component of the magnetic field intensity perpendicular to the target; $\Delta\mu$ is the drop of the magnetic permeability inside of the plasma sheath of thickness Δx; t the time the ion remains within this plasma sheath; N the ion density; and m_i the mass of one ion. From this increase of velocity, one can calculate the increase of the kinetic energy of the ions due to the plasma expansion velocity originating from the magnetic field. The simple formula for $\Delta\varepsilon$ is therefore

$$\Delta\varepsilon = \left[\frac{1}{2}(\Delta v_i)^2 + v_i \Delta v_i\right] m_i \qquad (2)$$

Combining this with Eq. (1), results in a final approximate expression for the energy increase $\Delta\varepsilon$ of an ion energy ε. Assuming that the ion density (plasma density) is proportional to the laser intensity

$$N = A\varepsilon \qquad (3)$$

A being a constant.

We arrive at an expression for the ion energy increase

$$\Delta\varepsilon = \frac{1}{2m_i \varepsilon^2}\left(\frac{H^2}{4\pi A}\frac{\Delta\mu}{\Delta x}t\right)^2 + \sqrt{\frac{2}{m_i}}\left(\frac{H^2}{4\pi A}\frac{\Delta\mu}{\Delta x}t\right)\varepsilon^{-1/2} \qquad (4)$$

If one assumes further that the permeability μ has dropped to 0 after the ion has passed through the plasma sheath of thickness Δx during the time t, one can set $\Delta\mu = 1$. Also $(t/\Delta x)^2$ might be assumed to be inversely proportional to the ion energy ε which then modifies Eq. (4) to:

$$\Delta\varepsilon = \frac{1}{\varepsilon^3}\frac{H^4}{64\pi^2 A^2}\left(1 + 16\varepsilon^2 \frac{\pi A}{H^2}\right) \qquad (5)$$

The second term in the parenthesis of Eq. (5) can be neglected for most of our experimental results so that one arrives at the simple relationship

$$\varepsilon^3 \Delta\varepsilon = \text{const} \cdot H^4 \qquad (6)$$

which is valid for magnetic fields up to 2000 gauss. In Figure 3 we have summarized all the measurements for the four different power levels and plotted down the relation as expressed in Eq. (6). One can say that all the points fall practically on one curve. Also, in Table I, we have given the results for the four different power levels as measured and calculated from Eq. (6).

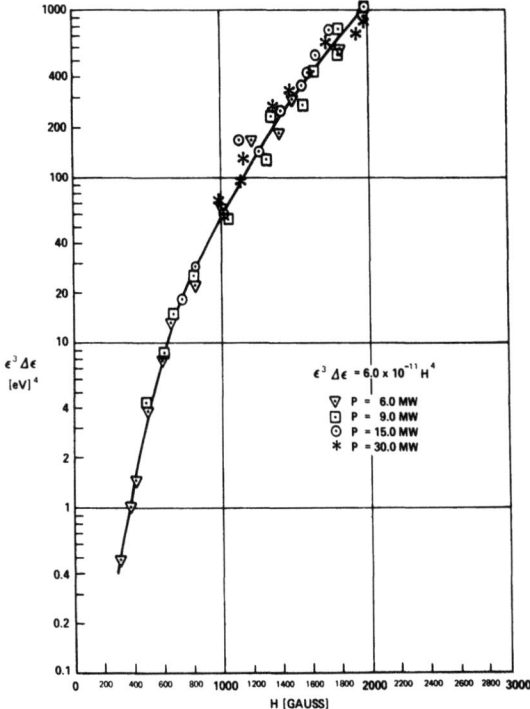

FIG. 3 - Relation between $\varepsilon^3 \Delta\varepsilon$ (ε = ion energy; $\Delta\varepsilon$ - increase of ion energy) and magnetic field H as measured at 4 different laser powers.

P [MW]	ε [ev]	H [Gauss]	$\Delta\varepsilon$ [ev]	$\varepsilon^3 \Delta\varepsilon$ [ev]^4	$\dfrac{\varepsilon^3 \Delta\varepsilon}{H^4}$ [ev/Gauss]^4 in 10^{-11} units
6.0	1.7	1000	12.0	60	6.0
		1500	58.0	290	5.8
		2000	190.0	940	5.9
9.0	2.5	1000	3.8	59	5.9
		1500	18.5	287	5.7
		2000	61.0	945	5.9
15.0	4.2	1000	0.8	62	6.2
		1500	3.9	290	5.7
		2000	13.5	1000	6.2
30.0	8.5	1000	0.1	61	6.1
		1500	0.5	305	6.0
		2000	1.5	920	5.8

TABLE I: Relationship between ion energy ε, ion energy increase $\Delta\varepsilon$ and magnetic field H.

The first column shows the laser power in megawatt, which was varied by changing only the optical density of a filter for the laser light before entering the vacuum and the focusing system; the second column gives the ion energy of the thermal group without magnetic field; and in the third column, the magnetic field strengths were applied which resulted in an energy increase $\Delta\varepsilon$ of this thermal group listed in the fourth column. The fifth and sixth columns are numerical evaluations.

REFERENCES

1. H. Schwarz, <u>Laser Interaction and Related Plasma Phenomena</u> (H. Schwarz and H. Hora, Ed.) Plenum, New York 1971, p.207.
2. H. Schwarz, S. Namba and P. H. Kim, Trans. of 8th Internl. Conf. on Phenomena in Ionized Gases, p. 59, Vienna, Aug. 27-Sept. 2, 1967.
3. H. Schwarz, S. Namba, P. H. Kim, T. Itoh and T. Arai, Sci. Papers I.P.C.R. (Scientific Papers of the Institute of Physical and Chemical Research), Vol. 60, #4, 101-106, Dec. 1966.

THEORETICAL ASPECTS OF ION ENERGY INCREASE IN LASER PRODUCED

PLASMAS DUE TO STATIC MAGNETIC FIELDS[*]

Heinrich Hora[+]

Rensselaer Polytechnic Institute-Hartford Graduate Ctr.

275 Windsor St., Hartford, Connecticut 06120

ABSTRACT

The increase $\Delta\varepsilon$ of the ion energy ε of a laser produced plasma due to an additional applied static magnetic field H oriented perpendicular to the stratified plasma boundary for moderate laser intensities led experimentally to the relation $\varepsilon^3 \Delta\varepsilon / H^4 =$ const. The ponderomotive force equation including the gasdynamic pressure and the electromagnetic fields from the laser and the static magnetic field reproduces the relation exactly, and results in a reasonable number for the constant consistent with the assumption of an ion density of about 10^{17} cm^{-3} in the diamagnetic boundary layer.

INTRODUCTION

Plasmas produced by focused laser irradiation of compact metal targets (discs) in vacuum showed some remarkable properties, especially in the transition range of moderate incident laser powers between 5 and 50 MW ruby laser pulses of about 20 nanosecond duration.[1] The increase of the maximum ion energies of some hundred eV changed nearly quadratic with the incident laser power[2] if the laser pulse profiles were kept constant, in contrast to higher laser powers with a sublinear increase[3] - however, with not always constant laser pulse profiles. Another point was the measurement

---

[*]Presented at the Second Workshop on "Laser Interaction and Related Plasma Phenomena" at Rensselaer Polytechnic Institute, Hartford Graduate Center, August 30 - September 3, 1971.
[+]Also from Max-Planck-Institut für Plasmaphysik, Garching, Germany

of the increase of the ion energy when applying a magnetic field perpendicularly to the target surface[2,4], as confirmed later on by other authors[5].

The increase of the ion energy in the range of moderate laser intensities was relatively very high and interpretation on the basis of the ponderomotive force equation[6] seemed to be possible[4]. Here we explore this increase of the ion energy following the parameters of applied magnetic field strengths and laser powers. The numerical agreement of the relations of the ponderomotive forces with the measurements indicates some reasonable consistency of the theoretical model. For example, a good numerical value of the ion density can be calculated from the evaluation of the measurements reported in Ref. 1, where the ion energies were derived from time-of-flight measurements.

The ion signals showed two maxima which could be explained as being due to a nonlinear fast group and a thermal slow group of plasma.[4] We are aware of the difficulty of interpreting such probe signals; especially negative biasing of the collector may change the second maximum of the oscillograms according to an acceleration of the ions of the thermal group. The consistency of the interpretation of the thermal group[4] with other optically identified thermal and nonlinear groups[7] all with the verified two groups of high energy plasmas[8] may assist our interpretation of the signals of the disc probe similar to the results of Papoular[9] of two thermodynamically uncorrelative groups of plasma after a very detailed discussion of the probe problems. The results of the measurements of H. Schwarz[1] are summarized in Table I.

P [MW]	ε [ev]	H [Gauss]	$\Delta\varepsilon$ [ev]	$\varepsilon^3 \Delta\varepsilon$ [ev]^4	$\dfrac{\varepsilon^3 \Delta\varepsilon}{H^4}$ [ev/Gauss]^4 in 10^{-11} units
6.0	1.7	1000	12.0	60	6.0
		1500	58.0	290	5.8
		2000	190.0	940	5.9
9.0	2.5	1000	3.8	59	5.9
		1500	18.5	287	5.7
		2000	61.0	945	5.9
15.0	4.2	1000	0.8	62	6.2
		1500	3.9	290	5.7
		2000	13.5	1000	6.2
30.0	8.5	1000	0.1	61	6.1
		1500	0.5	305	6.0
		2000	1.5	920	5.8

TABLE I: Relationship between ion energy ε, ion energy increase $\Delta\varepsilon$ and magnetic field H.

The first column shows the laser power in megawatt, which was varied by changing only the optical density of a filter for the laser light before entering the vacuum and the focusing system; the second column gives the ion energy of the thermal group without magnetic field; and in the third column, the magnetic field strengths were applied which resulted in an energy increase $\Delta\varepsilon$ of this thermal group listed in the fourth column. The fifth and sixth columns are numerical evaluations. While we are very much surprised at the magnitude of the increase of the ion energy[2], it can now be noticed that the ion energies increased much less at higher laser intensities. It is remarkable that the value for $\varepsilon^3\Delta\varepsilon/H^4$ listed in the sixth column of Table I is for all four laser powers up to a magnetic field H = 2000 gauss constant of approximately:

$$\text{const} = 6 \times 10^{-11} (\text{ev/gauss})^4 \qquad (1)$$

$$\text{or} \quad \text{const} = 1.02 \times 10^{-37} (\text{ev})^2 \text{cm}^6$$

where the relation $\text{gauss}^2/4\pi = 6.24 \times 10^{11} \text{ev/cm}^3$ was used.

THEORY

A remarkable consistency in the measurements of $\Delta\varepsilon$ and of the calculated constant of Eq. (1) was found with a theoretical treatment using the ponderomotive force equation. These relationships underline the advantage of this theory without going into detail of the physical meaning of each step derived. The force density $\underline{f}$ (the ponderomotive force) with a gasdynamic pressure tensor $\underline{p}$ and with the presence of electric and magnetic fields $\underline{E}$ and $\underline{H}$ is given by a general stress tensor $\underline{\underline{\sigma}}$ following Landau and Lifshitz's[10] treatment on the basis of Lorentz's theory:

$$\underline{f} = \nabla \cdot \underline{\underline{\sigma}} - \frac{\partial}{\partial t} \frac{\underline{E} \times \underline{H}}{4\pi c} \qquad (2)$$

Following a special interpretation of the plasma density and its dielectric properties given by the complex dielectric constant $\tilde{n}$ and by the magnetic properties determined by the permeability μ, we could successfully correlate the results of nonlinear forces of laser-plasma interactions with that derived on the basis of a two-fluid model or with the single particle model[6]. The final formulation of this force density following Eq. (2) is then

$$\underline{f} = -\nabla \cdot \underline{\underline{p}} + \nabla \cdot \underline{\underline{T}} - \frac{1-\tilde{n}}{4\pi} \underline{E}\underline{E} - \frac{1-\mu}{4\pi} \underline{H}\underline{H} - \frac{\partial}{\partial t} \frac{\underline{E} \times \underline{H}}{4\pi c} \qquad (3)$$

where $\underline{T}$ is the Maxwellian stress tensor in vacuum. The usefulness of this equation can be seen for the special case of a stratified plasma with a pressure p(x) depending only on coordinate x and a parallelly oriented static magnetic field (E = 0; $H_x = H_z = 0$; $H_y = H$), for which follows immediately from Eq. (3)

$$\frac{\partial}{\partial x}\left[p(x) + \frac{H_y^2(x)}{8\pi}\right] = 0 \qquad (4)$$

with the well known integral

$$p + \frac{H^2}{8\pi} = \text{const} \qquad (5)$$

More generally, we find for static magnetic fields and pressures using the Kronecker symbol $\delta_{\phi\chi}$

$$\nabla_\phi\left[p_{\phi\chi} + \frac{H_\sigma H_\sigma}{8\pi}\delta_{\phi\chi} - \mu\frac{H_\phi H_\chi}{4\pi}\right] = 0 \qquad (6)$$

For the case of the interaction of laser radiation incident perpendicular on a stratified plasma in the presence of a static magnetic field which is perpendicular to the plasma ($H_{sy} = H_{sz} = 0$), the ponderomotive force density can be separated

$$\underline{f} = \underline{f}^P + \underline{f}^L + \underline{f}^F \qquad (7)$$

where $\underline{f}^P$ results from the gasdynamic pressure, $\underline{f}^L$ from the laser fields, and $\underline{f}^F$ from the static field. Using Eqs. (3) and (6) we find

$$\underline{f}^F = \nabla_\phi\left[\mu\frac{H_\phi H_\chi}{4\pi} - \frac{H_\sigma H_\sigma}{8\pi}\delta_{\phi\chi}\right] \qquad (8)$$

We are now able to calculate the increase of the ion energy $\Delta\varepsilon$ due to the static magnetic field superimposed on the acceleration by the laser field. We have only to use the x-component of the force density $\underline{f}^F$:

$$f_x = \frac{1}{4\pi} H_x^2 \frac{\partial \mu}{\partial x} + \frac{2\mu-1}{8\pi} \frac{\partial H_x^2}{\partial x} \qquad (9)$$

The x-component of the ion velocity can be calculated now from this equation since we may neglect the last term of this Eq. (9), because this term causes only a small deconfining acceleration. The force is assumed to act during a time t which should be equal to the duration of the ion drift inside the plasma boundary region of thickness Δx. This time is also assumed to be the same for lowering the permeability to zero. In other words, $\Delta \mu = 1$. An increase of velocity due to the static magnetic field can be calculated from the simple relationship:

$$f_x = \frac{d}{dt}(Nm_i v_i) \simeq \frac{1}{4\pi} H_x^2 \frac{\partial \mu}{\partial x} \qquad (10)$$

which results in an approximate expression for the velocity increase:

$$\Delta v_i = \frac{H_x^2}{4\pi} \frac{\Delta \mu}{\Delta x} \frac{t}{Nm_i} \qquad (11)$$

The increase of the ion energy as given by the simple relationship

$$\Delta \varepsilon = m_i \left[\frac{1}{2}(\Delta v_i)^2 + v_i \Delta v_i \right] \qquad (12)$$

results then in

$$\Delta \varepsilon = \frac{1}{2m_i} \left[\frac{H^2}{4\pi} \frac{\Delta \mu}{\Delta x} \frac{t}{N} \right]^2 + \sqrt{\frac{2\varepsilon}{m_i} \left[\frac{H^2}{4\pi} \frac{\Delta \mu}{\Delta x} \frac{t}{N} \right]} \qquad (13)$$

At this point, we use an assumption which may be allowed for moderate laser intensity. We assume a linear increase of the plasma density N on the laser power and therefore, a proportionality with the laser produced ion energy

$$N = A\varepsilon \qquad (14)$$

A being a constant.

Substituting N from Eq. (14) into Eq. (12) leads to

$$\varepsilon^2 \Delta\varepsilon = \frac{1}{2m_i}\left(\frac{H^2}{4\pi A}\frac{\Delta\mu}{\Delta x}t\right)^2 + \sqrt{\frac{2}{m_i}}\left(\frac{H^2}{4\pi A}\frac{\Delta\mu}{\Delta x}t\right)\varepsilon^{3/2} \qquad (15)$$

For ion energies ε below kV and for magnetic fields H below 2000 gauss, the last term in Eq. (15) can be neglected. Using $\Delta\mu = 1$ (along the diamagnetic acceleration region for the ions to the static magnetic field) in identifying the time t with that of the acceleration by the laser light occurring only within the depths Δx, which results in $\varepsilon = \frac{m_i}{2}\left(\frac{\Delta x}{t}\right)^2$, we find from Eqs. (14) and (15)

$$\varepsilon^3 \Delta\varepsilon = \frac{H^4}{64\pi^2 A^2} \qquad (16)$$

This general relation reproduces the measured behavior of Eq. (1). Furthermore, we find from Eqs. (1) and (16) the identity

$$A = \frac{1}{8\pi\sqrt{\text{const}}} = 1.25 \times 10^{17} (\text{ev})^{-1} \text{cm}^{-3} \qquad (17)$$

which lies astonishingly in the right order of magnitude, since it leads to an ion density of approximately $10^{17} \ldots 10^{18}$ cm^{-3}.

CONCLUSION

The emperically found[1] proportionality between $\varepsilon^3\Delta\varepsilon$ and H^4 could be reproduced from the ponderomotive force theory for evaluation of the additional energy $\Delta\varepsilon$ which the ions gain by the magnetostatic field due to the interaction of the laser with the plasma. The measured[1] proportionality constant agrees well with the density

REFERENCES

1. H. Schwarz and H. Hora, "Magnetic Field Enhanced Energy Increase of Ions Emitted from Laser Irradiated Solid Targets" Laser Interaction and Related Plasma Phenomena (H. Schwarz and H. Hora, eds), Plenum, New York, these Proceedings.
2. S. Namba, H. Schwarz, Proc. IEEE Symposium on Electron, Ion and Laser Beam Technology, Berkeley, May 1967, p. 861; S. Namba P.H. Kim and Helmut Schwarz, Trans. 8th Internat. Conf. on Phenomena in Ionized Gases, Vienna, Austria, Aug. 27-Sept. 2, 1967.

3. D. W. Gregg and S. J. Thomas, J. Appl. Phys. 37, 4313 (1966); H. Opower, W. Kaiser, H. Puell, and W. Heinicke, Z. Naturforsch. 22a, 1392 (1967).
4. H. Schwarz, Laser Interaction and Related Plasma Phenomena, (H. Schwarz and H. Hora, eds.) Plenum, New York 1971, p. 207.
5. P. E. Faugeras, M. Mattioli, and R. Papoular, AIAA Fluid and Plasma Dynamics Conference, Los Angeles, Calif., June 1968. E. Fabre and H. Lamain, Physics Letters 29A, 497 (1969); G. Tonon e.a. as reported by F. Floux, Laser Interaction and Related Plasma Phenomena (H. Schwarz and H. Hora, eds.) Plenum, New York, these Proceedings.
6. H. Hora, Phys. Fluids 12, 182 (1969); Laser Interaction and Related Plasma Phenomena (H. Schwarz and H. Hora, eds.) Plenum, New York, these Proceedings.
7. A. G. Engelhardt, T. V. George, H. Hora, and J. L. Pack, Phys. Fluids 13, 212 (1970).
8. K. Büchl, K. Eidmann, P. Mulser, H. Salzmann, R. Sigel and S. Witkowski, Proc. 4th. Conf. on Plasma Physics and Controlled Nuclear Fusion, Madison, Wisconsin, June 1971; IAEA, Vienna 1971 C. Yamanaka, Second Workshop Laser Interaction and Related Plasma Phenomena (H. Schwarz and H. Hora, eds.) Plenum, New York, these Proceedings.
9. P. E. Faugeras, M. Mattioli, and R. Papoular, Plasma Phys. 10, 959 (1968).
10. L. D. Landau and E. M. Lifshitz, Electrodynamics of Continua Media (Pergamon Press, Oxford, 1966), p. 242.

SUMMARY OF DISCUSSION

(IV. Laser Produced Plasma Interacting
with Gases and Magnetic Fields)

In the discussion of Stamper's presentation on the spontaneous generation of magnetic fields in laser produced plasmas expanding in field free vacuum, it was mentioned that J. R. Wilson, Livermore[1] has done a theoretical study of similar properties which resulted in a magnetic field strength up to a few megagauss due to conductivity and heat flow effects.

One problem was brought up in connection with A. F. Haught's proposal to confine a plasma produced by a laser and a particle beam in a magnetic field of the minimum-B type. In early papers[2], it was reported that the plasma expands only according to the classical confining pressure of the magnetic field, while other authors have found a faster expansion perhaps due to instabilities or other plasma-magnetic field interaction. Haught observed faster expanding plasma groups but only in the case of minor fields or for minimum-B fields below 5 kgauss. The question of whether the plasma follows a classical diffusion across the magnetic field is still not answered. However, D. K. Bhadra's results of the instable vortex motion seems to have some bearing on this problem.

Following the contributions of Schwarz and Hora on the increased ion energy of laser produced plasmas expanding in a magnetic field, S. B. Segall mentioned the problem of interpretation of the probe signals measuring the time-of-flight of the ions. Despite the basic problems arising while applying these probes, as mentioned in the discussions of Chapter III, the measurements as reported by Schwarz can be substantiated by the fact that they are consistent with very general relationships derived from the ponderomotive forces including the static magnetic fields. The question of whether two or more ion energy groups exist in the plasma can be answered by comparing them with optical diagnostics as carried out by Yamanaka et al[3] and A. G. Engelhardt et al[4], who also observed two groups where no electric probes were involved.

References
1. J. R. Wilson, Private Communication 1971
2. A. F. Haught, D. H. Polk and W. J. Fader, Phys. Fluids 13, 2842, (1970).
3. T. Yamanaka, N. Tschuchimori, T. Sasaki, and C. Yamanaka, Technol. Progress Report, Osaka University 18, 155 (1968).
4. A. G. Engelhardt, T. V. George, H. Hora, and J. L. Pack, Phys. Fluids 13, 212 (1970).

ANOMALOUS ABSORPTION OF INTENSE RADIATION

W. L. Kruer and J. M. Dawson

Plasma Physics Laboratory, Princeton University

Princeton, New Jersey 08540 USA

ABSTRACT

Anomalous heating of a plasma by an electric field oscillating near the plasma frequency is considered. The large field excites the oscillating two-stream and the ion-acoustic decay instabilities. Numerical simulation of these instabilities has confirmed the linear theory for their growth and has shown that when the instabilities saturate a strong anomalous heating occurs. This heating results in the production of energetic electrons. A simple nonlinear theory gives results in reasonable agreement with the numerical calculations. These predictions could be applied to laboratory experiments.

---

*Presented at the Second Workshop on "Laser Interaction and Related Plasma Phenomena" at Rensselaer Polytechnic Institute, Hartford Graduate Center, August 30-September 3, 1971.

I. INTRODUCTION

There is substantial interest in the problem of anomalous ac resistivity in a plasma driven by a large field oscillating near ω_{pe}. This is an anomalous resistivity process for which we have a basic theoretical understanding.[1,2] The anomalous heating was first observed in computer simulations.[2,3] Since then a growing number of experiments have pointed to its existence. Gekker and Sizukhin,[4,5] Eubank,[6] Hendel and Chu,[7] and Dreicer et al[8] have all observed an anomalous absorption of intense microwaves in low-density plasmas. Shearer et al[9] have found an anomalous production of very energetic particles in intense laser experiments. Cohen et al,[10] and Wong et al[11] have observed results consistent with an anomalous absorption of intense radio-frequency waves in wave propagation experiments in the ionosphere.

In view of this substantial interest, we have carried out a large number of computer simulations to investigate in some detail the dependences of the anomalous resistivity on pump amplitude and frequency, the electron-ion mass ratio, and the number of dimensions. A simple nonlinear theory reasonably explains the simulation results.

II. NUMERICAL TECHNIQUES

The plasma is simulated by means of an electrostatic particle-pushing model of a type[12-14] widely used in computational plasma physics. Basically we follow the motion of a large number of ions and electrons in their self-consistent and any given external fields. We apply the external field by simply giving to each particle at each time step a force of specified amplitude and frequency, i.e.,

$$E_{ext} = E_o \cos \omega_o t .$$

We choose the cosine function since this field gives rise to no dc drift of the plasma at $t = 0$. Equivalently one could adiabatically turn on the pump field.

In the one-dimensional simulations[13] we generally used 25,600 ions and 25,600 electrons in a 512 λ_D periodic system with an ion-electron mass ratio of 100. For the weaker external fields and the large mass ratio simulations (an electron-ion mass ratio of 2000), we generally doubled both the number of particles and the size of the system. In two-dimensional simulations we used 20,000 ions and 20,000 electrons in a doubly periodic $64\lambda_D \times 64\lambda_D$ system. For these latter simulations the ion-electron mass ratio was again 100. The ions were always taken as relatively cold ($T_e/T_i \simeq 30$).

For these simulations we employed finite-size particles to reduce short-wavelength collisional effects. We start the simulation with the plasma uniform in space and with a Maxwellian distribution of velocities. The initial velocity distribution has been specified either by a random loading or by a quiet start[15] with no significant changes in simulation results. In the first case the velocity distribution is specified by assigning particle velocities via a random number routine. In the second case (usually employed in the one-dimensional simulations), we represent the Maxwellian velocity distribution by a large number of distinct velocity classes. We generally used a time step of $0.2\,\omega_{pe}^{-1}$. The results have been spot-checked by changing (1) the method of specifying the initial conditions, (2) the size of the system, (3) the number of particles, and (4) the time step.

III. A TYPICAL RESULT

Let us proceed to the physics with a brief survey of the phenomena that we are investigating. Figure 1 shows the results of a typical one-dimensional simulation. Here the strength of the driving field is

$$\frac{eE_o}{m\omega_{pe} v_{T_e}} = 0.5,$$

and its frequency is $\omega_o = 1.04\,\omega_{pe}$. The top of the figure shows the self-consistent wave energy of the plasma versus time, while the bottom shows the total energy of the plasma versus time. First, the plasma slowly joule heats[16] according to essentially classical two-particle collisions. Meanwhile, the ion and electron fluctuations are exponentiating. When these fluctuations become large and saturate, a strong anomalous heating of the plasma sets in.

The external driver has begun to efficiently couple energy into the plasma. A convenient way to discuss this heating is by

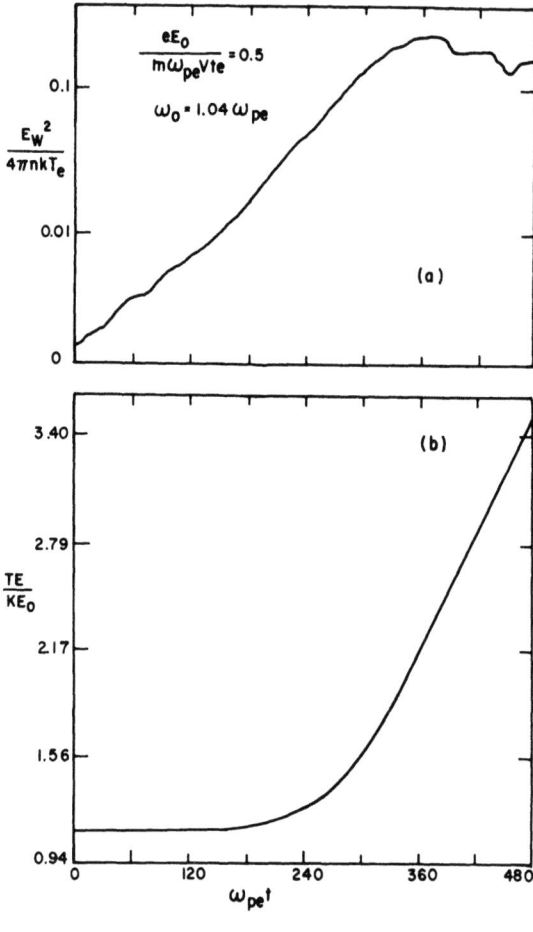

Fig. 1

The evolution of a) the self-consistent wave energy and b) the total plasma energy versus time (averaged over a plasma oscillation).

introducing an effective collision frequency defined in terms of the heating rate. Let us consider the equation for the spatially homogeneous component of the electron current (J_o),

$$\frac{\partial J_o}{\partial t} = \frac{n_o e^2}{m} E_o \cos \omega_o t - \nu^* J_o .$$

The first term on the right of this equation is due to the pump field which drives the oscillating current, and the second term represents the effective drag due to the interaction of the electron and ion fluctuations. Actually, the nonlinear interaction also gives a small component in phase with the pump field which can be considered a slight renormalization of the electron mass. We solve this equation for J_o and dot with E_o to find the net dissipation of energy by the external pump into the plasma,

$$\frac{d}{dt}(\text{total plasma energy}) = J_o \cdot E_o$$
$$= \frac{\nu^* E_o^2}{8\pi}.$$

We then measure ν^* by observing the rate at which the energy of the plasma increases. These measurements have been occasionally spot-checked by directly observing the phase shift between E_o and J_o. We emphasize that the heating is due to a collective process (the interaction of electron and ion waves) and ν^* should not be interpreted as simply an increase in the ordinary electron-ion collision frequency.

IV. LINEAR THEORY

We begin by discussing the instabilities responsible for the exponentiation of the ion and electron fluctuations. There are two hydrodynamic instabilities[17,18] for a plasma driven by a large field oscillating near the electron plasma frequency; the oscillating two-stream and the ion acoustic decay. Both these instabilities are parametric in nature and can be readily derived from the fluid equations as Nishikawa[18] has done.

It is instructive to derive the basic equations via simple physical arguments. In the absence of the external driving field, we have two uncoupled equations describing the evolution of the electron density at the Bohm-Gross frequency (ω_{ek}) and the ion density at the ion-acoustic frequency (ω_{Ac}). The large driver will clearly couple these high- and low-frequency oscillations. For example, consider an ion fluctuation. This ripples the background plasma density on the slow (ion) time scale, since the electron density follows along. The external driver now produces high-frequency charge fluctuations since it oscillates regions of high electron density into regions of low electron density and vice versa. Indeed, the magnitude of this coupling is

$$\delta n_{ek}^{h} = n_{ik}(x+x_o) - n_{ik}(x)$$

$$\simeq ikx_o n_{ik} \qquad (x_o = eE_o/m\omega_o^2),$$

where x_o is the excursion length of an electron in the external field and n_{ik} is the magnitude of the ion fluctuation with wavenumber k. Hence, the equation for the high-frequency electron fluctuation becomes

$$\frac{\partial^2 n_{ek}^h}{\partial t^2} + \omega_{ek}^2 n_{ek}^h = \frac{ikeE_o}{m} n_{ik}.$$

The superscript H denotes the high-frequency component of the electron fluctuations, and n_{ik} is assumed to be purely low frequency.

Similarly, the high-frequency electron-density fluctuations couple with the high-frequency pump field to drive ion fluctuations. It is well known that the electron pressure provides a force (∇p_e) which allows a charge imbalance on the ion-acoustic time scale. Let us add to the electron pressure the electric field pressure $E^2/8\pi$, where E is the total field in the plasma. The gradient of this latter pressure can also drive a low-frequency charge imbalance.[19] The magnitude of this force is

$$\nabla \frac{E^2}{8\pi} \simeq \frac{ikE_o E_k^h}{4\pi}$$

$$= -eE_o n_{ek}^h.$$

Here E_o^h is the high-frequency component of the self-consistent electric field, and Poisson's equation has been used to relate this field to the high-frequency electron density fluctuations. With this additional force the equation for ion fluctuations becomes

$$\frac{\partial^2}{\partial t^2} n_{ik} + \omega_{AC}^2 n_{ik} = - \frac{ikeE_o}{M} n_{ek}^h.$$

These two coupled equations describe the instabilities; indeed we readily obtain Nishikawa's dispersion relation (by also including a small damping term).

ANOMALOUS ABSORPTION OF INTENSE RADIATION

$$\left[\omega^2 + 2i\omega\Gamma_1 - \omega_{Ac}^2\right]\left[(\omega+i\Gamma_2)^2 - \delta^2\right] + \frac{k^2 e^2 E_o^2}{4mM}\frac{\delta}{\omega_{ek}} = 0.$$

Here $\delta = \omega_o - \omega_{ek}$, and the Γ's are the damping rates (collisional or Landau) of the oscillations.

Two different instabilities are predicted, depending on the sign of δ. For $\delta < 0$, high-frequency electron fluctuations and purely growing ion fluctuations are driven unstable. This is commonly called the oscillating two-stream instability since it goes over into the usual dc two-stream instability if one lets the driver frequency go to zero keeping E_o/ω_o constant. For $\delta > 0$, high-frequency electron oscillations and low-frequency ion-acoustic oscillations are driven unstable. We call this the ion-acoustic decay instability since the high-frequency driver decays into two normal modes of the plasma.

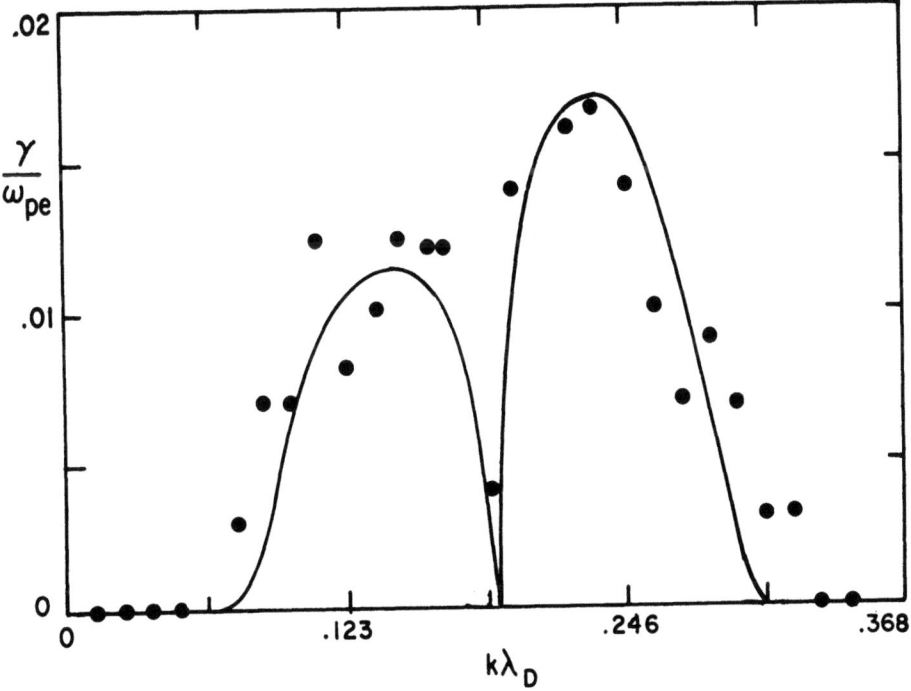

Fig. 2

The growth rate versus wave number ($\omega_o = 1.04$ and $eE_o/m\omega_{pe}v_{Te} = 0.5$).

We have solved this dispersion relation numerically to find the linear behavior of our system. The curve in Fig. 2 shows the theoretical prediction for the growth rate versus the wave number. The dots denote the growth rates observed in the simulation. For this example, $eE_o/m\omega_{pe}v_{Te} = 0.5$ and $\omega_o = 1.04\,\omega_{pe}$. We see the two instability branches. The portion of the curve to the left represents the ion-acoustic decay instability, and the portion to the right is due to the oscillating two-stream instability. It should be noted that in computing the theoretical curve we have incorporated a small correction due to the finite size of our simulation particles.[12]

V. ANOMALOUS HEATING

The parametric instabilities, then, readily account for the exponentiation of the ion and electron fluctuations. When the fluctuations become large, a strong anomalous heating of the plasma sets in. This enhancement of the high-frequency resistivity was predicted many years ago by Dawson and Oberman.[20] Basically, the (relatively light) high-frequency external driver oscillations are efficiently scattered by the (relatively massive) ion fluctuations into plasma oscillations. This energy is subsequently Landau damped from the plasma oscillations into the particles.

This basic nature of the heating process is confirmed by Fig. 3, which shows the electron and ion distributions after substantial anomalous heating has occurred. The parameters are the same as for the simulation shown in Fig. 1. We see the energy increase is primarily due to the creation of energetic tails on the electron distribution function. This is characteristic of heating by large amplitude electron plasma waves. It should be noted that these tails are not necessarily small. After substantial heating occurs, the tails become heavily populated.

The ions are heated much less (by about two orders of magnitude). The ion fluctuations act as sources or sinks of momentum in the basic coupling process, and the energy flow is principally from the external driver field into electron plasma oscillations. The relative magnitude of the ion and electron heating is consistent with simple estimates from the linear theory of the energy partition between the ion and electron fluctuations.

An interesting question is how much heating can ultimately occur. We have determined ν^* by measuring the heating rate shortly after the time of saturation. In some of our larger pump cases, we

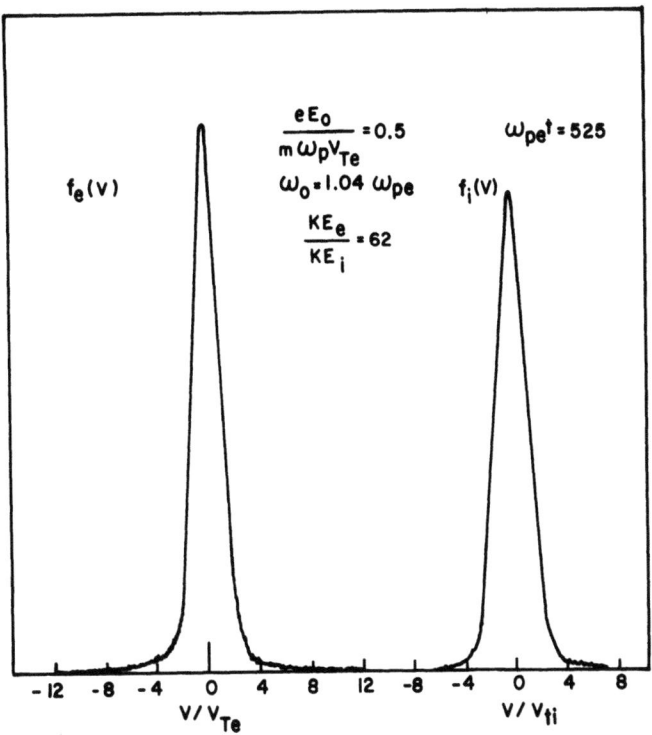

Fig. 3
The electron and ion velocity distribution function.

we have observed the plasma to heat up by several orders of magnitude. Computer movies show that as the plasma heats manifold, the plasma oscillations into which energy is scattered have larger and larger wavelength. This is as predicted since these plasma oscillations are those whose Bohm-Gross frequencies are near the driver frequency. As the effective thermal velocity increases, these waves have longer and longer wavelength. This presents numerical complications in determining the ultimate heating, since a very large number of modes (very large system sizes) is then necessary to follow the evolution of the plasma over many orders of magnitude in energy. In real plasmas the net amount of heating will probably be determined by density gradient effects such as energy flow out of the unstable region.

We can quantitatively estimate the magnitude of the anomalous resistivity by some simple energy balance arguments. There is an energy flow from the external driver into electron plasma oscillations, and an energy flow from the electron plasma oscillations into the plasma particles. Schematically we have

$$\text{External Driver} \rightarrow \begin{array}{c}\text{Electron}\\ \text{Plasma}\\ \text{Oscillations}\end{array} \rightarrow \begin{array}{c}\text{Plasma}\\ \text{Particles}\end{array}.$$

The instability theory allows us to estimate the energy transfer from the external driver to the electron plasma oscillations as $2\gamma E_W^2/4\pi$, where γ is a typical growth rate and $E_W^2/4\pi$ is the energy associated with the plasma oscillations. The transfer of energy to the particles is given by our definition of the anomalous heating rate as $\nu^* E_o^2/8\pi$. When the plasma oscillations saturate, these energy flows balance. Hence we estimate ν^* as

$$\nu^* \simeq 4\gamma \frac{E_W^2}{E_o^2} \quad,$$

where E_W^2 is the value of the wave energy at saturation.

VI. NONLINEAR THEORY

To proceed further we need a nonlinear theory for the instability. A detailed nonlinear theory for pump fields significantly larger than threshold does not exist and in general is quite difficult.[21,22] However, here the power of computer simulations becomes evident. Namely, simulations give us the saturation values and in general indicate what the dominant effects are. So guided, we have constructed a simple nonlinear theory which accounts reasonably well for the simulation results.

The computer simulations show that there are two regimes depending on whether $E_o^2/4\pi nkT_e$ is much less than one or is of order one. The distinguishing feature between the two regimes is the absence or presence of strong electron trapping by the electron plasma oscillations at time of saturation. Figures 4 and 5 illustrate this point. These figures show a portion of electron phase space at time of saturation for two different pump-field amplitudes. In Fig. 4, where $E_o^2/4\pi nkT_e = 0.09$, we see that some pulling out of the tail of the distribution has occurred, but there are no well-defined vortices such as are associated with particle trapping. In Fig. 5, where $E_o^2/4\pi nkT = 1.0$, we see very well defined vortices. These phase spaces clearly indicate the importance of strong electron trapping for the large pump-field cases.

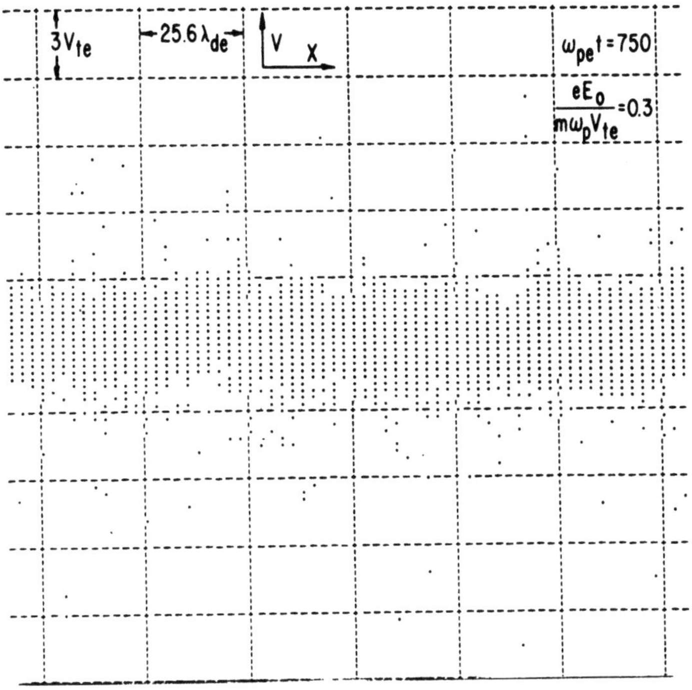

Fig. 4

A portion of electron phase space near time of saturation ($eE_o/m\omega_p v_{Te} = 0.3$).

In the weak pump-field regime, we propose a simple argument for the saturation. Namely, we expect saturation to occur when the perturbation in a typical particle's velocity becomes equal to the zero-order oscillating velocity responsible for the instability. In other words,

$$v_W \sim v_o,$$

where $v_W = eE_W/m\omega_p$, and E_W is the root-mean-square of the self-consistent electric field in the plasma. This then predicts $E_W \sim E_o$ and hence

$$\nu^* \sim 4\gamma.$$

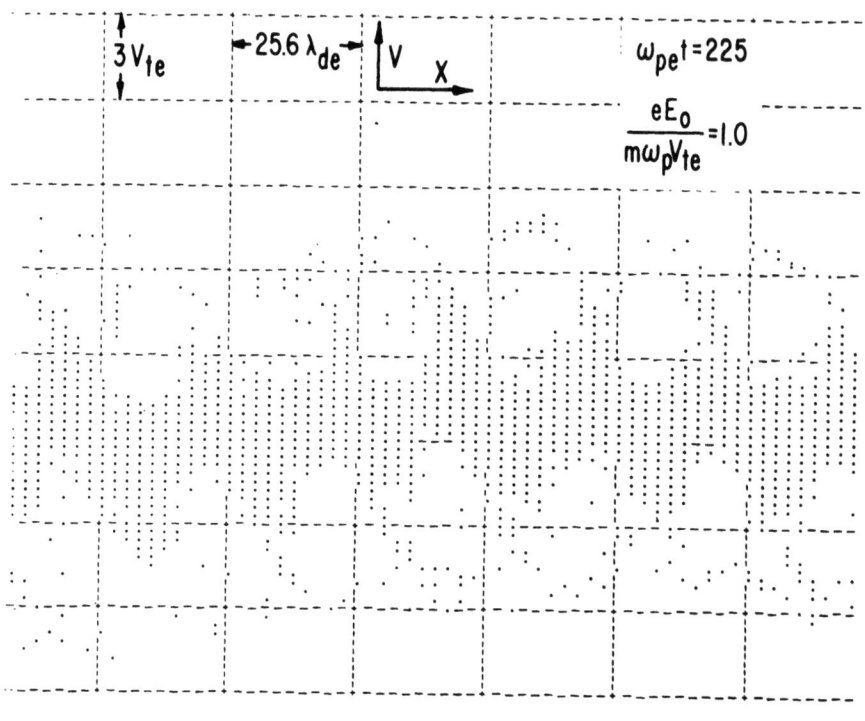

Fig. 5

A portion of electron phase space near the time of saturation ($eE_o/m\omega_p v_{Te} = 1.0$).

Other simple arguments are consistent with these estimates. For example, the excited plasma oscillations can themselves drive parametric instabilities as we have previously shown in computer simulations.[2] When $E_W \sim E_o$, these secondary instabilities can become as strong as the instabilities due to the external pump field. In other words, the excited plasma oscillations are also scattered by ion fluctuations. However, they are scattered less efficiently than the external pump, since they have wavelengths comparable to those of the ion fluctuations. A naive estimate for the rate of scattering is ν^*/α, where ν^* is the rate at which energy is scattered by ion fluctuations from the external pump and α is a factor of order 3 to allow for the less efficient scattering. The plasma oscillations then saturate when the energy lost by scattering from the ion fluctuations balances the energy input from the instability, i.e., $\nu^* \approx 2\alpha\gamma$.

The transition to the second regime readily follows when the wave fields at saturation become large enough to strongly trap electrons. A simple way to estimate the effect of particle trapping is via the concept of the trapping velocity.[23] This is simply the largest velocity a particle can have relative to a wave and still be trapped by it. By simple energy considerations the trapping velocity is

$$v_{TR} = \left(\frac{eE_W}{mk}\right)^{1/2},$$

where E_W is the electric field and k is the wave number of the large wave. Strong electron trapping occurs when a wave becomes large enough to trap many particles from the distribution function. A reasonable criterion for this condition is

$$v_{TR} \sim \text{Maximum velocity relative to the wave of an appreciable number of particles.}$$

Let us apply these arguments to our simulations. First, since we must stabilize a strong hydrodynamic-type instability, an appreciable number of particles must be trapped. Hence we allow the wave to reach back into the distribution function and trap particles with velocities at least several thermal velocities. In addition let us allow for the fact that as the external field increases in magnitude (the instability becomes stronger), more particles must be trapped in order to stabilize the instability. That is, the wave must reach even farther back into the distribution. A simple way to mock up this effect (guided by the data) is to demand that the wave trap particles back to a distance $(2v_{Te} - v_o)$ in the distribution. Hence when we get to fairly strong driving fields

$$\frac{E_o^2}{4\pi nkT_e} \sim 4,$$

we can trap essentially all particles. The strong trapping condition for stabilization then becomes

$$v_{TR} \simeq v_p + v_o - 2v_{Te},$$

where v_p is the phase velocity of the large wave. This condition is applied to the most unstable wave predicted by the linear theory.

These two regimes readily account for the simulation results. Figure 6 shows a plot of the wave energy at saturation and the heating rate versus the magnitude of the driving field. The points are results of the computer simulations, and the solid lines are the theoretical predictions. The formulae for these curves are

$$E_W^2 = \text{smaller of} \begin{cases} E_o^2 \\ \dfrac{m^2 k^2}{16\,e^2} [v_p + v_o - 2v_{Te}]^4 \end{cases}$$

and

$$\nu^* = 4\gamma\,\frac{E_W^2}{E_o^2},$$

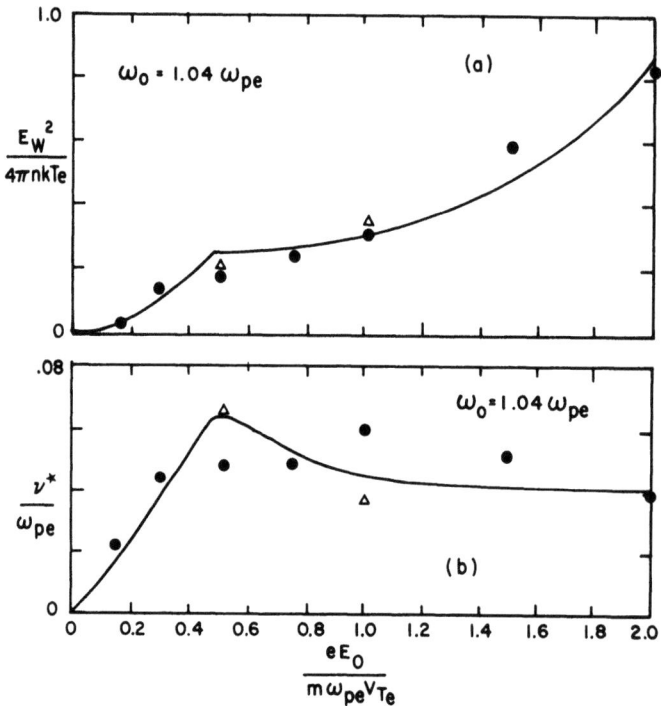

Fig. 6

The saturation values of the wave energy and the anomalous resistivity versus the magnitude of the external driver field ($m/M = 1/100$).

where v_p and γ have been taken to correspond to the most unstable mode. (Again, a small correction due to the finite-size of our simulation particles has been incorporated when computing γ and v_p.)

We see that there is reasonable agreement showing that our simple theoretical arguments capture the dominant physical effects. The triangles in this figure are the results of two-dimensional simulations. We will shortly discuss these in more detail, but we should note that the two-dimensional results are not significantly different from the one-dimensional ones.

VII. DEPENDENCE OF THE ANOMALOUS RESISTIVITY ON THE DRIVING FIELD FREQUENCY

Our basic picture of the anomalous resistivity predicts that for the second regime the enhancement of the resistivity will decrease when the driving frequency becomes significantly larger than the electron plasma frequency. The ion fluctuations couple energy from the driving field into electron plasma oscillations. The excited plasma oscillations are those whose Bohm-Gross frequencies are in the neighborhood (within roughly a growth rate) of the driving frequency. As ω_o increases from ω_{pe}, these plasma oscillations have larger wave numbers and smaller phase velocities. Their Landau damping rapidly increases, making them harder to excite; and when unstable they are saturated at lower amplitudes, since their lower phase velocities enable them to strongly trap particles at these lower levels.

Figure 7 confirms this behavior. Here we plot the saturation value of the wave energy and the anomalous resistivity versus the pump-field frequency for a fixed magnitude of pump field ($eE_o/m\omega_{pe}v_{Te} = 1.0$). The points are the results of the computer simulations, and the lines are the predictions of our simple theory.

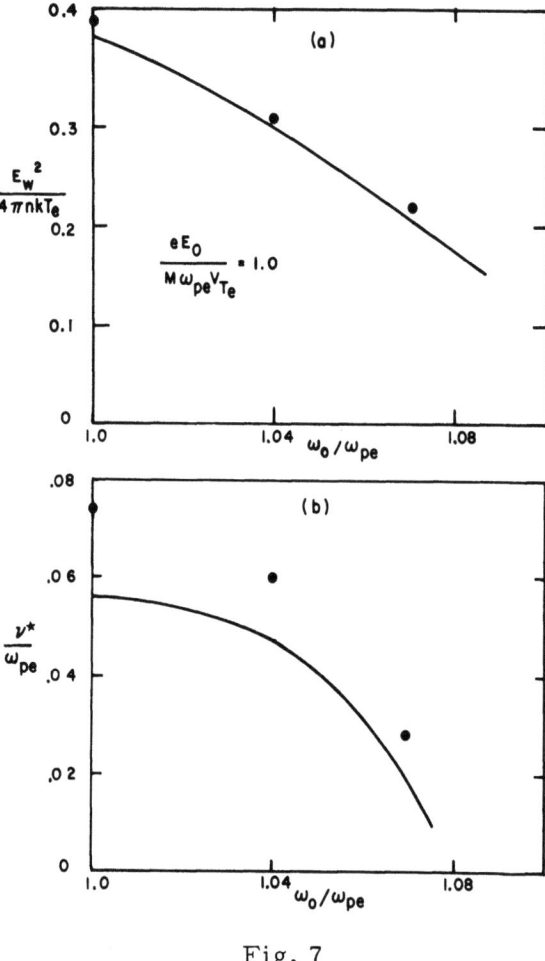

Fig. 7

The saturation values of the wave energy and the anomalous resistivity versus the external pump-field frequency ($eE_0/m\omega_p v_{Te} = 1.0$).

VIII. DEPENDENCE OF THE ANOMALOUS RESISTIVITY ON THE ELECTRON-ION MASS RATIO

Before discussing the two-dimensional simulations, let us indicate the dependence of these results on the electron-ion mass ratio. As is often the case in computer simulations, we have carried out the bulk of our simulations with an artificial mass ratio (in this case $m_i/m_e = 100$). The physics demands that the ion and electron time scales be well separated, and this we have achieved by a factor of 10. Nevertheless, it is important to indicate how the results scale with mass ratio. Hence we have carried out a few simulations with realistic mass ratios ($m_i/m_e = 2000$).

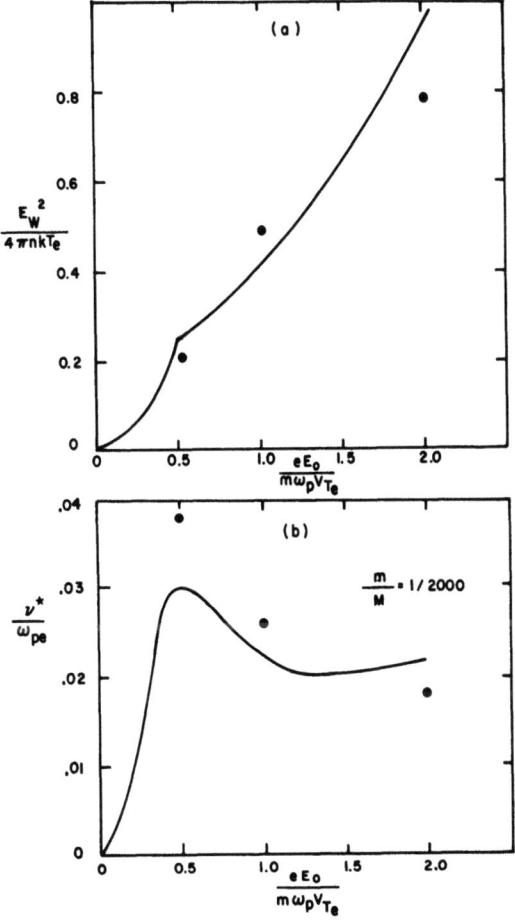

Fig. 8

The saturation values of the wave energy and the anomalous resistivity versus the magnitude of the external driver field ($m/M = 1/2000$).

Figure 8 shows the plot of the wave energy at saturation and the anomalous heating rate versus pump amplitude for several simulations with a mass ratio of 2000. Again the points are the results of computer simulation and the lines are the simple theoretical predictions.

The agreement between the simulations and theory is quite reasonable. A fair rule-of-thumb estimate is that the anomalous collision frequency varies roughly as the cube root of the electron-ion mass ratio. No simulations are presented for the weaker pump fields, since these would involve much larger systems and much longer runs on the computer.

IX. TWO-DIMENSIONAL SIMULATIONS

Finally, we have carried out some two-dimensional simulations of the anomalous heating. The instability is essentially one dimensional in character, but waves at an angle to the direction of the pump field can also grow. The growth rates fall off as $\cos^2 \Theta$, where Θ is the angle between the pump field and the wave vector of the excited oscillation. One consequence of this growth of off-angle modes will be a heating in the direction transverse to E_o. Furthermore, the off-angle modes allow richer possibilities for various mode-coupling processes which could saturate the instabilities at lower amplitudes. Hence it is of interest to examine the heating in a two-dimensional simulation.

The results of a typical two-dimensional simulation are shown in Fig. 9. The driving field parameters are the same as those for the sample one-dimensional simulation we have shown. It is apparent that the general behavior is quite the same — an exponentiation of the wave energy and a strong anomalous heating when the wave energy becomes large and saturates. Indeed the saturation values of the wave energy and the anomalous heating rates are reasonably the same as in the one-dimensional case, as we have noted previously.

The agreement of the two-dimensional and one-dimensional simulations is quite reasonable on the basis of our simple theory, when we note that the magnitude of the pump field used in the two-dimensional simulations puts in the second theoretical regime. Here strong particle trapping by the most unstable wave determines the saturation. Particle grapping from the distribution function does not depend sensitively on one or two dimensions.

Our first regime then, $E_o^2/4\pi nkT \ll 1$, is the one in which sizeable two-dimensional modifications of the saturation levels may be possible. We do not yet have two-dimensional simulations in this regime. Here the growth rates are significantly smaller (involving much longer runs) and the width of the unstable region in wave

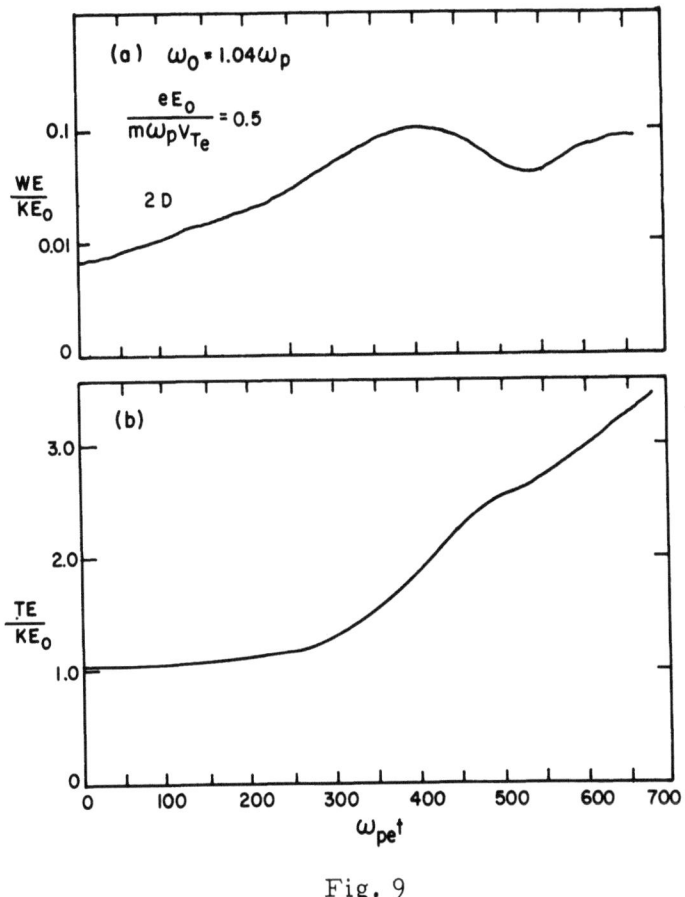

Fig. 9

The evolution of a) the self-consistent wave energy and b) the total plasma energy versus time (averaged over a plasma oscillation).

number is much narrower (necessitating the use of much longer systems). Nevertheless in this regime, our simple theory based on the one-dimensional results certainly gives useful upper bounds on the saturation amplitudes and heating rates to be expected.

The two-dimensional simulations yield additional very useful information. Namely, they show us what heating to expect in the direction transverse to the pump field. Figure 10 shows a plot of the distribution functions of the x and y velocities after anomalous heating has onset. The heating is dominantly along the pump field (x direction), but there is sizeable heating in the transverse

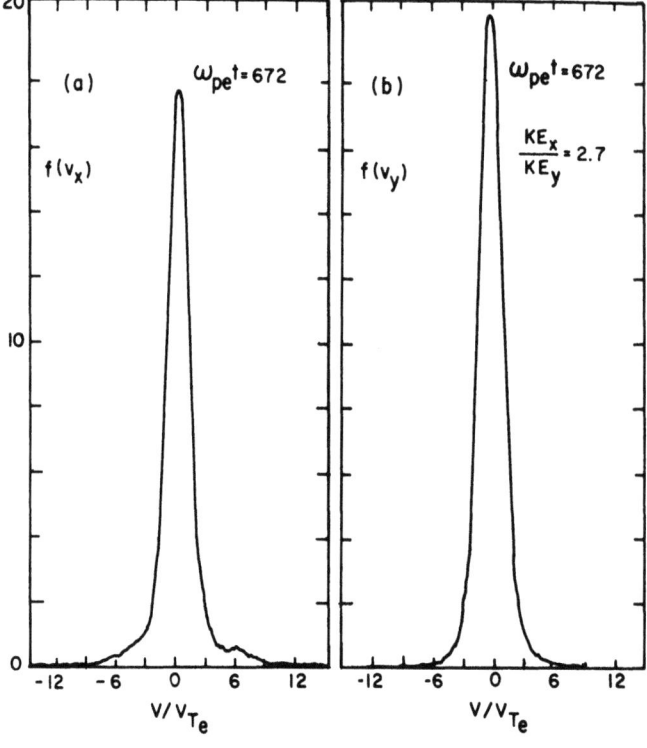

Fig. 10

The distribution functions of x and y electron velocities.

direction. The magnitude of the heating in this direction is less by a factor of about 3. This is reasonable since the off-angle modes grow less rapidly and the component of their phase velocities in the y direction is much greater than the component in the x direction. The anisotropy in the heating could lead to some interesting secondary instabilities. For example, the Weibel instability,[24] which is driven by a temperature anisotropy, might result and lead to a relaxation of the anisotropy.

X. SUMMARY

In conclusion, we have presented one- and two-dimensional computer simulations of anomalous high-frequency resistivity in a plasma driven by a large external field oscillating near the electron plasma frequency. The large pump field excites parametric instabilities which allow an efficient coupling of energy from the pump field into the plasma. We have investigated the saturation levels of the plasma fluctuations and the anomalous heating rates as a function of the pump field magnitude and frequency, and as a function of the ion-electron mass ratio. A very simple nonlinear theory reasonably accounts for the simulation results.

ACKNOWLEDGMENTS

We are happy to acknowledge useful discussions with Drs. B. Rosen, C. Oberman, P. Kaw, E. Valeo, H. Hendel, J. DeGroot, J. Byers, S. Bodner, and Mr. J. Iannucci. Messrs. R. Kluge, Y.C. Sun, and H. Fallon provided able numerical support.

This work was supported jointly by U.S. Atomic Energy Commission Contract AT(30-1)-1238, and Office of Naval Research Laboratory Contract N00014-67-A-0151-0021. Use was made of computer facilities supported in part by National Science Foundation Grant NSF-GP 579.

REFERENCES

1. P. K. Kaw and J. M. Dawson, Phys. Fluids 12, 2586 (1969).
2. W. L. Kruer, P. K. Kaw, J. M. Dawson, and C. Oberman Phys. Rev. Letters 24, 987 (1970); W. L. Kruer and J. M. Dawson, Phys. Rev. Letters 25, 1174 (1970); For anomalous heating by a large transverse field at ~ $2\omega_{pe}$, see W. L. Kruer and J. M. Dawson, Phys. Fluids 14, 1003 (1971).
3. Simulations of the anomalous heating by very large pump fields have been carried out by J. DeGroot and J. Katz at the Lawrence Radiation Laboratory (Livermore, California).
4. I. R. Gekker and O. V. Sizukhin, ZhETF Pis. Red. 9, 408 (1969) [JETP Lett. 9, 243 (1969)].
5. P. K. Kaw, E. Valeo, and J. M. Dawson, Phys. Rev. Letters 25, 430 (1970).
6. H. P. Eubank, Phys. Fluids 14, (1971).
7. H. Hendel and T. K. Chu (private communication).

8. H. Dreicer, D. Henderson, and J. Ingraham, Phys. Rev. Letters (to be published).
9. J. W. Shearer (private communication).
10. R. Cohen and J. D. Whitehead, J. Geophys. Res. 75, 6439 (1970); F. W. Perkins and P. K. Kaw, J. Geophys. Res. 76, 282 (1971).
11. A. Wong (private communication).
12. J. M. Dawson, C. G. Hsi, and R. Shanny, in Plasma Physics and Controlled Nuclear Fusion Research (International Atomic Energy Agency, Vienna, 1969) Vol. I, p. 735; W. L. Kruer and J. M. Dawson, presented at the Third Conference on Numerical Simulation of Plasmas, Stanford, Calif. (1969).
13. B. Rosen, W. Kruer, and J. M. Dawson, presented at the Fourth Conference on Numerical Simulation of Plasmas, Washington, D.C. (1970).
14. C. K. Birdsall and D. Fuss, J. Comp. Phys. 3, 494 (1969); R. Morse and C. Nielsen, Phys. Fluids 12, 2418 (1969); J. P. Boris and K. V. Roberts, J. Comp. Phys. 4, 552 (1969).
15. J. Byers and M. Grewal, Phys. Fluids 13, 1819 (1970); J. Denavit and W. L. Kruer, Phys. Fluids 14, 1782 (1971).
16. C. K. Birdsall and J. M. Dawson, in Computers and Their Role in Physical Science, edited by S. Fernbach and A. Taub (Gordon and Breach, New York, 1970) p. 247.
17. V. P. Silin, ZhETF 48, 1679 (1965), [JETP 21, 1127 (1965)].
18. K. Nishikawa, J. Phys. Soc. Japan 24, 916, 1152 (1968); J. R. Sanmartin, Phys. Fluids 13, 1533 (1970).
19. S. Aihara, S. Takamura, and K. Takayama, Nagoya University (Japan) Institute of Plasma Physics Report No. IPPJ-93 (1970).
20. J. M. Dawson and C. Oberman, Phys. Fluids 5, 517 (1962); 6, 394 (1963).
21. E. Valeo and C. Oberman have a systematic kinetic theory treatment of this problem for pump fields near threshold and for $T_e \sim$ few T_i. They then find that scattering of the unstable plasma oscillations by ion fluctuations is the dominant physical process for the saturation of the instability (private communication).
22. V. V. Pustovalov and V. P. Silin ZhETF 59, 2215 (1970).
23. J. M. Dawson and R. Shanny, Phys. Fluids 11, 1506 (1968).
24. E. S. Weibel, Phys. Rev. Letters 2, 83 (1959).

A CLASSICAL RESONANCE IN THE OPTICS OF THIN FILMS AT THE PLASMA FREQUENCY *

R. P. Godwin

University of California, Los Alamos Scientific Laboratory, Los Alamos, New Mexico 87544

Consider p light with electric vector E_o incident from vacuum on a dielectric film at an angle of incidence θ as shown in Fig. 1. The familiar boundary conditions of the continuity of the tangential component of the electric vector and the normal component of the displacement at the surface of incidence give

$$\tilde{E}_{||} = E_o \cos\theta (1-\tilde{r}), \text{ and} \tag{1}$$

$$\tilde{\varepsilon}\, \tilde{E}_{\perp} = E_o \sin\theta (1+\tilde{r}). \tag{2}$$

The reflectance $R = |\tilde{r}|^2$. The electromagnetic energy density just inside the surface is

$$U = \frac{1}{8\pi} |\tilde{E}|^2 = \frac{1}{8\pi} |E_o|^2 \{\cos^2\theta |1-\tilde{r}|^2 + \frac{\sin^2\theta}{|\tilde{\varepsilon}|^2} |1+\tilde{r}|^2\}. \tag{3}$$

In the special case of small thickness the transmittance $T = |\tilde{t}|^2 \to 1$ and the reflectance $R \to 0$. At the plasma frequency ε_1 (the real part of the complex dielectric constant $\tilde{\varepsilon} = \varepsilon_1 + i\, \varepsilon_2$) vanishes and we will assume $\varepsilon_2 \ll 1$. In this case we find the energy density in a thin film with p polarized light incident at the plasma frequency to be

$$U \simeq \frac{1}{8\pi} |E_o|^2 \frac{\sin^2\theta}{\varepsilon_2^2}. \tag{4}$$

The absorbed power per unit volume is

$$\frac{dU}{dt} = \frac{1}{8\pi} \varepsilon_2 \omega |\tilde{E}|^2 \simeq \frac{\omega_p}{8\pi} |E_o|^2 \frac{\sin^2\theta}{\varepsilon_2}. \tag{5}$$

Since $\varepsilon_2 \ll 1$ there is an "anomalous" enhanced absorption at the plasma frequency of a thin film which can be explained by ordinary linear optics.

Reference: R. P. Godwin, Springer Tracts Mod. Phys. __51__, 1 (1969).

1. W. L. Kruer and J. M. Dawson, "Anomalous Absorption of Intense Radiation," preceding contribution, p. 317

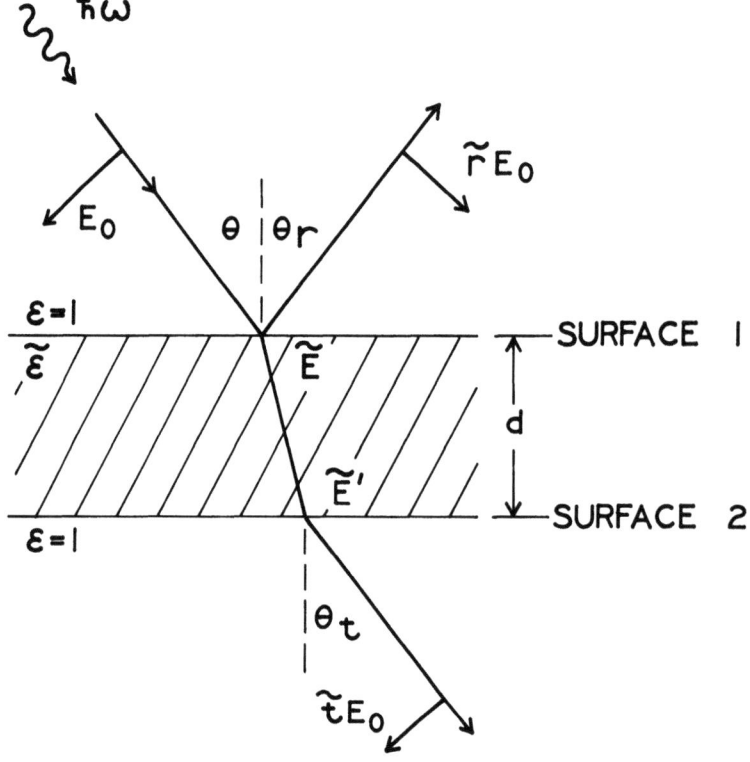

FIGURE 1

---

*Comments on the preceding paper[1] presented at the Second Workshop on "Laser Interaction and Related Plasma Phenomena" at Rensselaer Polytechnic Institute, Hartford Graduate Center, August 30 – September 3, 1971. See also R. P. Godwin, Phys. Rev. Letters __28__, 85 (1972).

NONLINEAR FORCES IN LASER PRODUCED PLASMAS[+]

Heinrich Hora

Max-Planck-Institut für Plasmaphysik
Euratom Association, Garching, Germany
and Rensselaer Polytechnic Institute, Hartford,
Conn., U.S.A.

ABSTRACT

The very recent observation of a correlation of the increase of reflectivity from laser produced plasmas with the onset of nuclear fusion reactions indicates a nonlinear mechanism. Numerical calculations of Shearer, Kidder and Zink of the plasma dynamics on interaction with light that allow for the nonlinear force, described before, on the basis of the ponderomotive interaction for collisionless dispersion effects indicated the increase of reflectivity due to this force. A review on this force is given and an exact numerical example of 10^{15} W/cm^2 laser intensity (Nd glass) in a plasma with a temperature of 10^3 eV and with a WBK-like density profile shows the predominance of the nonlinear force over the thermokinetic force. More detailed calculations of the threshold intensities I for the predominance of this force for a net plasma acceleration around $I^* = 10^{14}$ W/cm^2 are reported. These include the calculation of Steinhauer and Ahlstrom, where this threshold was given even for lower intensities but

---

[+] Presented at the Second Workshop on "Laser Interaction and Related Plasma Phenomena" at Rensselaer Polytechnic Institute, Hartford Graduate Center, August 30 - September 3, 1971.

so that temperatures exceeding 1o keV are then needed. The non-WBK-like density profile as treated analytically by Lindl and Kaw, resulted in the same net acceleration of the layer as in the WBK case, and the acceleration towards the nodes of the created standing wave was again very similar to the WBK case treated before by the author. For oblique incidence, where a Brillouin turbulence can occur, Lindl and Kaw found the possibility of a polarization dependent increase of the nonlinear force due to a Ginzburg-Denisov coupling of electromagnetic and electrostatic waves. Numerical calculations for non-WBK cases are reviewed, showing the predominance of the net acceleration and the increase of reflectivity. The forces in standing waves can exceed the thermokinetic forces even at intensities appreciably below I^*, as Mulser found from a dynamic numerical calculation. This fact allows the assumption that this mechanism may be one reason for the correlation of reflectivity and neutron production.

INTRODUCTION AND SYNOPSIS

Laser produced fusion plasmas show many very unexpected properties compared with the plasmas generated with lower laser powers than about 10^{14} W/cm^2, in which no fusion neutrons are produced. While the optical reflectivity of plasmas produced by laser intensities between 10^9 and 10^{13} W/cm^2 is remarkably low (only a few per cent although the radiation interacts with an overdense plasma)[1-4], a reflectivity of more than 3o % was reported from the first neutron producing plasma created by picosecond pulses[5]. Furthermore, it was observed that in the fusion experiment with nanosecond pulses of Floux et al.[6] the reflectivity was also always remarkably high when neutrons were produced, otherwise the reflectivity was small[7]. The same increase of reflectivity was correlated with other measurements of fusion neutrons, and a further correlation with the measurement of a temperature[8] of about 2 keV in addition to the usual temperature of about o.5 keV was found. The same increase of the reflectivity with the neutron production was also observed by Swain et al.[9], where the high reflectivity was observed only when a laser pulse impinged a few nanoseconds before the main pulse.

The increase of the reflectivity with increasing intensity could be seen from simplified numerical calculations using a rectangular laser pulse[7,8], and linear absorption processes only, but the experiment of Swaine et al. demonstrated that the preheating of the plasma and its creation of a slightly inhomogeneous plasma surface should not increase the reflectivity on the basis of the linear theory. It is to be underlined that a few months before Floux et al.[7] measured the correlation between reflectivity and neutron production for the first time, J. Shearer[10] et al. found from their numerical model WAZER that a remarkable increase of the reflectivity results if at high laser intensities the nonlinear force[11-13] is included in the calculations. At present it is too early to say whether this result completely explains the observed correlation, but it seems to be a first obvious indication of the action of nonlinear processes in high intensity laser produced plasmas. Indeed, there may be a number of other nonlinear processes involved, such as the nonlinear change of the collision frequency[13-15], the relativistic correction of the plasma frequency[16], coupling of the transverse waves with longitudinal waves[17], self-focusing with a low-power threshold[12,18] or with a high intensity threshold[19], two stream instabilities[20], or anomalous absorption at the cut-off density[21]. Nearly all of these processes have their threshold at about 10^{14} W/cm^2 and it would be puzzling if plasmas irradiated at higher intensities than 10^{14} W/cm^2 should be described only in terms of linear thermokinetic processes.

This contribution presents a review of the results for the theory of the nonlinear force, following the detailed description of the basic theory at the preceding Workshop[12]. From the outset it is to be pointed out that this type of nonlinear force we are treating is essentially connected with the strong changes of the absolute value of the complex refractive index $\tilde{n}$ insofar as the actual electrical field strength of the laser light $\underline{E}$ is given by the field $\underline{E}_v$ in vacuum as

$$\underline{E} = \frac{\underline{E}_v}{\sqrt{|\tilde{n}|}} \qquad (1)$$

or somehow modified by absorption or standing wave processes. Only these variations of the fields can create forces of the kind $\nabla(E^2/|\tilde{n}|)$ which give a net acceleration of the whole plasma layer against the laser light and a recoil to the plasma interior by an increased radiation pressure (increased possibly by factors of a hundred and more). Other types of nonlinear forces such as the ponderomotive, electrostrictive or magnetostrictive type have been well known for a long time, e.g. radiation pressure or plasma confinement by standing waves[22] or similar configurations[23]. With the exception of Ref.[12] there is no case known where the variation of the refractive index was included generally. All that is important for the net acceleration and is mainly responsible for the Brillouin turbulences (see Ref.[12] p. 408), the increase of reflectivity[10] and the polarization dependent increase of the net acceleration oblique incidence[24]. The standing wave process modified by the dispersion effects also seems to be important, as will be shown.

After sketching the derivation and the basic equations of the nonlinear force, the progress achieved in determining the range of validity is reported, including the paper of Steinhauer and Ahlstrom[25], where the predominance of the nonlinear force over the thermokinetic force is proved for laser intensities even below I^* if the electron temperature of the plasma exceeds 10^4 eV; the same treatment can be used to prove the predominance for $I > I^*$ at electron temperatures $T > 30$ eV. The properties of the force are demonstrated by the special example of a plasma of constant temperature where the density profile satisfies the WBK condition. The analytic treatment of Lindl and Kaw for a collisionless plasma not fulfilling the WBK condition proves the agreement of the calculated nonlinear force with the value of the WBK case where the latter is valid, and beyond this range the nonlinear force is even larger than the formal value of the WBK case. At oblique incidence a resonance-like increase of the force is derived. Finally, the numerical results of Mulser and Green[26] are reported for non-WBK-like profiles of a collisional plasma, the nonlinear force being compared with the thermokinetic force. Even at an intensity of neodymium glass lasers only a little higher than the threshold ($I = 5 \times 10^{14}$ W/cm^2) and at low plasma temperature (T_e 100 eV), the net nonlinear forces are 2 to 3 times as high as the thermokinetic forces. The forces due to standing waves are 12 times as high as the thermokinetic forces. All these cases were treated for the very special condition of static density profiles. The more interesting case of dynamics was treated for the first time by Shearer, Kidder, and Zink[10], some results of whose will be reported at the end of the paper. It is to be mentioned that the following short contribution of P. Mulser[27] reports on some further steps of a dynamic treatment.

BASIC CONCEPT OF THE NONLINEAR FORCE

The nonlinear force is given by direct interaction of the electromagnetic radiation with a plasma, thus causing the plasma to accelerate without any conversion of the radiation energy into thermal energy. There are different possible methods of derivation, always leading to the same expression for the force. The theory is based on the two-fluid model of plasma physics[11] or a combination of the single particle motion of the electrons in the plasma whose electric and magnetic fields $\underset{\sim}{E}$ and $\underset{\sim}{H}$ are determined by the collective properties of the plasma[12]. The most general derivation based on the nonrelativistic Lorentzian theory starts from the general equation of the ponderomotive force density[28]

$$\underset{\sim}{f} = \nabla \cdot \underset{\approx}{\sigma} - \frac{\partial}{\partial t} \frac{\underset{\sim}{E} \times \underset{\sim}{H}}{4\pi c} . \qquad (2)$$

By this definition of Landau and Lifshitz both forces in the plasma are ponderomotive forces, the force generated by the gas dynamic pressure tensor $\underset{\approx}{p}$ and any force given by the electromagnetic field strengths $\underset{\sim}{E}$ and $\underset{\sim}{H}$, determining the stress tensor $\underset{\approx}{\sigma}$ in Eq.(2). The tensor components are

$$\sigma_{ij} = -p_{ij} - \left[\frac{E^2}{8\pi}(\tilde{\eta}^2 - \rho\frac{\partial \tilde{\eta}^2}{\partial \rho}) + \frac{H^2}{8\pi}\right]\delta_{ij} + \tilde{\eta}^2 \frac{E_i E_j}{4\pi} + \frac{H_i H_j}{4\pi}$$

where ρ denotes the density, and the complex refractive index $\tilde{\eta}$ of electromagnetic waves with a frequency ω is

$$\tilde{\eta}^2 = 1 - \frac{\omega_p^2}{\omega^2 + \nu^2}\left(1 + i\frac{\nu}{\omega}\right) . \qquad (3)$$

Here the plasma frequency ω_p

$$\omega_p^2 = 4\pi e^2 n_e / m_e \qquad (4)$$

is given by the charge e, the density n_e and the mass m_e of the electrons. The relativistic change of the electron mass[16] at neodymium glass laser intensities exceeding 10^{16} W/cm^2 will be neglected here because these intensities are about 100 times as high as these considered here. The collision frequency ν in Eq.(3) is given by the theory of the inverse bremsstrahlung[29] or by the Dawson-Oberman theory[30], or derived[31] from Spitzer's d.c. plasma resistivity[32], nearly the same values always being obtained. The collision frequency of the collision theory is

$$\nu = \frac{\omega_p^2 \pi^{3/2} m_e^{1/2} Z e^2 \ln \Lambda}{8\pi \gamma_E(Z)(2kT)^{3/2}} \qquad (5)$$

with the Boltzmann constant k, the ion charge Z, with Spitzer's correction factor γ_E (being 0.2 ... 1) and the Coulomb logarithm with Λ

$$\Lambda = \frac{3}{2Ze^3} \left(\frac{k^3 T^3}{\pi n_e}\right)^{1/2} \tag{6}$$

The electron temperature T in Eq.(5) is such that the electron motion determines the dispersion relation, which means that the temperature is given by the random thermal motion according to a temperature T_{th} and by the coherent motion of the electrons in the electromagnetic field, given by the actual oscillation energy of the electrons ε_e^{osc}

$$T = T_{th} + \varepsilon_e^{osc}/k \tag{7}$$

In this way the nonlinear change of the collision frequency at high light intensities is included.

With an appropriate interpretation of the density factor in Eq.(3), Eq.(2) can be written

$$\underline{f} = -\nabla \cdot \underline{\underline{p}} + \nabla \cdot \left[\underline{\underline{U}} - (\tilde{\eta}^2 - 1)\underline{E}\underline{E} - \frac{\partial}{\partial t} \frac{\underline{E} \times \underline{H}}{4\pi c}\right] \tag{8}$$

where the gas dynamic part is called the thermokinetic force

$$\underline{f}_{th} = -\nabla \cdot \underline{\underline{p}} \tag{9}$$

and therefore

$$\underline{f}_{NL} = \underline{f} - \underline{f}_{th} \tag{10}$$

is essentially nonlinear owing to the quadratic forms of the electromagnetic field components. In the following the Poynting term (last term in Eq.(2)) can be neglected because of the much slower change of the radiation intensities compared with the laser periods. Averaging Eq.(8) over the time of one laser period neglects all effects occurring during these short times and results in the interesting long-term forces.

Some essential properties of the nonlinear force can be seen from the special case of perpendicular incidence of the laser radiation along the x-coordinate with linear polarization of the E-vector in the y-direction onto a stratified plasma. The time averaged nonlinear force is from Eqs.(8) and (10)

$$\overline{\underline{f}}_{NL} = \frac{\underline{i}_x}{8\pi} \frac{\partial}{\partial x} (\underline{E}_y^2 + \underline{H}_8^2) \tag{11}$$

This force is well known from standing waves in a thin plasma[22], where the particles are driven towards the nodes. If then the refractive index is always very close to unity, no spatially averaged net motion of the plasma results. Any net motion can be expected only by dispersion effects, when the absolute value of the refractive index differs appreciably from unity, which only occurs in the neighborhood of the cut-off density.

In the following sections the nonlinear force is evaluated for general density profiles, of which the quantities E_y and H_z are especially calculated. To gain a better insight into the mechanism of the force, we first discuss the case of the WBK approximation which forces also agree with those derived from general non-WBK cases[24].

The WBK solution for a reflection-free penetrating wave with an amplitude E_v in vacuum is

$$\underset{\sim}{E} = \underset{\sim}{i}_y \frac{E_v}{\tilde{\eta}^{1/2}} \exp(iF) \exp(\mp \frac{K(x)}{2} x); \quad \underset{\sim}{H} = \underset{\sim}{i}_3 |\underset{\sim}{E}| \tilde{\eta} \qquad (12)$$

where

$$F = \omega \left(t \mp \int^x \frac{\mathrm{Re}(\tilde{\eta}(\xi))}{c} d\xi \right) \text{ and } K(x) = \frac{1}{x} \frac{\omega}{c} \int^x \mathrm{Im}(\tilde{\eta}(\xi)) d\xi. \quad (12a)$$

The condition of the WBK approximation is

$$\theta = \frac{c}{2\omega} \frac{1}{|\tilde{\eta}|^2} \frac{\partial |\tilde{\eta}|}{\partial x} \ll 1 . \qquad (13)$$

Using Eq.(12) we find from Eq.(11)

$$\bar{\underset{\sim}{f}}_{NL} = \underset{\sim}{i}_x \frac{E_v^2}{16\pi} \frac{1-|\tilde{\eta}|^2}{|\tilde{\eta}|} \exp(-Kx) \frac{\partial}{\partial x} |\tilde{\eta}| + \underset{\sim}{i}_x \frac{E_v^2}{16} \frac{1-|\tilde{\eta}|^2}{|\tilde{\eta}|} \frac{2\omega}{c} \mathrm{Im}(\tilde{\eta}) \exp(-Kx) \qquad (14)$$

At densities below the cut-off density, only the first term of the right-hand side dominates, which is the essential part of the nonlinear force, as can be seen if Eq.(13) is written for a collisionless plasma ($\nu = 0$, $K = 0$; $\mathrm{Im}(\tilde{\eta}) = 0$) using Eqs.(3) and (4)

$$\bar{\underset{\sim}{f}}_{NL} = \underset{\sim}{i}_x \frac{E_v^2}{16\pi} \frac{\omega_p^2}{\omega^2 \tilde{\eta}^2} \frac{\partial \tilde{\eta}}{\partial x} , \qquad (15)$$

while at densities exceeding the cut-off density n_{eco}

(given by Eq.(4) if $\omega_p = \omega$) the second term of the right-hand side of Eq.(14) dominates. This term describes the usual radiation pressure, but increased by the refractive index variations.

NUMERICAL EXAMPLE OF THE WBK CASE

In Fig. 1 we report on the numerical calculation of the conditions and force densities of a special static case where a neodymium glass laser pulse with an intensity of 10^{15} W/cm^2 is perpendicularly incident on a plasma with local constant temperature of 10^3 eV and with a density profile such that the radiation can penetrate reflection-free according to the WBK condition (13). The term "static case" means that the evaluated light intensities, pressures and force densities are valid only during a time where the density profile can be considered constant. If this profile were an initial condition, the further dynamic development would obviously change the profile and the assumed local constance of the temperature.

Firstly, the electron density n_e was calculated on the basis of a value $\theta = 0.1$ in Eq.(13) with a monotonic increase along the depth x. This was performed by starting from a density difference Δn_e and calculating $|\tilde{\eta}|$ and $\Delta|\tilde{\eta}|$ from Eqs.(3), (4) and (5) using in Eq.(7) $T = T_{th}$ and then Δx from Eq.(13). The approximation $T = T_{th}$ in Eq.(7) causes a lower bound of the final values of the forces because the implication of the electron oscillation causes higher temperatures, hence lower values $|\eta|$, and then stronger force densities. The result in Fig. 2 is a very slow increase of $n_e(x)$ around the cut-off density, which for neodymium glass laser radiation is around 10^{21} cm^{-3}. The second step was to calculate the intensity

$$I = E^2 = \frac{E_v^2}{|\eta|} \exp(-K(x)x) \qquad (16)$$

following the numerical integration according to Eq.(12a). The strong increase of $(E/E_v)^2$ with depth up to values beyond 20 proves that the change of the $|\eta|$ values occurs much earlier than the decrease of I is influenced by the integral absorption constant K(x). Therefore, the maximum values of $|\eta|^{-1}$ are higher than 20. This results in an increase of the effective wavelength $\lambda = \lambda_o/|\eta|$ compared to the vacuum wavelength λ_o by more than 20 as well. From this point of view, the nearly pathologically flat density profile from 25 to 65 μm (Fig. 1) is only over about four effective wavelengths.

The next step was to calculate the thermokinetic force density $f_{th} = -\nabla p = -kT\nabla n_e$, which is given by the pointed curve in Fig. 1. The nonlinear force f_{NL} was negative up to a depth of 52 μm and is drawn positive in Fig. 1. First we have to remark that the nonlinear force is larger than the thermokinetic force for a depth larger than 13 μm and can be 9 times as large as the maximum of the thermokinetic force. To use one scale in the logarithmic diagram we drew the total force $f = f_{NL} + f_{th}$ and its components with negative sign, meaning an acceleration of the plasma in the negative x-direction. For depths larger than 53 μm the nonlinear force f_{NL} is positive (directed in the +x-direction), and because its absolute value is very high the total force f is positive up to a depth of 68 μm, where the nonlinear force decreases strongly because of the absorption of the radiation. The depth from a little more than 53 μm to 68 μm is nearly the skin depth, which is so large because the effective wavelength is at some depth larger than 2o vacuum wavelengths. We continued to increase n_e monotonically for higher depth in an arbitrary manner. A detailed dynamic calculation may also result in a decreasing n_e within the highly shocked material below the radiation zone, but this question is irrelevant at this point.

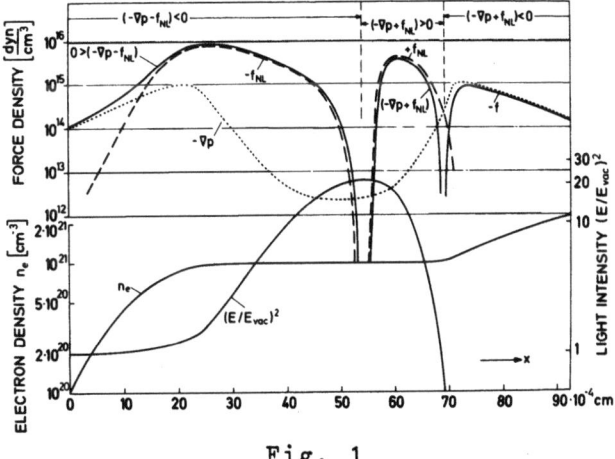

Fig. 1

Numerical calculation of a WBK-like density profile $n_e(x)$ of a plasma of temperature T = 1 keV. The radiation of 10^{15} W/cm^2, incident from the left-hand side, produces an amplitude E of the electric field of the penetrating light wave compared with the amplitude E_v in vacuum before incidence. The resulting thermokinetic force f_{th} and the nonlinear force f_{NL} are calculated on the basis of the linear collision frequency.

RANGE OF PREDOMINANCE OF THE NONLINEAR FORCE

The example of the preceeding section demonstrated the predominance of the nonlinear force over the thermokinetic force at intensities of neodymium glass lasers of 10^{15} W/cm^2. This agrees with our former calculations[12], where the threshold of predominance was found to be near 10^{14} W/cm^2. The evaluation of this threshold was given in more detail by an iteration method[13] and was discussed by Steinhauer and Ahlstrom[25]. We review these treatments in three steps, namely a simplified consideration, the iteration procedure and the estimation of Steinhauer and Ahlstrom[25].

Simplified Consideration

For a plasma where the absorption is small up to the cut-off density, as shown in the preceding section, the oscillation energy of the electrons

$$\varepsilon_e^{osc} = \frac{\varepsilon_{ev}^{osc}}{|\tilde{\eta}|} \qquad (17)$$

is given by its value in vacuum, but increased by the changes of the refractive index $\tilde{\eta}$. The maximum value of

$$\frac{1}{|\tilde{\eta}|}\bigg|_{max} = \frac{T^{3/4}}{a} \qquad (18)$$

has already been derived[12,13], where

$$a = \left[\frac{\omega_p^2 \pi^{3/2} m_e^{1/2} Z e^2 \ln\Lambda}{8\pi\omega_E^2 (Z)(2k)^{3/2}} \right]^{1/2} (eV)^{3/4} \qquad (19)$$

has values around 2...4 $(eV)^{3/4}$ for neodymium glass lasers at temperatures around $10^2...10^4$ eV. A simplified integration of the equation of motion was possible to evaluate the final energy ε_i^{trans} of the translational motion of the ions with a charge Z after the nonlinear acceleration where the details of the special density profile cancelled.

$$\varepsilon_i^{trans} = Z \, \varepsilon_e^{osc}\bigg|_{max} = Z \, \varepsilon_{ev}^{osc} \cdot \frac{T^{3/4}}{a} \qquad (20)$$

The problem is to use the electron temperature T of Eq.(7). The temperature where the random electron energy is equal to the energy of coherent motion

$$T_{th} = \varepsilon_{ev}^{osc} \cdot \frac{T_{th}^{3/4}}{k\,a} \tag{21}$$

leads with (at $\omega \approx \omega_p$)

$$\varepsilon_{ev}^{osc} = \frac{E_v^{1*\,2}}{16\pi n_{eco}} \frac{\omega_p^2}{\omega^2} \sim I^* \tag{22}$$

to

$$E_v^{1*\,2} = 16\pi n_{eco}\, ak\, T_{th}^{1/4} \tag{23}$$

In this way we find from Eq.(20)

$$\varepsilon_i^{trans} = \begin{cases} Z\, \dfrac{E_v^2}{16\pi n_{eco}} \cdot \dfrac{T_{th}^{3/4}}{a}, & \text{if } \varepsilon_e^{osc} \ll T_{th} \\[2ex] Z\, \dfrac{E_v^8}{(16\pi a n_{eco} k^{3/4})^4}, & \text{if } \varepsilon_e^{osc} \gg T_{th} \end{cases} \tag{24}$$

which shows resonance-like increase of the ion energy due to the nonlinear force at intensities a little higher than I^*. The damping mechanism can be explained qualitatively from the fact that at the start of this resonance the absorption due to the nonlinear force also acts and will result in an asymptotic limiting value of the ion energy ε_i^{trans}. The evaluation of Eq.(23) gives the threshold values with $[T] = $ eV

$$E_v^{1*} = \begin{Bmatrix} 2.83 \times 10^8 \\ 1.67 \times 10^8 \end{Bmatrix} \cdot T_{th}^{1/8}\ \text{V/cm};\quad I^{1*} = \begin{Bmatrix} 2.08 \times 10^{14} \\ 7.5 \times 10^{13} \end{Bmatrix} T_{th}^{1/4}\ \text{W/cm}^2$$

$$\text{for } \begin{Bmatrix} \text{ruby} \\ \text{Nd glass} \end{Bmatrix} \tag{25}$$

This first approximation is indicated by the index 1 of the threshold values.

Iteration Method with WBK Approximation

The most direct way to compare the nonlinear force with the thermokinetic force is to go back to Eq.(8) and compare the gasdynamic pressure with the momentum flux density of the electromagnetic field (after subtraction of its vacuum value) before performing the spatial differentiation. With the WBK approximation we have the following relation, with the general threshold indicated by an asterix:

$$n_e(1-1/Z)kT_{th} = \frac{E_v^{*2}}{16}\left[\left(\frac{1}{|\tilde{\eta}|} + |\tilde{\eta}|\right)\exp_o - 1\right]. \quad (26)$$

The exponential function given by the damping is to be determined separately and will be included as an open parameter between 1 and $1/e^2$. The fact that $|\tilde{\eta}|$ is to be used in the form of Eq.(17) with a temperature given by Eq.(7) causes the solution of Eq.(26) by an iteration only. For a set of E_v values with $\exp_o$ as a parameter we solved as a first step Eq.(26) for $|\tilde{\eta}| = a/T_{th}^{3/4}$ reaching $T_{th}^{(1)}$. With this value we calculated by iteration higher approximations $(n+1)(n=1,2...)$

$$n_e(1+1/Z)kT_{th}^{(m+1)} = \frac{E_v^{*2}}{16\pi}\left[\left(\frac{\left(T_{th}^{(m)} + \frac{E_v^{*2} T_{th}^{(n)3/4}}{16\pi a n_{eco} k}\right)^{3/4}}{a} + \frac{a}{\left(T_{th}^{(n)} + \frac{E_v^{*2} T_{th}^{(n)3/4}}{16\pi a n_{eco} k}\right)^{3/4}}\right)\exp_o - 1\right] (27)$$

When $T_{th}^{(n+1)} - T_{th}^n$ was less than 0.5 eV the iteration was finished. The results are shown in Figs. 2 and 3. We see that the first approximation I^{1*} is very close to the case of $\exp_o = 1$ to $1/2$, reproducing again the thresholds near 10^{14} W/cm^2 for neodymium glass lasers.

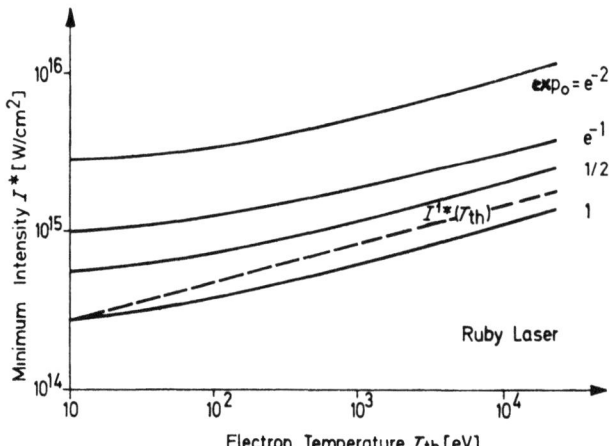

Fig. 2

Minimum intensity I^* for ruby laser to create larger nonlinear deconfining forces f_{NL} than thermokinetic forces $f_{th} = -\nabla n_e kT_{th}$. The undefined exponential function $\exp_o$ of collision induced attenuation is used as a parameter.

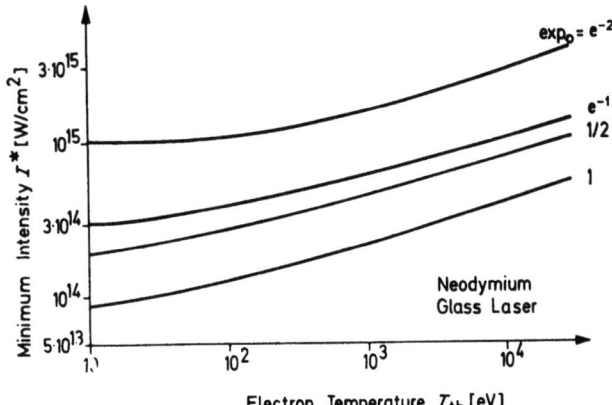

Fig. 3

Minimum intensity I^* for neodymium glass laser to create larger nonlinear deconfining forces f_{NL} than thermokinetic forces $f_{th} = -\nabla n_e kT_{th}$. The undefined exponential $\exp_o$ of collision induced attenuation is used as a parameter.

Steinhauer and Ahlstrom's Method

Steinhauer and Ahlstrom[25] compared the nonlinear force for the cases where no nonlinear changes of the collision frequency[12] take place, i.e. for neodymium glass laser intensities below 10^{14} W/cm^2. For a one-dimensional model, the thermokinetic pressure p due to the thermalization of the radiation is given by

$$\frac{dp}{db} = -\frac{2}{3} \nabla \overline{\underline{E} \times \underline{H}} \qquad (28)$$

Using approximations for the time of an acoustic wave to travel one absorption length and using averaging procedures which were recognized to include somewhat artificially the dependence of the thermokinetic force on the electron-ion collision time (as was pointed out by Lindl and Kaw[24]), the ratio

$$\frac{|\underline{f}_{th}|}{|\underline{f}_{NL}|} \gtrsim \frac{1.14 \times 10^5}{T^{5/4}} Z^{1/2} \quad ; \quad [T] = eV \qquad (29)$$

was given for the neodymium glass lasers. This means that the nonlinear force can exceed the thermokinetic force at low laser intensities only at temperatures exceeding 10^4 eV.

The temperature T in Eq.(29) is the quantity describing the energy of electrons with respect to the refractive index. Therefore, for high light intensities we have to use the value from Eqs.(7) and (17)

$$T = T_{th} + \frac{e \bar{e}_v^{osc}}{k|\tilde{\eta}|} \qquad (30)$$

Using the relation (18) with $T = T_{th}$ from Eq.(23) we get for $E \gg E^*$ a predominance of the second term on the right-hand side of Eq.(30). With such a T the relation (29) reads[13]

$$\frac{|f_{th}|}{|f_{NL}|} \geq \left(\frac{7.2 \times 10^8}{E_v}\right)^8 Z^{1/2} ; \quad [E_v] = V/cm \qquad (31)$$

This means that at intensities a little higher than I^*, a resonance-like increase of the predominance of the nonlinear force can be expected. The assumptions for Eq.(31) did not include such types of absorption which cause the nonlinear collisionless acceleration of the plasma. This mechanism will again cause a damping of the resonance.

Though the condition of Steinhauer and Ahlstrom[25] was not derived in the most general way it reconfirms the predominance of the nonlinear force for intensities exceeding I^* of about 10^{14} W/cm^2 and additionally proves a predominance at lower intensities as well if the temperature exceeds 10 keV. This result agrees also with the special numerical calculation of the preceding section, where at intensities of 10^{15} W/cm^2 the nonlinear force was more than 8 times as high as the thermokinetic force and also agrees with the following numerical results of non-WBK-like profiles of plasmas with collisions.

ANALYTICAL NON-WBK CASE OF LINDL AND KAW

An analytical treatment of the nonlinear force was performed by Lindl and Kaw[24] where a special restriction to an analytically well known case[33] was necessary, with a linear density profile at the surface of the plasma. The solutions of the Maxwellian equations for E_y and H_z for evaluating the nonlinear force in the case of a one-dimensional geometry, Eq.(11), can begin by solving the wave equation

$$\frac{\partial^2 E_y}{\partial x^2} + \frac{\omega^2}{c^2} \tilde{\eta}^2 E_y = 0 \qquad (32)$$

where the complex refractive index $\tilde{\eta}$ (Eq.(3)) has the following spatial variation

$$\eta^2 = -ax + is \qquad (33)$$

where

$$a \approx \frac{1}{|\tilde{\eta}|} \frac{\partial |\tilde{\eta}|}{\partial z} = \frac{2\theta\omega}{c} \quad ; \quad s = \frac{\nu}{\omega} \ll 1 \qquad (34)$$

the definition of θ being taken from Eq.(13). The collision frequency ν is assumed to be very small and the density profile $n(x)$ follows from Eq.(33) and (34)

$$\frac{4\pi e^2 n(x)}{m} = \omega_p^2(x) = \omega^2 \cdot \frac{x}{x_1} \qquad (35)$$

The layer has a depth x_1 and is distributed from $x = 0$ to x. At $x = x_1$ is the reflection point where a collisionless plasma creates total reflection of the incident wave. The solution of the wave equation (32) is given using the substitution

$$\zeta = \left(\frac{\omega}{ca}\right)^{2/3} \cdot (-ax + is) \equiv \varrho^{2/3} \xi \quad ; \quad \varrho = \left(\frac{\omega}{ca}\right)^{2/3} \qquad (36)$$

where the Eq.(32) reduces to

$$\frac{d^2 E_y}{d\zeta^2} + \zeta E = 0 \, , \qquad (37)$$

which results in the well known Airy function a_i. In expressions of the Bessel functions of the order $1/3$ we find

$$E_y = 3A a_i(-\zeta) = \begin{cases} A\zeta^{1/2}\left[J_{1/3}\left(\frac{2}{3}\zeta^{3/2}\right)+J_{-1/3}\left(\frac{2}{3}\zeta^{3/2}\right)\right], \text{ if Re}\zeta>0 \\[2mm] A(-\zeta)^{1/2}\left[I_{-1/3}\left(\frac{2}{3}(-\zeta)^{3/2}\right)+I_{1/3}\left(\frac{2}{3}(-\zeta)^{3/2}\right)\right], \text{if Re}\zeta<0 \end{cases}$$
(38)

with

$$A = \frac{2}{3}\pi^{1/2}\left(\frac{\omega}{ca}\right)^{1/6} E_v \exp\left(+\frac{i\omega}{c}\int_{-1/a}^{+is/a}\tilde{\eta}^{1/4}\,dz - \frac{i\pi}{4}\right) \qquad (39)$$

The solution of the magnetic field strength is derived from the Maxwell equations

$$H_z = + i\varrho^{-1/3}\left(\frac{dE_y}{d\varrho}\right) \qquad (40)$$

From Eq.(38) it can be seen[24] that for $x \geq 0$ there exists a totally reflected (standing) wave (exactly verified only in a collisionless plasma $\nu = 0$), as is to be expected, and for $x < 0$ a strongly damped wave occurs. The (time averaged) nonlinear force results simply by substituting Eqs.(38) and (40) in Eq.(11)

$$\bar{f}_{NL} = i_x \frac{\omega}{c} E_v^2 \exp(-2\rho s)\{(a_{iR} a'_{iR} + a_{iI} a'_{iI})$$

$$+ \rho^{-2/3} [a'_{iR}(\zeta_R a_{iR} - \zeta_I a_{iI}) + a'_{iI}(\zeta_I a_{iR} + \zeta_R a_{iI})]\} \quad (41)$$

where the real parts and the imaginary parts of the Airy functions are denoted by the indices R and I. The prime denotes differentiation with respect to the arguments.

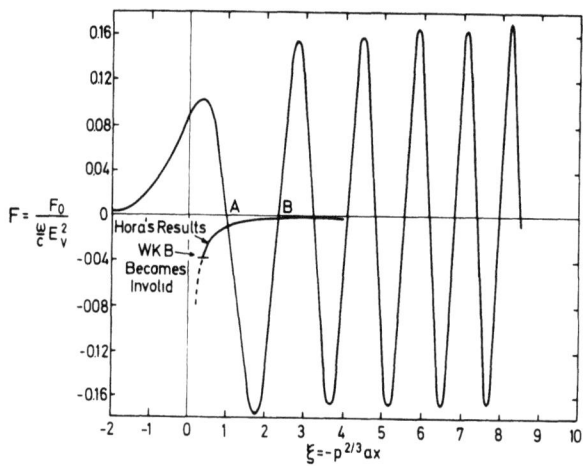

Fig. 4

Nonlinear force for a collisionless plasma of linear density profile at light incident from the right-hand side. The resulting standing wave creates the locally oscillating force, whose net acceleration is given by the double value of the monotonic force close to the abscissa. Comparison of the WBK case[12] with the rigorous case of the linear profile is given as derived by Lindl and Kaw[24].

Figure 4 reports on the numerical result of the evaluation of f_{NL} from Eq.(41) (oscillating curve). Very similar to our result of a standing wave with the WBK approximation[12] we find for the general case of Fig. 4 that the nonlinear force oscillates according to the acceleration of the plasma towards the nodes of the standing wave. In the neighborhood of the turning point ($x = x_1$; $\zeta = 0$) the effective wavelength increases and the maxima and minima of the force decrease monotonically. The spatially averaged net force acting on the whole surface region of the plasma is double the monotonic curve close to the abscissa. This superposition of the net nonlinear force was also observed in the WBK case[12].

If Lindl and Kaw[24] use the incident wave only, they find the nonlinear force

$$\overline{\underset{\sim}{f}}_{NL} = \underset{\sim}{i}_x \frac{E_v^2}{16\pi} \frac{\omega_p^2}{\omega^2 \tilde{\eta}^2} \frac{d\tilde{\eta}}{dx} \qquad (\frac{\nu}{\omega} \ll 1) \qquad (42)$$

in agreement with our expression Eq.(15). A more detailed evaluation of Eq.(41) of a collisionless plasma ($\nu = 0$) was given by Kaw[34]. The force at the deepest maximum of Fig. 4 is then

$$f_{NL}^{Kaw} = \frac{0.1}{36} \frac{E_v^2}{x_1} (\frac{\omega x_1}{c}) \left[1 + (\frac{c}{\omega x_1})^{2/3} \cdot 0.6 \right] \qquad (43)$$

This can be compared with the force in the WBK case, which by substituting Eq.(13) in Eq.(15) or (42)

$$f_{NL}^{WBK} = \frac{E_v^2}{16\pi} \frac{2\omega}{c} \theta \qquad (44)$$

because $\omega_p \approx \omega$. The ratio of Eqs.(43) and (44) is

$$\frac{f_{NL}^{Kaw}}{f_{NL}^{WBK}} = \frac{0.1}{36} \frac{16\pi}{2 \theta} \left[1 + (\frac{c}{\omega x_1})^{2/3} \; 0.6 \right] \qquad (45)$$

It is seen that at $(c/\omega x_1) \ll 1$ both forces are equal if the WBK parameter θ is 0.07. We note that for values $(c/\omega x_1) \gg 1$, in which case the WBK solution will become invalid, the rigorous solution of Kaw can even become very much larger than the formal (but invalid) value of the WBK solution. While the WBK solution will remain finite, in principle, as can be seen by combining Eqs.(13) and (15), Kaw's case could formally go to infinity but many other plasma physical conditions will then become critical.

Lindl and Kaw[24] also evaluated the solutions of the wave equations for the case of oblique incidence of the radiation on a plasma with linear increase of the electron density. The expressions for the ponderomotive force were taken from our former derivation of the case of oblique incidence[11]. While it was found for field strength derived from the WBK approximation that there was no strong difference of the plasma acceleration for the different cases of incidence, the linear density profile can cause an increase of the nonlinear force by a factor of 50 as high as for perpendicular incidence, if the electric vector of the light oscillates parallel to the plane of incidence. The reason is the influence of a certain function Φ, which is the result of the solution of the wave equation for this case, given by Denisov[35]. In this case a strong coupling of the electromagnetic waves and the longitudinal electrostatic waves occurs owing to finite temperature and plasma inhomogeneity in the interaction region, causing $|\tilde{\eta}| \ll 1$. There the collision frequency ν should be replaced by[33,35]

$$\nu_{eff} \simeq \nu + \omega \left(\frac{kTa^2}{m\omega^2} \right)^{1/3} \tag{46}$$

which, however, is a crude approximation. This strong increase of the nonlinear force at an optimum angle of incidence around 4 - 5° should be one of the possibilities to distinguish the nonlinear force from the thermokinetic processes.

The increased momentum P_{inh} transferred to the inhomogeneous layer was derived by Lindl and Kaw[24] on the basis of our relation[11,12]

$$P_{inh} = \int dy \, dz \int_{t_1}^{t_2} dt \int_{x_1}^{x_2} dx \, \underset{\sim}{f}_{NL} \tag{47}$$

where for oblique incidence with the $\underset{\sim}{E}$ vector in the plane of incidence

$$P_{inh} \approx -\frac{3}{2} P_o \frac{\exp(-2\rho s)}{\rho s^2} \tag{48}$$

where P_o is the momentum of the incident electromagnetic wave pulse at the times between t_1 and t_2

$$P_o = \int dy \, dz \int_{t_1}^{t_2} dt \, \frac{E_v^2(y,z,t)}{8\pi} \tag{49}$$

The quantity $s = (\nu/\omega)$ can become very small, especially at high electron temperatures or - as we have to point out - at high laser intensities following Eq.(7) at $I > I^*$. This results in another way in a very remarkable increase of the usual radiation pressure compared with the case treated before[36].

Lindl and Kaw[24] mentioned that the involved finite temperature and nonlinear effects may lead to enhanced absorption of the laser energy[20,21]. This may influence the nonlinear force positively or negatively.

NUMERICAL CASES

The numerical calculation of the field strengths $\underline{E}$ and $\underline{H}$ for a given density profile of a plasma permits numerical calculation of the nonlinear force according to Eq.(8) and (1o) or, for perpendicular incidence of the radiation, Eq.(11). At the beginning only static solutions could be treated in the same sense as the analytic treatment of the preceding section was a static case where the force was calculated for a special density profile and no attention was paid to the temporal development of the density profile and of the forces, as a dynamic treatment takes into account.

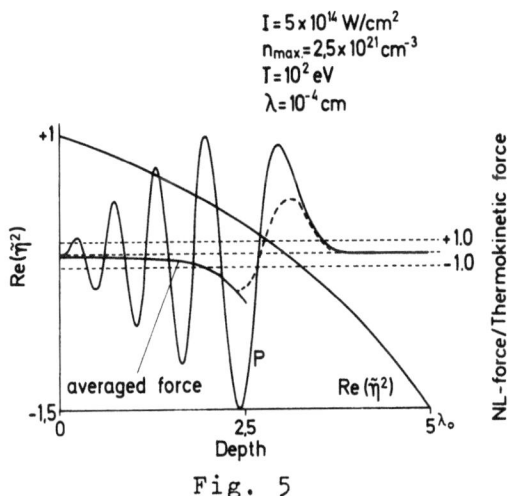

Fig. 5

Numerical calculation of the solutions of the wave equations for a density profile whose real part of the refractive index is parabolic, given by Mulser and Green[26]. The resulting standing wave at a plasma temperature T = 1oo eV for neodymium glass lasers of 5×10^{14} W/cm^2 incident intensity (from the left-hand side) creates nonlinear forces whose value is given as ratio to the actual thermokinetic force. The net force is given as a spatially averaged value. It exceeds the thermokinetic force by a factor higher than 2.5.

Figure 5 demonstrates the result of Mulser and Green[26], where a parabolic profile of the electron density n(x), given by Eq.(3) and (4) is used

$$R_e(\tilde{\eta}^2) = 1 - \frac{4\pi e^2}{m(\omega^2+\nu^2)} \cdot n(x) \qquad (50)$$

this value changing within the depth of five vacuum wavelengths from 1 (vacuum) to -1.5. The plasma is assumed to have a constant temperature of 10^2 eV and the intensity of the incident radiation is 5×10^{14} W/cm² of neodymium glass lasers. The oscillating curve of Fig. 5 gives the ratio of the nonlinear force to the thermokinetic force, each calculated for the actual conditions. The oscillation is due to the fact that a depth of the cut-off density at less than 3 vacuum wavelengths below the vacuum can obviously cause nearly total reflection of the wave only. As can be seen from Fig. 5, the force may be nearly twelve times the gasdynamic force. Besides the forces driving the plasma towards the nodes of the standing wave, there is a net force accelerating the whole plasma layer, which is calculated by averaging over the spatial oscillation period Δx

$$\bar{\underset{\sim}{f}}(x + \tfrac{1}{2}\Delta x) = \frac{1}{\Delta x} \int_x^{x+\Delta x} f(\xi) d\xi . \qquad (51)$$

We find that this averaged nonlinear force also exceeds the thermokinetic force by more than a factor of two. Averaging of the kind in Eq.(51) cannot be performed for $Re(\tilde{\eta}^2) < 1$, where only the aperiodic decrease of the nonlinear force results owing to absorption within the skin depth. Applying the Δx of the last period before this case, we get approximately the dashed curve in Fig. 5.

The larger net nonlinear force compared with the thermokinetic force is not unexpected, as can be seen from Fig. 3, where the threshold for the nonlinear force at 100 eV is around $I^* = 3 \times 10^{14}$ W/cm² at reasonable exponential factors e^{-1}, which is only a little less than the calculated case[36] with $I = 5 \times 10^{14}$ W/cm².

We also report a preliminary result of a dynamic numerical calculation of Shearer, Kidder and Zink[10], where numerical simulation of the interaction of plasma with laser pulses of some picoseconds is treated. Figure 6 (top) shows the time dependence of a 150 psec neodymium glass laser pulse of 10^{16} W/cm² intensity interacting with a plasma with a density profile of moderate slope over 150 vacuum wavelenghts crossing the cut-off

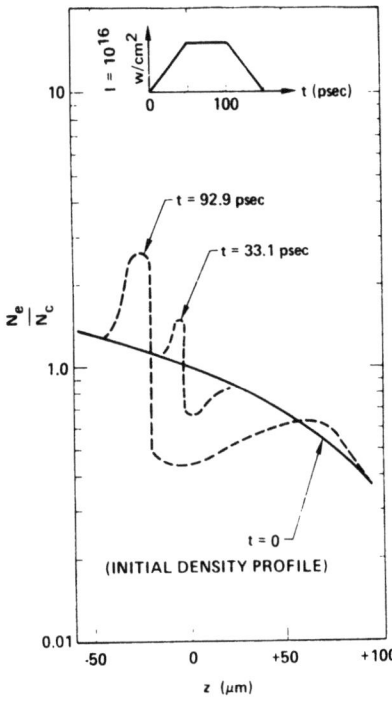

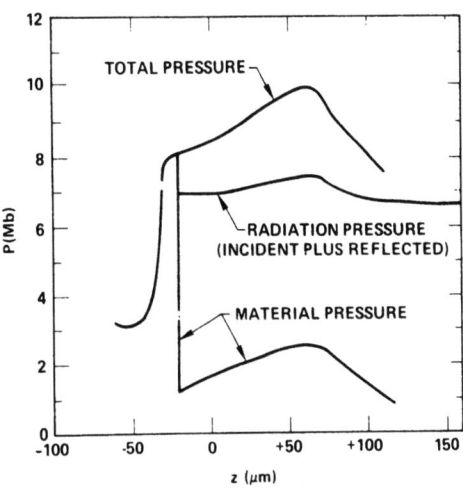

Fig. 7

Pressure vs distance at t =9o.9 psec for the WAZER computer run [38], see Fig. 6.

Fig. 6

Density profile at different times at a laser intensity described in the upper part of the figure in a dynamic computer program WAZER as given by Shearer[10,38]. The low density maximum is caused by the nonlinear force.

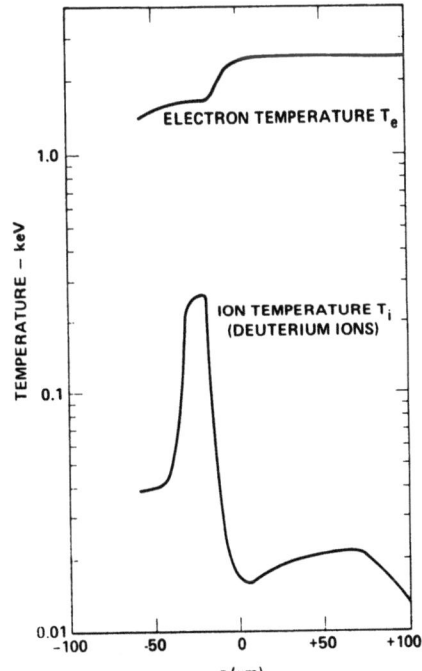

Fig. 8

Electron and ion temperature vs distance at t = 92.9 psec for WAZER computer run[38] of Fig. 6.

density at z=0. Figure 7 describes the pressure profile at the time of 9o.9 psec, where the material pressure p is distinguished from the electromagnetic energy-momentum-flux density (radiation pressure). The resulting temperature profiles at the time 92.9 psec (Fig. 8) demonstrates a state far away from thermal equilibrium between electrons and ions.

The result with respect to the nonlinear force was that the laser pulse is too short to generate a pronounced net motion of the plasma owing to this force. But it was established - and one can be sure that this is not due to numerical instabilities, which were certainly eliminated before the earlier steps of the program[37] - that the density profile was very strange at the times 33.1 psec and 92.9 psec (Fig. 6), showing strong maxima and minima. It is well known that the electron density can be monotonic or only have one maximum if only the thermokinetic force is taken into account. The strong bending of the profile causes an increase of the reflectivity of the laser light.

A further dynamic numerical treatment was performed by Mulser, the details of which will be reported separately[27]. It should merely be noted that the nonlinear force at the low intensity of 10^{14} W/cm^2 for neodymium glass lasers creates in a plasma no strong net acceleration, as expectable from the higher threshold I^*, but maxima in the standing waves which are forty times higher than the thermokinetic forces.

CONCLUDING REMARKS

The nonlinear force causing a net acceleration of the plasma has a threshold around 10^{14} W/cm^2 or at temperatures of 10^4 eV at lower intensities for ruby and neodymium glass lasers, as could be demonstrated analytically and by numerical examples of a WBK case (Fig.1) and of a non-WBK-like hyperbolic density profile (Fig.5). At oblique incidence and parallel orientization of the $\underline{E}$-vector to the plane of incidence, a strong increase of the nonlinear force is possible owing to coupling of transversal and longitudinal waves in the plasma. It is still an open question how the nonlinear increase of the absorption at the cut-off density (Kaw and Dawson[20]) influences the nonlinear force. It should be pointed out that positive changes of $|\tilde{\eta}|$ also cause strong gradients $\nabla |\tilde{\eta}|$ necessary for the nonlinear force.

The forces in standing waves, first treated under the aspects of changes of the refractive index under the conditions of the WBK case[12], more generally treated by Lindl and Kaw[24] and resulting in general numerical treatments[10,26,27], cause strong thermalization of the radiation and changes of the density profile. Therefore, the plasma is heated strongly and shows an increased reflectivity.

It is suggested that the standing-wave forces – which become effective even below the critical intensities I* – are the dominating mechanism for increased heating (and neutron production) if the maxima of the nonlinear force exceed the thermokinetic force because of the anticipating acceleration of plasma fronts at wave modes (Brillouin resonances). Under these strongly disturbed conditions, the reflectivity is increased synchroneously.

REFERENCES

1. R.W. Minck and W.G. Rado, J. Appl. Phys. 37, 355 (1966).
2. A.G. Engelhardt, T.V. George, H. Hora and J.L. Pack, Phys. Fluids 13, 212 (1970).
3. R. Sigel, Z. Naturforsch. 25a, 488 (1970).
4. H. Salzmann, K. Eidmann and R. Sigel, Verhndl. Dtsch. Phys. Ges. (VI) 6, 407 (1971).
5. N.G. Basov, P.G. Kriukov, S.D. Zakharov, Yu.V. Senatsky and S.V. Tchekalin, IEEE J. Quantum Electronics QE-4, 864 (1968).
6. F. Floux, Laser Interaction and Related Plasma Phenomena (H. Schwarz and H. Hora, Eds.) Plenum 1971.
7. F. Floux, J.F. Benard, D. Cognard and A. Saleres, Second Workshop Laser Interaction and Related Plasma Phenomena, these Proceedings, p. 409.
8. K. Büchl, K. Eidmann, P. Mulser, H. Salzmann, R. Sigel and S. Witkowski, Paper CN 28-D-11, 4th Conf. on Plasma Physics and Controlled Thermonuclear Fusion Research, Madison, Wisc., June 1971.
9. S.W. Mead, R.E. Kidder and J.E. Swain, Report UCRL-73356, August 17, 1971, IEEE J. Quantum Electronics (submitted).
10. J.W. Shearer, R.E. Kidder and J.W. Zink, Bull. Am. Phys. Soc. 15, 1483 (1970).
11. H. Hora, D. Pfirsch and A. Schlüter, Z. Naturforsch. 22a, 278 (1967); A. Schlüter, Plasma Physics 10, 471 (1968).

12. H. Hora, Phys. Fluids 12, 182 (1969);
    H. Hora, Laser Interaction and Related Plasma Phenomena (H. Schwarz and H. Hora Eds.) Plenum, New York 1971, p. 383.
13. H. Hora, Opto-Electronics 2, 2o1 (197o).
14. S. Rand, Phys. Rev. 136,B 231 (1964).
15. T.B. Hughes and M.B. Nicholson-Florence, J. Phys. A (2) 588 (1968).
16. R.Kidder, paper presented at the Varenna Summer School (July 1969), UCRL-Preprint 71775 (1969).
17. A. Caruso, A. de Angelis, G. Gatti, R. Gratton and S. Martellucci, Phys. Lett. 33A, 29 (197o).
18. H. Hora, Z. Physik 226, 156 (1969).
19. P.K. Kaw, Appl. Phys. Lett. 15, 16 (1969).
2o. P.K. Kaw and J.M. Dawson, Phys. Fluids 12, 2586 (1969).
21. W.L. Kruer and J.M. Dawson, Phys. Fluids 14, 1oo3 (1971); Second Workshop Laser Interaction and Related Plasma Phenomena, these Proceedings, p. 317.
22. see for example H. Motz and C.J.H. Watson, in Advances in Electronics and Electron Physics (L. Marton, Ed.) Academic Press, New York, Vol. 23, p. 153.
23. A.V. Gorbunov and M.A. Miller, ZhETF 34, 242; 751 (1958) (Sov. Phys. JETP 7, 168; 515 (1958)).
24. J. Lindl and P. Kaw, Phys. Fluids 14, 371 (1971).
25. L.C. Steinhauer and H.G. Ahlstrom, Phys. Fluids 13, 11o3 (197o).
26. B. Green and P. Mulser, Verhandl. Dtsch. Phys. Ges. (IV) 6, 4o5 (1971).
27. P. Mulser, Second Workshop Laser Interaction and Related Plasma Phenomena, these Proceedings, p. 381.
28. L.D. Landau and E.M. Lifshitz, Electrodynamic of Continuous Media (Pergamon Press, Oxford, 1966) p. 242.
29. see e.g. C.W. Allen, Astrophysical Quantities, Athlon Press, London 1955.
3o. J.M. Dawson and C. Oberman, Phys. Fluids 5, 517 (1962).
31. H. Hora and H. Wilhelm, Nuclear Fusion 1o, 111 (197o).
32. L. Spitzer,Jr., Physics of Fully Ionized Gases, Interscience, New York (1956).
33. V.L. Ginzburg, The Propagation of Electromagnetic Waves in Plasmas (Addison-Wesley, Reading, Mass. 1969), pp. 193-198 and 213-228.
34. P. Kaw, (private communication, July 1969).
35. N.G. Denisov, ZhETF 31, 6o9 (1956); Sov. Phys. JETP 4, 544 (1957).

36. H. Hora, Ann. Physik (7) <u>22</u>, 4o2 (1969).
37. J.W. Shearer and W.S. Barnes, <u>Laser Interaction and Related Plasma Phenomena</u> (H. Schwarz and H. Hora Eds.) Plenum, New York 1971, p. 3o7.
38. J.W. Shearer, Report Livermore Rad. Lab. UCID-15745 (Dec. 7, 197o).

STIMULATED SCATTERING AND SELF-FOCUSING PROCESSES IN DENSE PLASMAS*

A. Jay Palmer

Memorial University of Newfoundland

St. John's, Newfoundland

INTRODUCTION

During the years 1964 to 1966, there was a considerable amount of activity devoted to the investigation of optical non-linearities in plasmas[1-7]. At this time, attention was mostly confined to effects associated with non-linear excitation of the electron plasma mode. The stimulated scattering frequency mixing processes which were found to occur are analogous to the Raman type non-linearities which occur in neutral fluids and gases. It was also recognized at this time that, while the frequency mixing processes had promise of being observable, the stimulated scattering from the plasma mode was too weak to be of experimental importance[4].

Hydrodynamic optical non-linearities in plasmas associated with fluctuations in the plasma mass density analogous to stimulated Brillouin and stimulated thermal scattering were not looked into at this time. One reason for this was, evidently, that the plasmas of interest were generally of densities under 10^{18} cm^{-3} where collisional mean free paths are too long to permit hydrodynamic responses on the scale of an optical wavelength.

In the case of dense plasmas of the type created by high power laser interactions with dense media, where collisional mean

---

*Presented at the Second Workshop on "Laser Interaction and Related Plasma Phenomena" at Rensselaer Polytechnic Institute, Hartford Graduate Center, August 30 - September 3, 1971.

free paths are short relative to an optical wavelength, one can compare the non-linear response of the plasma for the Raman and hydrodynamic type of excitation. The non-linear excitation of the plasma mode in the Raman type process is essentially via the Lorentz force, and the damping is due to collisions. The resonant amplitude of the electron density fluctuation can be written (see Ref. 4 or 6):

$$\delta n_{e(\text{Raman})} \sim (m_e c^2)^{-1} (\omega_p \nu_c)^{-1} \left(\frac{\omega_p^2}{8\pi}\right) |E|^2 \ . \tag{1}$$

ω_p is the plasma frequency, ν_c is the ion-electron collision frequency, m_e is the electron mass, c is the speed of light, and E is the electric field amplitude. On the other hand, the non-linear excitation of ion acoustic waves in, for example, the hydrodynamic stimulated Brillouin process, is via ponderomotive forces, and the damping is due to ion viscosity. The resonant amplitude of the electron density fluctuation (which follows the ion fluctuation in dense plasmas) can be written (see Eqs. (7) and (8) below):

$$\delta n_{e(\text{Brillouin})} \sim (m_i e^2) (\omega_B \gamma_B)^{-1} \left(\frac{\omega_p^2}{8\pi}\right) |E|^2 \ . \tag{2}$$

m_i is the ion mass, ω_B is an acoustic frequency, and γ_B is the viscous damping coefficient (defined by Eq. (21) below). With the help of a gas-kinetic expression for the viscosity, the ratio of $\delta n_{(\text{Raman})}$ to $\delta n_{(\text{Brillouin})}$ (which is proportional to the ratio of the corresponding optical gains) can be reduced roughly to the form:

$$\delta n_{e(\text{Raman})}/\delta n_{e(\text{Brillouin})} \sim \left(\frac{m_i}{m_e}\right)^{1/2} \left(\frac{\omega_B}{\omega_p}\right) \left(\frac{\ell_{\text{coll}}}{\lambda_L}\right)^2 \tag{3}$$

where ℓ_{coll} and λ_L are the collisional mean free path and incident laser wavelength, respectively. The value of this ratio is on the order of 10^{-6} in typical laser produced plasmas. Thus, it is apparent that the hydrodynamic process is dominant, and it is the purpose of this paper to make order of magnitude determinations of the threshold conditions for the occurrance of hydrodynamic stimulated scattering and self-focusing processes in dense laser produced plasmas.

COUPLING MECHANISMS AND GOVERNING EQUATIONS

As in neutral fluids and gases, the two most important forces which drive hydrodynamic optical non-linearities are ponderomotive forces and thermal deposition. The general formula for the time averaged ponderomotive force on a dielectric fluid of dielectric constant, ε, is given in Landau and Lifshitz' book[8] as:

$$\vec{f} = \frac{(\varepsilon-1)}{8\pi} \vec{\nabla} \overline{(E^2)} + \frac{(\mu-1)}{8\pi} \vec{\nabla} \overline{(H^2)} + \frac{(\varepsilon\mu-1)}{4\pi} \frac{\partial}{\partial t} \overline{(E \times H)} \tag{4}$$

where μ is the magnetic permeability and $\overline{E^2}$ and $\overline{H^2}$ are the time averaged electric and magnetic field intensities. $\mu = 1$ for optical fields making the second term vanish and the last term can be neglected for interaction times longer than the "turn on time" for the fields. The remaining first term gives the well-known electrostrictive coupling on neutral systems which acts to compress the dielectric towards regions of greater field strengths. For a fully ionized, collisionless plasma with dielectric constant given by:

$$\varepsilon = 1 - \frac{\omega_p^2}{\omega_L^2} , \tag{5}$$

where ω_p and ω_L are the plasma frequency and laser frequency, respectively, the ponderomotive force is seen to act oppositely (for under-dense plasmas) tending to expell plasma from the high field regions[9]. Strictly speaking, Eq. (4) is valid only for non-dispersive dielectrics, its derivation being based on the assumption that no dissipation is present[8]. Its application to a plasma, however, appears to be justifiable, providing collisional frequencies remain less than the frequency of the incident field.

The force due to thermal deposition or absorbtive heating by the field gives rise to a pressure which may be written:

$$P_{th} = \left(\frac{\beta}{K_s}\right)(T - T_o) \tag{6}$$

where β is the coefficient of thermal expansion ($\beta = \frac{1}{T}$ for an ideal gas), K_s is the compressibility, and the temperature, T, satisfies the thermal diffusion equation (Eq. (9) below).

Under conditions typical of laser produced plasmas, the Debye length is small relative to the smallest length characteristic of the induced density fluctuations ($\sim \lambda_{laser}$). Under these conditions, while the coupling to the laser field is through the electrons, the inertial properties will be due to the ions. (We approximate the plasma as fully ionized.) Also, the ion-electron thermalization time is short relative to the transient times associated with the stimulated scattering and self-focusing processes. Thus, we can take $T_e = T_i$, and the specific heat, C_p, will be due to both electrons and ions. The thermal conductivity, K, will, however, be due to the electrons, while the viscosity, η, is due to the ions. Thus, we no longer have the condition, $\frac{K}{\eta C_p} = 1$, which is applicable to a neutral ideal gas and gives[10] $\frac{\gamma_B}{\gamma_R} = 1$, where γ_B and γ_R are the line-widths for spontaneous Brillouin and thermal Rayleigh scattering, respectively.

Under the above conditions, the response of the plasma mass density to the laser field is governed by the Navier-Stokes equation

$$\left[\frac{\partial^2}{\partial t^2} - v_s^2 \nabla^2 - \frac{\eta}{\rho} \frac{\partial}{\partial t} \nabla^2 \right] \rho = \nabla^2 [P_{pond} + P_{th}] . \qquad (7)$$

The ponderomotive pressure, P_{pond}, is defined in accordance with Eq. (4) as

$$P_{pond} = (\varepsilon - 1) \overline{E^2} . \qquad (8)$$

P_{th} is given by Eq. (6) and the thermal diffusion equation,

$$\frac{\partial}{\partial t} \cdot T = \frac{\alpha c |\vec{E}|^2}{C_p \rho} + \frac{K \nabla^2 T}{\rho C_p} . \qquad (9)$$

Here, ρ is the mass density, v_s is the sound speed $= \left(\frac{K_s}{\rho} \right)^{1/2}$, η is the ion viscosity $= 2.2 \times 10^{-15} \frac{T^{5/2} A_i}{Z^4 \ln(\Lambda)} \frac{gm}{cm\ sec}$ (A_i = ion mass number and Λ is Spitzer's Λ[11]), α is the absorption coefficient due to inverse bremsstrahlung $= \frac{1.17 \times 10^{-8} Z n_e^2 \ln \Lambda}{3 \nu^2 (kT)^{3/2}}$. (Here, n_e is the electron concentration, Λ is Spitzer's Λ, ν is the incident laser frequency, and kT is in eV[11]), C_p is the plasma specific heat per unit mass $= 3/2 \times k/m_i$ (k is Boltzmann's constant), K is the thermal conductivity $= 4.7 \times 10^{-12} \left(\frac{T^{5/2}}{2 \cdot \ln(\Lambda)} \right) \left(\frac{cal}{(sec)(^\circ K)(cm)} \right)$[11], and c is the velocity of light in the plasma.

The response of the laser field to the plasma is governed by the wave equation:

$$\nabla^2 \vec{E} - \frac{1}{c^2} \frac{\partial^2}{\partial t^2} (\varepsilon \vec{E}) = 0 \qquad (10)$$

with the plasma dielectric constant given by[11]:

$$\varepsilon = 1 - \left[\frac{\omega_p^2}{\omega^2 + \nu_c^2}\right] \left[1 + i\frac{\nu_c}{\omega}\right] . \qquad (11)$$

The plasma frequency, ω_p, is $(4\pi e^2 n_e/m_e)^{1/2}$ and ν_c is the ion-electron collision frequency. The imaginary part of ε gives rise to the strong inverse bremstrahlung attenuation of laser light in the plasma which is the source of the absorptive heating term in Eq. (8). The spatial gains of the various non-linear processes we will be considering depend primarily on the real part of ε which, for most laser produced plasma situations, can be approximated as Eq. (5). The attenuation resulting from Im ε is brought in to determine the thresholds for the onset of the stimulated scattering and self-focusing instabilities through the requirement:

$$\text{gain} = \alpha \qquad (12)$$

where the gain refers to the reciprocal e-folding distance of the linearized component of the field. Growth occurring over distances greater than α^{-1} will not be dealt with and this justifies our use of the small absorption approximation in the absorptive heating term in Eq. (9) for order of magnitude validity for the threshold conditions.

Below, we will apply stationary theories of stimulated scattering and self-focusing in calculating the relevant thresholds for the onset of these processes in laser produced plasmas, i.e., we will characterize the processes with spatial rather than temporal gain. For the case of stationary interaction zones, the criterion for applying such a steady state theory is that the laser pulse duration be long relative to the sound transit time across distances on the order of the thickness of the interaction region. This is normally the case for interactions with liquids and gases and may also be applicable to interactions in a blow-off plasma derived from laser heating of solid surfaces in a vacuum, provided one takes proper account of the Doppler effect.

In plasmas derived from laser induced gas breakdown, however, the interaction zone of the beam with the plasma (thickness $\sim \alpha^{-1}$) is known to move rapidly toward the laser[12] and the time, τ, that it takes the zone to move through a distance equal to its thickness is, in general, much shorter than the laser pulse duration. Thus, in the above criterion, laser pulse duration should be replaced by the time, τ. Except in special cases[13], the absorption zone moves with a velocity close to the sonic velocity of the plasma. In this case, the transient times are comparable to the effective interaction time, τ, and the above criterion is not well satisfied. While a more precise theory should, therefore, include the transient response of the instabilities for spark plasmas, these conditions seem to permit order-of-magnitude validity of threshold calculations based on a steady state theory and this is all that we desire here.

STIMULATED SCATTERING AND SELF-FOCUSING INDUCED THROUGH PONDEROMOTIVE FORCES

We now consider the non-linear effects arising from the ponderomotive force and from thermal deposition separately. With

$$\rho = \rho_0 + \rho_1 \; ; \quad \rho_1 \ll \rho_0 \tag{13}$$

we may write the dielectric constant in the form:

$$\epsilon = \epsilon_0 - \frac{\omega_p^2}{\omega^2} \frac{\rho_1}{\rho_0} \tag{14}$$

where

$$\epsilon_0 = 1 - \frac{\omega_p^2}{\omega^2} \; . \tag{15}$$

For the ponderomotive force coupling, we need only linearize Eqs. (7) and (10) with respect to ρ_1 and e_1 where

$$\vec{E} = \text{Re}\{\vec{E}_0 \exp[i(k_L z - \omega_L t)]\} + \vec{e}_1 \; ; \quad |\vec{e}_1| \ll |\vec{E}_0| \; . \tag{16}$$

If we then use, for backward scattered light,

$$\rho_1 = \text{Re}\{\rho_q \exp[i(q \cdot x - \omega t) + (\text{gain})z]\} \tag{17}$$

$$\vec{e}_1 = \text{Re}\{\vec{e}_q \exp[i(q_S \cdot x + \omega_S t) - (\text{gain})z]\} \tag{18}$$

the resulting dispersion formula gives, in complete analogy with the case for neutral gases[14], the maximum gain occurring for $\vec{e}_1$, Stokes-shifted by an amount

$$\omega = \omega_S - \omega_L = -\omega_B = -2\omega_L \left(\frac{v_s}{c}\right) \quad \text{(for backward scattered light)}. \tag{19}$$

The maximum gain is

$$g_I = \frac{q_S \, q}{8\pi \, \rho_0 v_S} \left\{\frac{\omega_p^2}{\omega_L^2}\right\}^2 |\vec{E}|^2 \, \gamma_B^{-1} \tag{20}$$

where γ_B is the spectral width of the corresponding spontaneously scattered line, and is

$$\gamma_B = \frac{q^2 \eta}{\rho_0} \; . \tag{21}$$

Ignoring the local field correction factor, $\frac{\tilde{n}^2+2}{3}$, which is ~ 1, Eq. (19) is just the expression for the maximum gain for stimulated Brillouin scattering in a neutral gas[14] with the factor $(\tilde{n}^2-1)^2$ replaced by $\left[\frac{\omega_p^2}{\omega_L^2}\right]^2$ where $\tilde{n}$ denotes the refractive index of the gas.

With $\frac{\partial}{\partial t}$ set equal to zero in Eq. (7), it can be seen that the ponderomotive force coupling gives an intensity dependence to the refractive index, $\tilde{n}$, such that $\frac{\partial \tilde{n}}{\partial |\vec{E}|^2} > 0$, and can thus cause self-focusing. A linearized theory of self-focusing analogous to the linearized theories of stimulated scattering has been given by Bespalov and Talanov[15]. The procedure for calculating the e-folding distance of the linearized component of the field, $\vec{e}_1$, within a self-focusing filament of transverse dimension, $\sim 1/q$, is the same as in the case of stimulated scattering (in the forward direction) except that one keeps only $\vec{E} \cdot \vec{E}^*$ combinations of the fields which give a $\omega = 0$ response to the density, ρ_1. With the dielectric constant written in the form:

$$\varepsilon = \varepsilon_o + \varepsilon_1 \cdot |\vec{E}|^2 \qquad (22)$$

the resulting dispersion formula gives the maximum gain[15]:

$$g = k_L \frac{\varepsilon_1 |\vec{E}|^2}{2} \qquad (23)$$

for filaments of reciprocal transverse dimension:

$$q = (|\vec{E}|^2 k_L^2 \varepsilon_1)^{\frac{1}{2}} \qquad (24)$$

where k_L is the laser wavenumber. For the ponderomotive force mechanism,

$$\varepsilon_1 = \left[\frac{\omega_p^2}{\omega_L^2}\right]^2 \left(\frac{1}{8\pi}\right) \left[\frac{1}{\rho_o v_S^2}\right] \qquad (25)$$

and the maximum gain is,

$$g_{II} = \frac{k_L}{2} \left[\frac{\omega_p^2}{\omega_L^2}\right]^2 \left(\frac{1}{8\pi}\right) \frac{|\vec{E}|^2}{\rho v_S^2} \qquad (26)$$

Note that we can write,

$$\frac{g_{II}}{g_I} = \frac{\omega_B \eta}{\rho_o v_S^2} \sim \frac{\ell_{coll}}{\lambda_L} \qquad (27)$$

where ℓ_{coll} is the gas-kinetic mean free path. This last relation is obviously of general validity whenever self-focusing and stimulated Brillouin scattering arise from the same coupling mechanism.

THERMALLY INDUCED STIMULATED SCATTERING AND SELF-FOCUSING

We consider now the non-linear effects due to thermal deposition. We want an expression for the gain of the process analogous to stimulated thermal Rayleigh scattering in gases. In a plasma with refractive index due entirely to free electrons, such stimulated thermal scattering should properly be called stimulated thermal Thomson scattering[16]. Following the procedure for stimulated thermal Rayleigh scattering in gases[14], we write:

$$T = T_o + T_1 \; ; \quad T_1 \ll T_o \qquad (28)$$

with

$$T_1 = T_q \exp\{i[q \cdot x - \omega t] + \text{gain} \cdot z\} . \qquad (29)$$

One then uses $\frac{\partial}{\partial t} = 0$ in Eq. (7) and linearizes Eqs. (7), (9) and (10) with respect to T_1, ρ_1 and $\vec{e}_1$. The resulting dispersion formula gives the maximum gain occurring for Stokes-shifted light at $\omega_S = \omega_L - \frac{\gamma_T}{2}$, where,

$$\gamma_T = \frac{2Kq^2}{\rho_o C_p} \qquad (30)$$

is the spectral width of the spontaneous scattering. Note that for a plasma, $\frac{\gamma_T}{\gamma_B} = 20 \, z^3 \, (\frac{m_i}{m_e})^{1/2}$. The maximum gain is:

$$g_{III} = \frac{q_S}{8\pi} \left[\frac{\omega_p^2}{\omega_L^2}\right] (\frac{\beta c \alpha}{\rho_o C_p}) \, (|\vec{E}|^2) \, \gamma_T^{-1} \qquad (31)$$

which, in this case, is just the expression for maximum stimulated thermal Rayleigh gain[14] with the factor $(\tilde{\eta}^2-1)$ replaced by $-\frac{\omega_p^2}{\omega_L^2}$. This minus sign, which causes positive gain for Stokes-shifted light in contrast to the case of stimulated thermal Rayleigh scattering, results from the opposite response of the dielectric constant to density changes in stimulated thermal Thomson scattering.

As in the case of stimulated thermal Rayleigh scattering, we can define a critical absorption coefficient, α_{cr}, for which $g_{III} = g_I$. From Eqs. (31) and (20), we find,

$$\alpha_{cr} = \frac{1}{2} \left[\frac{C_p \omega_B}{\beta c v_s^2} \right] \left[\frac{\gamma_T}{\gamma_B} \right] \left[\frac{\omega_p^2}{\omega_L^2} \right] \quad (32)$$

$$\sim \left(\frac{\omega_B}{c} \right) \left[\frac{\gamma_T}{\gamma_B} \right] \left[\frac{\omega_p^2}{\omega_L^2} \right] \sim \left(\frac{\omega_B}{c} \right) \left(\frac{K}{\eta C_p} \right) \left[\frac{\omega_p^2}{\omega_L^2} \right] \quad (33)$$

which for a plasma is

$$\alpha_{cr} \sim 20 \, z^3 \left(\frac{m_i}{m_e} \right)^{1/2} \left(\frac{\omega_B}{c} \right) \left[\frac{\omega_p^2}{\omega_L^2} \right] \quad (34)$$

and we can write,

$$g_{III} = \left(\frac{\alpha}{\alpha_{cr}} \right) g_I \quad . \quad (35)$$

For $\frac{\partial}{\partial t} = 0$ in both Eqs. (7) and (9), it is seen that thermal deposition also produces a self-focusing refractive index in the plasma. In this case, ε_1 in Eqs. (22) is

$$\varepsilon_1 = \frac{1}{8\pi} \left(\frac{\omega_p}{\omega_L} \right)^2 \left(\frac{\beta c \alpha}{\rho_o C_p} \right) \gamma_T^{-1} \quad (36)$$

and the maximum gain for thermal self-focusing is, according to Eq. (21):

$$g_{IV} = k_L \frac{1}{16\pi} \left(\frac{\omega_p}{\omega_L} \right)^2 \left(\frac{\beta c \alpha}{\rho_o C_p} \right) \gamma_T^{-1} |\vec{E}|^2 \quad . \quad (37)$$

We have, therefore (within a factor of 2),

$$g_{III} \stackrel{\sim}{=} g_{IV} \quad . \quad (38)$$

SAMPLE THRESHOLD CALCULATIONS AND DISCUSSION

In the absence of a known functional dependence of the plasma parameters on laser power, we will write the gains and the threshold condition, Eq. (12), in terms of the laser power density, P, and electron concentration, n_e, at a chosen value for the plasma temperature, T, ion mass, M, and ion charge, Z. As a simple example, we choose hydrogen at T = 10 eV. In this case, with P in MW/cm^2, and n_e in cm^{-3}, the maximum gains become (for the 6943A ruby laser wavelength):

$$g_I \simeq 2 \cdot 10^{-45} \, P n_e^2 \, cm^{-1} \tag{39}$$

$$g_{II} \simeq 1 \cdot 10^{-27} \, P n_e \, cm^{-1} \tag{40}$$

$$g_{III} \simeq g_{IV} \simeq 3 \cdot 10^{-28} \, n_e \, \alpha \, P \, cm^{-1} \tag{41}$$

and the attenuation coefficient is

$$\alpha \simeq 7 \cdot 10^{-39} \, n_e^2 \, cm^{-1} \,. \tag{42}$$

At threshold, gain = α and we have:

$$P_I(\text{thresh}) \simeq 3 \cdot 10^6 \, MW/cm^2 \tag{43}$$

$$P_{II}(\text{thresh}) \simeq 5 \cdot 10^{-12} \, n_e \, MW/cm^2 \tag{44}$$

$$P_{III}(\text{thresh}) \simeq P_{IV}(\text{thresh}) = 3 \cdot 10^{27} \, n_e^{-1} \, MW/cm^2 \,. \tag{45}$$

A limit on the laser power in the above calculations is imposed by saturation of the respective non-linear processes. The ponderomotive mechanism will saturate when the pressure due to the ponderomotive force becomes comparable to the plasma kinetic pressure. This occurs when

$$P_{pond} \sim nkT_o \tag{46}$$

where n is the plasma number density. Similarly, the thermal effects saturate when the induced temperature fluctuations become comparable to the plasma temperature, and this occurs when (see Eq. (9)),

$$T_q = \frac{\alpha c |\vec{E}|^2}{K q^2} \sim T_o \,. \tag{47}$$

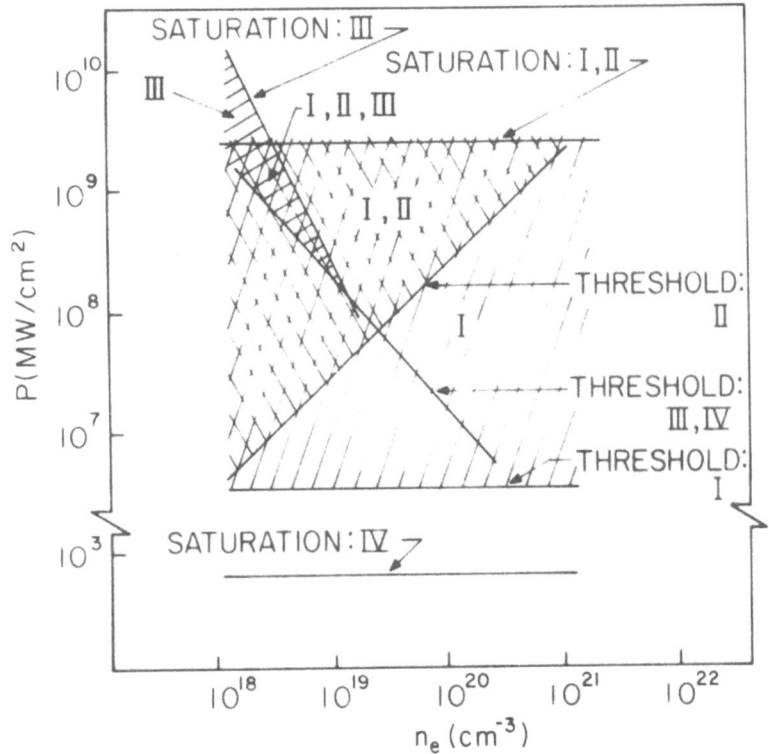

Fig. 1

Order-of-magnitude range of ruby laser power density and plasma density, within the limits of validity of the linearized theory used in the text, which permit a net positive gain for the indicated stimulated scattering and self-focusing process in a 10 ev, fully-ionized, hydrogen plasma. I: Stimulated Thomson-Brillouin scattering; II: Self-focusing induced through ponderomotive forces; III: Stimulated thermal Thomson scattering; IV: Thermally induced self-focusing.

In Fig. 1, the threshold conditions (43) - (45), the saturation conditions (46) and (47), and limits on n_e for which the calculations remain valid are used to define the approximate boundaries of regions in a P vs n_e plane where the stimulated scattering and self-focusing processes at the ruby laser wavelength may be expected to occur for the above chosen plasma parameters. The plasma density is limited to a range allowing simultaneous validity of the hydrodynamic theory ($\ell_{coll} < \lambda_L$) and of the slowly-varying approximation (gain$^{-1} > \lambda_L$).

Note that the self-focusing thresholds calculated here are the minimum thresholds for the onset of filamentary self-focusing instabilities, which may or may not lead to trapping of the beam as a whole. The threshold condition, Eq. (12), is far more restrictive in dense plasmas where strong absorption occurs than is the beam trapping condition used in Ref. (9) which requires merely that the self-focusing overcome diffractive spreading. It is, therefore, believed to be the more relevant threshold condition.

While the conditions indicated in Fig. 1 for the occurrence of ponderomotive self-focusing are not incompatable with conditions under which filamentation has been observed in laser induced spark plasmas, enough evidence has now accumulated[17] to indicate strongly that the observed filamentation arises during the breakdown process and not within the body of a fully ionized plasma as is assumed in these calculations. As can be seen from Fig. 1, thermal self-focusing occurs outside the range for which the threshold calculations are valid.

The plot in Fig. 1 illustrates that stimulated Brillouin scattering and stimulated thermal Thomson scattering should also occur within a realistic range of plasma conditions and laser power[18]. At this point, however, we must consider the question of the relative intensity of the stimulated scattered signal. The ratio of the power in the stimulated scattered signal to the incident laser power can be written approximately as:

$$\frac{P_{stim\ scatt}}{P_{incident}} \sim (\sigma_{Thom}) (n_e) \left(\frac{\Delta\Omega}{4\pi}\right) \left[\frac{\exp(\text{gain}\ \ell)}{\text{gain}}\right] \qquad (48)$$

where $\sigma_{Thom} = 6.7 \cdot 10^{-25}$ cm^2 is the total cross section for spontaneous Thomson scattering, n_e is the electron concentration, $\Delta\Omega$ is the solid angle subtended on the plasma by the incident beam, and the factor in brackets represents the integrated gain through the length, ℓ, of the interaction zone. For the spark plasmas, at threshold, gain $\sim \ell^{-1}$ and a nominal value for $\Delta\Omega$ is $\sim 10^{-2}$. Thus, for $n_e \sim 10^{19}$ cm^{-3}, we have, at threshold, $\frac{P_{stim\ scatt}}{P_{incident}} \sim 10^{-8}$. To bring this ratio up to 1% requires the gain to be on the order of 15 times the threshold value. From Fig. 1, we see that this is easily available below saturation for stimulated Brillouin scattering but is probably not possible for stimulated thermal Thomson scattering.

It is important to note that the spectral width of the stimulated Brillouin and stimulated thermal Thomson lines in a hydrogen plasma at 10 ev will be, respectively, $\sim 10^2$ and $\sim 10^5$ times the width of the corresponding stimulated Brillouin and stimulated thermal Rayleigh lines in a gas at room temperature (see Eqs. (21) and (30)) and will require, for their identification, detectors of much lower dispersion than have been used in the study of stimulated scattering from neutral gases and liquids.

Definitive observations of either of the stimulated scattering processes is also lacking. Recently, however, some interesting observations of non-linear backscattered radiation from a laser induced blow off plasma has been reported[19] which exhibits a sharp incident power density threshold of $\sim 1.5 \cdot 10^6$ MW/cm^2. This value is remarkably close to the $3 \cdot 10^6$ MW/cm^2 threshold predicted for stimulated Brillouin scattering in the sample calculation above. While the observed anti-stokes shift could be explicable in terms of the Doppler effect, it is more difficult to understand the the observed multiple line character of the backscattered spectrum.

ACKNOWLEDGMENTS

The author wishes to thank Professor C. W. Cho and the Department of Physics at Memorial University of Newfoundland for financial assistance in the form of a Visiting Assistant Professorship during the time this work was carried out. Helpful discussions with Drs. A. J. Alcock, N. D. Foltz, R. Papoular, and M. Mattioli are also gratefully acknowledged.

REFERENCES

1. P. M. Platzman, S. Buchsbaum and N. Tzoar, Phys. Rev. Letters 12, 573 (1964).
2. N. M. Kroll, A. Ron and N. Rostocker, Phys. Rev. Letters 13, 534 (1964).
3. D. F. DuBois and V. Gilinsky, Phys. Rev. 135, A995 (1964).
4. N. M. Kroll, Proceedings of the Physics of Quantum Electronics Conference, San Juan, Puerto Rico, 1965.
5. M. V. Goldman and D. F. DuBois, Phys. Fluids 8, 1404 (1965).
6. G. G. Comisar, Phys. Rev. 141, 200 (1966).
7. N. Bloembergen and Y. R. Shen, Phys. Rev. 141, 298 (1966).
8. L. D. Landau and E. M. Lifshitz, "Electrodynamics of Continuous Media" (Pergamon Press, Ltd., Oxford, 1966), pp. 68, 242.

9. The factor ($\varepsilon-1$) in the expression for the ponderomotive force is erroneously omitted in Hora's Treatment of Self-Focusing in Plasmas in: H. Hora, Ann. Physik 22, 402 (1969); H. Hora, "Laser Interactions and Related Plasma Phenomena", H. Schwarz and H. Hora, Eds. (Plenum Press, New York, 1971), p. 383; and H. Hora, "Opto-electronics" 2, 201 (1970). The author is grateful to Drs. R. Papoular and M. Mattioli for discussions related to this point.
10. D. R. Dietz, C. W. Cho, D. H. Rank and T. A. Wiggins, Applied Opt. 8, 1248 (1969).
11. L. Spitzer, Jr., "Physics of Ionized Gases" (Interscience, New York, 1950), Chap. 5.
12. Yu. P. Raizer, Usp. F.2. Nauk 87, 29 (1965) [Sov. Phys. Usp. 8, 650 (1966)].
13. The use of very high power and short focal length lenses can produce a considerably faster propagating "breakdown-wave"[12].
14. R. M. Herman and M. A. Gray, Phys. Rev. Letters 19, 824 (1967).
15. V. I. Bespalov and V. I. Talanov, Zhetf. Pis. Red. 3, 471 (1966) [JETP, Lett. 3, 307 (1966)].
16. We have retained the term, stimulated Brillouin scattering, for the stimulated scattering off ion-acoustic waves although it is, of course, also a type of stimulated Thomson scattering. "Stimulated Thomson-Brillouin Scattering" would also seem to be an appropriate label.
17. R. G. Tomlinson, IEEE J. Quantum Electronics QE-5, 591 (1969); A. J. Alcock, C. DeMichelis and M. C. Richardson, IEEE J. Quantum Electronics QE-6, 622 (1970); M. C. Richardson and A. J. Alcock, Applied Physics Lett. 18, 375 (1971).
18. Much larger gains for the stimulated scattering may also result from the self-trapped portions of the beam acting as the pump field.
19. P. Belland, C. De Michelis, M. Mattioli, and R. Papoular, Applied Physics Letters, 18, 542 (15 June, 1971).

A SELF CONSISTENT CALCULATION OF PONDEROMOTIVE FORCES IN THE LASER PLASMA INTERACTION[+]

B.J. Green, P. Mulser

Max-Planck-Institut für Plasmaphysik

Euratom Association, Garching, Germany

ABSTRACT

Previous estimates of the nonlinear ponderomotive forces which arise in the interaction of intense laser light with solid targets emphasized that they can dominate the plasma expansion. We treat a model problem solving the wave equation numerically and show for intensities up to 10^{15} W/cm^2 that the ponderomotive forces are important locally but the overall expansion of the created plasma is determined essentially by the thermal gas pressure.

As was first pointed out by H. Hora[1,2,3,4], the ponderomotive forces which act when an intense light beam is incident on a dense plasma, can be important long before the ordinary light pressure becomes dominant. For a laser produced plasma this means that above a certain intensity threshold the expansion of a hot plasma should no longer be determined by the hydrodynamic pressure alone. In[4] the interaction of strong laser radiation with dense plasma was investigated on the basis of the optical approximation (WKB) and it was shown that these

---

[+] Presented at the Second Workshop on "Laser Interaction and Related Plasma Phenomena" at Rensselaer Polytechnic Institute, Hartford Graduate Center, August 30 - September 3, 1971.

ponderomotive forces become very large in the vicinity of the reflection point (where the electron plasma frequency ω_p equals the frequency of the laser light ω. In addition this treatment claimed that for Nd laser radiation the ponderomotive force should dominate the gasdynamic pressure at light intensities $10^{14} - 10^{15}$ W/cm^2. Steinhauer and Ahlstrom[5] also attempted to evaluate these forces in a time dependent model. With a time averaging procedure and the no reflection approximation they calculated temperature-dependent intensity thresholds for the dominance of the ponderomotive forces.

It must be pointed out that in these treatments two difficulties arise: (i) in the neighborhood of the reflection point the optical approximation always breaks down, except for the case of the very special Kofink refractive index profiles[6], and (ii) in a nondynamic calculation there is no possibility to corectly relate the thermodynamic pressure to the incident radiation intensity.

To discuss the importance of ponderomotive forces we consider a one-dimensional model of linearly polarized laser light, perpendicularly incident on a plane target of solid hydrogen (particle density 5 x 10^{22}cm^{-3}). Because of the high frequency ω and the high value of light velocity, the radiation energy transport can be described by the stationary wave equation for the electric field

$$E'' + k^2 n^2(x) E = 0 ,$$

where $n^2(x)$ is given by the classical formula

$$n^2 = 1-(\omega_p/\omega)^2/(1+i\nu/\omega)$$

and ν is the electron-ion collision frequency. The plasma dynamics is described in terms of the conservation equations for mass, momentum and energy allowing for different electron and ion temperatures and electron heat conduction. A similar dynamical model has been discussed previously[7]. In the momentum and energy equations we include the ponderomotive force density as derived in[2] which, when averaged over the laser wave period ω^{-1} and transformed with the help of the stationary wave equation is

$$f_p = \frac{\varepsilon_0}{4} \left\{ (n^2-1)EE^{*'} + (n^{*2}-1)E^*E' \right\}$$

where * denotes the complex coniugate quantities. The selfconsistent calculations were performed for a Nd glass laser (vacuum wavelength 1.06 μ) with time-constant intensities between 10^{12} and 10^{15} W/cm². Due to electron heat conduction a plasma is formed which is largely overdense ($\omega_p > \omega$). Figure 1 is a snapshot after 5 nsec of such a dynamic calculation (light intensity 10^{14} W/cm²), taken in the neighborhood of the reflection point ($Re(n^2)=0$). Plotted as a function of the space coordinate x are: (i) the real part of the square of the refractive index $Re(n^2)$, (ii) the magnitude of the electric field $|E|$, and (iii) the local ratio of the ponderomotive to thermal force density f_p/f_{th}.

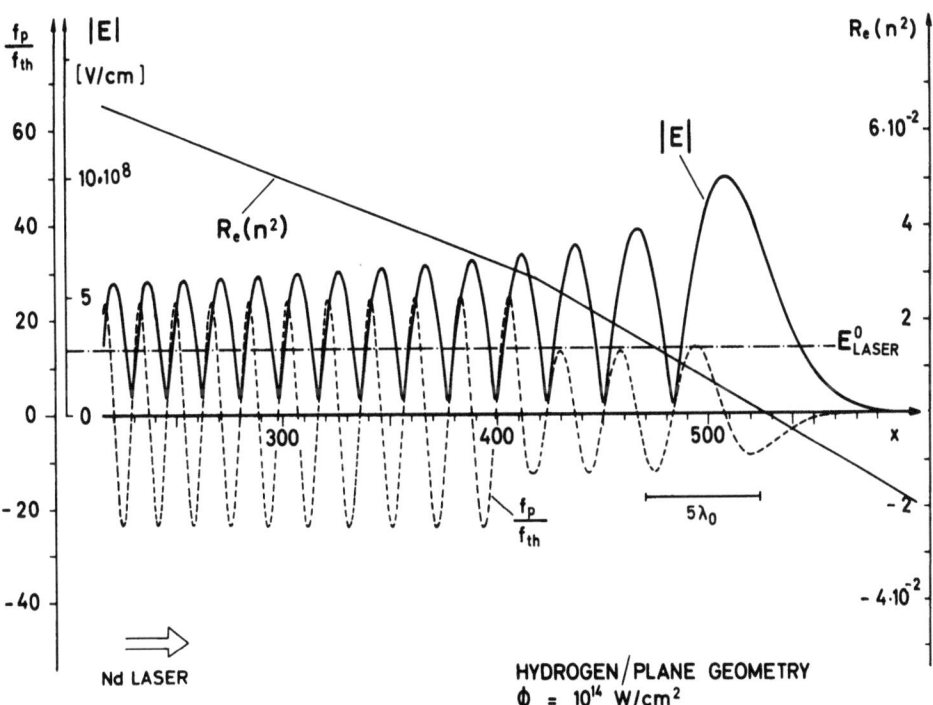

Fig. 1

Spatial distribution of (i) the real part of the square of the refractive index $Re(n^2)$, (ii) the electric field amplitude $|E(x)|$ (modulated by reflection), and (iii) the ponderomotive and thermal force density ratio f_p/f_{th}, in the vicinity of the reflection point ($Re(n^2)=0$) for a Nd laser. Note that in this region the electric field is increased above its vacuum value (indicated by -.-.-.-.-), although the laser energy flux is reduced by more than a factor of 5 due to absorption in the lower density plasma.

One of the characteristic features of the results is the oscillatory nature of the ponderomotive force, caused by the superposition of incident and reflected waves which modulate the electric field amplitude $|E|$. This aspect was pointed out in [7] and [8] and could not be observed, of course, in the no-reflection approximation [4]. As can be seen from the expression for f_p the modulation of $|E|$ produces very large forces with alternating sign. Already at an intensity of 10^{12} W/cm^2 the local ponderomotive force reaches values up to 12 times greater than the thermal forces; for 10^{14} and 10^{15} W/cm^2 this ratio is larger than 20, or 100 respectively (in Fig. 1 the knee in $Re(n^2)$ and the discontinuity in f_{th} and consequently in the ratio f_p/f_{th} are due to the fact that in the numerical calculation different space step sizes have to be chosen for dynamics and wave equation; the required values for $n^2(x)$ are obtained by linear interpolation between the calculated points for dynamics. Tests with different step sizes have been performed to show the accuracy of this procedure.

In order to investigate the g l o b a l effect of these forces, sets of two dynamical calculations were performed, one with the ponderomotive forces and one without. The resulting dynamical profiles were almost identical. However, in the case where ponderomotive forces are included perturbations in the plasma density profiles near the reflection point are observed. This means that also for intensities as large as 10^{15} W/cm^2 the overall expansion is determined essentially by thermal forces. From the numerical results therefore we conclude that for the laser intensities considered here the no-reflection approximation fails to determine the detailed structure of the forces and largely overestimates the magnitude of the electric field in the vicinity of the reflection point (by a factor of about 10) which leads to an overestimated net acceleration of the whole plasma. On the other hand our results confirm that ponderomotive forces can indeed be important, locally.

REFERENCES

1. H. Hora, D. Pfirsch, A. Schlüter, Z. Naturforschg. $\underline{22a}$, 278 (1967).
2. H. Hora, Phys. Fluids $\underline{12}$, 182 (1969).
3. H. Hora "Laser Interaction and Related Plasma Phenomena", H. Schwarz and H. Hora, Eds., Plenum Press, New York 1971, p. 383.
4. H. Hora, Opto-Electronics $\underline{2}$, 2o1 (197o).
5. L.C. Steinhauer, H.G. Ahlstrom, Phys. Fluids $\underline{13}$, 11o3 (197o).
6. L.M. Brekhovskikh, "Waves in Layered Media", Acad. Press New York-London, p. 23o.
7. P. Mulser, Z. Naturforschg. $\underline{25a}$, 282 (197o); for dynamics see also S. Witkowski "Laser Interaction and Related Plasma Phenomena", H. Schwarz and H. Hora, Eds., Plenum Press, New York 1971, p. 339.
8. J.D. Lindl, P.K. Kaw, Phys. Fluids $\underline{14}$, 371 (1971).

SUMMARY OF DISCUSSION

(V. Theory of High Intensity
Laser Interaction with Plasma)

P. Mulser commented on problems of numerical simulation of laser produced plasmas by computers for thermokinetic processes only. Codes for two-dimensional gasdynamical particles without collisions are very complicated. The results obtained with these codes are of limited value due to the omission of important physical properties such' as instabilities, nonlinear forces, relativistic effects and others. M. J. Lubin mentioned his detailed code of a spherical or three-dimensional plasma where he included relativistic effects, as well as the nonlinear force and instability effect of the energy dissipation. With this extended program, he could reproduce the process of neutron production due to shock waves of very high densities. Lubin stated that one computer run requiring three simultaneous CD 6600's for nine hours may be more expensive than performing the actual experiment.

Besides nonlinear effects of absorption - two stream instabilities and acceleration - some other instabilities must be considered as Silin, Bloembergen, Jackson, Volkov and others have done in wave-plasmon interactions in a general way. We recommend a review dealing with decay mode instabilities by D. F. Dubois[1].

The next step for studying the nonlinear forces further would be to include the instabilities where, however, more complicated analytical expressions will be needed.

The work of K. Nishikawa[2], as reported by C. Yamanaka, stimulated the question of the threshold for instabilities, since Nishikawa's values for the threshold were remarkably lower than those of other authors. In the discussion it was brought out that the lower thresholds are perhaps due to dielectric properties of the plasma, as in a similar way nonlinear forces may be due to relativistic effects.

References
1. D. F. Dubois, <u>Statistical Physics of Charged Particle Systems</u>, (R. Kubo and Taro Kihara, eds.) Benjamin Inc., New York 1969, p. 87-155.
2. K. Nishikawa, J. Phys. Soc. Japan, <u>24</u>, 916 (1968).

HEATING OF LASER PLASMAS
FOR THERMONUCLEAR FUSION*

N. G. Basov, O. N. Krokhin, G. V. Sklizkov

Lebedev Physical Institute, Academy of Sciences

Leninsky Prospect 53, Moscow, USSR

ABSTRACT

In this survey the experimental results of the investigation of the laser produced plasmas in a wide interval of energy and light pulse duration are reported. The processes taking place at a dense plasma heated by laser radiation for the obtaining of a thermonuclear yield have been analyzed. The directional motion for heating in the various schemes using cumulative effects is discussed.

The experimental results of plasma heating investigation are described. The laser description and the methods of investigation of plasma parameters are given. The results of measurements of neutron emission from deuterized polyethylene plasma are considered.

Spectroscopic and interferometric studies of gas-dynamical parameters in the process of heating are reported. The interferometric method of measurement of plasma pressure and momentum carried away by the heated plasma during laser pulse has been carried out. It has been observed that for a conventional laser the maximum of the pressure coincides with the beginning of a pulse.

---

*Presented at the Second Workshop on "Laser Interaction and Related Plasma Phenomena" at Rensselaer Polytechnic Institute, Hartford Graduate Center, August 30-September 3, 1971.

The application of multi-beam lasers for the target heating is discussed. The laser having a rectangular shape of light pulse with energy of 10^3 J within 2-16 nsec has been made. The efficiency of the beam energy absorption by the spherical target and the influence of laser beam parameters on this efficiency are determined.

1. INTRODUCTION

At present the idea of laser application for plasma heating seems to be prospective and is universally accepted. In Reference 1 it has been noted that in order to obtain a useful yield of a d-t thermonuclear reaction an amount of $\sim 10^5$ - 10^6 J of laser energy is required, and it seems to be quite a reasonable value. The same energy values are also presented in References 2 and 23. The same value was mentioned at the International Conference on Laser Plasma held in Moscow, November 1970. The obtaining of thermonuclear fusion by means of lasers can be realized nowadays only with the help of a Nd-glass Q-switched laser. One should remember that the efficiency of such lasers equals to about 1 per cent. Such a low efficiency will require a development of a new type of laser to be used in a real thermonuclear arrangement.

Now the investigators face two problems. On the one hand the enhancement of laser energy, and on the other hand the increase of neutron yield due to the improvement of the target and the elucidation of the phenomena taking place at laser interaction plasma. The up-to-date experimental studies of plasma heating are performed with the energy of several tens of Joules, see References 3-7, 22, and 27. Some works reported the neutron emission from the LiD target[5], solid deuterium[6], and deuterized polyethylene[7], the maximum number of neutrons equals to 10^4 - 10^5 neutrons per pulse. Thus, laser energy increase from 10^2 J to 10^6 J is required to enhance the neutron yield up to 10^{16} - 10^{17} neutrons per pulse in order to achieve a thermonuclear reaction. Speaking of the laser energy, we mean the radiative energy at the optimal duration (for various targets it is within 10^{-10} - 10^{-9} sec).

The usage of complex targets with mixtures having the atomic weight $A \sim 250$ can be promising when lasers of supershort duration are employed[8].

The present paper deals with the phenomena that take place in the heated plasma during a laser pulse. The usage of a multi beam laser resulted in the measurement of the heating efficiency of the spherically irradiated target at a level of 10^3 J energy.

2. TARGET AND LASER RADIATION PARAMETERS

In the case of the pulsed plasma heating the condition for a thermonuclear fusion is

$$n\tau \sim n \frac{r}{v_s} K \qquad (1)$$

where n is plasma density, r is the characteristic radius of a target for spherical irradiation, and v_s is thermal velocity. Hence, for $N\tau \geq 10^{14}$, $r = 10^{-2}$ cm, $v_s = 10^8$ cm/sec the required density of the heated substance should be more than 10^{24} cm^{-3}. One of the ways to achieve such densities is to use spherical hydrodynamic plasma implosion[24] or a statical compression of substance. The experiments on collisions of two laser flares showed that in the plane of the flare convergence an increase of plasma density and temperature is observed[9,10,25]. With the spherical geometry it seems to be reasonable to expect the density of two or three orders of magnitude. Such a substance compression is experimentally observed in the devices such as a plasma focus[11]. However, the use of targets with a high density requires the preliminary studies of the heating efficiency of plasma, the density of which is higher than the critical one.

The laser radiation parameters (divergence, duration, energy) are determined on the one hand by the target dimensions, and on the other, by the optical quality of laser elements and focusing system. The beam divergence α should be determined by an inequality

$$\alpha \leq \frac{r}{f}$$

(at $r \approx 10^{-2}$ cm and focus $f \approx 10^2$ cm, $\alpha \leq 10^{-4}$ rad). The use of objectives with $f \approx 10$ cm is possible only at the energy of about 10^3 J.

3. THE MEASUREMENT OF PLASMA PRESSURE

To understand the processes of plasma heating one should know the value of plasma pressure and its variation in the process of heating. In this section the value of pressure depending on time is determined using the data of high-speed interferometric measurements[4].

The experimental arrangement (Fig. 1) consisted of an Nd-glass laser used for plasma heating and a ruby laser that served as a pulse source for the illumination of an interferometer.

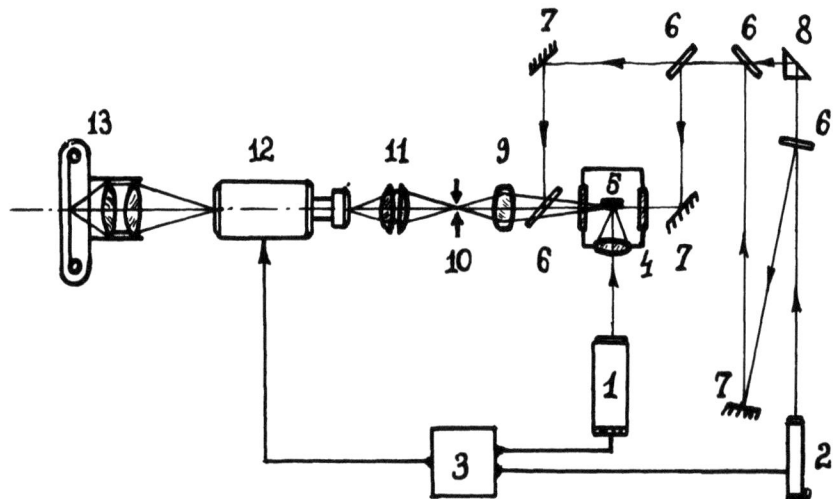

Fig. 1. Scheme of the arrangement for a time scanning of a flare interferogram using an image converter camera in streak mode. 1 - Nd-glass laser; 2 - ruby laser; 3 - controlling device; 4 - focusing objective; 5 - target; 6 - splitters; 7 - mirrors; 8 - full reflective prism; 9 - objective that is projecting the image of the interferogram on the slit 10; 11 - objective that is projecting the image of the slit on the photocathode of the electro-optical tube; 12 - image converter camera; 13 - photocamera.

The carbon target was placed in vacuum at pressure not higher than 10^{-5} mmHg. Nd-glass laser energy was equal to 8 J at pulse duration of 80 nsec at the level of 0.1 amplitude. Pulse oscillogram is shown in Fig. 2. The maximum beam divergence that corresponded to the maximum intensity was equal to 2×10^{-3} rad. During the rise time the divergence was increasing linearly with the intensity, and the focal spot radius varied from 0.05 mm to 0.2 mm (the radius of the beam at focal area due to abberation was equal to 0.05 mm).

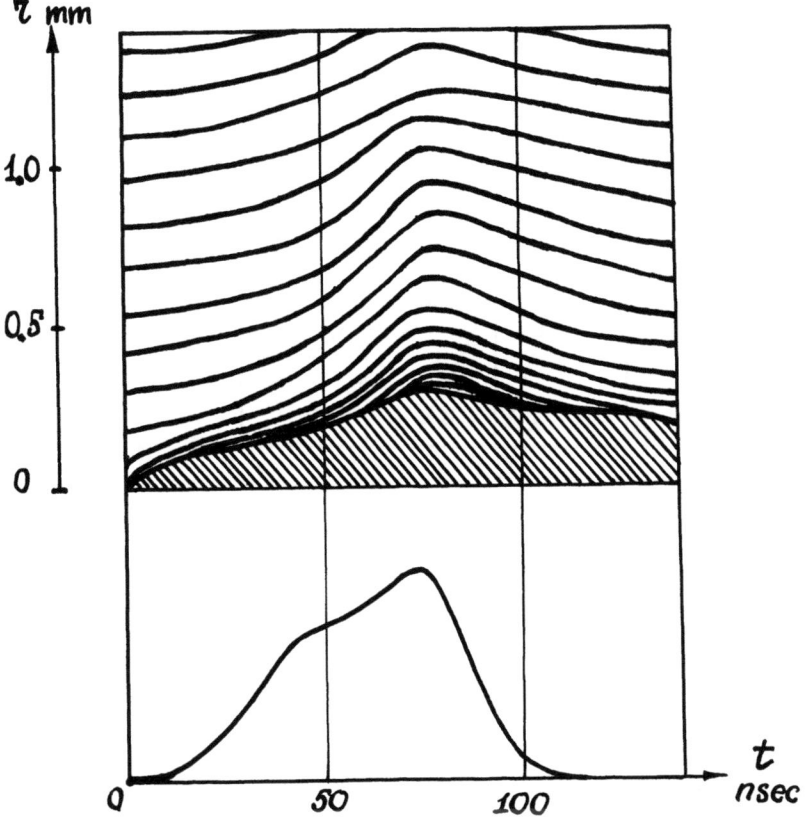

Fig. 2. Time scanning of the carbon interferogram which is correlated with the laser pulse (bottom curve). The slit of the camera coincides with the Nd-glass laser beam axis and is perpendicular to the target surface. Interference fringes in the zero position are parallel to the target. r - distance from the target, r = 0 - corresponds to the target surface. Dotted region is an opaque area.

The split image of the interferogram was scanned in time with an image converter camera in streak mode.

A typical streak interferogram is shown in Fig. 2. It is seen that an opaque zone is broadening with time, and it reaches a maximum value of 0.25 mm in 76 nsec after the beginning of the heating pulse. The electron density in the zone edge was equal to $5 \times 10^{19} cm^{-3}$. The profiles of the electron density for various time intervals have been determined by using the interferogram (Fig. 3). Interferograms have been unfolded by the parabolic method assuming an axis symmetry [21].

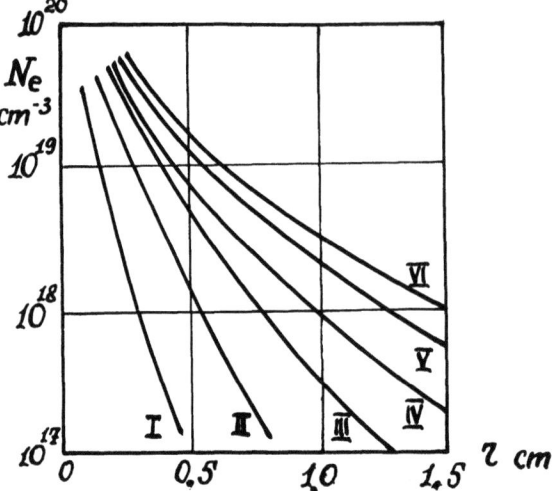

Fig. 3. Distribution of the electron density in the direction of the laser beam axis in various time intervals. Figures above the curves indicate the time from the beginning of a laser pulse in ns: I - 10; II - 16; III - 23; IV - 36; V - 56; VI - 76.

The full mass of plasma heated to a given moment was determined by integrating the density curves. Here the missing parts of curves were approximated by an expression:

$$n_e = \frac{10^{17}}{R^2} \frac{(1-\xi)^\alpha}{\xi^2} \qquad (2)$$

where $\xi = (r + 0.01)R^{-1}$, r - distance from the focal point, $R = v_u t$ - radius of the plasma edge, v_u - asymptotic velocity of plasma particles. Plasma expansion is assumed to be spherically symmetrical. The best approximation of all the curves is achieved at index $\alpha \approx 16$. The gradient of plasma density in the heated region is seen to be the largest and is reducing with time $\sim t^{-3}$.

The data of the density profile and of the velocity of various regions of plasma make it possible to calculate the momentum that is carried away by the plasma. Since the plasma acceleration occurs near the target the momentum value determines the plasma pressure on the target. Here we assume that plasma expands spherically symmetrically. During the laser pulse the plasma edge expands far away in comparison with the focal spot diameter. For the plasma momentum projection normal to the target surface we have

$$F(t) = \int \rho(r,t) v_n(r,t) dV \qquad (3)$$

where $\rho(r,t) = n_e m_i z^{-1}$, m_i - ion mass, z - effective charge, $v_n(r,t)$ - projection of the velocity to the normal, r - distance from the centre of the focal spot. Integration is over the whole of volume.

Hence the pressure in the hot part of a flare will be

$$p(\zeta) = (\pi r_0^2 \tau)^{-1} \frac{dF}{d\zeta} \qquad (4)$$

where r_0 - radius of the focal spot, τ - time of heating, $\zeta = t\tau^{-1}$ - dimensionless time.

The profile of the velocity $v(r)$ used in the calculations is presented in Fig. 4. The choice of such a profile is conditioned by the experimental data on the measurement of plasma velocity according to the Doppler shift[12], as well as on the results of time scanning of ion expansions of various changes in the light of corresponding lines[13]. Here v_0 - sound velocity, r_1 - 0.06 cm. Taking into account that $v_0 \sim q^{1/4}$ (q - radiation flux density), $r_0 = r_{0\tau}(x_1 + x_2 \zeta)$, $r_{0\tau}$ - maximum value of the focal spot at the moment of time; τ, x_1 and x_2 equal 0.25 and 0.75, respectively.

For the case of an Nd-glass laser with the radiation pulse shown in Fig. 2, and 8 J energy to the end of a pulse ($\tau = 76$ nsec), the temperature has been measured.

In our case we have temperature of $T_m \approx 50$ eV and an effective ion charge $z \approx 5$. Here $A_1 = 0.5 \times 10^7$ cm/sec, $v_{umax} = 2.5 \times 10^7$ cm/sec, $r_0 = 2 \times 10^{-2}$ cm. Laser beam divergence increases linearly with time during the largest part of the pulse front. Angular and space characteristics of the light beam were investigated by a slit-optical converter camera. To calculate the pressure the growing part of the pulse was approximated by a linear function. Noting the mentioned conditions the pressure is calculated by Formula (4). A force, acting on the target can be derived from the momentum

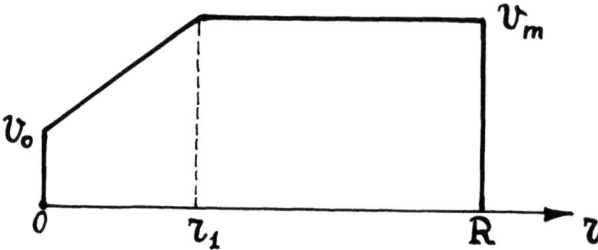

Fig. 4. Profile of the substance velocity in plasmas along the laser beam axis.

linear function. Noting the mentioned conditions the pressure is calculated by Formula (4). A force, acting on the target can be derived from the momentum

$$f(\zeta) = \frac{1}{\tau} \frac{dF(\zeta)}{d\zeta} \qquad (5)$$

Figure 5 illustrates the curves of the dependence in time of the force, momentum and plasma pressure on the target. The maximum pressure is seen to act at the beginning of a pulse due to the fact that the divergence is small at small intensity and the focal spot diameter is determined in this experiment only by the lens parameters. At later times (as is seen from Fig. 3) the density profile is rising. The plasma region where the absorption occurs, is shifting from the target and is growing. The mass of gas heated by laser is also increasing. The decrease of pressure indicates an attenuation of laser radiation. The temperature in the hot region falls and the larger part of the radiative energy is transformed into the kinetic energy of the expanding substance.

Thus, changing the time dependence of the radiation divergence one can shift the maximum pressure and use the laser energy optimally for the plasma heating in real conditions. The presence of the peak of pressure and hence, the maximum temperature at the beginning of the laser pulse can explain the experiments of Reference 14, the authors of which have observed the intensive peak of x-ray emission from the flare in the very beginning of a laser pulse with the radiative power much less than the maximum one.

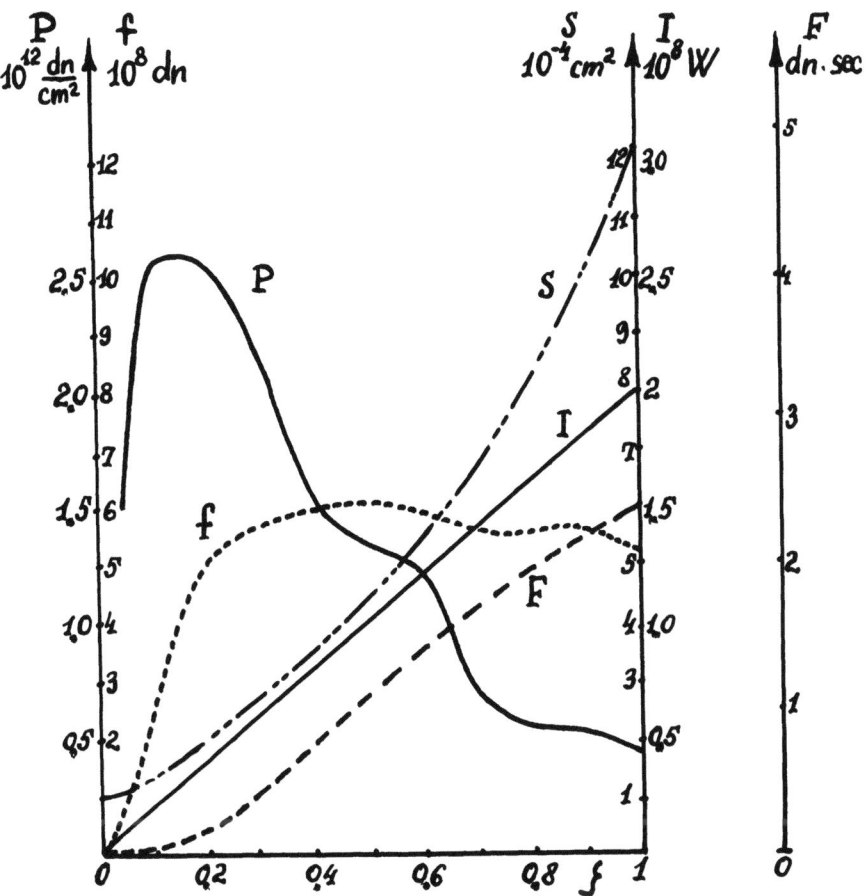

Fig. 5. Time variations of plasma pressure p in the "hot" region and momentum F carried away by plasma during the growing part of the laser pulse, the intensity of which is approximated by the linear function (line I). f is force, applied to the target, S is a focal square.

4. NEUTRON GENERATION

The improvement of an Nd-glass laser has made it possible to record neutron emission from the deuterized polyethylene target[7]. The presence of heavy ions in the deuterium plasma should lead to the increase of temperature and decrease of density[15] (in comparison to pure deuterium)

$$T = T_D \, Z^{2/3} \left(\frac{2}{Z+1}\right)^{2/3} \left(\frac{A}{2}\right)^{2/9}$$

$$N = N_D Z^{-1} \left(\frac{2}{Z+1}\right)^{1/2} \left(\frac{A}{2}\right)^{1/6}$$

(6)

where Z, A - average charge and ion mass, respectively, $T_D N_D$ - temperature and pressure in the case of pure deuterium.

In the experiment we used a laser with five amplifiers. The laser beam was focused by a lens $f = 100$ mm onto the massive target prepared from the powder-like polyethylene. At the beam divergence of 1.2×10^{-4} rad the heated area was equal to 10^{-4} cm^2. Maximum laser energy was of 80 J and the duration did not exceed 3.5 nsec at the 1% level.

Neutron emission was recorded by a photomultiplier with a plastic scintillator. The photomultiplier recorded the light flash of recoil photons.

Figure 6 illustrates the pulses from the neutron detector at distances from the scintillator to the target of 10 cm and 60 cm. Laser pulse from the coaxial photodiode recorded at the same trace served as a time mark. The minimum number of neutrons is easily determined because at the distance of 60 cm the neutrons are always recorded in each experiment (at energy not less than 14 J). It makes not less than 10^4 neutrons for the total number.

5. SPECTRAL MEASUREMENTS

To measure electron temperature T_e of the dense laser plasma there have been used the relative intensities of the spectral lines in the vacuum ultraviolet spectrum region, in the range 100 - 200 Å. They correspond to the transitions $2s2p^n - 2s2p^{n+1}$ of ions of the elements with charge of $Z = 20-30$, that are stripped at temperatures of 100 - 1000 eV[16].

Stark line broadening of the hydrogen-like ion CVI has been used to measure the electron density. The choice of the ion CVI of a rather high ionization potential (498 eV) permits the

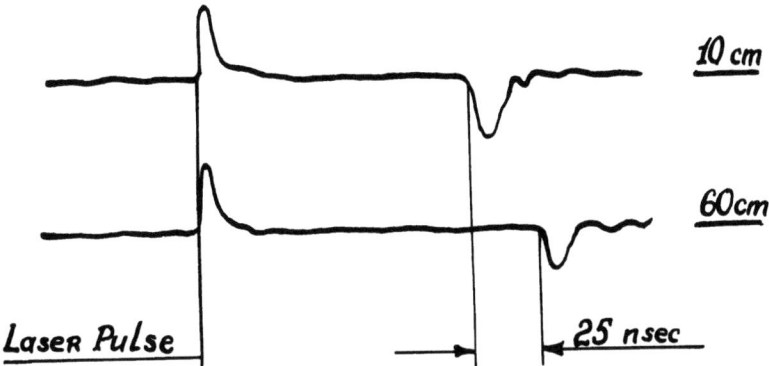

Fig. 6. Oscillograms of the pulses from the neutron detector distanced 10 cm and 60 cm from the target. A light pulse from the coaxial photodiode is recorded at the beginning of the traces.

evaluation of the density distribution in the hot flare region during the action of the laser pulse[19] by the integrated spectrograms.

The measurement of the profile of plasma velocity was performed by the analysis of data on time scanning of CVI ion lines in the visual spectrum region[13] and on the Doppler shift of absorption lines[17] of the resonance doublet of CIV ion (Fig. 7). The hot dense flare nucleus radiates in a continuous spectrum through an expanding cloud of the rarefied cold plasma. It leads to the appearance of the absorption lines similar to the "Fraunhofer lines" in the solar spectrum. The shift of absorption lines relative to the emission lines to the blue side of $\lambda \sim 1 \text{Å}$ permits the evaluation of the velocity of the directed motion of the "cold" envelope. Spectrogram of Fig. 7 shows that the multi-charged laser plasma that is leaving a dense "hot" region, accelerates due to the thermal electron energy up to the velocity $\sim 3 \times 10^7$ cm/sec.

6. MULTIBEAM LASER[26]

With the help of a powerful laser arrangement there has been measured the efficiency of radiation absorption in the target with spherical irradiation. The arrangement consisted of a successive-parallel system of amplifiers and it provided the obtaining of

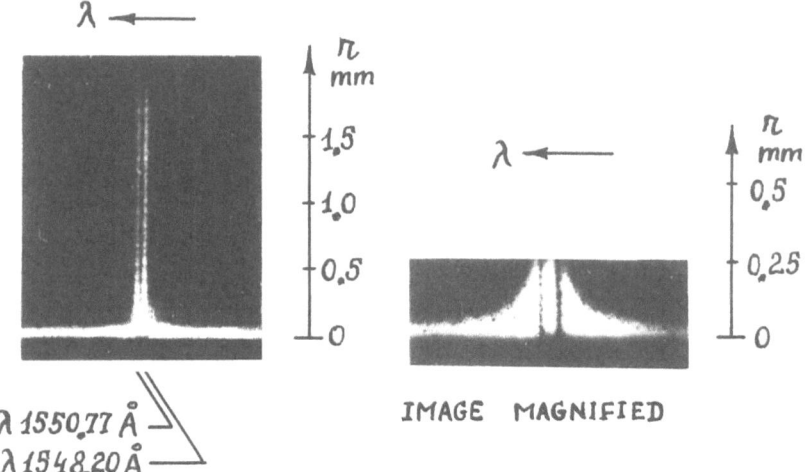

Fig. 7. Spectrograms of the lines λ = 1550.77 Å, λ = 1548.20 Å, r - distance from the target. Absorption lines of the cold envelope are shifted on a $\Delta\lambda$ = 1.2 Å relatively to the emission line.

about 10^3 J energy at the controllable pulse duration from 2 to 16 nsec. Nine beams were formed at the output. Divergence of each beam did not exceed 2×10^{-4} rad. The background energy did not exceed 0.1 percent of the useful energy. The energetic characteristics of the laser arrangement are listed in Table I. E - total output energy in Joules (with an accuracy of about 10%), B - radiative brightness, q - flux density on the target, η - efficiency relating to the electric energy of the condensor. From Table I one can see that with the pulse reduction from 16 to 2 nsec the energy falls about a factor of two. Accordingly, the space distribution of intensity and coherent properties of the beams do not vary, in fact.

A preamplified laser beam was directed to the powerful amplifying cascade (Fig. 8) where the light beam obtains the whole energy despite a rather small amplification coefficient ($\sim$10). Amplification proceeds into two stages. In the first stage a beam is split into three parts, each of which is amplified by a rod after the corresponding isolating shutter. In the second stage each beam is again divided into three, and each is again amplified. As the

τ nsec	E J	$B \cdot 10^{-16}$ W/cm^2 stread.	q W/cm^2	η %
2	600	4.3	1.5	0.15
4	800	2.9	1	0.2
8	1000	1.8	0.6	0.24
16	1300	1.2	0.4	0.3

TABLE I

active elements Nd-rods are used, 4.5 cm diameter and 60 cm pumping part length.

The arrangement provided a flux density on the target that equaled 10^{16} W/cm^2 (for a flat target) and 2×10^{15} W/cm^2 (for a spherical target). In the case of spherical irradiation the time derivative of the specific energy yield in the heated plasma can achieve 10^{18} W/cm^3, in the condition of full radiation absorption by the plasma. For $r_o = 2.5 \times 10^{-2}$ cm an energy yield of 5×10^{15} W/cm^3 has been experimentally achieved. The latter value can be substantially increased, using heavy targets of smaller dimensions[8].

Compensation of the optical paths of each beam was realized by optical compensators that provided an accuracy of synchronization of light pulses coming to the target not greater than 3×10^{-11} sec.

The formation of a rectangular light pulse is performed by means of a shutter controlled by a spark gap with laser triggering[18]. The shutter provided a contrast of 0.5×10^5. The oscillograms of light pulses after the shutter are given in Fig. 9.

After amplification all nine beams are directed to the focusing objectives mounted in the walls of a vacuum chamber. The focusing system provided a practically uniform irradiation of the

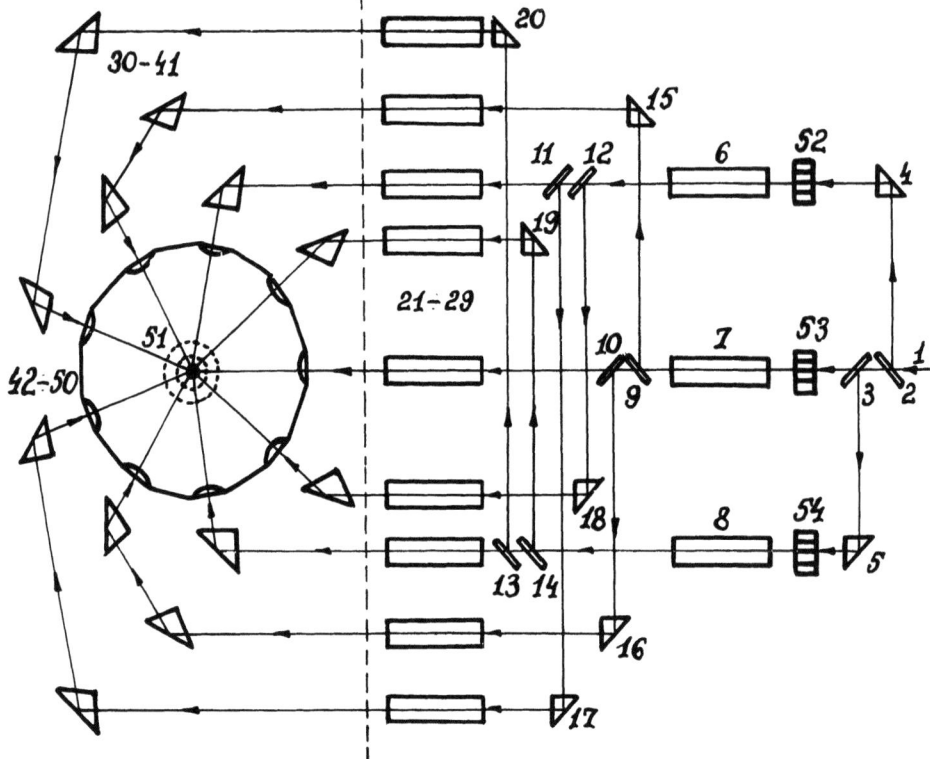

Fig. 8. Scheme of the powerful amplifying system. 2, 3, 9, 10, 11 ÷ 14 - dividing elements; 4, 5, 15-20, 30-41 -- prisms of the full internal reflection; 6-9, 21 ÷ 29 - Nd-rods; 42-50 -- focusing objectives; 51 - spherical target; 52-54 -- isolating shutters. Calorimeters and coaxial photodiodes are not shown in the scheme. In some beams the light delays are used to compensate optical ways (they are not shown in the scheme).

spherical target, 0.1 - 0.5 mm diameter.

To photograph high temperature processes that occur at the target a system of a high speed laser photography has been used[19]. Frame exposure equaled 0.4 ns, a number of frames - 7.

7. EFFICIENCY OF THE TARGET HEATING

Light radiation was focused on the spherical target dimensioned from the condition $r_o \approx v\tau$, where v - thermal velocity of expansion at temperature T. At $T \approx 1$ KeV and pulse duration $\tau \approx 2 \times 10^{-9}$ sec we have $r_o \approx 2.5 \times 10^{-2}$ cm.

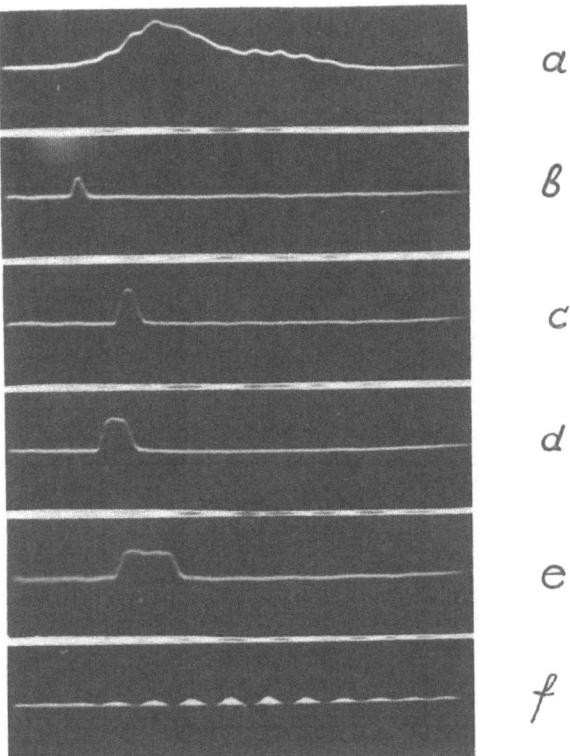

Fig. 9. Radiation light pulses of the powerful laser. a - a pulse after the oscillator; b, c, d, e - radiative pulses after a forming shutter and a first three passes amplifier for the regimes, respectively: 2, 4, 8 and 16 nsec; f - time marks of 10 nsec.

The energy, absorbed in the target, has been measured by observation of a gasdynamic expansion of the spherical shock wave in the residual gas. The motion of the wave is described by formulae of the spherical instant point explosion[20] when the mass of the shock driven gas is sufficiently greater than the mass of the heated target. In our case it occurs when the radius of the shock wave equals $R \approx 20\ r_o$.

Figure 10 illustrates a seven-frame shadow photograph of the shock wave formed in air at pressure of 17 torr. Due to the spherical irradiation of the target the shock wave has good spherical symmetry. The long duration of the shock wave formation (first

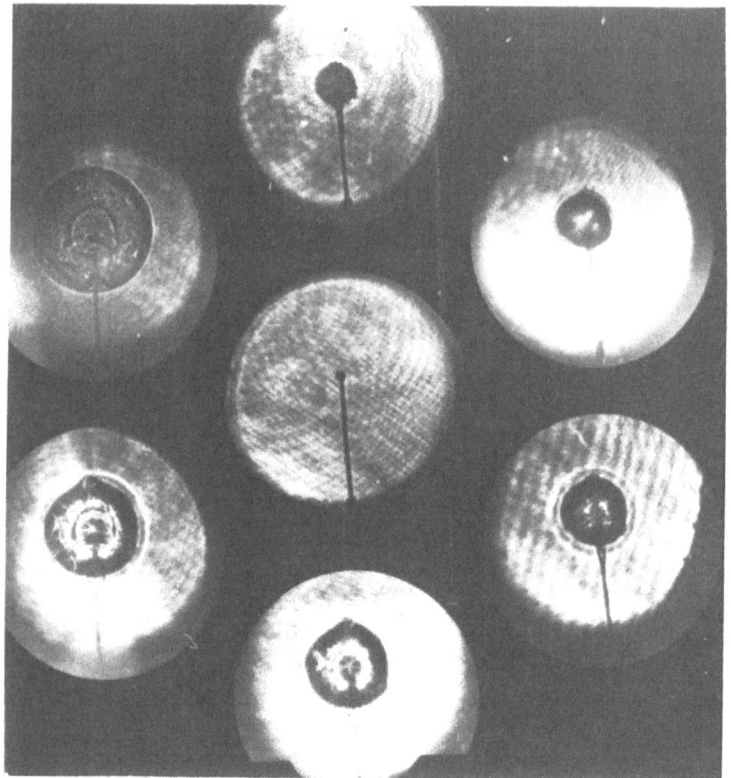

Fig. 10. Shadowgram of the spherical shock wave. Time intervals correspond to the following delays from the beginning of heating. 1 - 2.0; 2 - 20; 3 - 45; 4 - 70; 5 - 140; 6 - 210; 7 - 460. Still diameter equals 5.0 cm. A polyethylene target has $r_o = 2.5 \times 10^{-2}$ cm. Laser energy is equal to 560 J. Pulse duration - 2 nsec.

three frames) compared to the time of heating, 2 nsec, is explained by high target density versus the density of the surrounding gas.

The law of the shock wave front motion is defined by the expression

$$R = \left[\frac{a(\gamma) E_e t^2}{\rho_o} \right]^{1/5} \tag{7}$$

HEATING OF LASER PLASMAS FOR THERMONUCLEAR FUSION

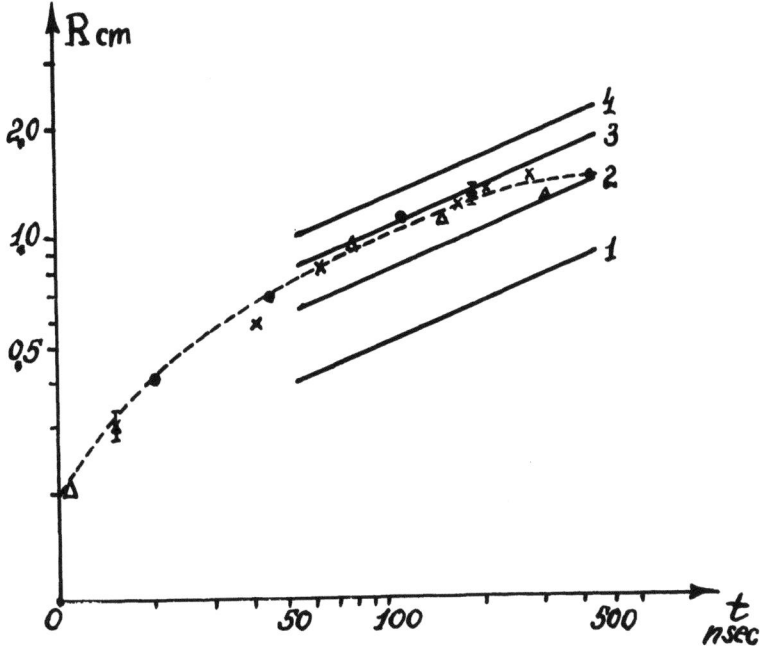

Fig. 11. Experimental dependence of the shock wave radius on time - a dotted curve. The marks o, x, Δ correspond to the three separate experiments in the mode of τ = 2 nsec. Theoretical curves correspond to the explosion energies: 10, 10^2, 3.5 x 10^2; and 10^3 J. Gas pressure (air) before the wave equals 15 mm Hg.

where $a(\gamma)$ - coefficient that depends on the adiabatic index, which in our case is equal to 1, E_e - energy of the explosion, which equals the absorbed laser light energy neglecting the plasma radiation, ρ_o - density of unperturbed gas.

Figure 11 shows the dependence of R(t) for three experiments at the same conditions. The same figure illustrates R-t diagrams that are calculated by Formula (7) at the explosion energies of 10, 100, 350, 1000 Joules. Comparing theoretical and experimental curves one can conclude that the explosion energy equals 300 J. This is the minimal value for the laser radiation energy absorbed by the plasma. Taking into account the energy losses on the optical elements of the converging and focusing systems one can derive that a light energy of about 450 J falls on the target surface, more than 70 percent of which is absorbed in the target.

After 250 nsec the dependence of radius on time changes from $R \sim t^{0.4}$ to $R \sim t^{0.2}$. This can be explained by the losses due to recombinative radiation for which the plasma is transparent behind the shock wave front in the considered time interval.

Note that the reduction of the energy yield (see the Table) for shorter pulse length cannot yet be explained. It can turn out to be an important point when designing a thermonuclear system, because for a laser with 10^6 J energy the optimal pulse duration is equal to $10^{-10} - 10^{-9}$ sec. If the energy reduction is connected with the saturation of the lower nonradiating transition then for the decrease of the pulse to 10^{-10} sec the energy reduction won't be more than about two times. However, this assumption needs a special experimental verification.

REFERENCES

1. N. G. Basov, O. N. Krokhin, "Laser Application for Thermonuclear Fusion", Vestnik Ac. Sci. USSR, N 6, 55 (1970).

2. H. Hora, "Application of Laser Produced Plasmas for Controlled Thermonuclear Fusion", p. 427 in <u>Laser Interaction and Related Plasma Phenomena</u>, Plenum Press, New York-London, 1971.

3. N. G. Basov, V. A. Gribkov, O. N. Krokhin, G. V. Sklizkov, "Investigation of High-Temperature Phenomena Induced by Powerful Laser Radiation Focused onto the Solid Target", JETF <u>54</u>, N 4, 1073 (1968).

4. N. G. Basov, V. A. Boiko, V. A. Gribkov, S. M. Zakharov, O. N. Krokhin, G. V. Sklizkov, "Gasdynamics of Laser Plasma During Heating", JETP <u>61</u>, N I, 154 (1971).

5. N. G. Basov, P. G. Kriukov, S. D. Zakharov, Yu. V. Senatsky, S. V. Tschekalin, IEEE J. Quant. Electron., <u>QE-4</u> 864 (1968). G. W. Gobeli, J. C. Bushnell, P. S. Peercy, E. D. Jones, Phys. Rev. <u>188</u>, N I, 300 (1969).

6. F. Floux, D. Cognard, L. Denoeud, G. Piar, D. Parisot, J. Bobin, F. Delobeau, C. Fauquignon, "Nuclear Fusion Reactions in Solid Deuterium Laser Produced Plasma", Phys. Rev. A (General Physics), <u>I</u>, N 3, 821 (1970).

7. N. G. Basov, V. A. Boiko, S. M. Zakharov, O. N. Krokhin, G. V. Sklizkov, "Neutron Emission from Laser Plasma Heated by Nanosecond Pulses", JETP Letters, 13, N 12, 691 (1971).

8. Yu. V. Afanasiev, E. M. Belenov, O. N. Krokhin, I. A. Poluektov, "About Possibility of Obtaining of the Intense Neutron Source at Laser Plasma Heating", JETP Letters, 13, N 5, 257 (1971).

9. N. G. Basov, O. N. Krokhin, G. V. Sklizkov, "Investigation of Dynamics and Heating and Expansion of Plasma Created by Laser Radiation Focused onto the Substance", Trydi FIAN, 52, 171 (1970) "Quantum Radiaphysics".

10. V. A. Gribkov, V. Ya. Nikulin, G. V. Sklizkov, "Plasma Density Increase at Laser Flare Collision", Short Communications in Physics, FIAN, N 2, 45-49 (1971).

11. N. V. Filippov et al, "Experimental and Theoretical Investigation of Pinch Discharge Like 'Plasma Focus'", Proc. of the 4th Conference on Controlled Thermonuclear Fusion, 1971, USA, Madison, Report CN-28-D6. N. J. Peacock, M. G. Hobby and P. O. Morgan, "Measurements of the Plasma Confinement and Ion Energy in the Dense Plasma Focus", Report CN-28-D3.

12. E. V. Aglitzky, V. A. Boiko, S. M. Zakharov, G. V. Sklizkov, "Determination of Electron Density Profile in Laser Plasma by Stark Spectral Line Broadening", Preprint N 143, FIAN, Moscow, 1970.

13. N. G. Basov, V. A. Boiko, Yu. A. Drozhbin, S. M. Zakharov, O. N. Krokhin, G. V. Sklizkov, V. A. Yakovlev, "Investigation of Initial Stages of Gasdynamical Laser Flare Plasma Expansion", DAN, 192, N 6, 1248 (1970).

14. G. L. Bobin, F. Floux, P. Langer, M. Pignerol, "X-rays from a Laser Created Deuterium Plasma", Physics Letts., 28A, 398 (1968).

15. O. N. Krokhin, "High-Temperature and Plasma Phenomena Induced by Laser Radiation", Proc. of the Intern. School of Physics, Enrico Fermi, Course XLVIII, Academic Press NY-L, 1971, 278-305.

16. N. G. Basov, V. A. Boiko, Yu. P. Voinov, E. Ya. Kononov, S. L. Mandelshtam, G. V. Sklizkov, JETP Letters, 5, 179 (1967); 6, 849 (1967). V. A. Boiko, Yu. P. Voinov, V. A. Gribkov, G. V. Sklizkov, Optics and Spectroscopy, XXIX, 1023 (1970).

17. E. V. Aglitzky, V. A. Boiko, S. M. Zakharov, G. V. Sklizkov, "Determination of Velocity and Electron Density Profile in Laser Plasma by Measurements in UV Spectral Region", Short Communications in Physics, FIAN, N 6, p.3 (1971).

18. A. H. Guenther, J. R. Bettis, "Laser Triggered Switching", p. 131-172 in Laser Interaction and Related Plasma Phenomena,' ed. by H. Schwarz and H. Hora, Plenum Press, NY-L, 1971. N. N. Zorev, G. V. Sklizkov, S. I. Fedotov, A. S. Shikanov, "Investigation of Spark Channel Triggered by Laser Radiation", Preprint FIAN, N 56, Moscow 1971.

19. N. G. Basov, O. N. Krokhin, G. V. Sklizkov, "Laser Application for the Production and Diagnostics of Pulse Plasma", Appl. Optics $\underline{6}$, N 11, 1814 (1967).

20. Korobeinikov, N. S. Melnikov, E. V. Ryazanov, "Theory of the Point Explosion", Fizmatgiz, Moscow (1961).

21. V. A. Gribkov, V. Ya. Nikulin, G. V. Sklizkov, "Interferometric Investigation of Laser Plasma Collision", Preprint FIAN N 153, Moscow (1970).

22. G. V. Sklizkov, "Kinetic and Ionization Phenomena in Laser Produced Plasmas", p. 235-257 in Laser Interaction and Related Plasma Phenomena, Plenum Press, NY-L, 1971 (ed. by H. Schwarz and H. Hora).

23. P. P. Pashinin, A. M. Prokhorov, "Obtaining of High Temperature Dense Plasma at Laser Heating of a Special Gas Target", JETP, $\underline{60}$, 1630 (1971).

24. J. W. Daiber, A. Hertzberg, C. E. Wittliff, "Laser Generated Implosions", Phys. of Fluids, $\underline{9}$, N 3, 617 (1966).

25. H. Puell, H. Opower, H. G. Neusser, "Experiments with Two Laser Produced Interpenetrating Plasmas", Phys. Letts. $\underline{31A}$, N 1, 4 (1970).

26. N. G. Basov, O. N. Krokhin, G. V. Sklizkov, S. I. Fedotov, A. S. Shikanov, "Powerful Laser Installation with a Successive-Parallel System of Amplifiers for Plasma Heating", JETP, N 1, (1972).

27. M. P. Vaniukov, V. Venchikov, V. I. Isaenko, P. P. Pashinin, A. M. Prokhorov, JETP, To be translated.

NUCLEAR DD REACTIONS IN SOLID

DEUTERIUM LASER CREATED PLASMA*

F. FLOUX, J.F. BENARD, D. COGNARD and A. SALERES

Commissariat à l'Energie Atomique, Centre d'Etudes de Limeil

B.P. n° 27 - 94-VILLENEUVE-SAINT-GEORGES - FRANCE

I - INTRODUCTION

Research in high density and high temperature plasmas has lead the development of more and more powerful lasers. The past nine years have been devoted to the ruby and neodymium glass lasers, while the future devices will probably look like molecular or chemical ones.

Among the neodymium glass lasers, two main classes can be distinguished : the picosecond mode locked laser which was first developped at the Lebedev Institute by Prof. Basov and his team ; the nanosecond laser which is more common and widely used in our laboratories. The corresponding laser produced plasmas are also to be distinguished because, while the purpose of the experiments remains the same , the mechanisms of heating and the characteristic lengths and times of the produced plasma are basically different.

The problem to create a high density and high temperature plasma has been continuously studied at Limeil with nanosecond Nd^{3+} glass lasers. This obstinacy is not due to the heavy technology of the research programs which are to be developed to reach significant steps but it mainly results from an overall examination of the experimental results achieved in laser produced plasmas.

---

* Presented at the second workshop on "Laser Interaction and Related Plasma Phenomena" at Rensselaer Polytechnic Institute, Hartford Graduate Center, August 30 - September 3, 1971.

We have considered, in fact, that the hydrodynamic behavior of the plasma was necessary to reach very high temperatures when using infrared optical generators. For this purpose, we have given great attention to the plasma behavior during the absorption of laser light [1] and we have been lead to modify the pulse shape during the successive steps of our research program [2,3]. Then, at the same time that the laser output power was increased, first from 1 to 4 GW, and then from 5 to its present level of 30 GW, the rise time of the pulse was shortened, first from 100 to 5 nsec and then from 5 to 1 nsec.

In 1969, during the first workshop [4]. I took this opportunity to present our experimental results obtained when a 40 nsec, 1 GW, Nd^{3+} laser was focused onto a solid deuterium ice. The measured plasma electron temperature was closed to 400-500 eV. The theoretical interpretation of such results lead us to propose the radiative deflagration model [5] in which the temperature of the heated matter varies as the 2/3 power of the laser flux density impinging onto the target.

In August 1969, the laser pulse was modified (P $\sim$ 4 GW ; $\Delta t \sim 7$ to 8 nsec ; $\tau_m \sim 5$ nsec) and a sharp focusing on the deuterium ice gave us neutrons produced by nuclear fusion reaction taking place inside the heated plasma [6]. Let me recall briefly the main results of such experiments :

- First, the new pulse shape was obtained by inserting a Pockels cell between the oscillator stage and the amplifier cascade. The total energy ranged between 20 and 40 J.

- The electron temperature was formed to be 500 to 700 eV when hydrogen or deuterium ice were used.

- The detected signals were accurately identified as corresponding to neutrons created inside the laser produced deuterium plasma. They did not occur when using hydrogen instead of deuterium while the electron temperature was nearly the same.

- Time of flight measurements were performed with phosphors located 1.50 from each other and the neutron energy was 2.45 MeV as expected for nuclear DD reactions. The neutron emission is simultaneous with the laser pulse except for some scattering from the experimental items (30 %) and the floor of the lab (15 %). This scattering had been evidenced by static measurements with a conventionnal DD neutron source.

- A theoretical attempt using computer calculations [7], showed to us why these DD reactions were achieved in the plasma. In fact, the low increase of electron temperature between the two experiments was probably less efficient than the steepness of the rise time of the laser pulse. Because of this latter improvement an electron conduction mechanism is allowed to develop between the deflagration front where the laser energy is absorbed ($n = 10^{21}$ cm^{-3}) and the shock wave propagating inside the ice ($n_0 \sim 5.10^{22}$ cm^{-3}). This region increases by a factor 4 or 5 the efficiency of ions heating and we know, presently, that neutrons are produced inside this region. The increase of neutrons output is, then, strictly dependent on the efficiency of the photon-electron transfert through the deflagration front.

Roughly speaking, we can say that the laser output power gives the maximum boundary temperature and then the pulse length (so the energy) sustains the heating mechanism.

We shall deal, now, with our experimental and theoretical attempts made, since 1970, to achieve larger neutron output yields when using such optical energy source.

II - EXPERIMENTAL SET-UP

1. The laser system

A constant 3.5 nsec half-width laser pulse has been used in this experiment [3]. The rise time, of 1.2 nsec, is achieved by inserting two Pockels cells (PC) between, the Q - switched oscillator and the seven stages amplifier cascade. (see fig. 1). The end rod diameter of 64 mm, allows 120 J energy output which may be varied from 20 to 85 J on the interaction chamber by using neutral filters.

The rise time of such a pulse is determined to be the time necessary to grow from 10^{-5} P_M to P_M, P_M being the maximum output power.

The function of the successive Pockel's cells used in such a device is depending on their location. Starting from the oscillator stage, the two first PC shape the required laser pulse, while the two following ones prevent the first stages of the laser from being broken down by the backwards reflected energy, coming from the interaction chamber. Moreover, the third PC prevents us from a possible superradiance effect due to the high gain amplifier cascade. At last, a Faraday cell has been recently used, in order to improve the protection against backwards reflected energy.

The output divergence of the beam is ranging from 2 to 4 mrad.

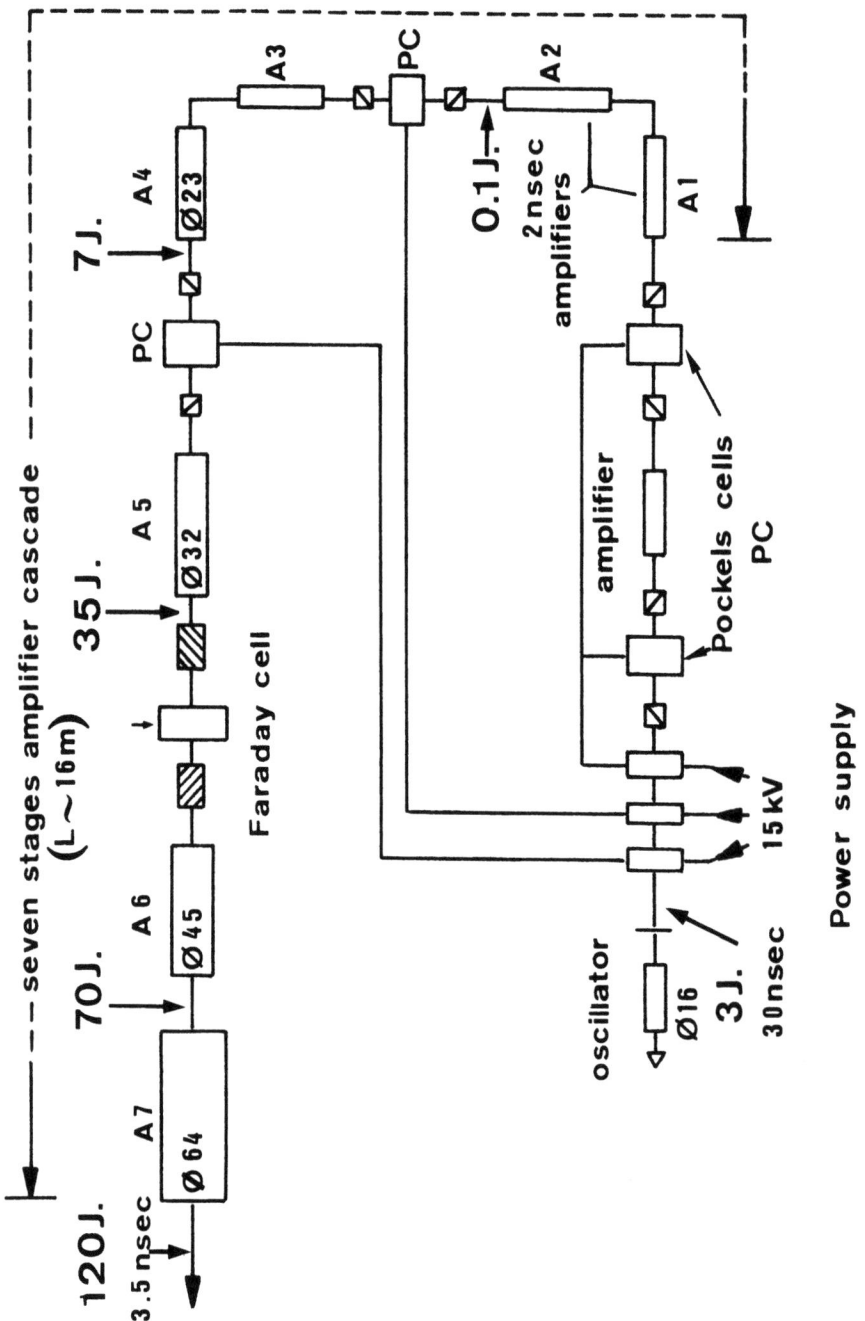

Fig. 1 : Limeil laser system (C.G.E.)

2. Focusing of the laser beam

A plane aspheric large aperture lens [8] focuses the beam on the surface of a 1 mm thick, square section, deuterium ice. The aperture is F/0.9 and the focal length 70 mm. So, the focal diameter is less than 30 μm in vacuo and can be estimated to 100-150 μm inside the plasma.

The focusing lens is located by a static alignment and, into operation, the cryogenic device is moved with respect to the focal point. For this purpose, an image of the ice is made on a screen 5 m away from the chamber. This system allows us to focuse the beam with an accuracy better than 20 μm, value which is necessary to optimize the interaction phenomena.

We have used four diagnostics, which are respectively :

- Reflected energy measurements
- STL streak camera recording
- Neutron counting by BF 3 gas detectors
- Neutron fluxes versus angle of emission, using phosphors.

3. Reflected energy measurements

When striking solid targets with powerful laser beams, we know that some amount of energy is backwards reflected. This effect is important to be studied, at least for three reasons :

a) First, we have to prevent the laser itself to be broken down by the reflected energy ;

b) Secondly, in order to know the absorbed energy, we have to measure this backwards effect ;

c) At last, reflected power can be of interest in the knowledge of the phenomena occuring inside the plasma (absorption, hydrodynamic behavior).

This experiment has been carried out with a laser beam axis perpendicular to the target. The absorption of the experimental system (entrance window and focussing lens) has been previously determined by replacing the deuterium stick by a concave mirror. The reflected energy is only measured along the beam axis, since reflected signals 45° away from it have been found more than hundred times smaller. The transmitted energy through the ice is less than 3 % of the incident one, and is essentially due to the light passing the central hole of the focussing lens [8].

Calibrated calorimeters pick up the incident and reflected energies while the corresponding pulses are gathered on the same CVHC 20 fast photodiode. Signals are recorded on a Ferisol OZ 100 oscilloscope. The delay accuracy between the two signals has been found close to 300 psec.

When the incident laser pulse is kept constant, we have found a maximum reflected signal, depending on the focussing conditions and sharply connected with the maximum neutrons emission. Fig. 2 gives account of such results and one can see that hundred microns away from the focal point, the two data have decreased by a factor 5. In the same time, the absorbed energy has a little increased. This fact is particularly important, in order to optimize the neutron yields versus energy.

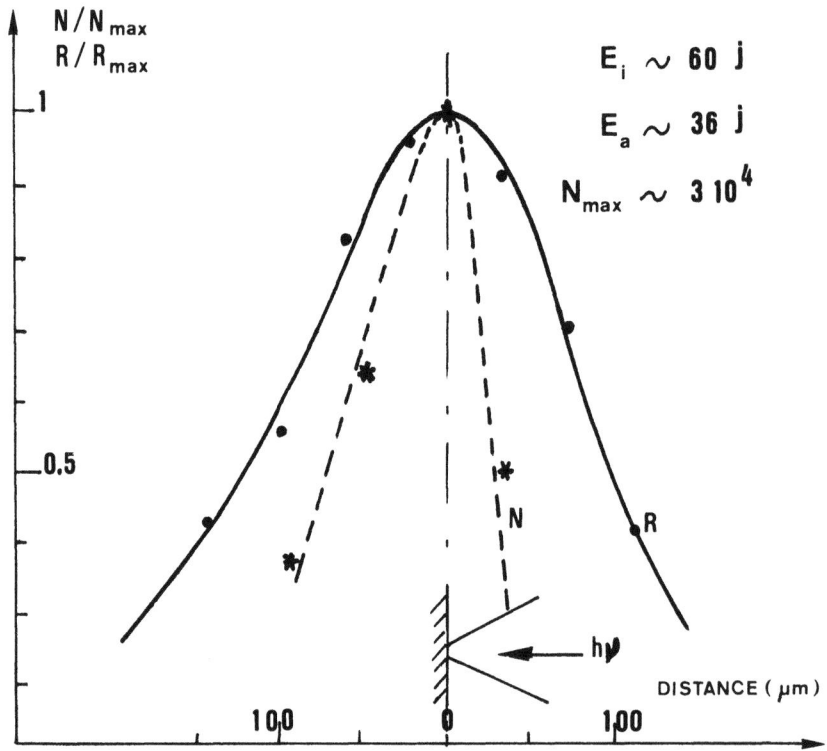

Fig. 2 : Influence of the focussing accuracy upon reflection and neutron output.

We think that this effect closely depends on the temperature : As a matter of fact, a high temperature gives a steep density gradient and then, it occurs a strong reflection in the same time that neutrons are emitted. Furthermore, an exact analytical treatment of linear absorption process inside an exponential density gradient has been achieved by A. Salères [9] and gives account of these results.

On Fig. 3, we have plotted a typical reflected signal, P_r, correlated with the incident power P_i for a total incoming energy of 44 J (The neutron flux is then of 2.10^4). The difference P_a gives quite well the laser energy absorbed into the plasma. The P_r curve shows two different stages in the interaction process :

- During the first 2 nsec, the reflection coefficient R, is varying with the rise front of the incident laser pulse until

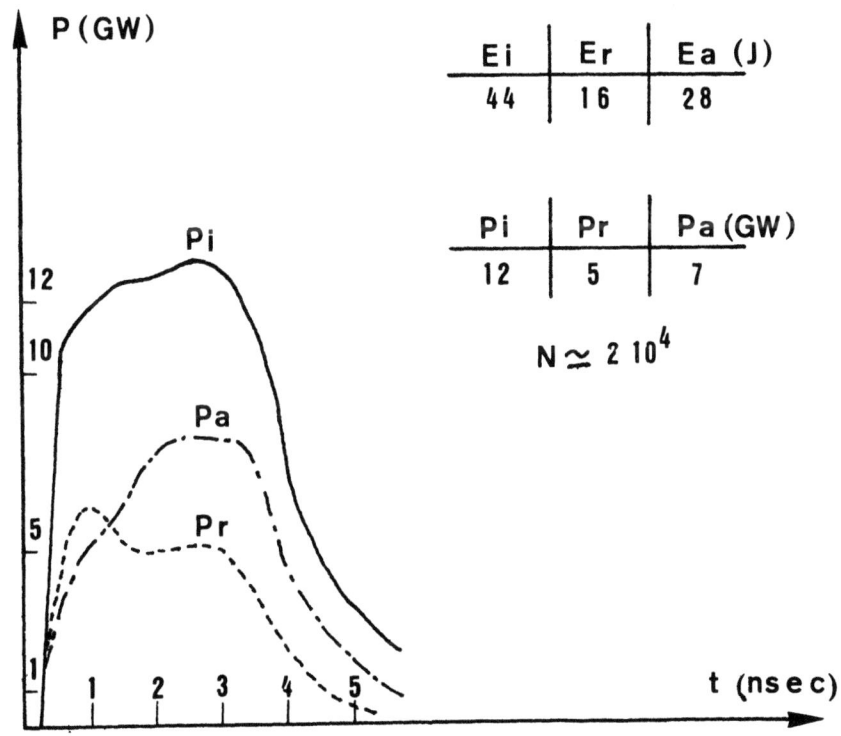

Fig. 3

a nearly constant value P_a has been reached. We interpret this first stage as corresponding to the time which is necessary to create a smooth density profile and to develop an heating mechanism, inside the plasma, by non-linear electron conduction. We shall confirm these conclusions with computer results given in section III.

- The second stage of the process shows a nearly constant R value, which can be assumed to correspond to a stabilization of the density profile in the same time that P_i and P_a are quite constant. This stage may be a sustainment of the heating process, while the maximum value of the temperature has been reached at the end of the first stage. If such a conclusion can be extended to picosecond heating, one may easily conclude how will be important a prepulse heating in such a regime.

The variation of the reflection coefficient R, has been obtained in the 20-85 J energy range. Fig. 4 gives account of the corresponding results, where R ($\sim E_i^{1/6}$) varies from 28 to 42 %. In these results, the value of R has been averaged over the total pulse length. These high values for R give explanation of the necessary Faraday an Pockel's cells, inserted into the amplifier cascade.

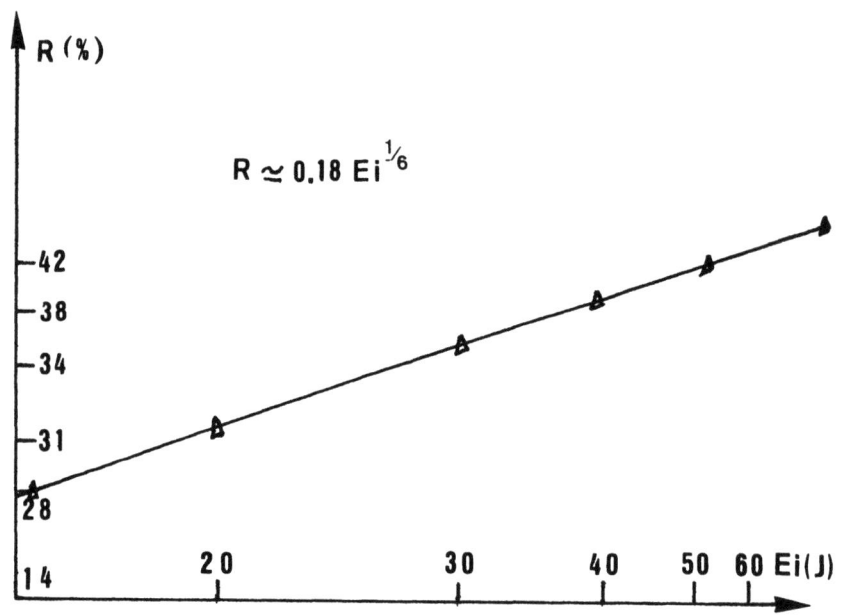

Fig. 4

4. Streak camera recording

Using a conventional STL streak camera, we have plotted in Fig. 5, the evolution of the luminous created plasma, correlated in time with incident laser power. Quantitative measurements cannot be significant because the luminous intensity, recorded on the polaroïd film, is depending on $n^2 T^{1/2}$ (Bremsstrahlung emission) and on the luminous emission distribution, as a function of the wave length. In spite of that fact, this picture shows that, after 2 nsec, the dimension of the heated region is decreasing, while the laser is still running. The same effect occurs for lower output power, but for time up to 4 and 5 nsec. In this later case, one may think that the decrease of laser energy is the main reason for the plasma to disappear.

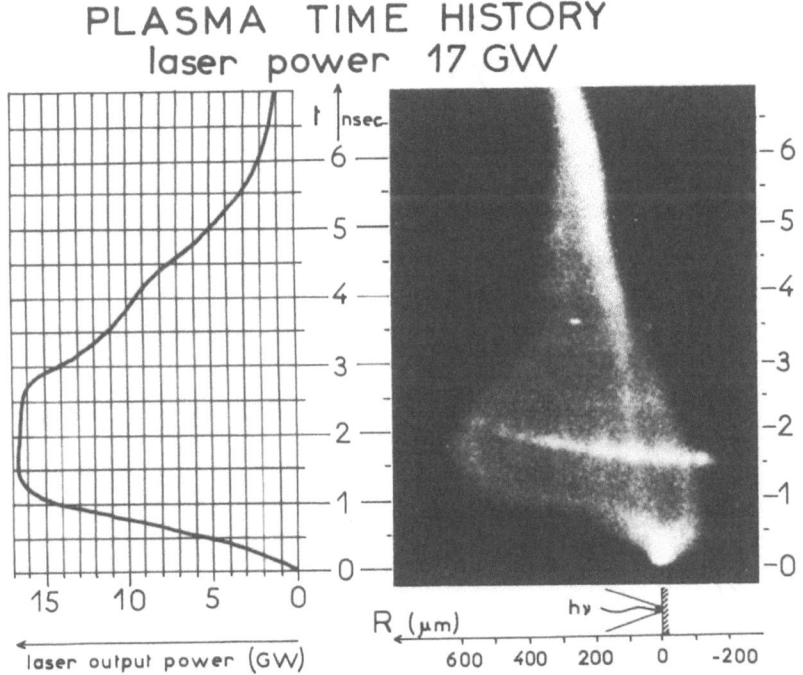

Fig. 5 : Streak camera recording

We have chosen this 17 GW case, in order to show a possible effect of tridimensionnal expansion. This idea is confirmed by the scaling law found for the neutron output yields versus absorbed energy.

5. Neutron fluxes measurements

Neutron detection has been carried by using 2 GMBF3 counters, the calibration of which has been achieved as in [6]. When located at 50 cm from the plasma, these counters give one pulse for respectively 750 and 1200 neutrons, emitted in the total solid angle. So, our 10 to 20 pulses, recorded on each oscillogram, allow us a detection accuracy of 20 to 30 per cent. On Fig. 6 has been reported the total number of recorded neutrons, versus the laser absorbed energy.

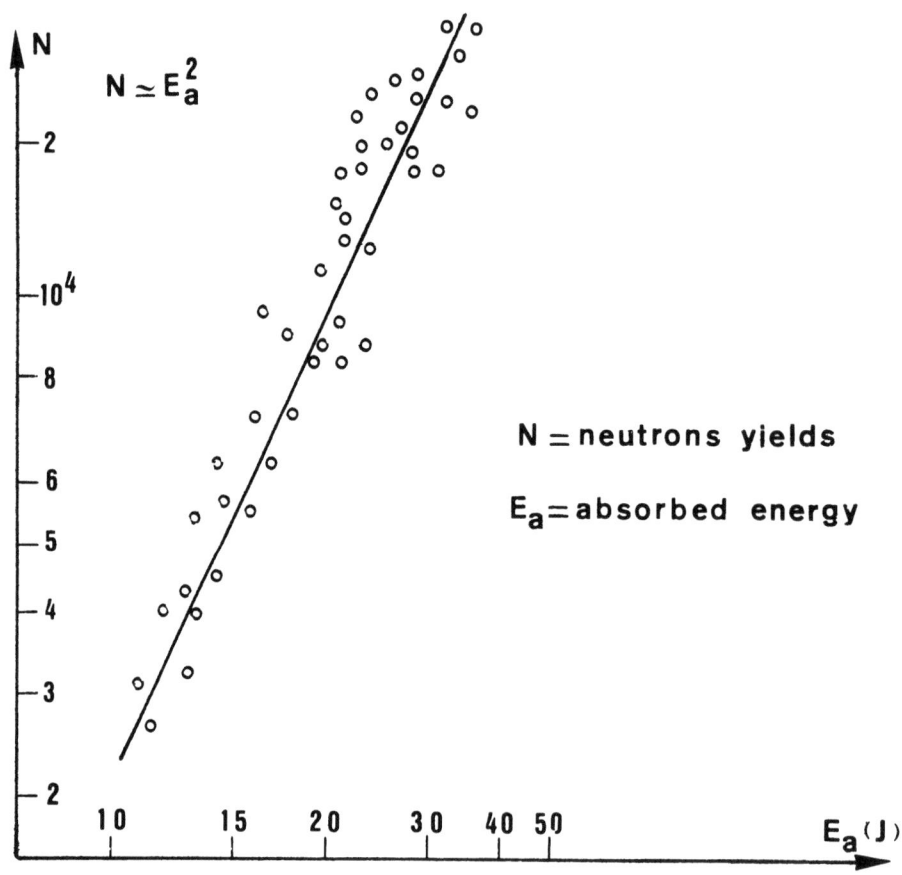

Fig. 6

Each experimental neutron value is averaged over 10 to 20 laser shots at the same energy, in order to balance the counter statistics. The total number of neutrons, N, has been found to vary as E_a^2 in the explored energy range ($20 < E_1 < 60$ J). This result has been confirmed over 500 to 600 laser shots, and it looks quite different from earlier reported theoretical calculations [10, 11], which assume a plane, one dimensional plasma expansion during the heating mechanism. On the contrary, we think that our experimental law is mostly due to an influence of a three-dimensional expansion into vacuum, which induces losses of heated particles away from the high density region. For this reason, we plan now to perform interferometric measurements, in order to see the plasma expansion shape.

When using large plastic phosphors, it is possible to look at the neutron yields versus the angle of emission. The origin of angle is taken to be the laser axis. Two such counters have been located respectively at 45° and 135° as mentionned in Fig. 7.

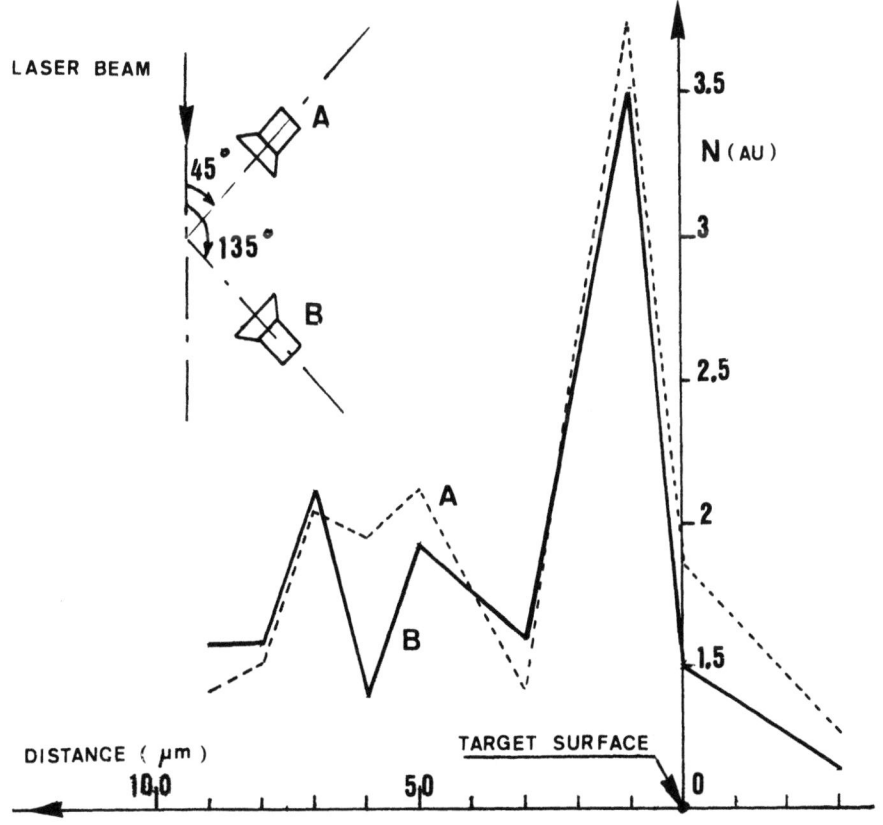

Fig. 7

For each laser shot, the area of each neutron recorded pulse is determined by graphic integration and plotted on the y-axis (arbitrary units), while the focussing conditions have been reported on the x-axis (in microns). These curves have been drawn for about hundred laser shots for which the neutron output was close to 10^4 in the x = 0 position. So, each experimental point has been averaged over a tenth of shots. We can conclude at a good agreement between the results of the two counters.

Nevertheless, we have performed more recently the same experiment using three phosphors located at 45°, 135° and 225° with respect to the laser beam axis. The total number of shots has been of about 70 with an average emission of 8×10^3 neutrons for one single shot. The counters located at 135° and 225° are in very good agreement each other ($\Delta N \simeq 5 \%$), but a discrepancy of 20 to 25 per cent has been observed with the phosphor located at 45°. (This last one giving the stronger value for N). We don't know, presently, the origin of this disagreement : it may be due to counter statistics, to a calibration defect, or to the neutron emission itself. Further experiments, with higher neutron fluxes, have to be achieved before we shall be able to conclude.

III - THEORETICAL DATA

The problem of heating with nanosecond laser pulses had been widely investigated before nuclear DD reactions were evidenced in the experiments [12, 13, 14]. Then, each author agrees with the conclusion that neutrons are created by heat conduction mechanism. If the picosecond laser heating is possible to be explained by analytical solution [15], due to the fact that plasma expansion is not taken into account, the nanosecond electron conduction heating requires the complete solution of the non-linear system of fundamental equations. For this reason, computer calculations appear necessary to investigate such a problem. The main characteristics of such a computer code will be to take into account :

- Electron conduction heating connected with plasma expansion and laser light absorption.
- Two different temperatures for electrons and ions..
- Reflection of laser light on the expanding plasma.

This system is derived from the Waser code which has been used in Livermore Laboratory [16]. A conical geometry is introduced in the one dimensional spherical calculation, in order to give account of the strong focussing of the laser beam on the deuterium target. The laser pulse is chosen as close as possible to the experimental one, and the reflection coefficient R, has been taken to be zero in the expanding plasma, except for the cut-off density ($n_c \simeq 10^{21}$ cm^{-3}) where R is equal to 1 (complete reflection). Such a treatment assumes a classical light reflection and non linear phenomena are not at all taken into account. We shall see that such a situation is in agreement with experimental data for power flux densities ranging at least from 10^{12} to 10^{14} W/cm^2.

Moreover, the thickness of each light absorbing space step is not allowed to grow up to 1.5 or 2 photon mean free paths (a few microns), in order to give account of classical inverse bremsstrahlung mechanism. Our earlier calculation [7] has shown how this system is efficient in order to get temperatures in agreement with experimental measurements.

1. Scheme of the flow

The calculations have been achieved for laser powers of 5, 10, 15 and 30 gigawatts. Fig. 8 gives density and temperature profiles for 10 GW power ($\phi \simeq 4.5 \; 10^{13}$ W/cm^2) at time t = 3.2 nsec. The R = 0 position corresponds to the surface of the ice on which the laser beam is focussed (focal diameter equal to 150 μm).

Due to the high density of the solid material ($n_0 = 5.10^{22}$ cm^{-3}) the cut-off density ($n_c = 10^{21}$ cm^{-3}) is reached into the expanding plasma and located 150 to 200 μm in front of the target. This result is important, because the increase of the absorbing area leads to a power flux density limitation. Moreover, the hot plasma, in front of the target, is not bounded by solid walls and can expand in an half-sphere geometry leading to an escape of hot ions from the electron conduction region. We can notice that this effect is stronger with small focal spot diameter.

The heat conduction mechanism spreads from the cut-off density up to 0.5 ρ_0, and the ion temperature is maximum for $\rho \simeq 2 \rho_c$ (T_i max $\sim 6.10^6$ °K). Neutrons are emitted in this region which is located near the surface of the target. On the contrary, a large difference can be seen between T_e and T_i in the free expanding plasma and through the absorbing region (0.5 $\rho_c \leq \rho_A \leq \rho_c$) ; the maximum electron temperature has been found to be 10^7 °K in this case. So the ratio $(T_i/T_e)_{max}$ is close to 0.6, that is to say 2 times greater than the value found in picosecond heating by the authors of [20].

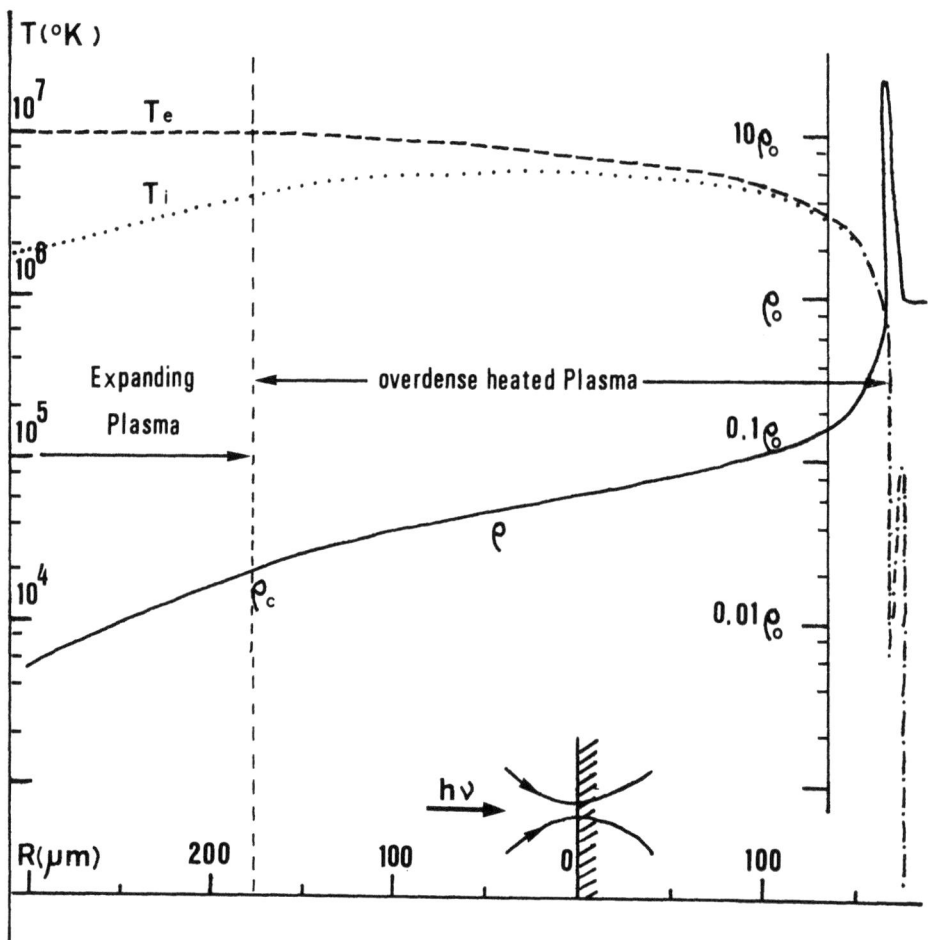

Fig. 8 : Computed profiles into the heated plasma (t = 3.2 nsec)

Table 1 gives values of Φ, T_e, T_i and $(T_i/T_e)_{max}$ for the reported laser powers.

At last, we can compare our results with those obtained by Mulser [17] in plane one dimensional computer calculations. In this work, the author finds that neutron are emitted into the expanding plasma for $\rho \simeq 0.5 \rho_c$, where the ion temperature is maximum. Electron heat conduction is not relevant for neutron production. Moreover, this region is located a few millimeter in front of the target. The discrepancy of such results shows how much geometrical effects may be important, and how much experimental data are necessary to manage computer calculations.

P_i (GW)	$\emptyset$ in 10^{13} W/cm²	T_e maximum in 10^6 (°K)	T_i maximum in 10^6 (°K)
5	3.2	8	4.6
10	4.5	10	6
15	4.8	11	6.5
30	8.3	14	7.5

Table 1

2. Reflection of laser light

This result is, presently, the single diagnostic for which an accurate comparison with experimental data can be made. On Fig. 9 and 10 are plotted the incident and reflected theoretical pulses for laser powers of respectively 5 and 10 GW.

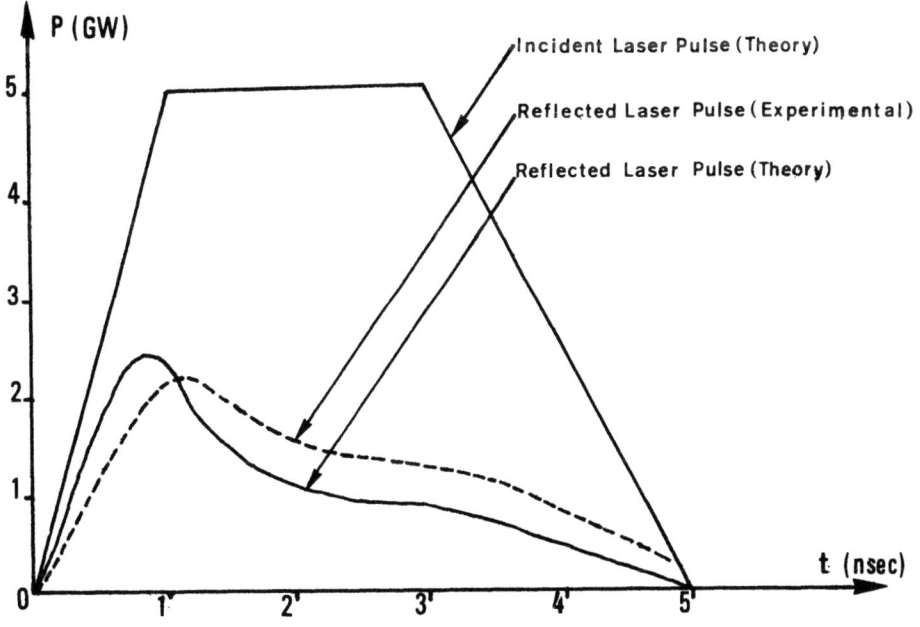

Fig. 9

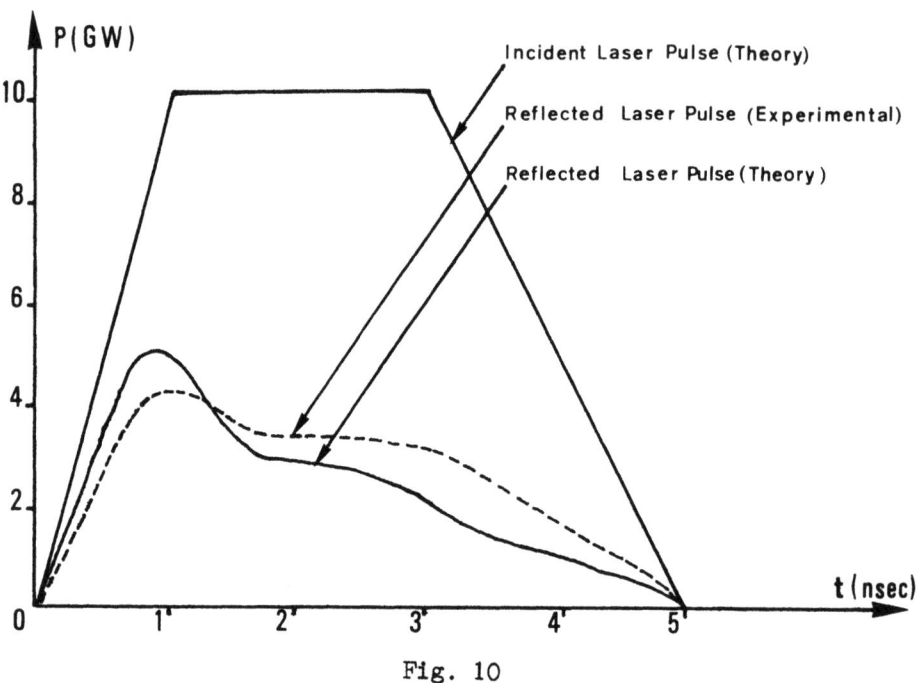

Fig. 10

On each graph, the experimental reflected pulse is drawn in dotted line. A very good agreement may be found between these two data, owing to the little difference between the incident laser pulses used in the experiment and the computer code. Except for the first nanosecond where the initial conditions of the calculations appear somewhat artificial, the differences upon R values are quite low : 23 and 29 % for calculations to be compared with 30 to 35 % for experimental data. In addition, we have found also a very good agreement for P = 15 GW between the calculated value of R (38 %) and the experimental data (42 %). An interpretation of the experimental reflection measurement has been given in section II.3. It is confirmed by such calculations which show a nearly constant density profile after the two first nanoseconds of the interaction process.

3. Sequence of the interaction process

Two stages can be distinguished during the five nanoseconds of the laser pulse duration.

a) During the first 1.5 or 2 nsec, duration which is dependent on the power flux density, the boundary electron temperature increases in the same time as the absorbing region is put away in front of the target. This expansion leads to the creation of an electron conduction region over 100 to 150 μm located from the cut-off density layer to the shock wave, which is increasing in density. The temperature reaches its maximum value at the end of this stage.

b) The 3 to 3.5 following nanoseconds are spent by a sustainment of the boundary electron temperature and an increase of the total overdense heating region. But this mechanism is mainly due to the propagation of the shock wave inside the solid material. In the same time, a decrease in density of the shocked region feeds the hot plasma.

At this stage of the process, we can imagin a situation where lateral expansion may occur in the expanding overdense region as well as in the shock wave front. The first region looses particles while the shock wave decreases by spherical expansion. So the sustainment can be shorter in experiment, where such a lateral expansion is a direct consequence of the high temperature value. The streak photograph of the plasma, shown in Fig. 5, can be interpreted by such a limitating process. The result of calculated neutron output is also in agreement with this conclusion.

4. Neutron emission

As mentionned in section III.1, neutrons are created in the overdense plasma ($n_i \simeq 2 - 3 \ 10^{21}$ cm^{-3}) where the ion temperature is maximum. This result is basically different from computer data obtained by Mulser [17] in which the neutron emission is occuring into the underdense expanding plasma ($n_i \simeq 5.10^{20}$ cm^{-3}). This fundamental difference may be explained by the two geometries choosen for the calculations : plane geometry for Mulser, conical one for us. Moreover, the agreement or the discrepancy on the calculated and measured reflection coefficients may be due to the same origin.

The emission starts after 1 or 1.5 nsec that is to say for an absorbed energy threshold of 4 or 5 J and lasts until the laser pulse has been entirely delivered (t = 5 nsec). In our experiment, some amount of neutrons (30 %) are detected with a delay time ranging from 20 to 100 nsec but we have determined that they were neutrons scattered by experimental items [6].

On Fig. 11 have been plotted our computer data (dotted line) dealing with neutron yields versus the calculated absorbed energy. We have reported on the same scale the experiments achieved in different laboratories.

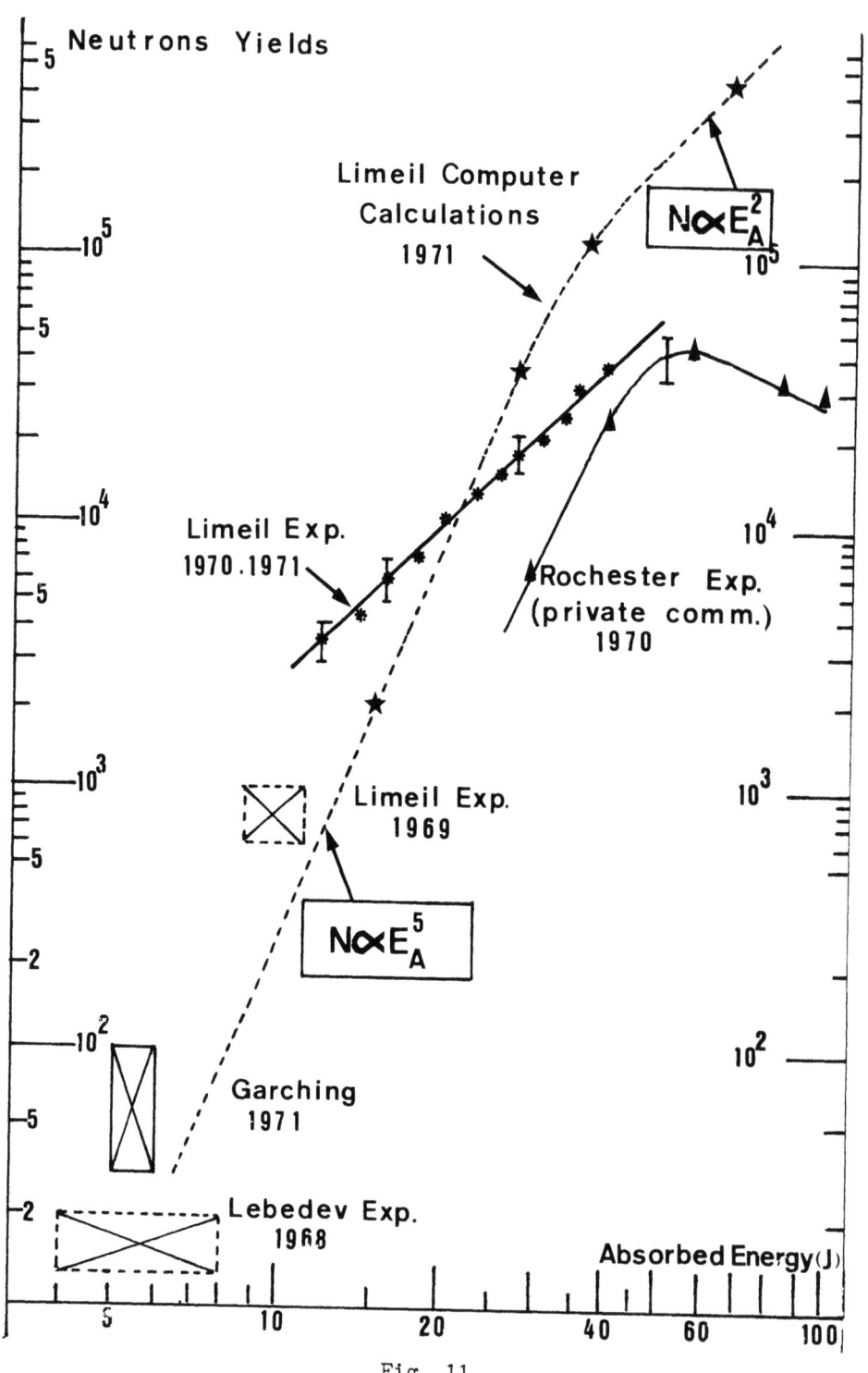

Fig. 11

For 5 and 10 GW incident powers, the scaling law looks like E_A^5 that is to say the same as found by J. Shearer 10 in picosecond heating computer calculations and S. Zakharov and alii 11 by analytical attempts. Moreover, some of the reported experiments are not far from such a scaling law.

On the contrary, for 15 and 30 GW laser powers, the growth of neutrons varies as E_A^2, scaling law which is in agreement with our 70-71 experimental data and on the other hand with a recent analytical calculation made by Zakharov and alii 18. Such a situation may be understood as follows :

- In the computer code, the f/0.9 conical geometry is topped by a right cylinder, the diameter of which has been taken to be 150 μm. So, the expansion of the heated plasma is successively plane and spherical shaped. When the incident power is not too high, the plane geometry plays an important role and leads to the E_A^5 scaling law. If the laser power, in our case, overcomes 10 GW the spherical portion of the nozzle influences strongly the neutrons yields and leads to the E_A^2 scaling law. We have verified this conclusion in connection with the location of the maximum ion temperature compared to the change in our geometry. We can easily deduce that if the focal spot is taken to be 100 μm or less, the spherical expansion will occur early.

- In the experiments, we can imagine that lateral effects increase the possibility to reach a spherical situation for the expanding plasma. For laser absorbed energies lower than 10 J, the experimental neutron emission seems to grow as E_A^5 (the caracteristic plasma lengths are shorter than the plasma diameter). On the contrary, above 10 J absorbed energy, we have found the E_A^2 scaling law with a disappointing reproducibility.

In any way, the agreement between experimental and theoretical absolute values is good (a factor 2.5 to 3) and confirms the results upon the reflected energy. This little difference may be probably reduced if the focal spot diameter is taken to be 100 or 120 μm.

In order to conclude this section, we shall make a remark upon the experimental reported data from Rochester University (Dr. Moshe Lubin) 19. The first part of the reported curve varies as expected from our computer data. Then, the neutron flux decreases for energy greater than 60 J.

The author concludes that for a particular size droplet (220 μm in this experiment) an increased absorbed energy does not necessarily mean more neutron yield as the hot electrons cause an expansion which is too rapid for more burning.

5. Heating efficiency

We shall restrict presently our conclusions to instantaneous efficiencies because some amount of heated matter is lost into the expanding plasma ($n \leq 5.10^{19}$ cm^{-3}) in order to shorten the time of calculation (the time step is of the order of 0.3 psec).

About 50 % of the laser power impinging the cut-off density layer is converted into electron conduction. This conduction flux is varying with time as the power flux density delivered by the laser.

Into the heat conduction region (overdense plasma) 85 % of the energy is converted into thermal energy, 45 % of which concerns the ion heating. Into the expanding plasma (underdense region) the absorption of photons leads to an heating efficiency which is less than 50 %.

The ratio T_i/T_e has been found nearly constant and equal to 0.58 in the four investigated cases. The values which are considered in such a ratio are the maximum ones for T_i (which occurs in the overdense plasma) and T_e (which occurs on the cut off density layer). So, such a ratio can be considered as pessimistic value, but it is of practical interest for extrapolation purposes. Such constancy of ion heating efficiency can be understood in considering that with increasing electron temperature, the overdense region is enlarged in the same time that ion velocities are increased.

This result has to be compared with those obtained by Zakharov and alii [20]. The authors deal with picosecond heating and find that the ratio T_i/T_e is no larger than 0.30 and shows an optimum value depending on the power and the duration of the laser pulse. They conclude that nanosecond laser heating should increase this efficiency.

IV - CONCLUSIONS

1. We have reported in this paper, the results of experiments carried out with 3.5 nanosecond laser pulse. For output powers ranging from 5 to 20 gigawatts, a few 10^4 nuclear DD reactions are created into the heated plasma provided that the focussing accuracy will be good enough.

We have given results about backwards reflected energy from the target and we have shown that the maximum neutron output was achieved in the same time that the reflection coefficient was maximum. Presently, this phenomenon may be interpreted with classical theory but non linear effects cannot be avoided, specially when the power flux density above 10^{14} W/cm^2 will be reached. The reflection coefficient shows a monotonic increasing law with the incident energy and a trick has to be discovered in order to use efficiently laser energy of a few hundred joules or more.

The neutron yields has been found to vary as E_A^2 in our explored energy range.

2. A computer calculation has been achieved and gives well account for most of the experimental data. We plan now to perform density gradient and electron temperature measurements in order to confirm such agreement.

The results of these calculations clearly show that the neutron output is depending not only on the laser power but overall on geometrical effects connected with the keV electron temperature achieved into the plasma.

3. So, in order to increase the output of nuclear reactions it appears important to balance the lateral plasma expansion in the same time that electron and ion temperatures will grow with the output laser powers.

Such results can be probably achieved in three different ways [21] :

- The first one, is to control the plasma expansion. A CO_2 laser heating, connected with a 1 MG. magnetic field confinement, is presently studied in some laboratories in order to increase the absorption and confine the plasma blow off. Difficulties are, then, appearing on the high magnetic field technology and the physics of laser absorption into such a large volume of heated matter. Furthermore, geometrical effects and particles losses have to be taken into account if a few keV temperatures are desired. At last, electron-ion energy transfert has to be optimize by a long confinement time (several hundred nanoseconds).

- The second way may be to minimize the plasma expansion with the use of short wavelength laser pulse, specially in the UV range. Such a laser beam would be able to heat the solid density itself. Furthermore, a plane onedimensional mechanism could be probably achieved due to the fact that total dimensions of the heated matter will be a few tenths of microns. Difficulties, in this case, are appearing from the laser technology.

- At last, the third way is concerned with the pursuit of experiments carried on with the old neodymium laser. In this case, two possible situations can be imagined in order to increase the temperature and the neutron fluxes :

When using deuterium targets in which some amount of impurities has been added, one can increase the absorption ($\sim Z^2$) and minimize the plasma dimensions (plane one dimensionnal expansion). Targets as DL_1 or CD_4 appear well fitted for such a purpose. But, the increase of temperature of deutons may be balanced by the impurities atoms. Nevertheless, recent calculations from Lebedev Institute [22] seem to be hopeful for such attempts. Moreover, people from Livermore have performed experiments on polyethylene deuteride in which 10^4 neutrons are produced [23] ;

In order to increase the energy output from Nd^{3+} glass laser, a possible mean is to use multichain amplifier system. Such a device should allow an increase of energy absorption in order to balance the growth of the heated volume. Moreover, the shaping of the laser pulse appears reasonable to minimize the spherical expansion of the plasma. One can imagine some of the focused laser beams to be second harmonic generated in order to absorb the energy into the overdense heated region. But such ideas are only into study and perhaps will be concretized for the third workshop.

ACKNOWLEDGEMENTS

The authors are greatly indebted to R. AMAN, H. CROZO, G. FAUCHEUX, D. MEYNIAL and A. QUEFFELEC for their decisive contribution to experimental data. They express their thanks to A. LETEINTURIER-LAPRISE for computer calculation and to J.L. BOBIN for fruitful discussions.

REFERENCES

1. J.L. Bobin, F. Delobeau, G. de Giovanni, C. Fauquignon and F. Floux, Nuclear Fusion 9, 115, (1969).
2. F. Floux, D. Cognard, J.L. Bobin, F. Delobeau and C. Fauquignon C.R. Acad. Sciences Paris, 269, 697, (1969).
3. P. Langer, Paper presented at the last IQE Conference held in Kyoto (Japan), September 1970.
4. F. Floux, laser interaction and related plasma phenomena, Plenum Press, New York, 447, (1971), Eds. H. Schwarz and H. Hora.
5. C. Fauquignon and F. Floux, Phys. Fluids, 13, n°2, 386, (1970).
6. F. Floux, D. Cognard, L.G. Denoeud, G. Piar, .D. Parisot, J.L. Bobin, F. Delobeau and C. Fauquignon, Phys. Rev., A1, 3, 821, (1970).
7. F. Floux and A. Billebize, Rep. CEA-R-4148, February 1971
8. J. de Metz, Appl. Optics, 10, 7, 1609, (1971).
9. A. Saleres, Priv. comm., Not yet published.

10. J.W. Shearer and W. S. Barnes, Phys. Rev. Letters, $\underline{24}$, n°3, 92, (1970).
11. S.D. Zakharov, O.N. Krokhin, D.N. Kryukov and E.L. Tyurin, J.E.T.P. Letters, $\underline{12}$, n°1, 36-38, (1970).
12. A. Caruso, B. Bertotti and P. Guipponi, Nuovo Cimento, 45B, 176, (1966).
13. O.N. Krokhin, International school of physics Enrico Fermi, XLCIII course, Varenna (1969).
14. R. Kidder, same course as 13.
15. A. Caruso and R. Gratton, Plasma Physics, $\underline{11}$, 839, (1969).
16. R. Kidder and W.S. Barnes, International Report UCRL 50583 (1969).
17. K. Büchl, K. Eidmann, P. Mulser, H. Salzmann, R. Sigel and S. Witkowski, Fourth Conference on Plasma Physics and controlled nuclear fusion research, Madison, 17-23 June 1971.
18. S.D. Zakharov, O.N. Krokhin, P.G. Kriukov and E.L. Tyurin, Quantum Electronics, $\underline{2}$, 104, (1971).
19. M. Lubin, Private communication.
20. S.D. Zakharov, O.N. Krokhin, P.G. Kriukov and E.L. Tyurin, J.E.T.P. Letters, $\underline{12}$, n°2, 92-94, (1970).
21. F. Floux, to be published in the next issue of Nuclear Fusion.
22. S.D. Zakharov, O.N. Krokhin, P.G. Kriukov and E.L. Tyurin, Quantum Electronics $\underline{2}$, 102, (1971).
23. G. Kachen, Private Communication.

LASER HEATED OVERDENSE PLASMAS FOR THERMONUCLEAR FUSION[*]

M. Lubin, J. Soures, E. Goldman, T. Bristow,
and W. Leising
Laboratory for Laser Energetics
University of Rochester
Rochester, New York 14627

ABSTRACT

In this paper we report theoretical and experimental work on the vaporization and heating of spherical overdense plasmas. Details of the required laser optics are described, together with charged particle, x-ray, and neutron diagnostics. A series of numerical experiments have been conducted to investigate the dynamics and neutron production in dense spherically symmetric plasmas. These have led to a development of scalings for energy absorption, shock strength and optimum neutron production. Similar numerical work has also been used to study the conversion of input laser radiation to output soft x-rays from an oxygen doped spherical laser plasma. The fast risetime of the calculated spectra suggests a method of producing a population inversion.

Finally, we put forward one example of a breakeven configuration not requiring anomalous absorption.

---

[*]Presented at the Second Workshop on "Laser-Interaction and Related Plasma PHenomena" at Rensselaer Polytechnic Institute, Hartford Graduate Center, August 30-September 3, 1971.

INTRODUCTION

The production of multikilovolt, dense plasmas by the irradiation of solid targets with high peak power high brightness lasers has led to the consideration of this process as a possible means of producing controlled thermonuclear reactions.[1,2] Early theoretical and experimental results have indicated that in order to achieve favorable ratios of output energy in neutrons to input laser energies special care must be taken to insure high absorption during the expansion of the plasma, which initially constituted the target and/or increase the yield by volume compression. Indeed, the neutron yield, or burn(γ), scales as η^2 for the simplest process:

$$\gamma \simeq \kappa \alpha^4 \eta^2 \tag{1}$$

where η is the compression ratio and α the overall laser and radiation absorption efficiency. The assumption that anomalous and non-linear absorption mechanisms may be critical for high neutron yield from small solid targets has given a special importance to the theoretical work dealing with such processes at laser frequencies and near solid densities.[3,4,5]

We describe below work done by our group on the laser irradiation of spherical light element targets. The objective of this work is an understanding of the absorption physics. The majority of the results were obtained with spherical targets of LiD whose size and purity is closely controlled to include a specific percentage (1%-6%) of oxygen throughout the target volume. Use of oxygen allows the line radiation to be used as an independent measurement of temperature. The optimum energy deposition time for a single pulse absorbed in small spherical dense plasma targets is bracketed by electron-electron thermalization times ($\tau_{ee} \simeq 10^{-11}$ seconds at 10 Kev) and free expansion of the plasma to a point of transparency to the incident radiation ($\tau_a \simeq 10^{-9}$ sec). The neodymium-glass laser system described below has a pulse duration of 120×10^{-12} seconds. Our use of a "prepulse" to establish an absorption density profile surrounding the spherical plasma sphere is discussed along with a comparison of experimental results with and without prepulse. Absorption of main heating pulse energy content as high as 43% have been measured using a prepulse. These measurements are correlated with x-ray temperature determination.

Finally, we present a detailed numerical study of one spherical geometry which yields a "breakeven" energy release, utilizing a simple 10^{-10} second laser pulse, at 1.3×10^5 joules.

LASER PARAMETERS

These laser plasma experiments utilize a Nd-doped glass laser system capable of producing focused power densities on the order of 10^{17} w/cm^2 for a pulse duration of 120×10^{-12} seconds.

The system consists of a mode-locked Nd-glass oscillator, a pulse selection unit, a string of five rod amplifiers, and several large aperature zig-zag geometry slab amplifiers which are presently being added to the system, Fig. (1). The oscillator produces a train of several 120×10^{-12} sec. wide pulses separated by twice the cavity transit time (typically $\simeq$ 10nsec.) at a repetition rate of once every eight seconds. A pulse selector consisting of two pockels cells, four polarizers, and a laser-triggered spark gap, selects a single pulse for amplification with a contrast ratio of > 10^4.[6] The typical energy in the single switched out oscillator pulse is on the order of 10^{-4} joules. An amplifier chain consisting of five Nd-doped glass rod amplifiers is used to boost the energy to the 50-100 joule level. The pertinent parameters of the first five amplifiers are listed in Table I. Large aperture amplifiers have been developed to use in subsequent amplification of the 100 joule beam. The first amplifier constructed was a 15cm diameter open disc geometry similar to that developed at LRL.[7] As much as 700 joules was extracted from this unit in four passes. However, because of the low repetition rate required for cooling the discs in air and frequent failure of the 55" flashlamps used for excitation, a liquid cooled 11cm aperture amplifier was built in cooperation with the General Electric Research Laboratories. The new amplifier modules have three Nd-doped glass slabs at 45° to the incident beam which are separated by large BK-7 glass prisms. Cooling is done by means of an index matching liquid which flows between slabs and prisms. The overall length of each module is 18" and each module is pumped by eight FX-47B-18 flashlamps. A total of four such modules are planned as the second stage of the amplifier string to boost the pulse energy to 1000 joules. Measurements of

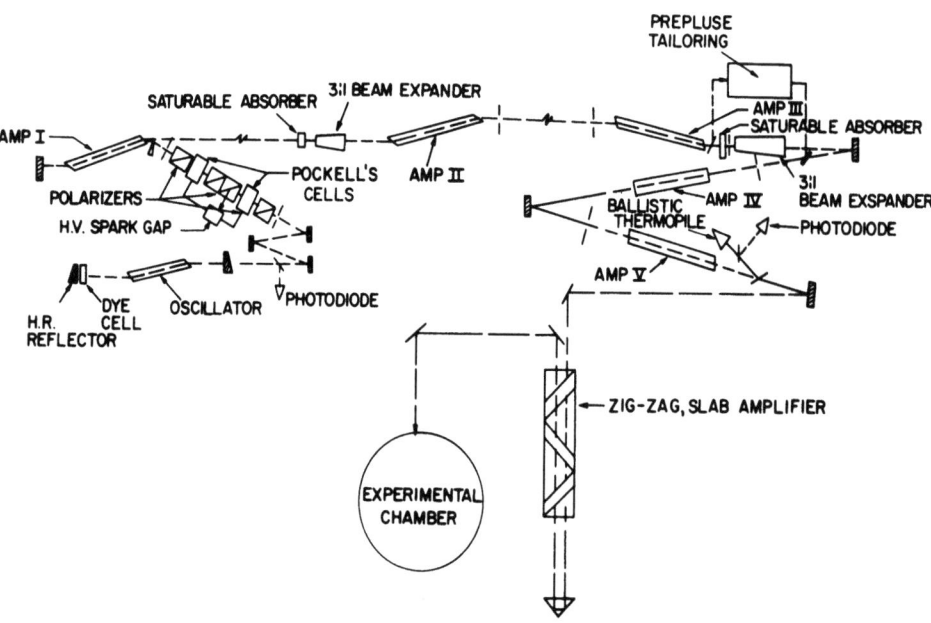

Fig. 1 Schematic of University of Rochester High Power Laser System.

TABLE I

	I	II	III	IV	V
Active Length	51 cm	51 cm	76 cm	45 cm	50 cm
Diameter	2.0 cm	2.5 cm	2.5 cm	5 cm	5 cm
Pump Energy	5 KJ	10 KJ	10 KJ	22.5 KJ	34 KJ
Max. Output Energy	2 J	8 J	25 J	75 J	200 J

Table I Characteristics of First Five Amplifiers in University of Rochester High Power Laser.

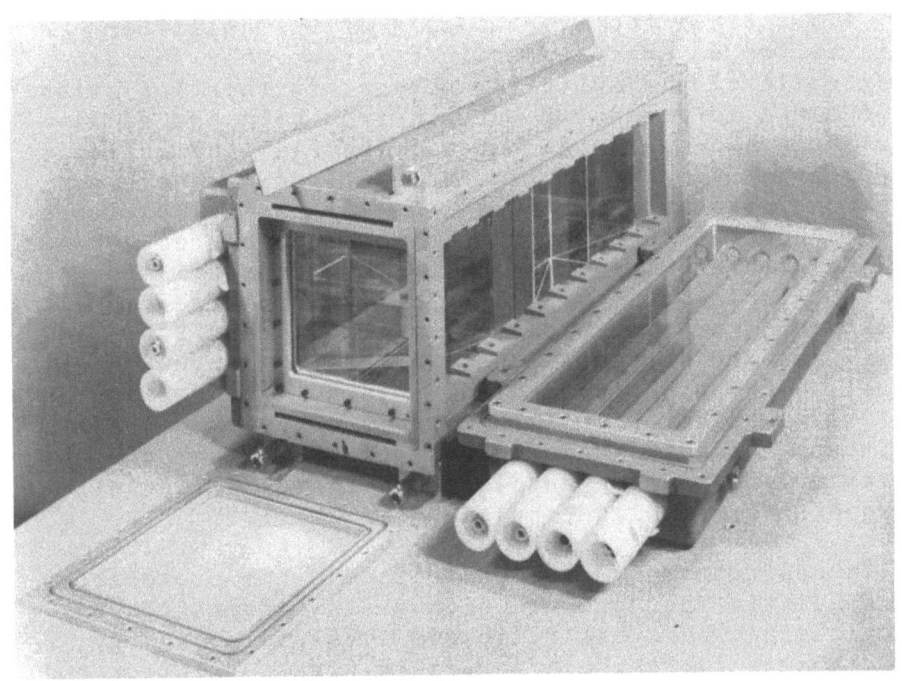

Fig. 2   Photograph of Zig-Zag, Liquid Cooled Laser Amplifier.

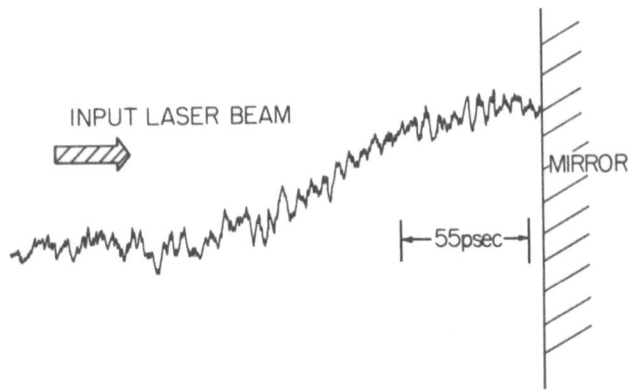

Fig. 3   Microdensitometer Recording of Typical Two Photon Florescence Camera Trace. The 120 psec Pulse Traverses a Cell Filled with Rhodamine Solution and is Reflected Back on Itself by the end Mirror.

the gain of the zig-zag amplifiers at the output of the present rod system show a storage efficiency of 2.5%, Fig. (2). Since the overall gain of the amplifier chain is two orders of magnitude larger than the contrast ratio of the switching optics, further isolation is required to prevent feedback, from both the amplifier chain and the target area. The required isolation is achieved by means of saturable absorbers and large distances between amplifiers themselves. Cells with Kodak 9860 or 9740 q-swich solutions are placed between the amplifiers. The typical low level transmission of these cells is between 10% and 20% while large signal (>50 mw/cm^2) transmission is between 50% and 80%. The amplifiers are spaced at least 5 meters apart with an overall beam path length of 50 meters. The portion in Fig. (1) labeled prepulse optics is a system of beamsplitters and adjustable delays for injecting pulses of variable amplitude and lead time into the target, ahead of the main laser pulse.

In addition to vacuum photodiodes, PIN photodiodes and ballistic thermopiles which yield information on pulse timing, contrast ratio and total pulse energy, a two photon florescence camera is used to monitor the pulse width. With the present oscillator, pulse widths between 30 and 120 x 10^{-12} seconds may be obtained. All the experiments reported on here were performed with 120 x 10^{-12} sec. wide pulses. A typical microdensitometer recording of a single amplified pulse, TPF camera exposure is shown in Fig. (3).

The quality of the output laser pulse is characterized by the beam divergence, and the beam energy density profile. The divergence of the beam at the output of the fifth amplifier has been measured to be less than 200μradian. The beam uniformity is determined mainly by the gain profile of the last two rod amplifiers. The 1% rod in the last amplifier exhibits a 15% drop in uniformity near the center. Fig. (4) shows the beam profile of the 1% doped rod in the fifth amplifier head.

Focusing Optics

The lenses used for focusing the laser beam are f/1 or f/2 aspherics with a focal length of 75 and 150 mm respectively. The aspheric surfaces are required to prevent secondary reflections inside the lens from forming damaging caustics. A small axial depression is

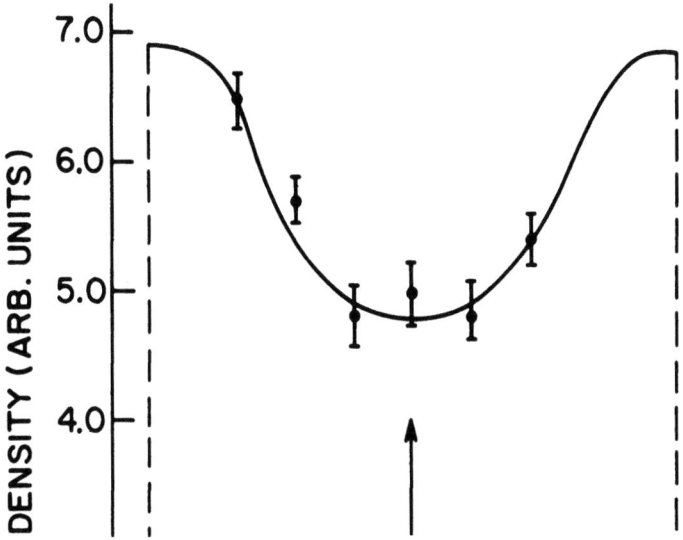

Fig. 4 Energy Density Profile of Laser Beam at Output of Last Rod Amplifier.

drilled in the center to prevent the lens from being damaged. Measurements of the focal properties of these lenses showed that at least 90% of energy in a 50mm beam is concentrated with 50μm at the focal spot.

A schematic of the experimental chamber presently being used is shown on Fig. (5). The laser beam is focused by lens L_1 onto spherical target T_1, suspended at the center of the vacuum chamber, by means of thin (2 to 5μm) glass whiskers. The targeting system is composed of a continuous YAG laser colinear with the main laser beam, a motor driven X-Y-Z translational mount for the target assembly and two TV camera systems for viewing and positioning the targets. The first viewing system, C_1, consists of a telemicroscope and TV camera. It is used for positioning the targets in a plane parallel to the laser optic axis. The second camera system, C_2, is infrared sensitive and views an enlarged image of the target through the focusing lens by means of a beamsplitter. The focal spot is determined to within 10μm using a 200 mesh copper screen. A check on the accuracy of the alignment and focal spot size is done by firing the laser at the copper

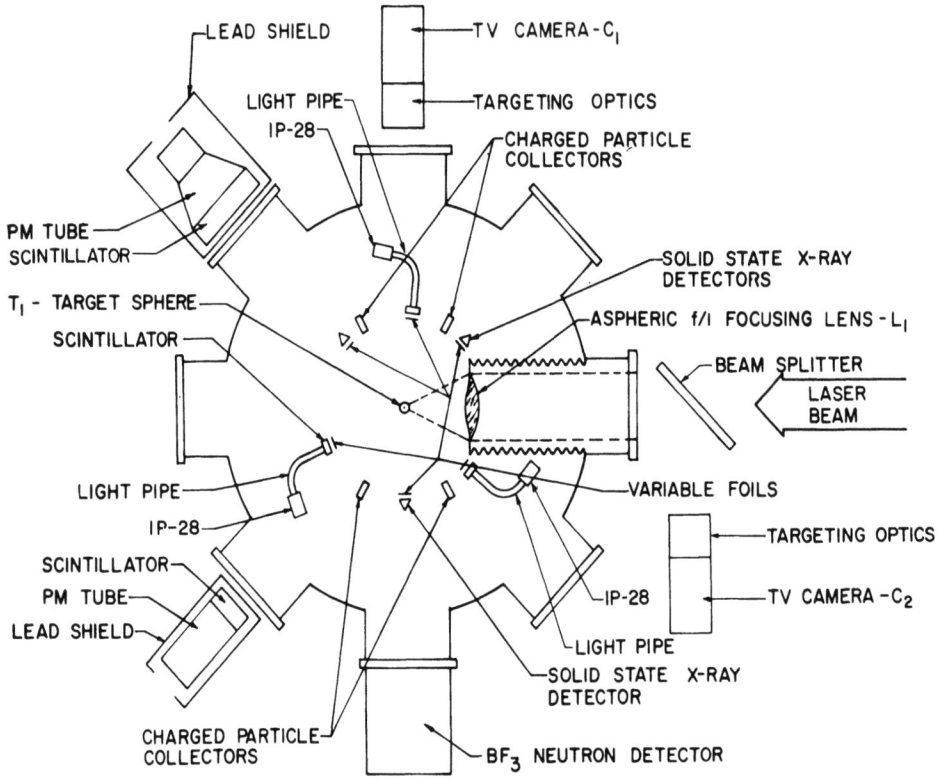

Fig. 5   Schematic of Vacuum Chamber for Laser-Plasma Interaction Experiments.

screen. Position and size of the hole burned on the screen is compared with the predicted focal spot. In this manner, it has been determined that this combination of laser and focusing element deposits 90% of the energy within 50μm. Once the position of focal spot is known, it is marked on the TV monitors and the targets are remotely positioned to coincide with it.

The energy in the focused laser beam is not uniformly distributed across the focal spot. Indeed, a consideration of the focusing properties of a perfect, aplanatic optical system, focusing a monochromatic, coherent light beam, reveals minima and maxima of electric field intensity in the focal region, as shown in Fig. (6).[8] From this figure it is readily seen that positioning of the target in the focal spot is a very critical part of any experiment. Furthermore, repeatability for fast lens systems can only be achieved by maintaining accuracy of 10 μm or better when the experiment relies on the strongest field strengths available. The highest degree of reliability in our case, (neutrons, x-rays, and energy coupling) have been achieved by displacing the spherical target a few tens of microns ahead of focal plane.

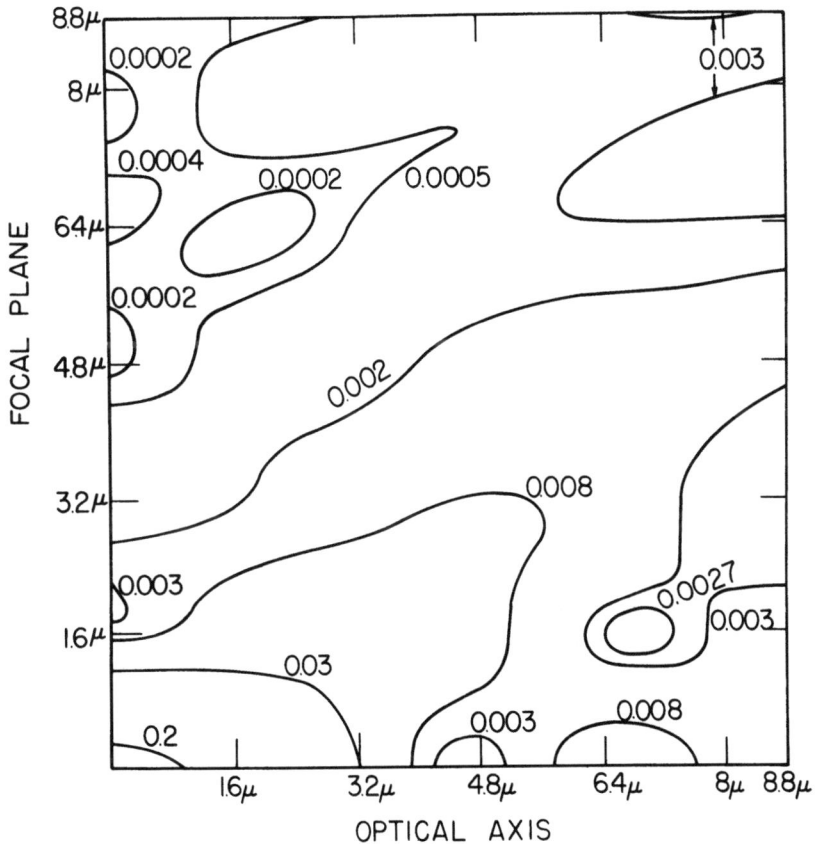

Fig. 6 Time Averaged Electric Energy Density Contours in the Vicinity of the Focus of an Aplanatic f/0.5 Lens. The Contours are Normalized with respect to the Value at the Origin.

PREPULSE IRRADIATION[9]

The transformation of the solid target into a cold dense plasma occurs on a time scale much greater than the optimum heating time of $\simeq 10^{-10}$ seconds. Consumption of a solid target via vaporization occurs on a time scale of:

$$t \geq \frac{d}{C_a} \qquad (2)$$

for intensities in the range of

$$W\sqrt{a/t} < I_{ab} < WC_a \qquad (3)$$

where: W - binding energy, a - thermal diffusivity, C_a - solid acoustic speed, t - duration of laser pulse, and I_{ab} - absorbed energy flux.

To better understand the prepulse vaporization regime a theoretical and experimental study has been undertaken utilizing a low energy ruby laser and aluminum wire targets. The initial radiation incident on the metallic solid interacts with the electrons of the conduction band. The electrons respond collectively to the incident field, and energy is absorbed via free-free electron transitions due to electron collisions with impurities, lattice imperfections, surface irregularities, and thermal fluctuations. Because of the distinct possibility that the collision frequency, laser frequency and effective plasma frequency of the electons in the lattice are comparable, the real and imaginary parts of the complex index of refraction, given as a function of frequency are

$$\hat{n} = n + ik \qquad \text{complex refractive index}$$

$$k^2 = \frac{\left(1 - \frac{\omega_p^{*2}}{\omega^2 + \frac{1}{\tau^2}}\right) \pm \sqrt{1 - \frac{\omega_p^{*2}}{\omega^2 + \frac{1}{\tau^2}} + \left(4\frac{\omega_p^{*2}}{\omega^2 + \frac{1}{\tau^2}}\right)\frac{1}{4\omega^2\tau^2}}}{2}$$

$$n = \left(\frac{\omega_p^2}{\omega^2 + \frac{1}{\tau^2}}\right)\frac{1}{\omega\tau\ 2k} \qquad (4)$$

where: $\omega_p^{*2} = \dfrac{4\pi N^* e^2}{M_e^*} \equiv \dfrac{4\pi\sigma(0)}{\tau}$ plasma frequency,

ω = incident laser frequency, τ = electron collision time, N^* = number density of electrons in conductor band, e = electronic charge, and m_e = effective electron mass due to lattice, $\sigma(0)$ = DC conductivity.

The skin depth is then given by

$$\delta = \lambda\ \text{vacuum}/4\pi k(\omega,\omega_p,\tau) \tag{5}$$

and the absorption is given approximately by

$$A \simeq 2/\omega_p^* \tau \tag{6}$$

For aluminum the collision frequency is near the laser frequency and less than the plasma frequency. In this region the index of refraction makes a transition from less than one for $\omega_c < \omega$ (plasma-like behavior) to greater than one for $\omega_c \approx \omega$ (dielectric-like behavior). Measured values corresponding to this type of theory are given by Haas and Waylonis[10] as $k \simeq 7.1$ - strong damping; $n \simeq 1.4$ - dielectric behavior; $\delta \simeq 150\text{Å}$ - 1/3 point for E field; $A = 10\%$ - moderate absorption. The nature of the skin depth absorption is illustrated in Fig. (7).

ABSORPTION MECHANISMS

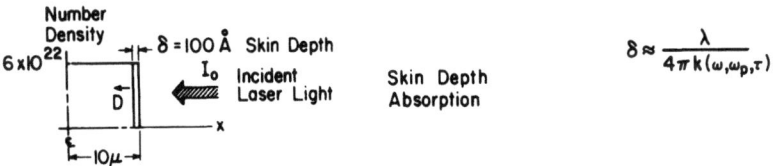

Fig. 7 Absorption Mechanisms Considered for Prepulse Vaporization, Skin Depth Absorption at Interface, Inverse Bremsstrahlung in Plasma, ω_1 vs. the Incident Radiation Frequency.

The skin depth region is heated very rapidly. Equation 3 shows that for an absorbed energy flux of 10^9 watts/cm^2, a latent heat of 10^4 joules/cm^3, and a thermal diffusivity of 1 cm^2/sec; only 10^{-10} seconds are required to overcome the lower bound limit. After this time a vapor interface is formed, and may move into the solid. The upper limit on this motion (for these low incident intensities, less than 10^{12} w/cm^2) is the acoustic velocity of the solid. To attain velocities higher than this requires a solid/shock to be formed. These require driving pressures exceeding 10^6 atmospheres[6], an exceedingly high value.

At the same time vapor begins to expand into the vacuum surrounding the target. Based on simple kinetic theory the vaporization cannot proceed on a large scale unless the temperature is such that kT≃latent heat; the vaporization rate:

$$e^{-W/kT} \simeq \text{vaporization rate}$$

and thus for kT much less than the latent heat, the rate becomes exponentially small.[11] The latent heat is of the order of 3 electron volts per molecule. From Saha equilibrium the material is partially ionized, >1%. This degree of ionization is sufficient to insure that absorption in the vapor is dominated by inverse bremsstrahlung. The absorption length $L \simeq C\tau_{e-i}$ is approximately 1 micron for a 3 ev plasma at cutoff density.

The expansion of the vapor coupled to the behavior of the interface has been considered by Afanasev and Krokhin.[12] Similarity solutions have been obtained for a number of cases applicable to the prepulse regime. The expansion of the vapor is taken as a centered rarefaction wave driven by the absorption of the incident radiation at or near the interface. The physics of the phase change governs the behavior of the interface, which serves as a boundary for the rarefaction waves.

The same governing equations have been solved numerically[13] using a one-dimensional Lagrangian code which includes aribtary interface motion. These solutions are not restricted to self-similar behavior. A comparison of the numerical and similarity solution for the same parameters is shown in Fig.(8). A complete computed time history is shown in Fig.(9). In the absorption region the non-linearity of the velocity

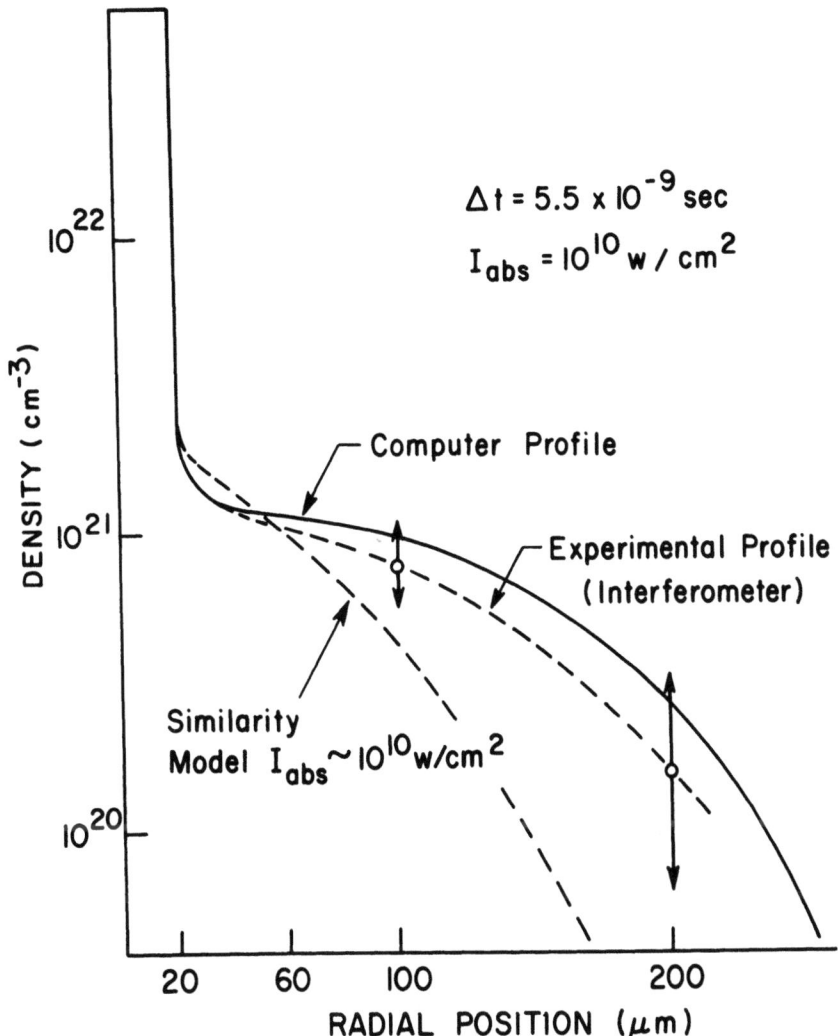

Fig. 8 Density Profiles for t=5.5 nanoseconds all profiles are for one-dimensional case. Computed profiles and similarity profile are for the same boundary interface conditions.

profile produces a significant deviation from similarity in the density distribution. The shape of this density profile is important with respect to subsequent heating by a short intense laser pulse. Two points are of interest in Fig.(9). First, the propagation of the vaporization front into the solid is approximately 1/10

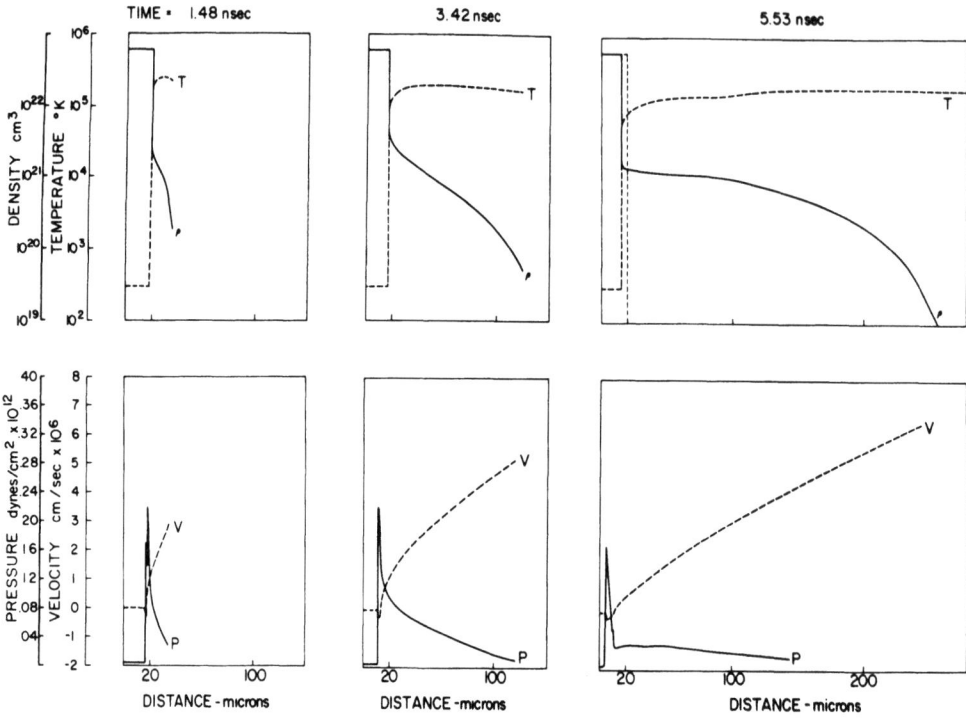

Fig. 9 Computed profiles for density, pressure, temperature, and velocity at three different times, t_1 = 1.5 nsec, t_2 = 3.4 nsec, t_3 = 5.5 nsec.

the acoustic speed C_a. Second, even at these modest intensities substantial pressures (10^4 atm) are generated at the solid interface.

An experimental determination of the density profile has been made using a microscope interferometer driven by a delayed portion of the incident laser pulse, and time of flight charge collectors. The target consisted of a 17 micron aluminum wire irradiated by a 100 micron spot size laser pulse with risetime of 2×10^{-9} seconds and a duration of 10^{-8} seconds. Peak incident intensity was less than 10^{11} watts/cm^2.

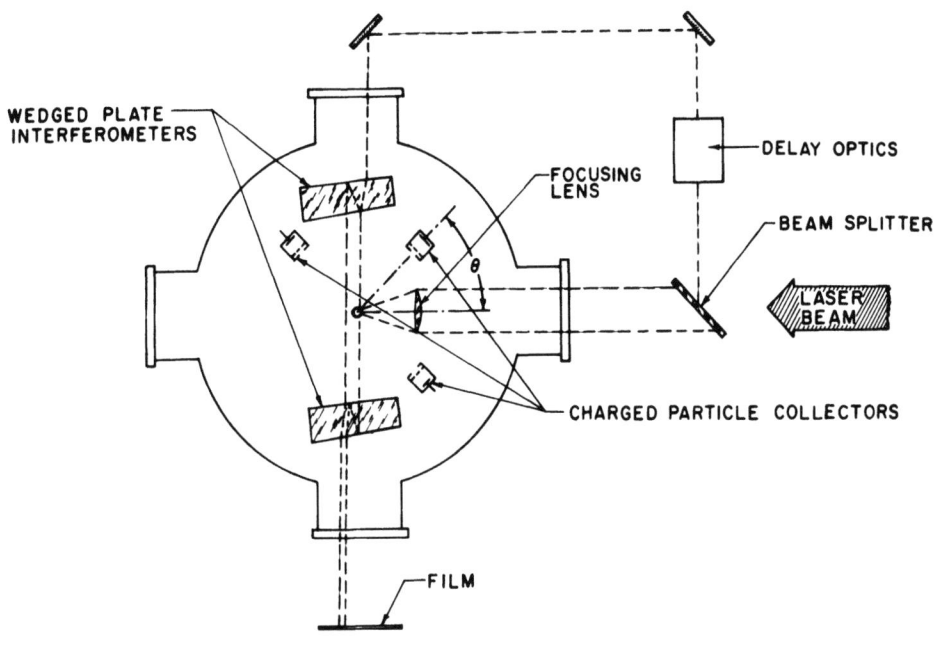

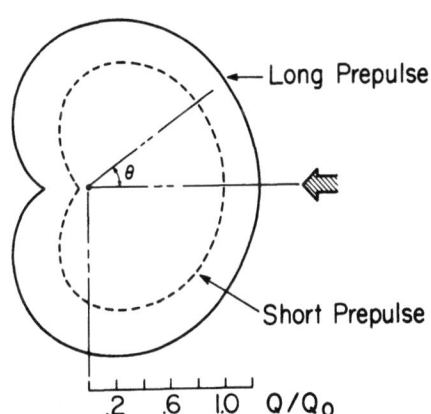

Fig. 10 Schematic of experiment, illustration of cardoidal symmetry exhibited by charge collection measurements.

A schematic of the prepulse experiment is shown in Fig. (10). With a fringe spacing of 100μ and a divergence limited minimum measurable shift of 1/5 of a fringe electron densities exceeding $10^{18} cm^{-3}$ could be detected over 100 micron regions. In view of the measured cardoidal symmetry for charge collection exhibited in Fig. (10), fringe shift profiles were computed using surfaces of $r = \cos\theta/2$ as surfaces of constant density. The best fit of the computed profile to experimental results is:

$$\delta = \frac{r_o^2}{x_o^2} \cos\theta/2 = \left(\frac{r_o^2}{x_o^2} \cos\theta/2 \right) \tag{7}$$

$$\rho = \frac{r_o^2}{x_o^2} \cos\theta/2 \, \rho_{1-D} \tag{8}$$

where $\tan\theta = x/y$; x = direction of incident laser beam; y = direction of incident analyses beam; z = axis of wire target; $x_o(r)$-x axis of cardoid passing turn r; ρ_{1-D} = one-dimensional computed density profile.

Under these conditions, the agreement between computed fringeshift and observed fringe shifts is good, Fig.(8). It is noted that the similarity densities profiles could not be fitted to the observed fringe pattern because the density gradient is too steep over the interferometer analysis region. A knowledge of this vaporization time history and profile allows us to consider in a self-consistent manner subsequent heating by an intense pulse tailored to plasma time scales.

SPHERICAL HEATING

We have conducted a series of numerical experiments to investigate the dynamics and neutron production on laser irradiated, dense, spherically symmetric plasmas. In particular, we are interested in developing scalings for the absorbed energy, shock strength and neutron production. This code (LAPP-I) contains a two temperature hydrodynamic description of the plasma dynamics along with relativistically corrected inverse bremsstrahlung absorption,[14] electron-ion energy coupling, separate electron and ion thermal transport, neutron production, alpha particle heating and continuum bremsstrahlung radiation.[15]

A prepulse from the laser converts a pellet of frozen deuterium into a spherical fully ionized plasma with a density profile shown in Fig. (11). The profile consists of a high density core of radius r_c surrounded by a low density tail of thickness r_t. This tail is necessary to promote energy absorption from the laser which is assumed to uniformly illuminate the surface of the plasma.

When the electrons in the tail have been heated to 0.1 - 1 Kev the ability of the plasma to absorb energy decreases rapidly. However, the tail does not become completely transparent as part of the electron thermal

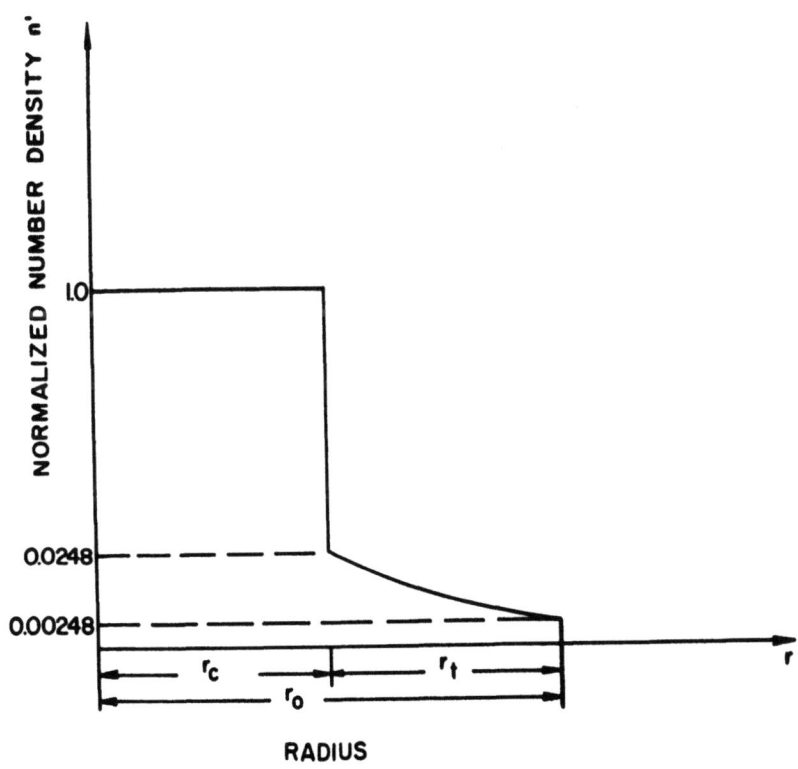

Fig. 11 Sketch of the initial normalized density profile as a function of radius. The core radius is r_c, plasma radius r_o, and tail thickness r_t.

energy is steadily drained away by collisions with ions. The energy in the electron thermal mode within the core drives an imploding spherical heat front. The energy transport across a unit area per unit time is

$$\varepsilon \simeq \frac{3}{2} nkT_e V_T \qquad (9)$$

where
$$V_T \simeq \frac{2}{3} \hat{C} T_e^{3/2} \frac{\frac{\partial T_e}{\partial r}}{nk} \qquad (10)$$

This thermal front can exist only as long as there is sufficient energy behind it to maintain $V_T > U_a$; where U_a is the isothermal sound speed of the mixture. When this velocity becomes acoustic, the density perturbation resulting from the velocity gradient at the thermal front, will steepen into a converging shock wave. The transition from thermal to shock behavior can be estimated from[24]

$$V_T/U_a = 2^{3/2} C[nk^{3/2} \ell_f]^{-1} T_e^2 = 4.1 \times 10^{-14} T_e^2 \qquad (11)$$

where n is the ion-electron number density ($4 \times 10^{22} cm^{-3}$), k Boltzmann's constant, ℓ_f a measure of frontal width (about 5μm), C the numerical constant in the electron thermal conductivity ($\lambda_e = CT_e^{5/2}$) and T_e the electron temperature. Electron energies on the order of 2 Kev are necessary to maintain the thermal front.

If the neutron production is plotted as a function of energy absorbed, Fig. (15), the curve increases rapidly to a peak and slowly decreases. For absorbed energies below 100 joules, the production of neutrons is dominated by the imploding shock wave formed by the decaying thermal front as the electron thermal energy decreases below 2 Kev. Above 100 joules the thermal front does not decay and fewer neutrons are produced as more energy is retained in the electrons.

Two important observations can be made from these numerical results. The first is that once part of the laser energy absorbed by the tail electrons has been transported into the dense core, a spherically converging heat front is formed. In accordance with the theory of non-linear heat conduction, the front may or may not decay into an imploding shock wave depending on the amount of electron thermal energy behind it. The dynamics of the plasma and its neutron production differ greatly under these two possibiliities. A second observation is that a plot of total neutron production as a function of the absorbed energy shows a peak in that portion of the curve which represents shock cases.

Typical time histories for plasma density, electron and ion temperature and neutron production are shown in Fig. (12) thru Fig. (16).

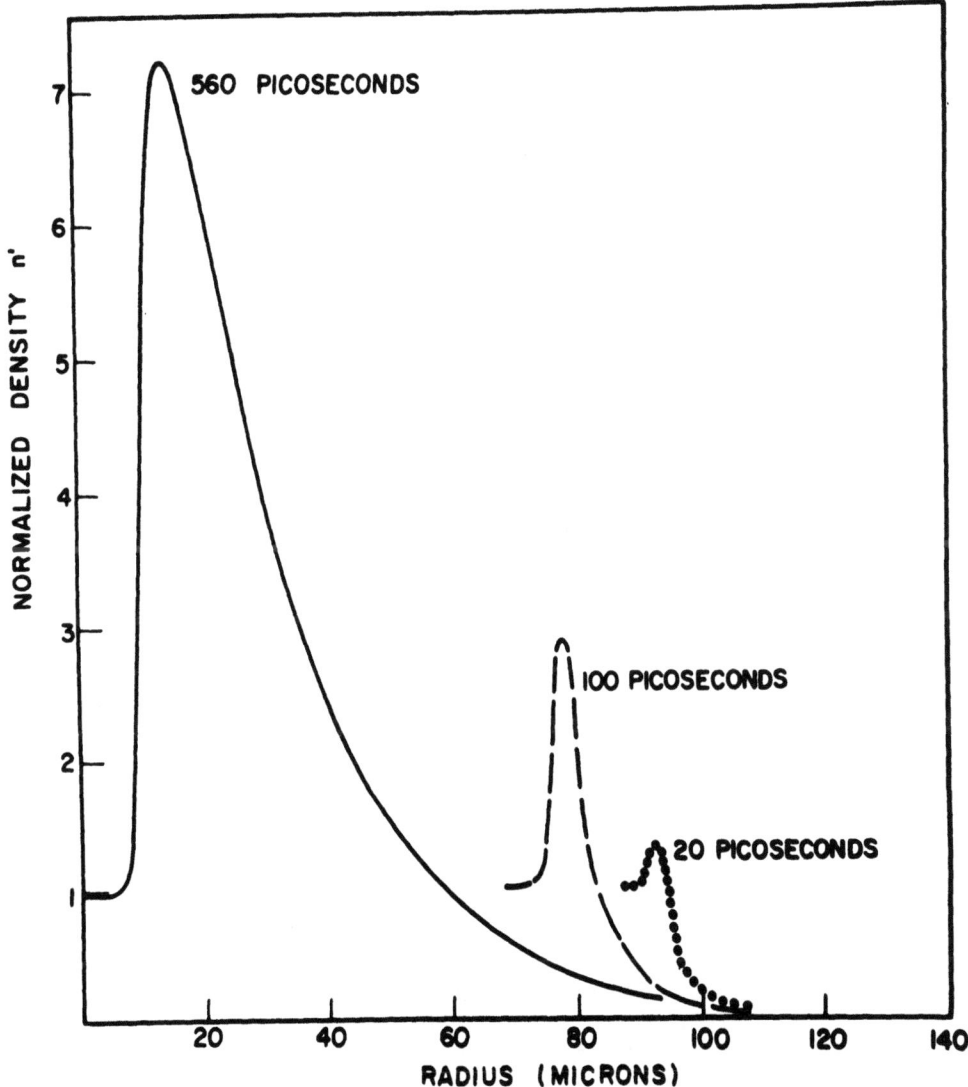

Fig. 12 Normalized density distribution as a function of radius at three different times. Peak laser power is 10^{13} watts with a rise time of 10^{-11} seconds.

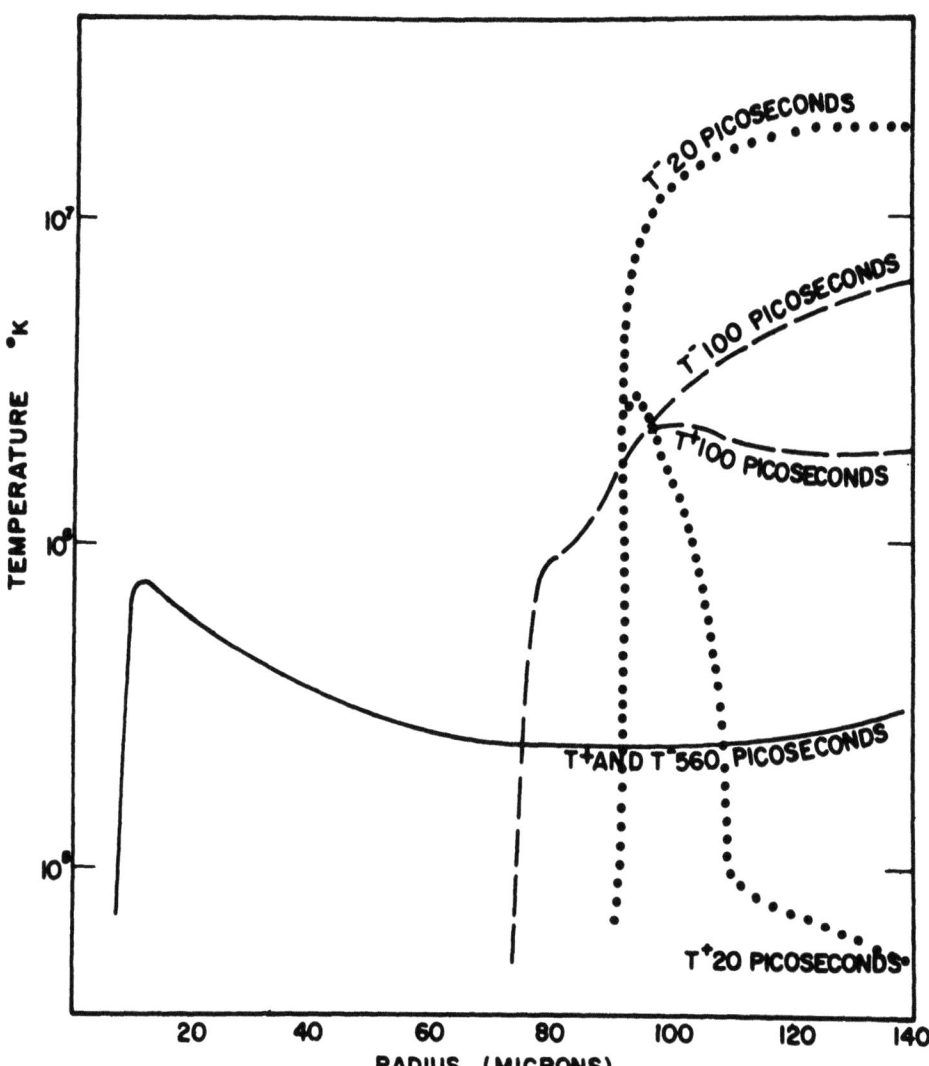

Fig. 13 Electron (T^-) and ion (T^+) temperature distributions as a function of radius at three different times. Peak laser power is 10^{13} watts with a rise time of 10^{-11} seconds.

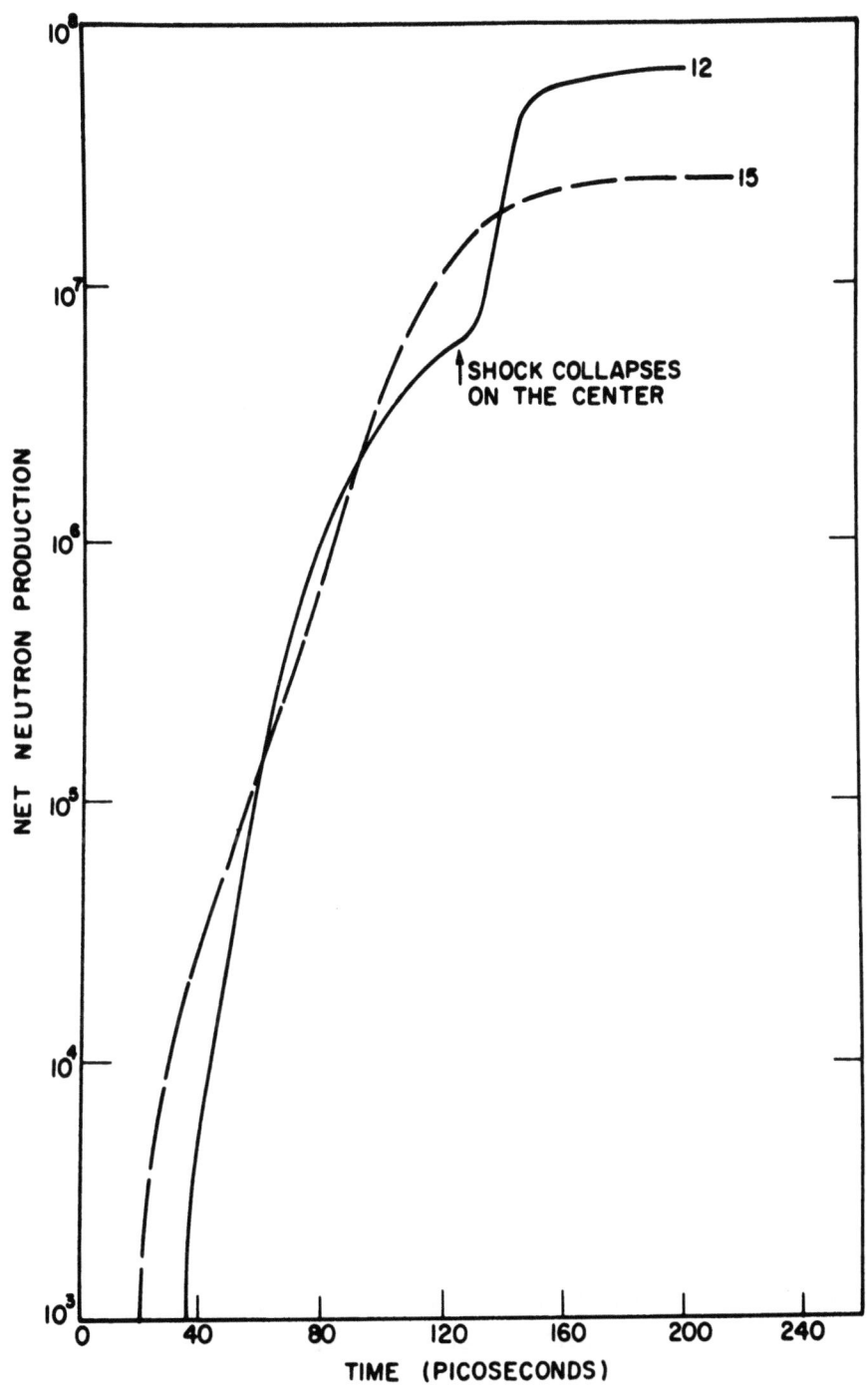

Fig. 14 Net neutron production as a function of time for the shock dominated case #12 (Peak laser power 4.5 x 10^{14} watts, 10^{-11} second rise time) and the thermally dominated case #15 (Peak laser power 9x10^{14} watts, 10^{-11} second rise time).

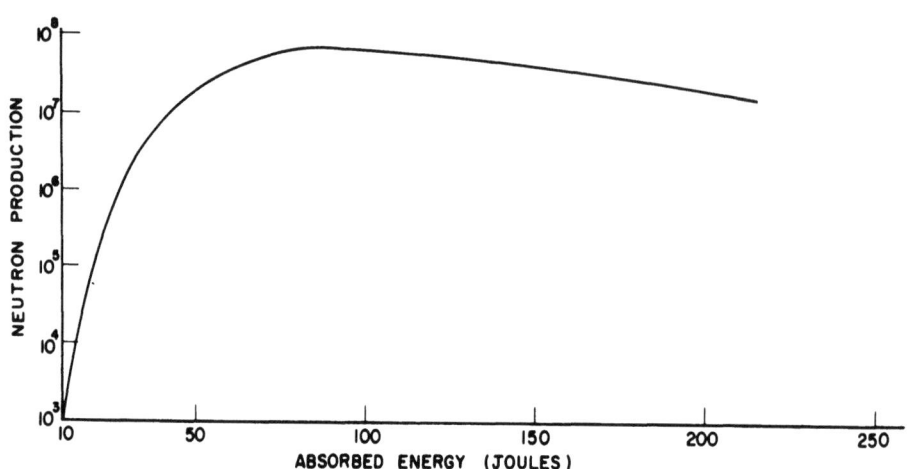

Fig. 15 Neutron production as a function of absorbed energy for cases with a 100μ radius core.

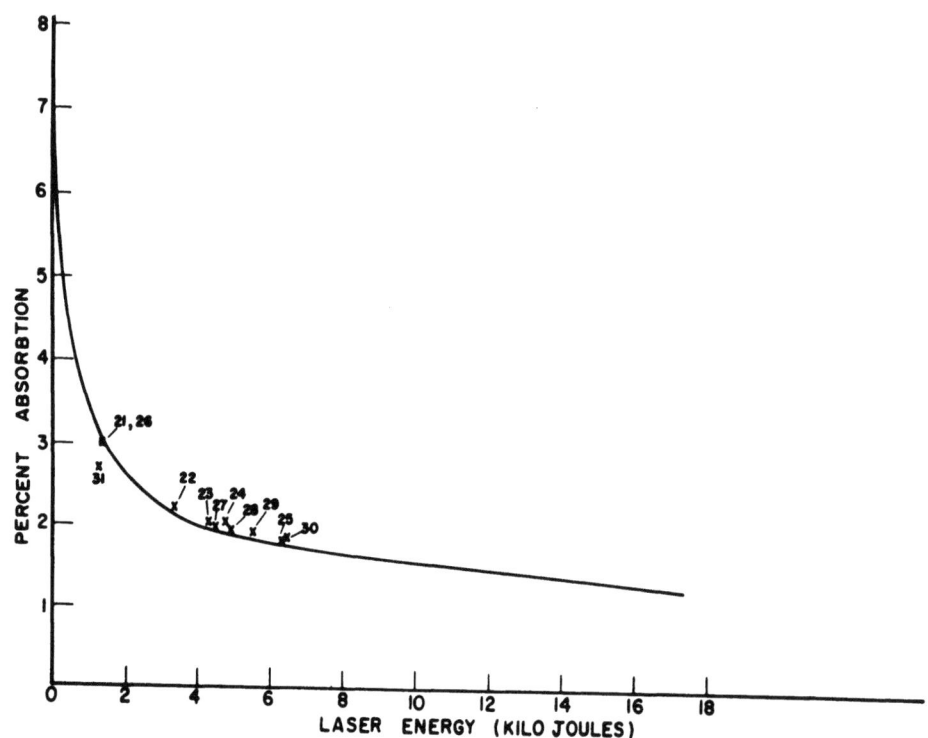

Fig. 16 Percent absorption as a function of the laser energy. The solid line is constructed from the 100μ/100μ cases while the reduced values for 200μ/200μ, 300μ/300μ and 700μ/700μ cases are shown as numbered crosses.

Experimental work on the irradiation of spherical pellets of LiD has been carefully carried out up to input energies of 50 joules. The objective of these experiments is to study the correlation of energy absorption (via charge particle collectors) neutron yield and temperature (x-rays) with input energy. Four distinct sets of conditions have been examined;
(i) a single heating pulse without a prepulse,
(ii) a single heating pulse with a prepulse,
(iii) two equal intensity pulses separated by 10^{-8} sec. without a prepulse, and
(iv) two equal intensity pulses with a prepulse.

The intensity of the prepulse is $\simeq 10^{-3}$ of the main pulse. The four cases are shown in Fig.(17). Absorbed energy is determined by integrating the total charge collected. Peak energies as high as 10 Kev per ion have been measured. Complete symmetry in charge collection was never achieved above incident energies of 2 joules. A ratio of collected current in the forward (toward the laser) direction to current normal to the beam in the plane of the target was typically 4:1, for spherical targets the size of the cross section of the focused laser beam at the point of heating.

Even a modest prepulse makes a significant difference in the absorbed energy. Clearly the second main heating pulse, in the two pulse case is more efficient than the first. Neutron yields $>10^4$ have been measured using a 10 cm. diameter scintillator 12 cm. from the focal spot enclosed in a lead shield. In the single pulse case, the absorbed energy, as determined from charged particles, x-rays, and neutrons is always more than that predicted by the spherically symmetric calculations (by as much as 50%). No neutrons have been observed without a prepulse. Calculations have not yet been carried out for multiple high energy pulses.

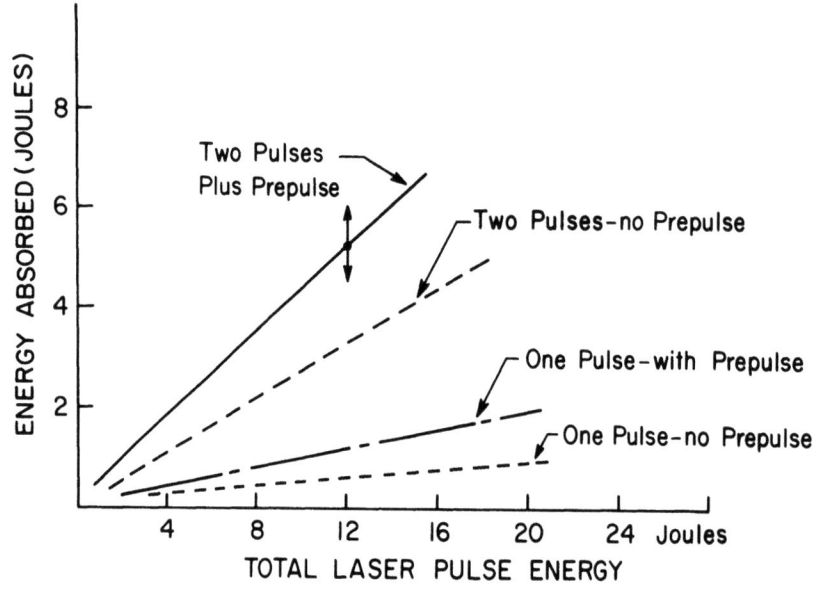

Fig. 17 Experimentally measured energy absorbed by laser plasma, produced from 150μm diameter LiD particle as a function of input laser pulse energy.

X-RAY RADIATION AND TEMPERATURE MEASUREMENTS

Numerical studies of x-ray radiation from low Z spherical targets lightly doped with heavy elements have been performed using a one-dimensional Lagrangian computer code.[23] Oxygen was chosen for these studies because the spectra is well-known, and because it is the main controlled impurity present in the experimental LiD spherical targets. This study involved both the continuum and line components of the x-ray radiation; in particular, the emission from the OVI, OVII, and OVIII species. The transitions used are the OVI $2s^2S - 3p^2P^o$ transition ($\lambda = 150.12$Å); the OVII $1s^2\ {}^1S - 1s2p\ {}^1P^o$ ($\lambda = 21.6$Å), $1s^2\ {}^1S - 1s3p\ {}^1P^o$ ($\lambda = 18.63$Å), $1s2s\ {}^1S - 1s3p\ {}^1P^o$ ($\lambda = 128.25$Å) transitions and the OVIII $1s\ {}^1S - 2p\ {}^2P^o$ ($\lambda = 18.97$Å) transition[14] The study was restricted to these configurations because the excitation cross sections and oscillator strengths for these lines are considerably larger than other cross-sections at temperatures characteristic of laser produced plasmas.

Absorption of the incident focused laser radiation by the spherical plasma is predominantly an inverse bremsstrahlung process, i.e., the absorption exhibits a Z^2 dependence.[15] Hence, the addition of high Z impurities to the plasma leads to a greater overall absorption with elevated electron temperatures. The radiation power density from the plasma consisting of both continuum (free-free and bound-free) and line components, is strongly Z dependent; i.e., the free-free and bound-free power densities exhibit a Z^2 and Z^4 dependence, respectively.[16] The line radiation from the impurity atoms exhibits a Z^6 dependence for a plasma in L.T.E. For a non-LTE plasma the dependence is given through the excitation potential.

In these studies, it is assumed that the collisional radiative model (CR) model is valid so that excitation of a bound level is by electron collisions and recombination of the lower levels is by spontaneous emission.[17] The upper levels of the ions are depopulated by collisions and so the density of these levels is given by the usual Boltzmann distribution. In addition, each ion species is dependent upon electron collisional ionization while recombination is by three-body and radiative processes. The maximum level for spontaneous emission is chosen so that the Einstein A coefficient is greater than the collisional de-excitation rate. In this manner the condition for a maximum level is dependent on the electron number density and temperature.[16] This condition is included in the numerical code, as each shell has a specific electron number density and temperature.

We describe the radiation dynamics of the impurity ions by a rate equation for the degree of ionization of a particular state given in terms of ionization and recombination coefficients. Reabsorption of the emitted radiation is not important in lightly doped laser plasmas because they are optically thin to this calculated soft x-ray spectra. The rate equations are of the form:

$$\frac{dq_j}{dt} = n_e q_{j-1} \langle\sigma v\rangle^i_{j-1} - n_e q_j \left(\langle\sigma v\rangle^i_j + \langle\sigma v\rangle^r_j\right)$$
$$+ n_e q_{j+1} \langle\sigma v\rangle^r_{j+1} \qquad (12)$$

where n_e is the electron number density in cm^{-3}, q_j is the fractional ionization of the jth species (for oxygen q_8 is the hydrogenic ion), and $\langle\sigma v\rangle^i_j$ and $\langle\sigma v\rangle^r_j$ are ionization and recombination cross-sections respectively, averaged over a Maxwellian distribution.[18] The coefficients $\langle\sigma v\rangle^i_j$ and $\langle\sigma v\rangle^r_j$ are available in the literature.[16,19] Recombination, as described by $\langle\sigma v\rangle^r_j$, consists of contributions from radiative and three-body recombination.

The power density from an excited state in the jth species is:

$$P^{ku}_{Lj} = n_e^2 q_j q_o \langle\sigma v\rangle^{ku}_{ex} h\nu \qquad watts/cm^3 \quad , (13)$$

where q_o is the fraction of impurity atoms (with respect to n_e); and

$$\langle\sigma v\rangle^{ku}_{ex} = \frac{6 \times 10^{-6}}{X_u(kT_e)^{1/2}} f_{ku} \exp(-X_u/kT_e) \qquad cm^3 sec^{-1} ,$$
$$(14)$$

where f_{ku} is the absorption oscillator strength for a u to k line transition, and X_u is the excitation potential of the uth state.

A typical calculated time history of an impurity doped spherical pellet heated by an incident 1.06 mm 265 joule laser pulse of 10^{-10} seconds duration is shown in Figs. (18) and (19). These calculations were carried out using a two-temperature hydrodynamic, spherically symmetric Lagrangian code which allows for complete energy transfer between species. Figs. (18a) and (18b) show the density profile and electron temperature vs. the radius of the plasma for various times of interest. In addition to the 100 micron core, initially at solid density, a low density tail is present which has been formed by a prepulse applied to

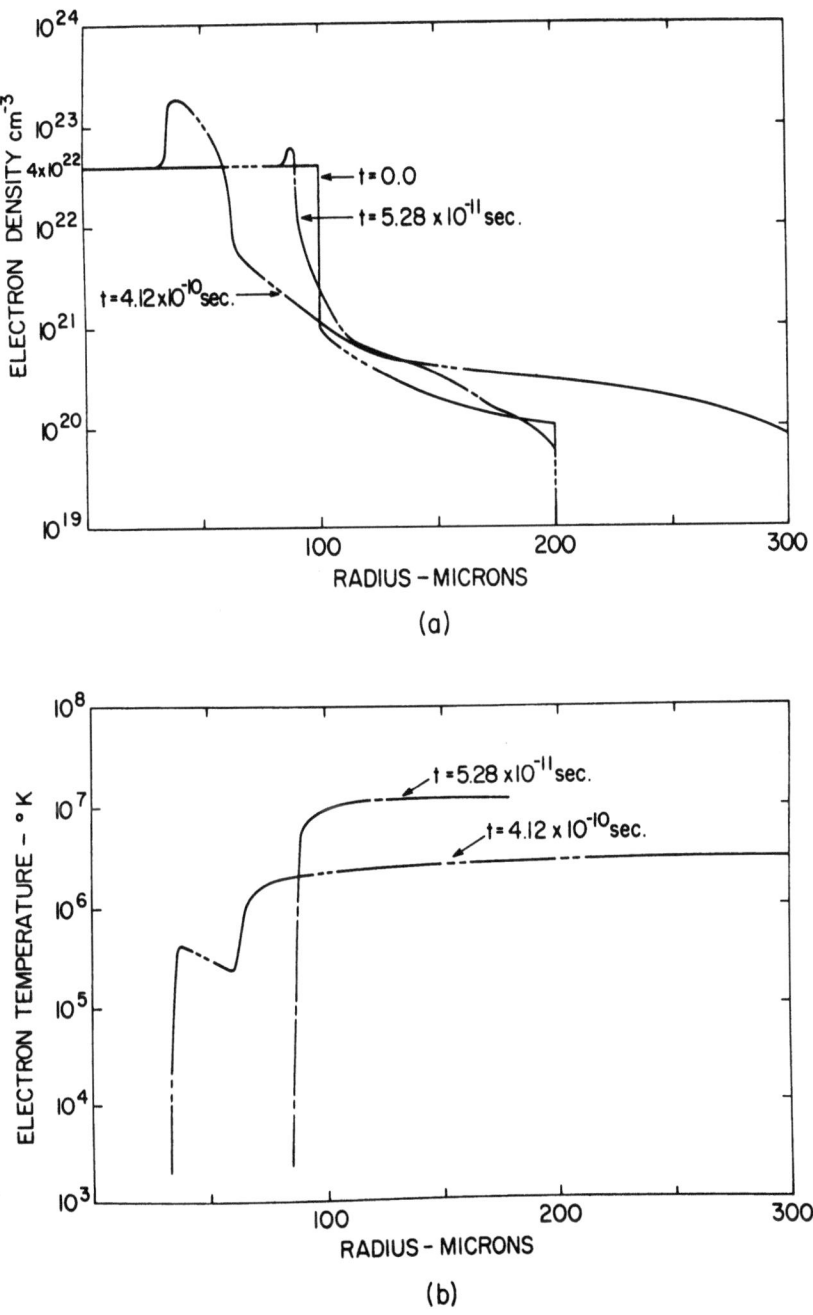

Fig. 18 Development of laser heated impurity doped plasma sphere, showing the density (a) and electron temperature (b) at various times. Incident laser energy ≈265 joules at $\lambda=1.06\mu m$ delivered in 10^{-10} seconds.

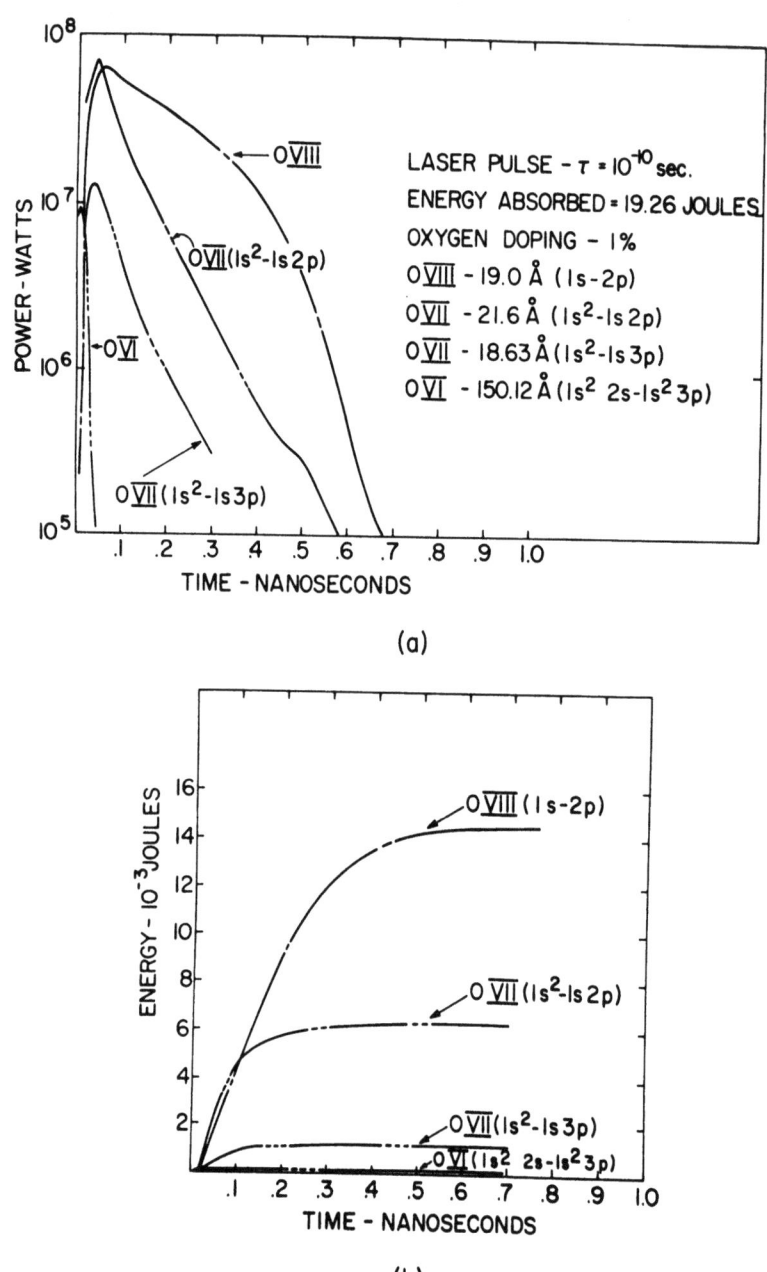

Fig. 19 Time history of line radiation showing the power (a) and energy (b) in each line. Impurity doping concentration is 1%. Rise time of each line is less than 10^{-11} seconds.

enhance the absorption process.

From Fig. (19a) the fast rise ($<10^{-11}$ sec) of the OVII and OVIII states (for 1% doping) is apparent during the laser heating of the plasma. As the electron temperature drops to 10^6 °K the line emission decreases, because the excitation cross-section is proportional to $\exp(-X_u/kT_e)$. The line radiation originates in the outer high-temperature low-density region of the plasma; the cold center core does not contribute to the radiation dynamics at these short times. Fig. (19b) gives the energy in each line vs. time. In this calculation the total energy absorbed by the plasma from the incident laser beam is 19.3 joules. The continuum radiation emitted from the deuterium plasma is 1.1 joules whereas the contribution from the impurity species to the continuum for 1% oxygen doping is 0.3 joules.

The experimental diagnostics included the measurement of the electron temperature by the x-ray foil transmission technique.[20] The x-rays were measured by solid state (PIN-Silicon) and PM-Scintillator detectors, each with partially absorbing aluminum or nickel foils. The results showed electron temperatures ranging from 700 ev to 3 Kev. Other results[21] have indicated the possibility of hard x-rays due to plasma instabilities, or due to energetic electrons striking the vacuum tank walls. Our results do not indicate the presence of such hard x-rays. In Fig. (20) the temperature from the aluminum foil measurements is plotted vs. the absorbed energy of the plasma for the cases in Fig. (17). These temperatures are based on the absorbing foil technique and the assumption of a Maxwellian velocity distribution.

The conversion efficiency of laser energy to soft x-rays has been suggested as an efficient means of generating continuum and line radiation.[22] Experimental confirmation of the high conversion efficiencies to soft x-rays (continuum) has been made in the case of 6% oxygen-doped LiD spherical targets subjected to a 2 joule 10^{-10} second laser pulse with a prepulse. A conversion efficiency of laser energy to soft x-rays is found to be 0.44%. The numerical code was used for the experimental case corresponding to 0.3 joules absorbed and the conversion efficiency was calculated to be 0.52%, in good agreement with the experimental results.

The calculated fast risetime of the line emission ($<10^{-11}$ sec) may lead to extreme non-equilibrium conditions since the electron-electron thermalization time is 10^{-11} sec at electron densities of $10^{20} cm^{-3}$ and electron temperatures of 1.2 Kev. Also the very rapid non-equilibrium electron impact excitation in oxygen-doped laser plasmas suggests a possible mechanism for population inversion. During early time, it should be possible to produce an inverted population on the 2s-3p transition of OVII and hence lead to stimulated emission at 128Å.[23]

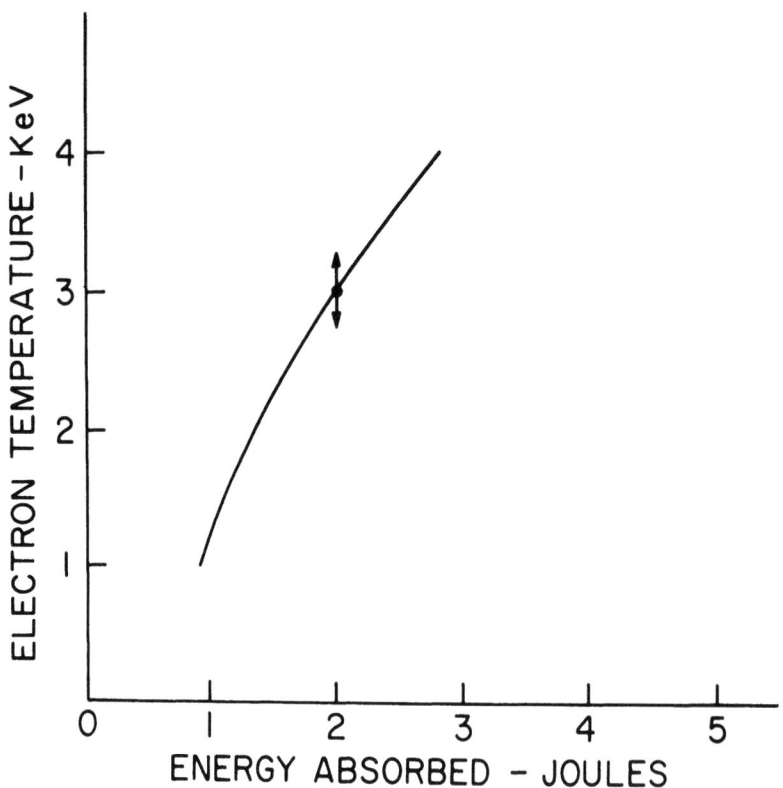

Fig. 20 Measured electron temperature vs. energy absorbed by plasma created from 150μm diameter LiD particle.

BREAKEVEN SPECULATION

In previous work a series of spherical heating experiments in which various size particles with dense cores surrounded by low density tails (to promote absorption) irradiated with single laser pulses of varying energy content have been conducted. Once part of the laser energy absorbed by the tail electrons has been transported into the dense core, a spherically converging heat front is formed. Extension of these studies to larger particles and higher energies gives no clear indication that breakeven yields can be obtained from this simple single pulse configuration.

In a different configuration, the spherical target has an initial density profile with a depression in the center, Fig. (21). The laser pulse is incident on the inner core at a wavelength of 1.06 microns and a pulse width of 10^{-10} seconds.

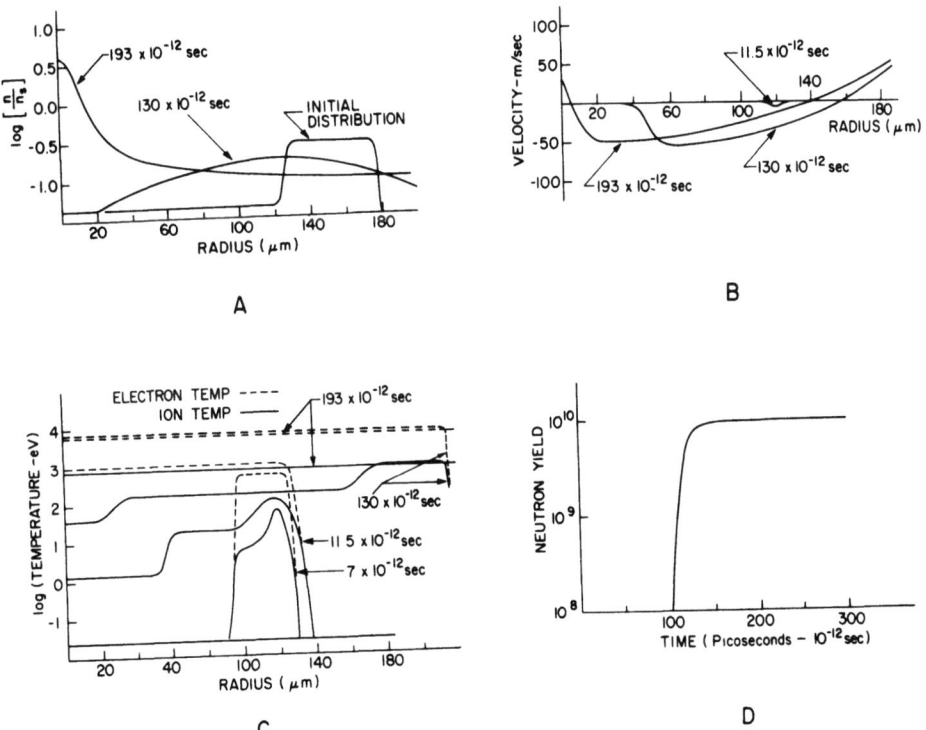

Fig. 21 Computed time development of (a) density, (b) velocity, (c) electron and ion temperatures and (d) neutron output, for the irradiation of a "castle" target. Laser pulse duration $T=10^{-10}$ sec., $\lambda = 1.06 \mu m$.

This "castle" profile is efficient for two reasons:
(i) the outer region being denser than the inner core prevents a rapid expansion of the inner portion which is absorbing the radiation, reducing the velocity of expansion by $\sqrt{n_2/n_1}$ hence, increasing the effective confinement, and
(ii) the absorption is virtually 100% as the outer layer is over dense.

From Fig. (21) it is possible to follow the time history of the density, electron and ion temperatures as well as the integrated neutron output. The target core is heated by a 10^{-10} second, 1.06 micron 100 joule, laser pulse. This initial density profile is readily established; for example, by applying a weak external compression pulse (10 joules) in such a manner as to leave a tunnel for the subsequent heating pulse. The initial heating occurs at the peak absorption region on the inner side of the high density layer in a spherically symmetric manner. Very rapidly (3.6×10^{-12} seconds into the pulse), electron thermal conduction spreads the heat inward, the electrons always being hotter than the ions. 11.5×10^{-12} seconds into the pulse the electrons have equilibrated above 1 Kev across the inner core while the ion temperature profile lags far behind. At 10^{-10} seconds significant neutron output is now apparent with the ion thermal front approaching the center of the core. By 150×10^{-12} seconds the electron temperature is starting to relax and the neutron yield is leveling off. The center core of ions is still heating up but the overall heated volume in which reactions can take place is now less dense. Electron and ion temperatures are very close together in the center and the target is now dissembling. The integrated neutron output levels out after 200×10^{-12} seconds.

The effectiveness of this configuration as measured by the production of neutrons via DT reactions is shown for various spherical target sizes in Fig. (22). The spherically symmetric single pulse external heating of a uniform solid density target of the same size is less effective by 5 orders of magnitude.

Around "breakeven" the peak is broad, indicating that a 10% variation in target size is tolerable. For a given laser pulse duration and energy, the output drops at smaller target sizes due to a lack of sufficient high temperature electron-ion thermalization before the target dissembles. On the other hand, when the target is too large, electron and ion temperatures are

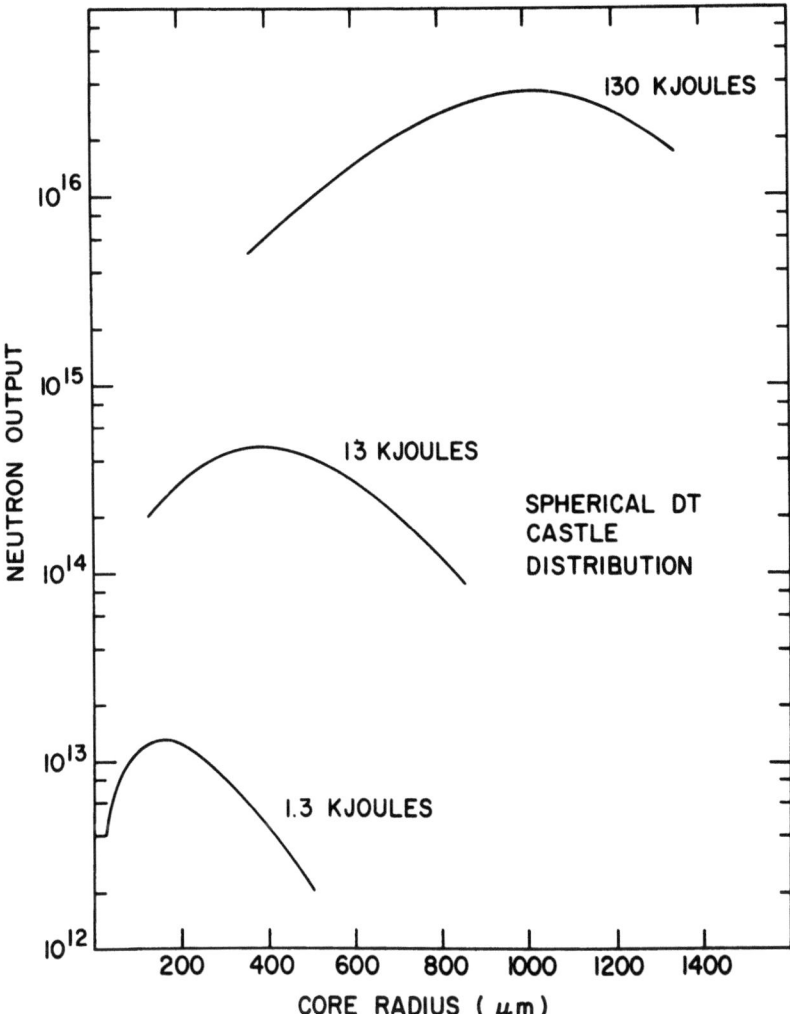

Fig. 22 Computed neutron output showing optimization of target size. These results are for a 10^{-10} second duration laser pulse at $\lambda=1.06\mu$. Breakeven occurs at 130 Kjoules.

too low for significant output. An optimum range of sizes exists for a particular laser pulse.

In light of the state of the art of present day high peak power Nd-glass laser systems, which is a delivery capability of a tunable pulse width at 10^3 joules in one amplifier string, we should anticipate near break-even experiments in the very near future.

REFERENCES

1. Yu.V. Afanasov, et al. JETP Lett. 13, 182 (1971).
2. R.E. Kidder, Proceedings of Esfahan Symposium on Fundamental and Applied Laser Physics, Esfahan, Iran (1971).
3. H. Hora, these Proceedings, p. 515.
4. D.F. DuBois and M.V. Goldman, Phys. Rev. Lett. 164, 207 (1967).
5. P.K. Kaw and J.M. Dawson, Phys. Fluids 12, 2586 (1969).
6. G. Dube', Appl. Phys. Lett. 18, 69 (1971).
7. S.W. Mead, et al., Appl. Optics 11, 345 (1972).
8. A. Boivin and E. Wolf, Phys. Rev. 138B, 1561, 1965.
9. O.N. Krokhin, High Temperature and Plasma Phenomena Induced by Laser Radiation, Physics of High Energy Density, Edited by Corso Academic Press, New York (1971).
10. G. Haas and J.E. Waylonis, J. Opt. Soc. Am. 51, 719 (1961).
11. N. Frenkel, "Kinetic Theory of Liquids", Oxford University Press, London (1947).
12. Yu. V. Afanasev and O.N. Krokhin, Sov. Phys. JETP 25, #4 (Oct. 1967).
13. E. Goldman, Numerical Modeling of Laser Produced Plasmas, One-Dimensional Fluid-In-Cell Model, Lab. for Laser Energetics Report No.6 and No. 12 University of Rochester, Rochester, New York.
14. W.L. Weise, M.W. Smith, and B.M. Glennon, Atomic Transition Probabilities, NSRDS-NBS4, Vol. 1, U.S. Gov't. Printing Office, Washington, D.C., 1966.
15. J. Dawson and C. Oberman, Phys. Fluids 5, 517 (1962).
16. H.R. Griem, Plasma Spectroscopy, McGraw-Hill, New York, 1964.
17. R.W.P. McWhirter, Spectral Intensities, in Plasma Diagnostic Techniques (R.H. Huddleston and S.L. Leonard Eds.) Academic Press, New York, 1965,p.201.
18. R.F. Post, Plasma Physics 3, 273 (1961).
19. R.C. Elton, Atomic Processes, in Methods of Experimental Physics (H.R. Griem and R.H. Lovberg Eds.) Academic Press, New York, 1970, Vol. 9A, p. 115.

20. R.C. Elton, *Determination of Electron Temperatures Between 50 eV and 100 Kev X-Ray Continuum Radiation in Plasmas*, Naval Research Laboratory, NRL Report 6738, Washington, D. C. (1968).
21. N.G. Basov, et al., ZhETF Pis. Red. **13**, 691 (1971); English translation JETP Lett. **13**, 489 (1971); and J.W. Shearer, et al., *Experimental Indications of Plasma Instabilities Induced by Laser Heating*, Lawrence Radiation Laboratory Report UCRL-73489.
22. M.J. Bernstein and G.G. Comisar, J. Appl. Phys. **41**, 729 (1970).
23. T.C. Bristow, et al., *High Intensity X-Ray Spectra and Stimulated Emission from Laser Plasmas*, University of Rochester Report No. 10, Laser Energetics Laboratory (1972).
24. E. Goldman, Lab. for Laser Energetics Report No. 8, University of Rochester, Rochester, New York (1972). (Numerical Modeling of Laser Produced Plasmas: The Dynamics and Neutron Production in Dense Spherically Symmetric Plasmas.)

NANOSECOND AND PICOSECOND LASER IRRADIATION OF SOLID TARGETS[†*]

E. D. Jones, G. W. Gobeli, and J. N. Olsen

Sandia Laboratories, Albuquerque, New Mexico

ABSTRACT

The design and operating characteristics of a large Nd^{+3} glass laser which produces large-energy, single mode locked pulses will be discussed. With rigorous care to minimize back reflections and precise optical alignment, pulses of up to 35 joules and 3 psec FWHM have been focused onto the surface of CD_2. Incident, transmitted, and reflected energies were monitored and for targets 0.04 cm thick, 0.025% transmission and 0.4% reflection were observed for a typical 10 joule psec pulse. These values remained essentially the same for a 35 J, 3.5 nsec pulse. Measurements of ion velocity distributions at several orientations relative to the laser beam indicate considerable anisotropies, velocities back toward the laser being substantially larger than those away from the laser.

I. INTRODUCTION

Recently, preliminary experiments on high energy laser irradiations of deuterium[1-3] and deuterated solids [4,5] have raised the possibility of utilizing such devices in the area of research in controlled thermonuclear reactions. Due to the limited energy (E < 1000 joules) currently available in short (< 10^{-8} sec) pulses, experiments germane to the problems involved in such research can be carried out only in very small dimensioned target volumes; and hence, it is concomitantly necessary to utilize very short duration pulses. Although some theoretical-calculational results[6-7]

[†]This work was supported by the U. S. Atomic Energy Commission.

[*]Presented at the Second Workshop on "Laser Interaction and Related Plasma Phenomena" at Rensselaer Polytechnic Institute, Hartford Graduate Center, August 30-September 3, 1971.

indicate the advisability of utilization of pulses in the 0.1-10 nanosecond regime, based on current plasma theories, there are alternate possibilities, e.g., much shorter time durations might be effective due to their much higher power levels.

The technique of generating a train of mode locked[8] pulses from a Nd^{+3} doped glass laser, isolating a single pulse from such a train[9,10] and amplifying it to substantial energy levels[4,5] allows pertinent experiments to be performed in the ultrashort (10^{-12}-10^{-11} sec) time regime and at adequate energy levels (up to 30 joules). This paper reports the details of the design and operating characteristics of a large glass laser oscillator-amplifier system which produces large-energy, single mode locked pulses; some problems encountered in the focusing of the energy into small areas on solid targets; some results concerning the partition of the incident pulsed laser energy into reflected, transmitted and scattered components; and some measurements of the kinetic energy of expansion of the resultant plasma ions.

II. LASER SYSTEM

A block diagram of the Nd^{3+} glass picosecond laser system is shown in Fig. 1. The mode locked oscillator consists of a 9 x 150 mm Brewster-Brewster rod which is mode locked by Eastman-Kodak 9860 bleachable dye. The oscillator optical cavity is plano-plano with an output reflectivity of 55%. The spacing between the individual mode locked pulses is about 14 nsec. Typical energy content in each mode locked pulse is approximately 2 millijoules and for a properly aligned optical cavity, the total pulse train duration is greater than 500 nsec.

The optical gate consists of a dual-crystal KD*P Pockels cell placed between two crossed Glan polarizers. With very careful alignment and no voltage applied to the Pockels cell, a rejection ratio (incident energy/transmitted energy) of 2000-3000 can be obtained. When the 10 nsec wide $\lambda/2$ voltage pulse (generated by a N_2 pressurized laser triggered spark gap) is applied at the appropriate time to the Pockels cell, essentially all of the energy of a single mode locked pulse is gated out and directed to the first amplifier.

The first and second amplifiers are identical in their construction which is as follows. These amplifier rods are 1.3 cm dia. by 66 cm long with both faces cut at 12°. The maximum pump energy is 12 kJ per rod and is applied to four linear flashlamps in a 500 μsec wide pulse. The resulting output energy after the first amplifier is approximately 0.1 joule and after the second amplifier is in the 5 J regime.

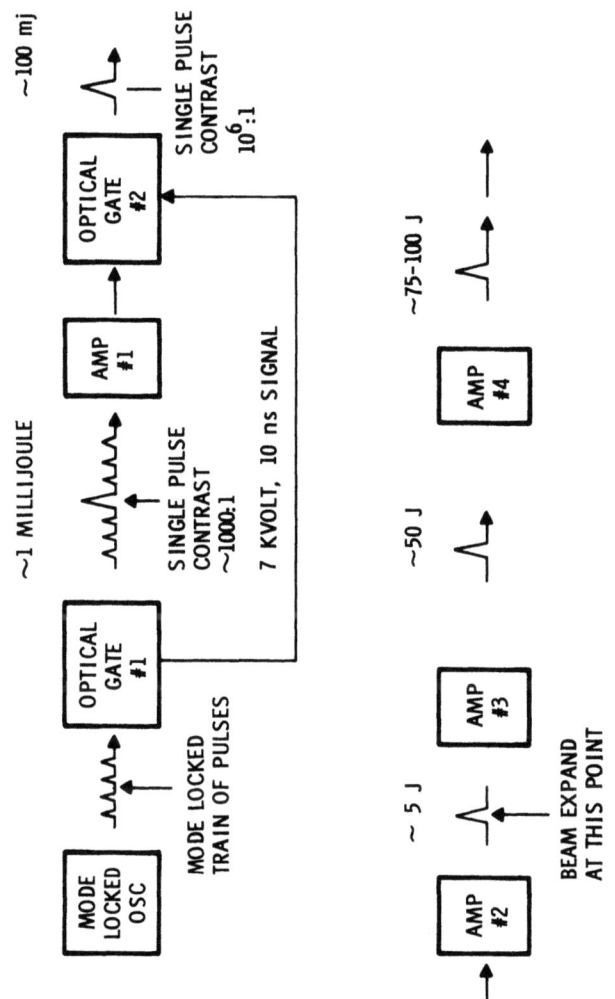

FIGURE 1. Block diagram of laser system.

In order to provide further isolation and also to improve the transmitted energy contrast ratio, a second optical gate is placed between amplifiers 1 and 2. This optical gate is similarly constructed to the one described above. The $\lambda/2$, 10 nsec voltage pulse which operates this gate is the same one which was used for triggering the first gate and is connected to the first gate by a coaxial cable whose length has been adjusted such that the $\lambda/2$ pulse arrives at the Pockels cell just as the amplified mode locked pulse arrives. In this manner, contrast ratios of $10^6/1$ are achieved. The transmitted mode locked pulse is then directed to the second amplifier for further amplfication to the nominal 2-5 joule level.

After two stages of amplification, the beam is expanded from $\frac{1}{2}"$ → $1\frac{1}{2}"$ by an aspheric lens system which has been designed to correct the measured 2 MR input beam divergence from the $\frac{1}{2}"$ amplifier system. This expanded beam is then amplified by a 38 x 1120 mm 12°-Brewster rod which is pumped by a 48 kJ capacitor bank. Typical output energies at maximum flashlamp energy levels is of the order of 50 J. Amplifier 4 is a 16 kJ capacitor bank pumped Brewster-Brewster 38 x 560 mm rod and for maximum pump energies, the output laser pulse energy is in the 75-100 J range. Typical measured beam divergence at this point is found to be 1.8 MR in the vertical (long axis) direction and 2 MR in the horizontal (short axis) plane. It is felt that the greater beam divergence in the horizontal direction is due to the difficulty of achieving correct alignment with a Brewster-Brewster system. This result is evidenced by the fact that beam divergences of 2 MR vertical and 5 MR horizontal have been observed from a deliberate slight misalignment of the Brewster-Brewster system.

Various beam diagnostics have been installed on this laser system which facilitates the operation of the laser. These diagnostics are coupled to the laser system by thin film pellicle beam splitters. In particular, it has been found that installation of integrating photodiodes at various check points have proven invaluable. These photodiodes not only provide information about the timing of the system, but they also act as energy measurement devices and indicators for existence of long pulse operation (prelase caused by target feedback). In addition to the integrating photodiodes, other diagnostics include fast response time photodiodes for observation and monitoring of the laser pulse and its purity, i.e., other mode locked pulses, etc. Located after the second and fourth amplifiers are two TRG calorimeters which are used for total output energy measurements at these points. The maximum energy delivered to these calorimeters is kept below one joule in order to avoid destruction of these units.

The principal difficulties in the operation of this laser system are as follows. The repeatability of obtaining good mode locked pulses from the air cooled mode locked oscillator is unsatisfactory. Maintaining alignment of the system is difficult if large temperature variations occur in the room. However, by far the most difficult problem to overcome is the prelase problem when experiments demand a focal spot size irradiation area. Because the target is at the focal spot of the lens, any target feedback to the lens is automatically directed back to the laser system. For this reason, the maximum energy that could be delivered to the target without prelasing was approximately 30 J.

For nanosecond operation, the same laser system is used with the following two modifications. The mode locked oscillator is replaced by a standard 40 nsec Pockel cell Q-switched oscillator system. The only other modification needed was to change the laser triggered spark gap from giving a 10 nsec $\lambda/2$ voltage pulse. After amplification to the 40 J level, the output pulse width was measured to be 3.5 nsec. Because of the increased energy content in the 3.5 nsec pulse from the oscillator compared to a single mode locked pulse, the amplifiers need not be pumped at their maximum levels. However, when calibration shots on the system were made, output energies of 90-100 J were easily obtained.

III. FOCAL SPOT SIZE

An estimate of the focal spot size achieved by the laser-focusing lens combination as well as a determination of the proper focusing condition to achieve maximum surface incident energy density was established by utilizing a cw YAG:Nd^{+3} laser. The output of the cw laser was injected into the properly aligned glass laser oscillator rod and allowed to traverse the entire amplifier train including the optical gates.* An output beam of full aperture, 10 cm^2, and proper beam divergence, 2 milliradian, was achieved and was incident upon the carefully aligned F/5.0, 60 mm dia. single element dual aspheric focusing lens.**

A 50 micron diameter nichrome wire which had been carefully stretched over the surface of the target was then maneuvered into the focal volume of the lens via a x-y drive for the wire and a z drive on the lens itself for focal depth positioning. The transmitted cw laser energy was monitored and a x, y, z location of the wire-lens combination which maximized obstruction of the incident laser energy was determined. The average of many (> 6) such determinations indicated that 94% to 97% of the energy was blocked

---
*High transmission through the finally aligned gates was achieved by simply inserting a 1.06 micron half-wave plate between the crossed polarizers.
**Designed and fabricated for best focusing of a 1.06 micron wavelength, 2 milliradian beam divergence, coherent laser beam by Space Optics, Inc., Chelmsford, Mass.

by the optimum wire position. Displacing the z location of the lens by 200 microns resulted in 65% maximum obstruction. For optimum focusing conditions, the lens was advanced toward the target by 25 microns from the established focal position and the experiments were conducted. More quantitative measurements of the focal volume utilizing a precision bilateral slit in place of the 50-micron diameter wire indicate that 80% of the cw laser energy is concentrated into a spot 26 microns in diameter and that the final 20% extends to a diameter of greater than 38 microns. Assuming a focal spot of 40 microns for 90% energy content appears reasonable for the lens capability, giving about 5×10^{-5} cm^2 as a focal area. Thus, a 10 joule pulse gives an energy density of 2×10^5 joule/cm^2. For the 3.5 nanosecond wide pulse this yields an incident power density of 6×10^{13} watts/cm^2 and for the picosecond pulses, power densities in excess of 10^{16} watts/cm^2.

The possibility that the focal spot diameter for the laser under full-pump, pulsed operation might be different even though the proper beam divergence has been simulated for the cw laser measurements still remains. Measurements to ascertain the relation between the above described measurements and the full operational mode are currently in progress and will be reported in the near future.

IV. REFLECTION AND TRANSMISSION OF LASER ENERGY

Measurements of the incident, reflected, transmitted, and scattered laser energy were made utilizing identical fast response time photodiodes, each covered by a 1.06 micron band pass filter, as indicated in Fig. 2. The attenuator stacks were empirically adjusted until the photodiode outputs were approximately equal in amplitude. The output of the incident and reflected diodes were fed to a "cross" input to a 519 oscilloscope through appropriate coaxial cable lengths, which imposed a 15 nsec delay for the reflected diode signal and a 30 nsec delay for the transmitted diode signal. Thus, the three signals were displayed sequentially on the same trace as illustrated in Fig. 3. Figure 3 shows the results for a .010" thick CD_2 film irradiated by a 33 joule, 3.5 nanosecond pulse. The indicated transmission is 0.06% while the indicated reflection is 0.2%. The scattered 1.06 micron signal was of such low intensity that it was necessary to operate the diode in an integrating mode in order to observe a signal. Utilizing the diode sensitivity and integrator characteristics, it is estimated that about 0.1 - 0.13 millijoule is incident on the detector which, when corrected for solid angle, indicated that a maximum of 10% of the laser energy is scattered in a 4π solid angle. However, this photodiode could possibly be detecting scattered light off the walls of the vacuum chamber. Hence, an estimate of 10% scattered light could be too large by some unknown amount. The results for thick, 3 mm, CD_2 are the same except that the transmitted energy is down to 0.02%.

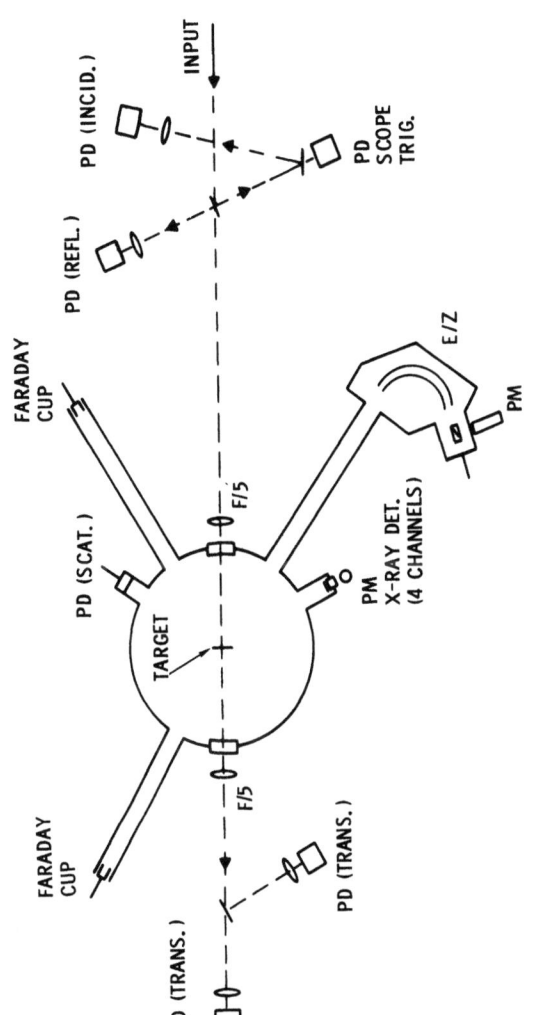

FIGURE 2. Schematic of experimental chamber.

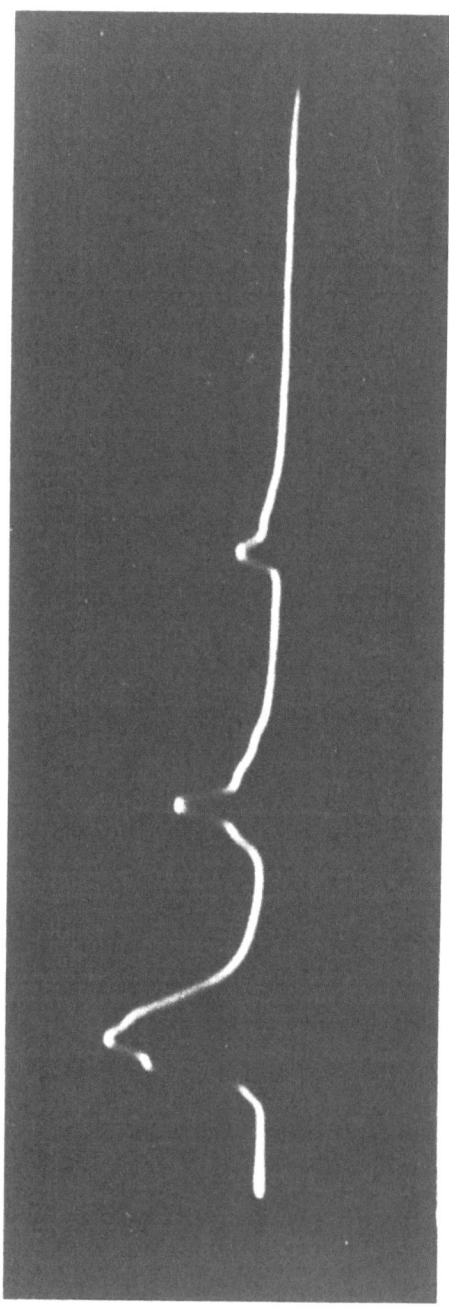

FIGURE 3. Oscilloscope pictures of incident reflected and transmitted laser energy. Total trace length is 60 nsec. See text for percentage of reflection and transmission.

Similar records taken from picosecond time scale irradiations indicate 0.3% - 0.4% reflection and 0.01% - 0.015% transmission, in substantial agreement with the nanosecond pulses.

These data are interpreted as arguing convincingly for absorption of the laser energy by the solid targets rather than transmission, reflection and/or scattering.

V. ION VELOCITY DISTRIBUTIONS

As indicated in Fig. 2, time of flight ion detectors were mounted at angles of 30° and 150° relative to the incident laser beam. A typical velocity distribution is shown in Fig. 4 for a 35 joule, 3.5 nanosec pulse on a thick CD_2 target, looking back toward the laser. Assuming the first peak to be due to H^+ (which might arise due to absorbed H_2O or more likely due to incomplete deuteration) one obtains an energy of 1.2 keV. Assuming the second peak to be due to D^+, one obtains an energy of 1-1.2 keV and assuming the third general peak area to be carbon in various ionized states gives an approximate energy of 1.5 keV. The ratio of H^+ to D^+ indicates about 75% deuteration which is comparable to the approximate value of 67% obtained by NMR analysis of the same sample.

Similar measurements for a picosecond pulse with energy of 14 joules gives H^+ and D^+ expansion energies of $\sim$ 1.9 keV and a large C^+ peak near the same energy. In all measurements, picosecond pulses consistently produced higher expansion energies than nanosecond pulses by substantial (> 2 times) factors. In some preliminary experiments on carbon particles suspended on a very thin plastic membrane (thickness $\sim$ 2 microns) expansion energies reached 5-6 keV with picosecond pulses of 15-20 joules amplitude.

The isotropy of ion velocities toward and away from the incident laser direction were measured at 30° and 150° as indicated. For a 0.04 cm thick CD_2 membrane target and a 36 joule, 3.5 nanosec pulse, front side ion expansion energies of 1.8 keV were recorded while simultaneously backside energies of 0.210 keV were obtained. For a 10 joule picosecond pulse backside energies of 0.4 keV and front side energies of $\sim$ 4 keV were obtained.

It should be pointed out that the 0.04 cm (400 micron) target thickness is greater than the focusing lens depth of focus ($\sim$ 200 microns for 33% energy density decrease) and hence this large front-to-back ratio of about 6:1 might well in part be due to a greater energy absorption toward the front of the sample. In this regard, the preliminary measurements on carbon particles on plastic membrane yield front-to-back ratios of 7:1 to 10:1. For example, a picosecond pulse of 18 joules gave ratios of 5 keV to 0.53 keV.

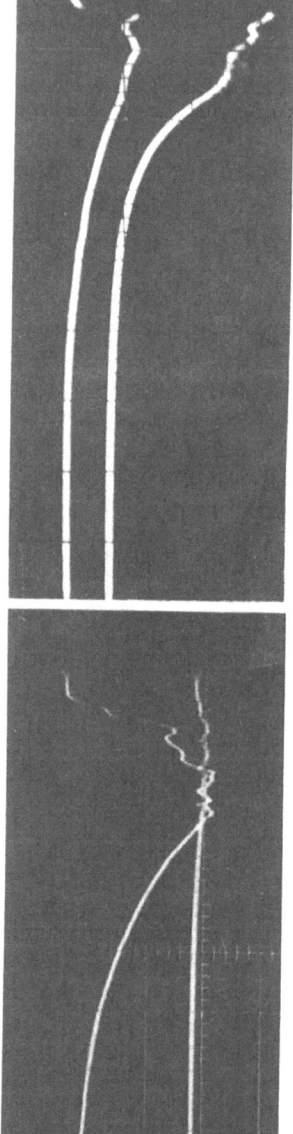

FIGURE 4. Typical time of flight velocity distributions obtained for thin CD_2. Part A upper trace is rear detector, 5 μsec/cm sweep. Part A lower trace, front detector, 2 μsec/cm sweep. Part B, ion signal obtained with picosecond pulse, 2 μsec/cm sweep.

In this regard, the necessity of precise description of the incident laser pulse is evidenced by the fact that a 16 joule picosecond pulse with 1.5 joule of prelase produced only 700 eV and 125 eV ions in the front/back directions, respectively. Further experiments on very thin targets and on membrane supported particles are being pursued and will be reported in detail in the near future.

Finally, it is possible to make a crude estimate of the total expansion energy of the ions by taking into account the front-to-back anisotropy, the collector Faraday cup area and its distance from the target. Assuming singly ionized H^+ and D^+ and an average charge state of 4+ for the carbon ions, the TOF curves indicate that at least 60% of the incident laser energy appears as kinetic energy of expansion. It should be noted here that preliminary data utilizing a 127° electrostatic deflection energy/charge, i.e., E/Z, analyzer indicate the choice of 4+ for the carbon charge state to be reasonable although possibly on the high side. Thus, these data indicate that the laser energy is well accounted for in the kinetic energy of expansion of the resultant ions.

The data on ion velocities as measured by the time-of-flight technique are interpreted as indicating substantial expansion energies resulting from picosecond and nanosecond irradiation of solid targets. Typically, ion velocities of 0.5 to several keV are obtained, with the higher energies resulting from picosecond pulses with no prelase.

ACKNOWLEDGMENTS

The authors wish to thank Lawrence Livermore Laboratories for furnishing the CD_2 samples, Dr. H. T. Weaver, Sandia Laboratories, for the NMR analyses of these samples, and G. C. Hauser and V. K. Sower for expert technical assistance.

REFERENCES

1. F. Floux, D. Cognard, L-G. Denoeud, G. Piar, D. Parisot, J. L. Bobin, F. Delobeau, and C. Fauquignon, Phys. Rev. A1, 821 (1970).
2. K. Büchl, K. Eidmann, P. Mulser, H. Salzmann, R. Sigel, and S. Witkowski, Proceedings on Fourth Conference on Plasma Physics and Controlled Nuclear Fusion Research (to be published), Paper CN-28-D-11.
3. A. Caruso, Proceedings of First Workshop on Laser Interaction and Related Plasma Phenomena (Plenum Press, New York, 1971), p. 289.

4. N. G. Basov, P. G. Kriukov, S. D. Zakharov, Yu. V. Senatzky, and S. V. Tschekalin, IEEE J. Quant. Electron. QE-4, 864 (1968).
5. G. W. Gobeli, J. C. Bushnell, P. S. Peercy, and E. D. Jones, Phys. Rev. 188, 300 (1969).
6. J. W. Shearer and W. S. Barnes, Proceedings of First Workshop on Laser Interaction and Related Plasma Phenomena (Plenum Press, New York, 1971), p. 307.
7. M. Lubin, private communication.
8. A. J. DeMaria, W. H. Glenn, Jr., M. J. Brienza, and M. E. Mack, Proceedings of First Workshop on Laser Interaction and Related Plasma Phenomena (Plenum Press, New York, 1971), p. 11.
9. G. Kachen, L. Steinmetz, and J. Kysilka, Appl. Phys. Letters 13, 229 (1968).
10. G. Dubé, Appl. Phys. Letters 18, 69 (1971).

THERMONUCLEAR FUSION PLASMA BY LASERS*

Chiyoe Yamanaka**

Institute of Plasma Physics, Nagoya University

Nagoya 464 JAPAN

ABSTRACT

As lasers have an ability to deliver a large amount of energy very rapidly to matters, one can produce thermonuclear temperature plasma. We observed the neutron yield from a solid deuterium target irradiated by the beam of glass laser, wave length of 1.06 µ, pulse width of 2 ns and power of about 10 GW.

The heating process by laser is one of the most interesting subject in plasma physics. The classical absorption becomes weak at high temperatures and the over-dense plasma is to perfectly reflect the laser beam. Considering the experimental results, anomalous absorption induced by the high electric field of laser beam seems to be introduced. The electron temperature was estimated from the soft x-ray measurement. The reflection intensity of laser beam from plasma and the time of flight of ions were investigated. These data have a close correlation with the event of the neutron yield. Until up to 200 eV of the electron temperature near the cut off density, the absorption is caused by the classical process and beyond this threshold the parametric instability, predicted by Nishikawa, begins to be functional. The electron temperature showed a steep rise with increase of the incident laser power after the

---

*Presented at the Second Workshop on "Laser Interaction and Related Plasma Phenomena" at Rensselaer Polytechnic Institute, Hartford Graduate Center, August 30 - September 3, 1971.
**Permanent address: Faculty of Engineering, Osaka University, Osaka 565 JAPAN; The members of project: T. Yamanaka, T. Sasaki, H. Kang, K. Yoshida and M. Waki.

threshold, 10^{13} W/cm^2. Near the threshold the oscillation of reflected beam was often observed. The fast component of ions began to appear.

In appendix, brief descriptions of damages in glass laser rods are presented as for the reference.

INTRODUCTION

With the development of high intensity laser, one can produce plasmas with high density and high temperature which are available to thermonuclear fusion research.[1] Recently, French group[2] has reported substantial neutron emission from a solid deuterium target irradiated by a 7 nsec, 4 GW Nd^{3+} laser.

In order to clarify the potentiality of laser produced plasma for a future fusion reactor, detail investigation of the properties[3] of these plasmas is very important. Especially the heating process by laser is one of the most interesting subject in plasma study. The classical absorption is weak at high temperatures and becomes inefficient as the temperature increases. This absorption length at near the cut off density of laser light is

$$\ell \simeq 5 \times 10^{39} \, T^{3/2}/n^2 \text{ microns,}$$

where T is the electron temperature in eV and n is the density in cm^{-3}. For a density 10^{21} cm^{-3} and a temperature 1 keV, the absorption length is 150 μ. This shows a situation of the absorption process becomes critical in a focal spot of the diameter 100 μ. According to the classical process, the laser beam would be perfectly reflected or perfectly transmitted, depending on whether laser frequency ω is less than or greater than the plasma frequency ω_p. Considering the experimental results, the anomalous absorption due to the high field strength of laser beam seems to be functional in plasma heating.

In this paper we report the properties of laser plasmas from solid deuterium, deuterated polyethylen and LiH, measured by various kinds of diagnostics. The neutron emission is observed from deuterium plasma.

EXPERIMENTAL ARRANGEMENT

The laser is composed of five stage amplifiers[4,5] constructed in our laboratory. Three oscillators are provided to it. Two of them are glass ones, used alternatively in nanosecond range and in picosecond range. The third is a YAG oscillator which has a narrow spectrum beam especially suitable for scattering spectroscopy. The nanosecond oscillator can deliver a pulse, the duration of which is variable[6] from 2 ∿ 10 ns and the rise time is 1 ns. The pulse forming system is composed by a laser trigger spark gap, a KDP Pockels cell and Glan prisms. An oscillator pulse can be amplified up to more than 30 Joules, corresponding to 15 GW for 2 ns pulse. This output is a suppressed value for a precaution of the hazard by the reflected beam from a target. The damages of the glasses have been completely studied[7] some of which are presented in appendix. To protect the reflection, we used a Faraday rotator, dye cells and a special uniguide slit switched by the reflected beam. These methods can attenuate the reflection by more than 10^4.

The picosecond oscillator can deliver a single pulse selected from the train of mode locked pulses by the pulse transmission mode method.[3] To amplify the picosecond pulse, $POCL_3$-Nd^{+3} liquid[8] laser seems to be suitable on account of larger transition cross-section than that of glass amplifier.

The YAG oscillator is used to SHG by a KDP crystal for light scattering diagnostics as well as to drive a main amplifier for plasma production. The YAG laser is Q switched by a Pockels cell and connected to a glass preamplifier. In this case the spectrum of laser beam is so narrow as 0.01 A in wave length.

TABLE I COMPOSITION OF HIGH POWER LASER SYSTEM

Amp.	I	II	III	IV	V
Rod Dimension	$\phi 20 \times 320^l$	$\phi 20 \times 320^l$	$\phi 30 \times 320^l$	$\phi 30 \times 320^l$	$\phi 40 \times 600^l$ mm
Nd_2O_3	3.5	3.5	3.5	3.5	3.5 wt%
Flash Lamp	4	4	6	6	10
Pumping Energy (Max.Pump.Ener.)	9 (20)	9 (20)	13.5 (30)	13.5 (30)	60 kJ (200)
Gain	7	5	4.5	4	3.5

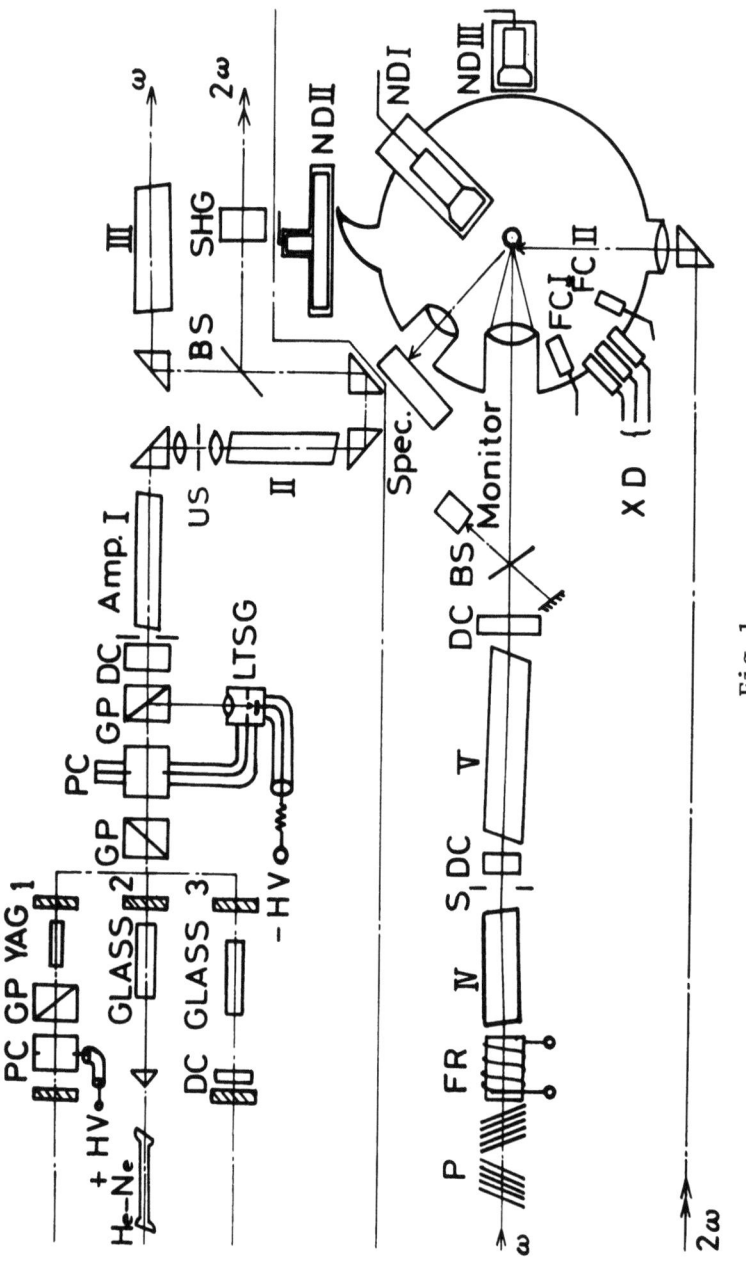

Fig.1

Experimental arrangement of laser plasma research. 1: Pockels cell Q-switched YAG laser and glass preamplifier (2 ns, 6 mJ), 2: Rotating Q-switched glass laser (2 ns, 9 mJ), 3: Mode locked glass laser (50 ps, 1 mJ), PC: Pockels cell, DC: Satulable dye cell, XD: X-ray detector, ND: Neutron detector, FC: Faraday cage, Spec: Spectrometer, US: Uniguide slit

As for a Booster the disc type laser is also constructed. The beam divergence of the systems is about 1 m rad. It is focused onto the target by either aspherical lens of focal length 50 mm, diameter 60 mm or Luboshez lens of focal length 60 mm, diameter 60 mm. The experimental arrangement is shown schematically in Fig.1. A cryostat of liquid helium can produce a solid deuterium stack, the diameter of which is 2 mm in vacuum of 10^{-7} torr. The whole view of apparatus is shown in Fig.2.

Fig.2

Whole and detail view of apparatus.

EXPERIMENTAL RESULTS

(a) Preliminary Results of Laser Plasmas

To clarify the properties of plasma due to the target we performed the experiments on a small suspended particle and a solid thick plate of LiH. Streak photographs of LiH particle 50 μ in diameter irradiated by a glass laser beam were taken using STL streak camera. The expansion velocity was about 5×10^7 cm/sec as shown in Fig.3. At the distance of 30 cm from the irradiated point, the Li^{1+} and Li^{2+} ions were disolved by an energy analyser, the velocities of which were 2.7×10^7 cm/sec and 5×10^7 cm/sec respectively.[9]

By our simple gas dynamical computation[10] under the following condition: glass laser rise time 2 ns, power 10 GW, the temperature reaches up to 300 eV within 1 ns. Applying a magnetic field to suppress the expansion, the temperature can be recovered to 600 eV after the expansion. Without magnetic field, inertia confinement seems to finish in 1 ns. The calculated expansion velocity accorded with the experimental one. These are shown in Fig.4.

When we used a thick target, the plasma formation had some structures[11] in space which would be caused by the self induced magnetic field. After the appearance of plasma, very high dense neutral atoms followed. They were measured[12] using Thomson and Rayleigh scatterings of a ruby laser light as shown in Fig.5.

(b) Experiments with Neutron Yields

<u>(i) Electron temperatures</u>. The electron temperature was measured from the soft x-ray by plastic scintilators with beryllium windows of different thickness (25 μ, 50 μ, 100 μ). The small type photo multiplier HTV-R292 has a time resolution of ∿10 ns. The experimental results with solid deuterium are shown in Fig.6. The influence of laser pulse[13] duration to the electron temperature was

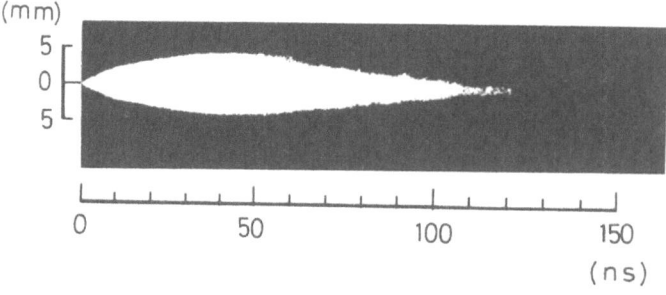

Fig.3
Streak photograph of LiH particle irradiated by glass laser.

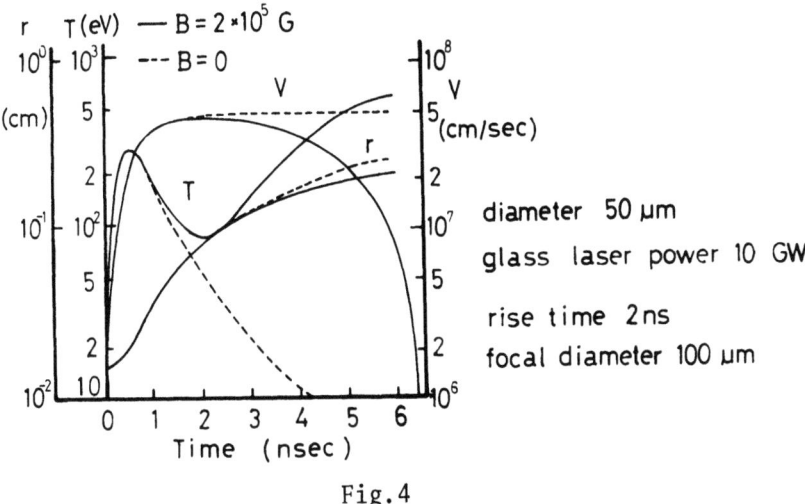

Fig. 4

Gasdynamical behaviors of laser irradiated LiH particle, laser power 10 GW, 2ns.

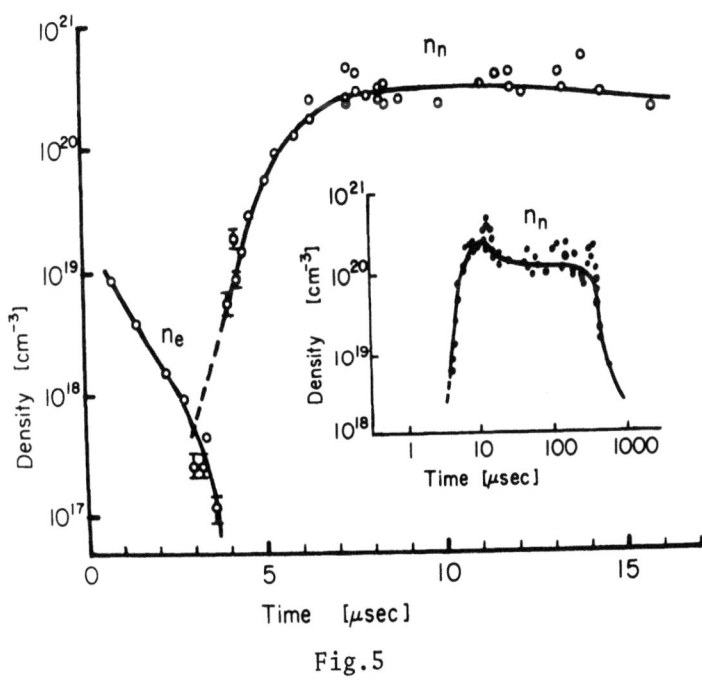

Fig. 5

Evaporation of dense neutral atoms from solid target by laser irradiation.

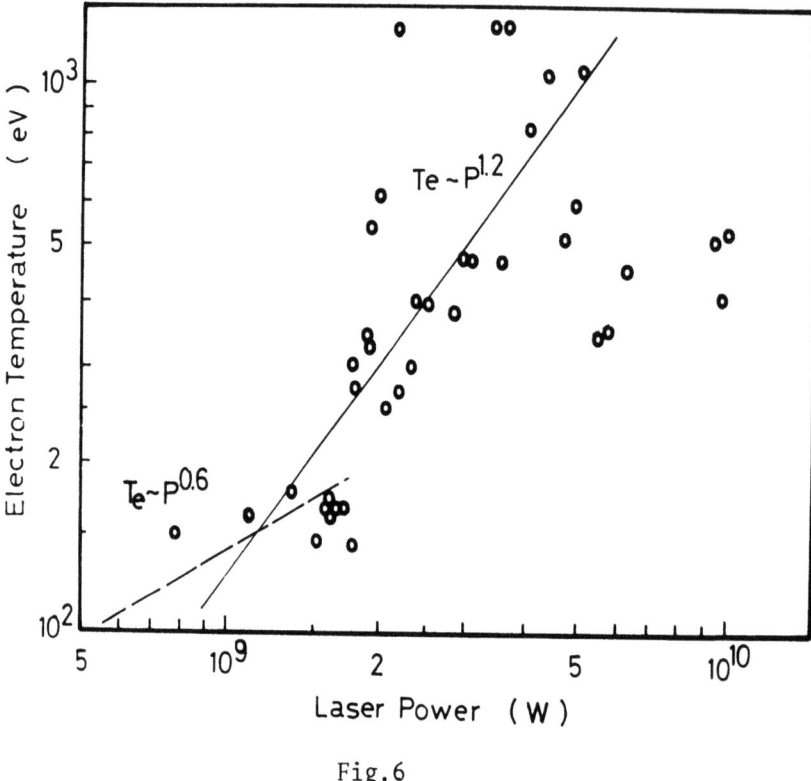

Fig.6

Electron temperature of deuterium plasma with incident laser power.

not observed in the cases of durations 2.3 ns, 4 ns and 10 ns. The electron temperature increased remarkably with increase of the laser power. The abrupt change seems to be introduced at the power of 1 GW. In the lower side of this point the electron temperature had a dependence on laser power p as $p^{0.6}$, and in the higher side it was proportional to $p^{1.2}$. In addition, a close correlation was observed between electron temperature, reflection intensity, particle velocity and neutron yield.

(ii) Laser beam reflection. A fairly amount of the incident laser energy has been reflected from the target. The pulse form of the incident and reflected laser light was measured by a same bi-planer photodiode HTV-R317. Typical responces of laser reflection are shown in Fig.7. The reflection intensity was rather scattered. The ratios of reflected energy to incident one with the electron temperature are indicated in Fig.8. At the low incident energy of about 10^{11} W/cm^2 the oscillation of the reflected laser light was observed near the electron temperature 200 eV. The frequency was

THERMONUCLEAR FUSION PLASMA BY LASERS

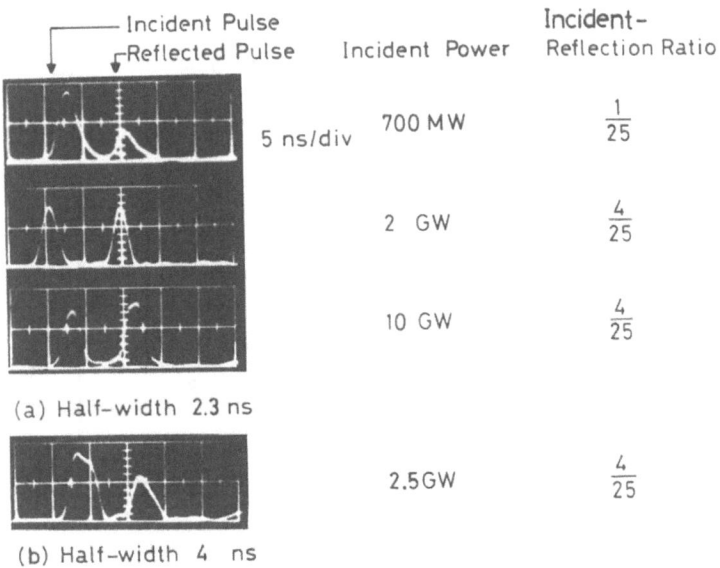

Fig. 7

Typical reflections of laser power.

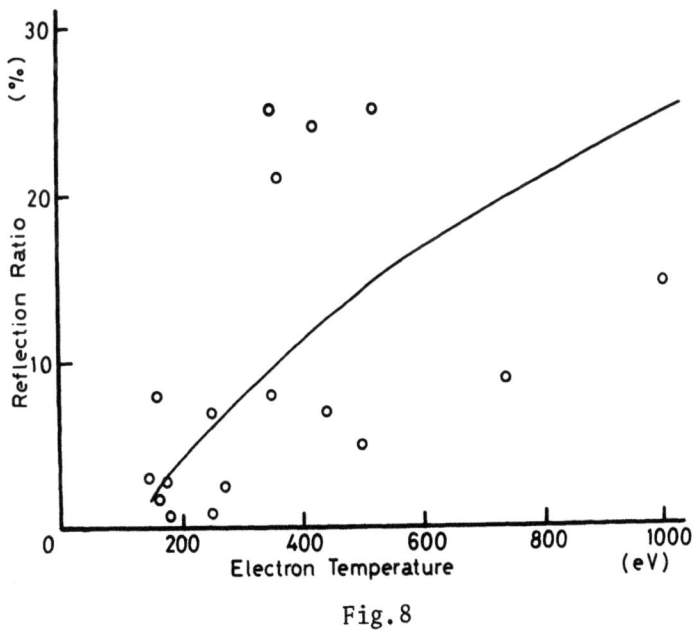

Fig. 8

Reflection ratio of energy with the electron temperature.

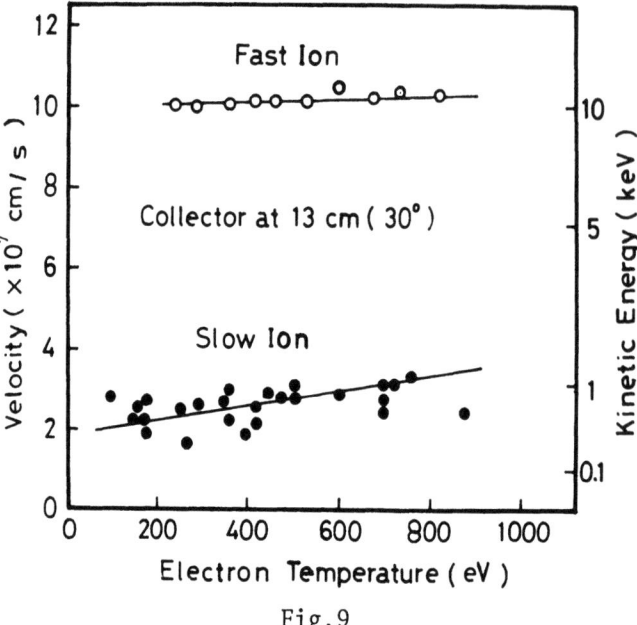

Fig. 9

Expanding velocity of deuterium plasma with the electron temperature

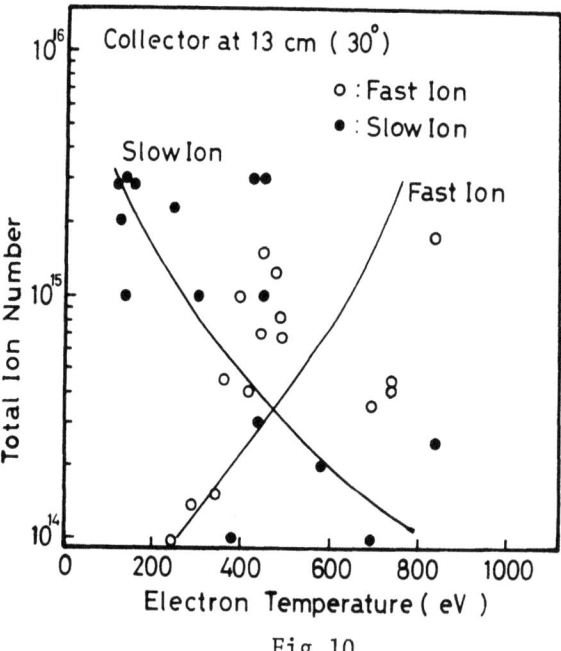

Fig. 10

Ion numbers of fast and slow component with the electron temperature.

10^9 Hz which accorded with the ion acoustic frequency in a focal region.

(iii) <u>Ion collection and time of flight</u>. As the previous report,[11,14] laser plasma is ejected from a target with the directional dependence. We measured the ion flux at the 30°, 45° to the incident beam. The time of flight measurement shows the two groups of ions, fast and slow components. The velocity dependence to the electron temperature is indicated in Fig.9. The velocity of slow component has an increase with the electron temperature, but that of the fast component is almost constant. The ion numbers of slow component show the decrease with the increase of electron temperature and these of fast component have reverse characteristics as shown in Fig.10. The neutron yield was always accompanied by the appearance of the fast ions. The total energy of the fast ions is up to 70 % and that of the slow ions is about 10 % of the incident energy from laser.

(iv) <u>Laser light scattering</u>. To get the ion temperature, we used a YAG oscillator to deliver SHG beam to the plasma as well as to send the fundamental frequency beam to plasma production. From these experiments the preliminary results indicate the ion temperature up to 1 keV, even if considering the non-Maxwellian plasma[16] behaviours.

(v) <u>Neutron yields</u>. We used three plastic scintillators which were set at the distance 5 cm, 10 cm and 40 cm from the target to detect the neutrons. Calibration was performed by a Am-Be neutron source. Threshold energy of laser for neutron emission was 5 J for 2 ns pulse and the electron temperature was about 500 eV. With laser energy of 10 Joule the neutron was observed in 40 % of the shots. The focal condition was very critical to produce the neutrons. Fig. 11 indicates the response of plastic scintillators for neutron events. The characteristics of neutron yield with laser energy are shown in Fig.12. The number of neutrons N depends strongly on the laser energy ε such as $\varepsilon^{4.5}$. The check method for experiment is as usual the replace of deuterium with hydrogen.

CONCLUSIONS

As the focal spot of laser was less than 100 μ in diameter, the incident laser flux was up to 10^{14} W/cm². From the theory of the anomalous absorption, Nishikawa[16] predicted the threshold laser flux for the instabilities near $\omega \approx \omega_p$.

For parametric instability: $\omega > \omega_k$

$$\frac{|E_0|^2}{4\pi n T_e} > \frac{2\sqrt{3}}{9} \frac{\nu_{ei}^2 \nu_{ii}}{\omega_p \Omega_k^2},$$

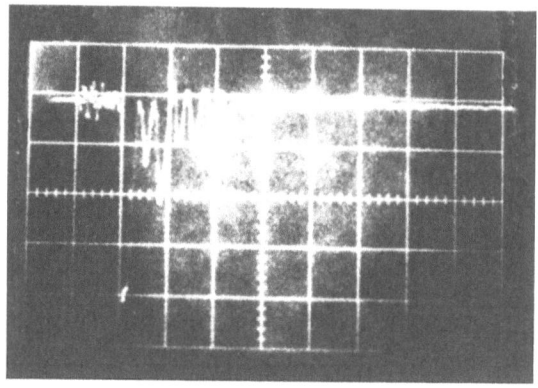

Fig.11

Neutron yield from solid dueterium irradiated by laser beam, 50 ns/div. laser energy 13.5 J.

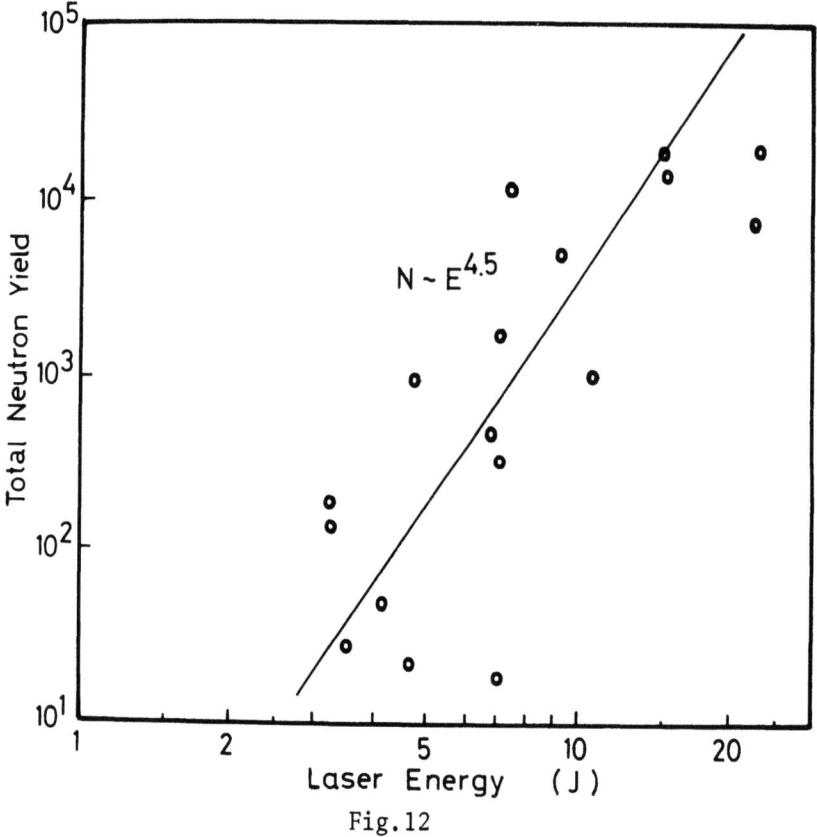

Fig.12

Dependence of neutron yield to laser energy.

Table II

Property of Plasma

Laser Pulse Width : 2.3 ns

Input Energy (J)	Electron Temp. T_e (keV)	Fast Ion E_k (keV)	Fast Ion T_i (keV)	Slow Ion E_k (keV)	Slow Ion T_i (eV)	Total Neutron Yield
3	~0.2	—	—	0.6	70	—
5	~0.5	9	0.8	0.9	80	~300
12	2	18	2.1	1.5	90	~5,000
20	4	27	4	2	140	~20,000

and for oscillating two stream instability: $\omega < \omega_k$

$$\frac{|E_0|^2}{4\pi n T_e} > \frac{\nu_{ei}}{\omega_p},$$

where $\omega_k = \omega_p[1 + \frac{3}{2} k^2 \lambda_D^2]$, λ_D is Debye length, ν_{ei} and ν_{ii} are collisional frequency between electron-ion and ion-ion, ω_p is plasma frequency, and $\Omega_k = k\sqrt{\frac{T_e}{M}}$ is ion acoustic frequency.

These threshold are estimated the balance of growth rate and damping of instabilities. The electron temperature of 200 eV is a critical point above which the parametric instability has smaller threshold than the oscillating two stream instability. Corresponding flux value is about 10^{13} W/cm^2. Experimentally, the oscillation of reflected light was observed at this electron temperature. At the temperature of 1 keV, plasma density of 2.3×10^{21} cm^{-3}, ν_e is 2×10^{12} sec^{-1}, the threshold flux of parametric instability is 10^{12} W/cm^2 and that of oscillating two stream instability is 10^{13} W/cm^2. Until when the electron temperature gets 200 eV, the heating process is mainly controlled by the classical absorption and after that is seems to be attributed to the anomalous heating of parametric instability. As shown before, the electron temperature measurement of soft x-ray endorses this situation. The growth rate of instability is known to be very fast above the threshold and the induced plasma waves tend to electron heating. This effect seems to be very effective in high temperature ranges.

At the yield of neutrons we noticed the appearance of fast ion component as well as high electron temperature and strong reflections. The neutron number depends strongly upon the laser energy as the power of about 5. The incident energy of laser is distributed mainly to fast ion component up to 70 % and slow component only have less than 10 %. Pulse duration of nanosec laser has no appreciable influence to the plasma properties of high temperatures.

The heating process of keV region[7] is very important subject for laser plasma research, which is mainly caused by the anomalous resistivity of plasma. The temperature measurement of ions by the YAG-SHG light scattering is going on to clarify this problem.

REFERENCES

1. N. G. Basov, IQEC, Miami, 1968
2. F. Floux, IQEC, Kyoto, 1.3, 1970
3. C. Yamanaka, T. Yamanaka, IQEC, Kyoto, 2.5, 1970 ; Int. Conf. on Laser Plasma, Moscow, 1971
4. T. Sasaki, T. Yamanaka and C. Yamanaka, Japan. J. Appl. Phys. 8, 1037 (1969)
5. C. Yamanaka, T. Yamanaka and T. Sasaki, IQEC, Kyoto, 20.3, 1970
6. K. Yoshida, T. Yamanaka, T. Sasaki, H. Kang M. Waki and C. Yamanaka, to be published in Japan. J. Appl. Phys.
7. C. Yamanaka, T. Sasaki, M. Hongyo and Y. Nagao, Conf. on Damage in Laser Materials, Boulder, 1971
8. C. Yamanaka, T. Sasaki and M. Hongyo, Conf. Laser Eng. and App., Washington, 1971
9. M. Onishi and C. Yamanaka, Technol. Rep. Osaka Univ. $\underline{20}$, 121 (1970)
10. C. Yamanaka and T. Yamanaka, Prog. Rep. of Plasma Electronics II, Osaka Univ. (1968)
11. T. Yamanaka and C. Yamanaka, Technol. Rep. Osaka Univ. $\underline{18}$, 155 (1968)
12. Y. Isawa and C. Yamanaka, Japan. J. Appl. Phys. $\underline{7}$, 954 (1968)
13. M. Waki, T. Yamanaka, H. Kang, K. Yoshida and C. Yamanaka, to be published in Japan. J. Appl. Phys.
14. H. Kang, T. Yamanaka, K. Yoshida, M. Waki and C. Yamanaka, to be published in Japan. J. Appl. Phys.
15. W. Kegel, Plasma Physics $\underline{12}$, 295 (1970)
16. K. Nishikawa, J. Phys. Soc. Japan $\underline{24}$, 916 (1968)
17. P. Kaw, J. Dawson, W. Kruer, C. Oberman and E. Valeo, MATT-817 (1970)

APPENDIX

INVESTIGATION OF DAMAGES IN LASER GLASSES

Chiyoe Yamanaka

Abstract

Experiments on laser glass damages were performed to clarify the damage mechanisms and to obtain the informations of the most preferable glass to a high power laser system. As for the testing method, passive and active tests were adopted by multi and single mode lasers. The plasma formations were always observed in internal and surface fractures. To investigate the plasma behavior, time resolved photographic and spectroscopic methods were used. The damage threshold on entrance surface was larger than exit one. The mechanism of fracture in glass laser seems to be due to the plasma formation by light which produces a shock wave to cause a cleavage in glass.

If one produces glass carefully without Pt inclusion, the damage threshold is mainly determined on surface. In this trend, the improvement of surface damage threshold becomes very important for the practical use.

Introduction

To construct a high power glass laser system the characteristics of laser glass are very important, especially high durability against damages. It is reported that present laser glasses can withstand over 1000 J/cm^2 with a long pulse of about 1 ms, however the energy passage of several ten J/cm^2 in a short pulse 10 ns may lead to fracture.

These results seem to show the fundamental process of fracture can be explained by the energy balance between the laser supplied energy and the energy lost in glass under irradiation. The laser light can produce the free electrons in glass through the multiphoton absorption process. If the energy balance between gain and loss becomes unstable, these electrons will be accelerated by the inverse blemsstrahlung and cause cascade yields of ionization. Once the plasma is produced, the shock wave due to laser supported deflagration will be driven into the solid which induces the cracks when the shocked stress becomes larger than the fracture threshold. In a longer duration of laser pulse, much more amount of energy may be necessary to heat up the plasma which is always growing in time. The mechanical strength of glass has also an important role to the growth of the damage in solids.

We performed the experiments on glass damage by Q-switched laser single and multi modes. As for the methods of damage testing, there are a passive test, that is performed only by laser irradiation of samples, and an active test that is done under pumping condition. Succeeding the previous report[1], in this paper the improvement of damage threshold is mainly described. The damage threshold on surface is increased by the treatment of hydro-fluoric acid. The chemical strengthen treatment has negative effect to prevent the damage growth on surface and coating treatment are discussed.

(a) Damage   (b) Stress

Fig. 1 (a) Exit surface breakdown of barium crown glass without inclusions. (b) Interference stress pattern on surface

Surface damage

Fig. 1 shows a picture of exit surface fracture. Laser beam came from left to right. The focal point was in a sample just near the exit. Increasing laser power, the plasma light on the entrance surface increased slightly, whereas that on the exit surface was more serious than that of entrance surface and produced a network of cracks around the plasma epicenter. Fig. 2 shows the fracture on both surfaces after several shots. As the plasmas tend to travel toward the direction of the insident laser beam, formation of shock wave on the exit surface seems to be much stronger to induce heavy fracture.

We used the barium crown glass samples with 15 x 15 mm^2 cross section and a lens of 6.5 cm focal length to collimate the peak power 30 MW, pulse duration 25 ns. Whenever the fracture was observed, plasma always occured. The damage threshold on the entrance surface was about 28 ± 5 J/cm^2 with multi mode and 88 ± 5J/cm^2 with single mode. The threshold was decreased by successive irradiation to the same site. The threshold depended upon materials and surface conditions especially microcracks and impurities on surface. The threshold on entrance surface was a litter larger than that of exit. If we used long samples with a long focal lens the threshold on the exit surface decreased due to self-trapping.

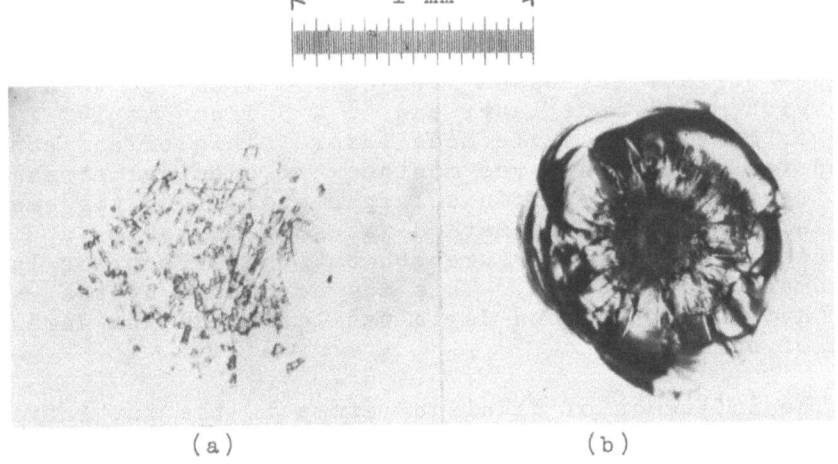

Fig. 2 (a) Fracture on entrance surface and (b) exit surface. (After several shots)

Fig. 3 indicates spectroscopic photographs of surface breakdown where, (a), (b) correspond to barium crown glass and light flint glass respectively. These were ten shots exposed. (c) and (d) are the spectrum of air breakdown by laser beam and the reference lines of Hg, respectively. All broad dim lines in (a), (b) were attributed to spectra of air. From this, it shows that the air is ionized by plasma of glass. Fig. 4 is the densitometric spectrograph of breakdown of barium crown glass. The line spectra of Ba^+ were observed intensely. Fig. 5 shows the time variation of Ba^+ (4,554 Å), Ca^+ (3,968 Å) and N_2 (5,011 Å). The rise time of the emission of N_2 was slow than that of metallic ions. This endorses the deflagration wave is driven to the laser beam which produces the air breakdown with time delay. Plasmas continued about 1 µsec at half width of maximum height.

Improvement of Surface Damage Threshold

When the glass was very carefully manufactured, most of platinum inclusions could be eliminated. Even if few inclusions makes fractures in laser glass, this is not practically serious. But the damage on surface is very severe limitation[2] for operation.

There seems to be several methods to raise the damage threshold. One is treating the glass either by a hydrofluoric acid[3] or by dimethyldichlorsilane[4]. In the case of the acid etch, it is considered that small surface irregularities and cracks are smoothed over, thus making the surface stronger to damage. We used the barium crown glass which was immersed for 10 minutes in 10 % hydrofluoric acid. The damage threshold on entrance surface increased from 28 ± 5 J/cm^2 to 40 ± 5 J/cm^2 with multi mode laser and 88 ± 5 J/cm^2 to 140 J/cm^2 or more with single mode laser. This effect continued for more than three monthes. The chemical strengthen treatment is to diffuse larger atoms into the small alkali matrix. This method was so remarkable to increase the mechanical strength but the effect for laser damage was negative. This may be caused by the fact that the strengthen depth was less than the damage depth of surfaces.

The influence of alkaline atoms to the initiation of ionization by light was investigated. The removal of alkaline oxide from the surface shows no meaningful difference with the original sample. Coating effect of the surface for the damage threshold were investigated.

INVESTIGATION OF DAMAGES IN LASER GLASSES

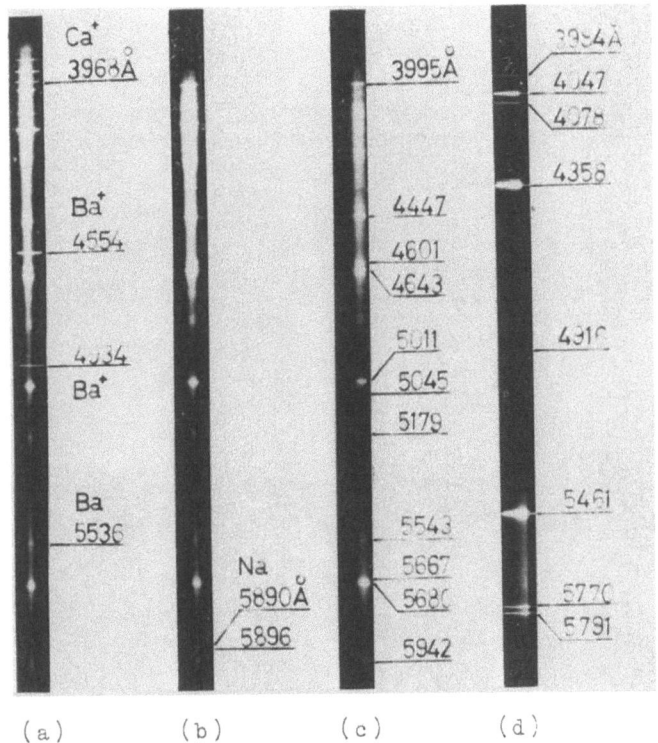

Fig. 3. The spectra of surface breakdown plasmas (a) Barium crown glass (b) Light flint glass (c) Air (d) Hg

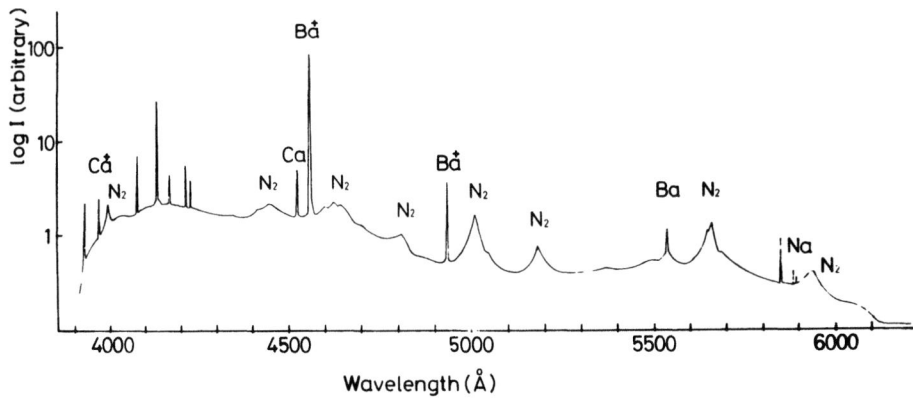

Fig. 4. Densitometric spectrograph of barium crown glass fracture

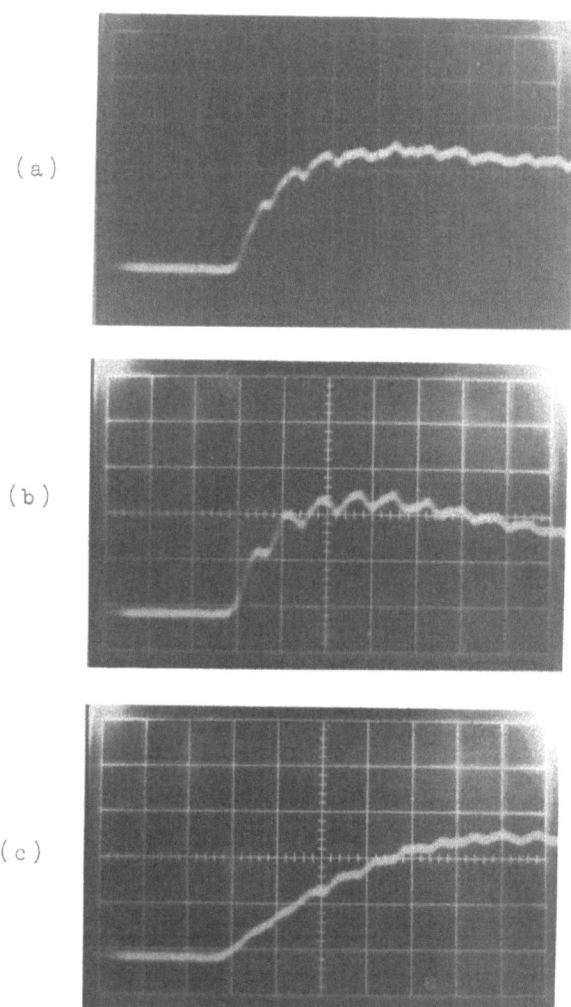

Fig. 5. Time trace of various line spectra.
(50 ns/div.)
(a) Ba$^+$ (4,554 Å)
(b) Ca$^+$ (3,968 Å)
(c) N$_2$ (5,011 Å)

As shown in Table 1, SiO_2 coating shows a slight decrease while MgF_2 coating reduces the damage threshold down to one third of the original value.

Sample	Damage Threshold	
Barium Crown Glass Original	37 ± 5 23 ± 5	J/cm^2
Ditto SiO_2 Coating (300mµ)	26 ± 5 30 ± 5	J/cm^2
Ditto M_gF_2 Coating (300mµ)	10 ± 5 9 ± 5	J/cm^2
Quarz	84 J/cm^2 over	

Table 1. Effect of Surface Coating
Test with multi mode laser. Damage threshold is five times larger with single mode laser.

Conclusion

We performed the experiments of glass damage by Q-switched laser beam to investigate the damage mechanisms. The main conclusion as follows.

(1) From the passive test we determined the damage threshold of platinum inclusion, platinum free and surface of barium crown glass. They were 12 ± 3 J/cm^2, 400 J/cm^2 and 28 ± 5 J/cm^2, respectively with multi mode laser.

(2) It is desirable to use the glass matrix which is mechanically strong in order to protect the growing of damage.

(3) Platinum included glass sample showed the difference of damage thresholds between passive and active test but platinum free glass had a same threshold.

(4) The plasma formation was always observed whenever the damage was induced. The mechanism of damage will be mainly due to the shock impulse from plasma.

(5) The fracture on the exit surface was more serious than that of the entrance surface. The shock wave caused by plasma on the exit surface will be more effective to produce damage.

(6) Time resolved photographic and spectroscopic methods were adopted to investigate the plasma behavior. The

invitation of plasma seems to be due to the ionization of alkaline atoms.

(7) The damage threshold on entrance surface increased from 28 ± 5 J/cm^2 to 40 ± 5 J/cm^2 with multi mode laser and 88 ± 5 J/cm^2 to 140 J/cm^2 with single mode laser by the treatment of hydro-fluoric acid.

(8) The chemical strengthen treatment will have negative effects on the damage growth on surface.

(9) Surface coating shows the serious reduction of damage threshold.

References

1. C. Yamanaka et al, IQEC 20-3, 404 (1970), Conf. on damage of laser materials, Boulder (1971).
2. J. E. Swain, Damage in Laser Glass, ASTM STP 469 69 (1969).
3. J. E. Swain, J. Quant. Elec. QE-4 362 (1968).
4. J. David, J. Decoux, J. Gautier and M. Souli, Revue de Physique Applique 3 118 (1968).

PLASMA PRODUCTION WITH A Nd LASER AND NON-THERMAL EFFECTS[+]

K. Büchl, K. Eidmann[++], P. Mulser,
H. Salzmann and R. Sigel

Max-Planck-Institut für Plasmaphysik
Euratom Association, Garching, Germany

INTRODUCTION

With the improvement of high power lasers it has become possible to produce plasmas of very high density and temperatures which possibly are of interest for thermonuclear research. Recently substantial neutron emission from such plasmas has been reported[1-6]. In order to investigate the future possibilities in this field of plasma physics detailed understanding of the properties of these plasmas is necessary. In this paper we report results obtained by using a variety of diagnostic techniques applied to plasmas produced within a wider range of laser power and show the evidence of non-thermal effects occurring in laser produced plasmas at moderate energies.

EXPERIMENTAL SETUP

The laser used is a multi-stage neodymium glass laser constructed in our laboratory. The nanosecond oscillator is Q-switched by a Pockels cell. A pulse shaping system consisting of a laser-triggered spark gap, a

---

[+] Presented at the Second Workshop on "Laser-Interaction and Related Plasma Phenomena" at Rensselaer Polytechnic Institute, Hartford Graduate Center, August 3o - September 3, 1971.
[++] on leave from University Marburg/Lahn

Kerr cell, and two crossed polarizers produce rectangular laser pulses of 1o nsec duration with a rise time less than 2 nsec. These pulses could be amplified up to 25 J by passing through five Nd glass rods, corresponding to a power of 2.5 GW. The laser pulses which have a beam divergence of about 1 mrad are focused onto solid hydrogen and deuterium targets by a spherical lens of 15 cm focal length and an aspherical one with f = 5cm[7]. Performance of the laser system without damage to the optical components was possible only by inserting two optical isolators (Faraday rotators consisting of Quarz rods together with sets of glass discs under the Brewster angle working as polarizers). Thereby the light reflected back from the target into the amplifier chain is attenuated by more than a factor of 10^4.

The different diagnostic methods which are partly described in previous papers[8,3] are shown schematically in Fig. 1.

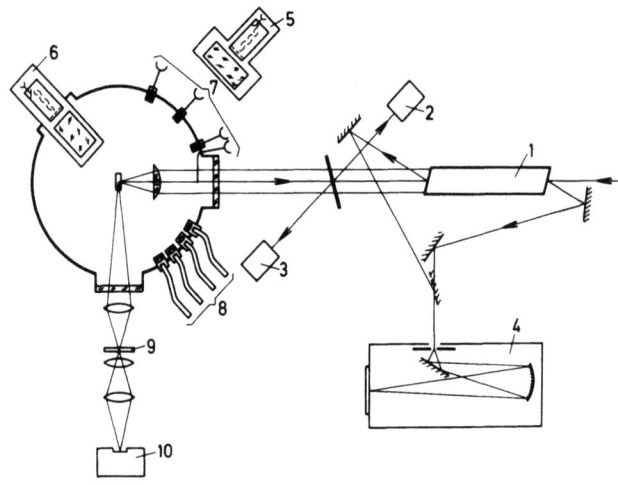

Fig. 1

Scheme of the diagnostics: 1 last amplifier rod, 2,3 photodiodes, 4 grating spectrograph, 5,6 neutron counters, 7 probes, 8 X-ray diagnostics (Be, Al foils, scintillators, light pipes), 9 streak slit, 1o image converter streak camera.

An image converter camera with a time resolution of 1 nsec was used to take streak photographs of the expanding plasma and to ensure that no plasma is produced by a forerunning laser pulse. A time mark generated by the incident pulse allowed the light emission of the plasma to be exactly pinpointed in time relative to the laser pulse. The pulse shape of the incident and reflected light was recorded using calibrated vacuum photodiodes. Part of the incident and reflected light was supplied to a 2 m grating spectrograph with a resolution of 0.5 Å.

The electron temperature was determined from the bremsstrahlungscontinuum using the well-known absorber method. The X-ray emission from the plasma was transmitted through four differently thick beryllium or aluminum windows respectively, then converted into visible light by plastic scintillators and fed by light pipes to four photomultipliers with a time resolution of about 1o nsec[9]. Extensive tests of this method were made with a carbon target under a wide range of conditions (pulse duration, laser energy, focal spot diameter). For time of flight and charge collection measurements up to six ion collecting probes under angles of 22.5° between two consecutive probes were used to measure the angular distribution of ion flux and asymptotic kinetic ion energy. The 0° probe was looking through a central hole in the focusing lens.

Two scintillation detectors with time resolution of 5 - 1o nsec monitored the neutron emission from the plasma. The counter located inside the vessel was at a distance of 4 cm from the target, the other detector was positioned outside at distances variable between 34 and 1oo cm. Calibration was performed by substituting the target by a Be-Pu neutron source. Both counters were shielded by 5 mm thick lead sheets. The different diagnostics were used simultaneously shot by shot.

EXPERIMENTAL RESULTS

When the laser pulse is impinging on the solid target a rapidly expanding plasma cloud forms. The measurements of the reflected light, the electron temperature, the ion currents and energies, and the number of neutrons were performed at laser energies varying from 5 to 25 J. The pulse shape and the intensity of the reflected light depend on the incident light flux. The highest degree of reflection and the most reproducible time behavior of the reflected intensity were observed at the highest light flux when the f = 5 cm aspherical lens was used. In some shots the reflectance reached values up to 3o %

at the beginning of the pulse. At the lower energy limit an irregularly spiking behavior of reflection signals appeared. The spectral resolution of the reflected light taken at 1o J laser energy (corresponding to a flux density of $5 \cdot 10^{12}$ W/cm^2) showed a line shift of 2.5 Å towards longer wavelengths (Fig. 2).

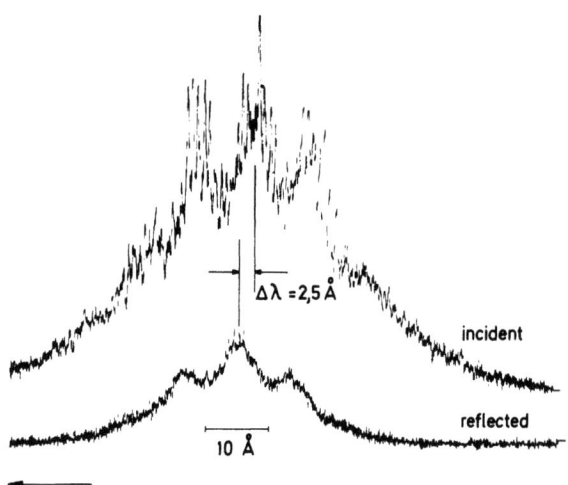

Fig. 2

Photometer traces of incident and reflected laser light (λ = 1.06 u) from a laser produced deuterium plasma. The incident laser line consists of a great number of lines grouped in three maxima; in the reflected light this fine structure has disappeared.

It was interpreted as a Doppler shift of the pulse reflected from the critical layer (where $\omega_p = \omega$) moving in the direction of the incident beam with a velocity of 3.5×10^6 cm/sec. (The observed red shift is not in contradiction to the blue shift measured by other authors. It is a question of the outstreaming velocity of the reflecting layer with respect to the shock velocity, i.e. of intensity and geometry, whether red or blue shift takes place.)

The number of ions and their asymptotic kinetic energy was determined by probe measurements. For typical experimental conditions (1o - 25 J focused with the f = 5 cm aspherical lens) the mean kinetic energy of the ions averaged over all directions is 1.5 - 2 keV, the mean kinetic energy of the 0° probe is 2.5 - 3.5 keV. However, fast ions with energies up to 2o keV are also observed.

A characteristic example of the angular distribution of particle number and their energy obtained with the spherical f = 15 cm lens is given in Fig. 3. From such measurements an energy balance can be established by integration over the halfspace. It was found that about 80 % of the incident laser energy is contained in the plasma. Taking into account the measured energy loss due to reflection of laser light and the amount of energy transferred to the compressed material by the shock wave in the target the whole incident laser energy is recovered[8].

Whereas the appearance of fast ions up to 20 keV could be considered as a first indication that non-thermal effects may occur in laser produced plasmas from solid targets the measurement of the electron temperature reveals clearly such non-thermal anomalies.

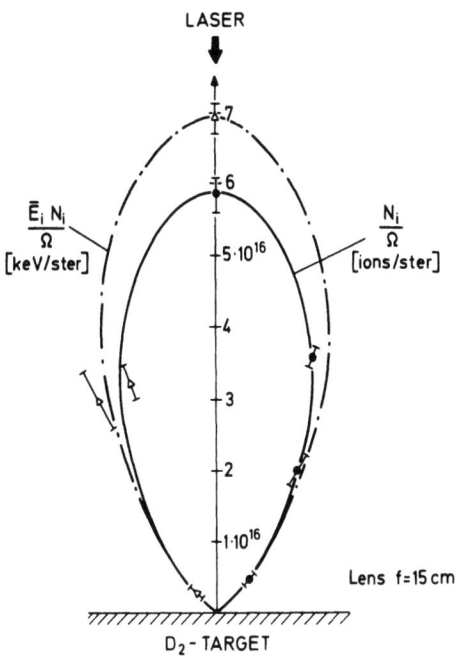

Fig. 3

Angular distribution of ion number and asymptotic kinetic energy measured by ion collecting probes. Laser energy 12 J in 10 nsec.

In Fig. 4 the X-ray intensities passing through four differently thick Be and Al foils are reported for a 20 J 10 nsec laser pulse. The measured values are lying between the two curves for T_e = 500 eV and T_e = 2 keV which are calculated under the assumption of a Maxwellian distribution. It results clearly that the slope of the dotted curve which should give the electron temperature of the plasma varies from point to point leading to temperature values from 500 eV up to about 2 keV. If we identify the lowest value of 500 eV with the electron temperature, this means that the X-ray emission is enhanced up to a factor of 10^5 towards the high energy end of the measured interval (15 keV). If the laser energy is lowered then this non-thermal radiation dominates the whole measuring range and simulates to high temperatures because in that case the dotted curve of Fig. 4 is shifted to the left hand side and the long wavelength X-ray radiation can no longer penetrate the thinnest foil. In addition, a strong correlation between the X-ray intensity and the intensity of the reflected light has been observed. The correlation is especially striking at low laser energies when the intensity and the pulse shape of the reflected light vary strongly from shot to shot.

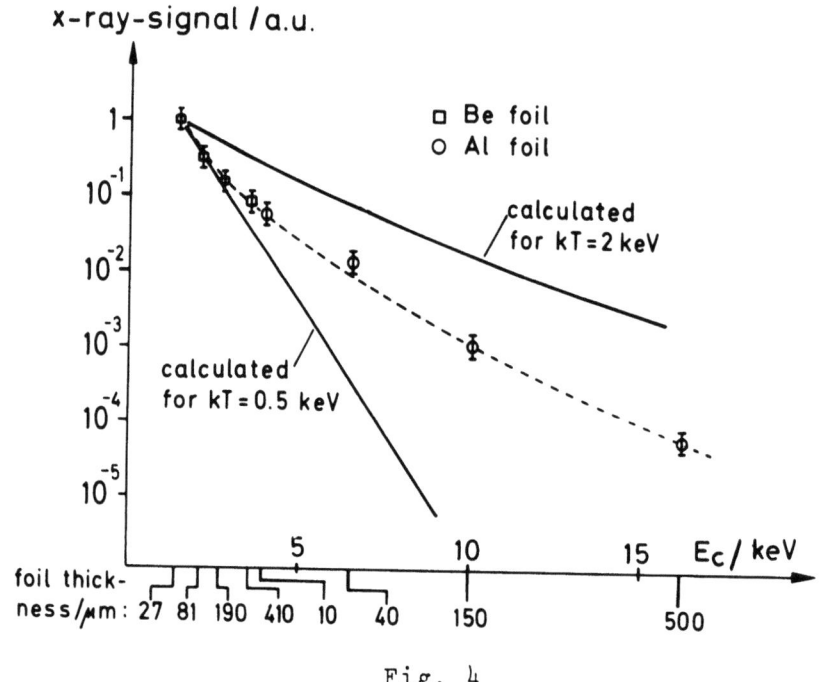

Fig. 4

Emitted X-ray intensities from the plasma as a function of cut-off energy E_c of four Be and Al foils of different thickness. The slope of the dotted line which fits the experimental points is lying between those of curves for T_e = 500 eV and T_e = 2 keV. Laser energy is 20 J delivered in 10 nsec.

Neutron emission from plasmas produced with our experimental device has also been observed. On both counters in more than 80 % of the shots neutron emission was registered. A typical oscilloscope trace of one of the counters is given by Fig. 5.

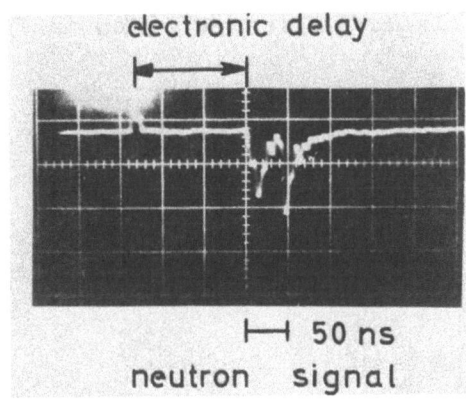

Fig. 5

Neutron signal of one of the counters

To be sure that the signals really monitor D-D fusion reactions the following facts can be considered as a proof:

(a) If solid hydrogen instead of solid deuterium was used as target material the signal disappeared under otherwise identical conditions.

(b) From time of flight measurements with the counter located outside of the vacuum vessel the neutron energy was determined to be 2.45 MeV.

(c) Neutron signals were never observed before the laser pulse was impinging on the deuterium target.

On the average a few hundred neutrons are emitted per shot; the maximal neutron number was 10^3. Only about 20 neutrons were registered in the first 10 nsec of pulse duration. For the time behavior of the neutron signals see[3], Fig. 4. The neutron emission threshold was as low as 5 J for the f = 5 cm aspherical lens and 10 J for the f = 15 cm conventional lens.

To see whether the detected neutrons were of thermal origin or not numerical calculations were performed in different one-dimensional geometries. Since most reasonable agreement with experiments was found for spherical geometry only results for this case are reported here. Two temperatures, T_e and T_i were introduced, and heat conduction and absorption of laser energy were taken into account locally. The spatial distribution for electron and ion temperatures T_e, T_i, particle density n_e, laser intensity $\phi.F$, and the neutron production rate over the mass coordinate or the radius respectively for a 1o J, 1o nsec laser pulse are given in Fig. 6 at the

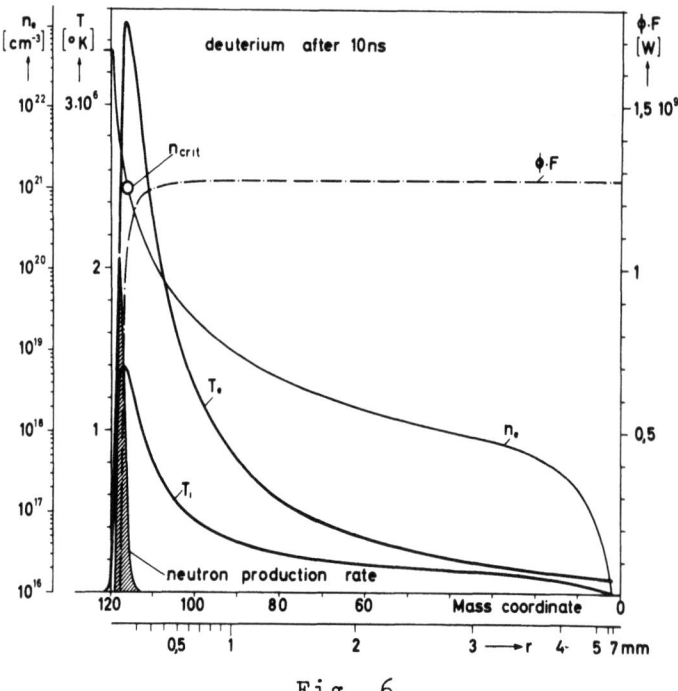

Fig. 6

Distribution of electron and ion temperatures T_e, T_i, electron (= ion) density n_e, light flux $\phi.F$ and neutron production rate (arbitrary units) over the mass (Lagrange) coordinate or the radius (nonlinear scale) at the end of a 1o J, 1o nsec Nd laser pulse.

end of the pulse. Whereas electron temperature and total particle number are in good agreement with the measured quantities for such a pulse ($T_{e,exp}$ = 42o eV, $T_{e,calc}$ = 35o eV, N_{exp} = 4 x 1o^{16}, N_{calc} = 3.6x 1o^{16}) a strong disagreement has been found for the neutron production since the hydrodynamic calculation gave less than 1o^{-2} thermal neutrons per shot.

To investigate further the non-thermal behavior of the plasma, we repeated the temperature and neutron emission measurements with the target surrounded by a low density gas rather than by a high vacuum. Tests were made to ascertain that the background gas had no influence either on the laser beam by absorption or deflection, or on the target itself. With neutral gas added the plasma behaviour was influenced in two respects: Firstly, the number of neutrons produced is rapidly reduced to zero with increasing gas pressure, as shown in Fig. 7a. Secondly, the hard X-ray intensity decreases. Already at a pressure of 3×10^{-2} torr the experimental points fit the calculated curve for a thermal plasma with $T_e = 420$ eV (see Fig. 7b). The measurements of Fig. 7a,b were performed in helium, however, deuterium as a background gas gave essentially the same results.

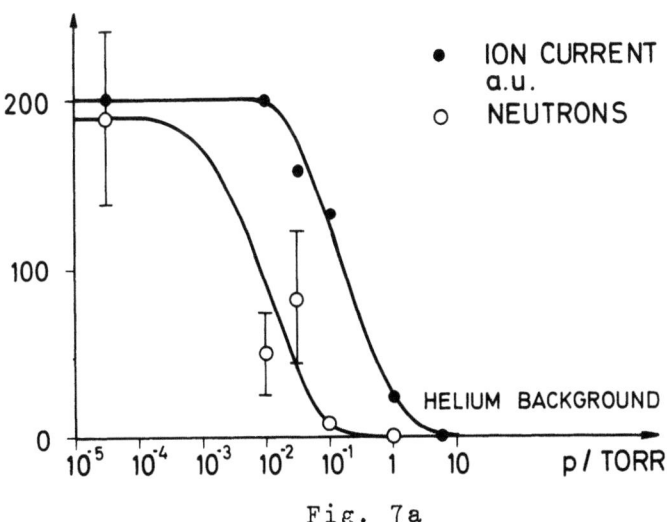

Fig. 7a

Mean number of neutrons per shot emitted into 4π and the ion current as functions of helium background pressure. Mean laser energy 11 J.

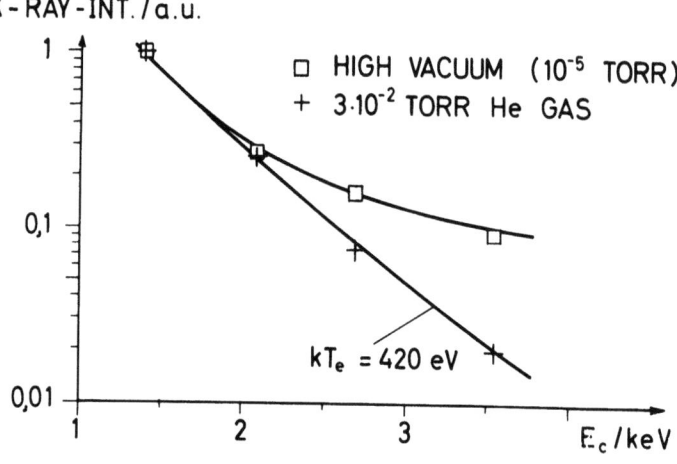

Fig. 7b

X-ray intensity versus photon cut-off energy of the beryllium absorber foils. High vacuum (10^{-5} torr) and helium background pressure 3×10^{-2} torr. Lower curve is calculated for thermal bremsstrahlung emission with T_e = 420 eV. Mean laser energy 11 J.

The nature of the processes which under high vacuum conditions lead to enhanced X-ray and neutron emission, and which are heavily damped in the presence of a background gas is not yet understood. Nevertheless, it is possible to decide from these observations whether the observed fusion reactions are of thermonuclear origin or not.

From gasdynamic theory we expect thermal fusion reactions to occur at densities as high as the critical density n_{crit} where the laser frequency equals the plasma frequency $n_{crit} = 10^{21}$ cm^{-3} for a wavelength of λ = 1.06 μ) or even in the more dense interior of the plasma, which is heated by electronic heat conduction (see Fig. 6). It is hard to imagine that this high density region of the plasma is influenced by the applied low density background gas. This is confirmed by the fact that the following measured quantities remain essentially unaffected:

(1) The total intensity of the low energy X-rays as measured with the thinnest absorber foil (see Fig. 7b).

(2) The electron temperature or, more precisely, the mean energy of the bulk of the electrons. This follows from the fact that the slope of the measured curve is conserved for low energy X-rays (see Fig. 7b).

(3) The total number and mean energy of the ions at least up to a gas pressure of 3×10^{-2} torr (see Fig. 7a).

(4) The reflection coefficient of the plasma for the incident laser light.

Therefore, the following conclusions can be drawn: The electron temperature has to be calculated from the curve slope of the thinnest foils. This is suggested, too, by the comparison with the ion probe measurements and the computational results.

The number of thermal fusion reactions in the high density region of the plasma is undetectably small compared with the total number of fusion reactions observed. This is in agreement with numerical calculations, which do not predict a measurable amount of fusion reactions under our experimental conditions. Thus, for the laser intensities applied here thermonuclear conditions are not achieved. A number of observations indicate that the origin of the observed fusion reactions must be sought in the fact that the plasma is not in a purely thermal state.

REFERENCES

1. F. Floux, D. Cognard, L.G. Denoeud, G. Piar, D. Parisot, J.L. Bobin, F. Delobeau, F. Fauquignon, Phys. Rev. $\underline{1}$, 821 (1970).
2. M. Lubin, Astronautics and Aeronautics, Nov. 1970, p. 42.
3. K. Büchl, K. Eidmann, P. Mulser, H. Salzmann, R. Sigel, S. Witkowski, Fourth Conf. on Plasma Physics and Controlled Nuclear Fusion Research, Madison, Wisconsin, June 1971, Paper CN 28-D-11.
4. N.G. Basov, V.A. Boiko, S.M. Zakharov, O.N. Krokhin, G.V. Sklizkov, JETP Lett. $\underline{13}$, 691 (1971).

5. C. Yanamaka, T. Yamanaka, Hyung-Boo Kang, K. Yoshida, in Progr. Rep. VI, Lab. for Plasma and Quantum Electronics, Aug. 1o, 1970, p. 6.
6. S.W. Mead, R.E. Kidder, J.E. Swain, LLL Rep. Livermore UCRL-73356, Aug. 17, 1971.
7. R. Sigel, H. Krause, S. Witkowski, J. Phys. E: Sci. Instrum. $\underline{2}$, 187 (1969).
8. R. Sigel, Z. Naturforschg. $\underline{25a}$, 488 (197o).
9. R. Sigel, S. Witkowski, H. Baumhacker, K. Büchl, K. Eidmann, H. Hora, H. Mennicke, P. Mulser, D. Pfirsch, H. Salzmann, Institut für Plasmaphysik, Report IV/9 (1971).

INFLUENCE OF FAST ION LOSSES IN INERTIALLY CONFINED

NUCLEAR FUSION PLASMA[+]

H. Hora[++] and D. Pfirsch

Max-Planck-Institut für Plasmaphysik

Euratom Association, Garching, Germany

ABSTRACT

The gain G of the thermonuclear fusion energy of very dense spherical plasmas heated by lasers in times of about 1 nsec is treated theoretically. Assuming as a pessimistic estimate that fast ions leave the plasma without contributing to fusion neutrons, we find that for G = 1 laser energies of 3×10^7 joules are needed. A kinetic theory for a one species gas starting with a Gaussian density and velocity profile yields almost the same result as found by a hydrodynamic model with adiabatic expansion and without fast ion losses taken into account. In the latter case the even-point of G = 1 is reached at the well-known value for the laser energy of 4.6×10^6 joules for a D-T reaction.

[+] Presented at the Second Workshop on "Laser Interaction and Related Plasma Phenomena" at Rensselaer Polytechnic Institute, Hartford Graduate Center, August 3o - September 3, 1971.
[++] Also from Rensselaer Graduate Center, Hartford, Conn. U.S.A.

INTRODUCTION

Two different ways of attaining thermonuclear conditions are mainly studied at present: one is to heat a low-density plasma rather slowly and confine it sufficiently long by magnetic fields; the other is to heat a high-density plasma very rapidly, which then expands freely, this being called inertial confinement. The latter might be possible by using laser pulses[1], as is discussed here, or, by means of intense electron beams[2]. The importance of fast heated, inertially confined high-density plasma for controlled thermonuclear fusion was pointed out by Linhart[7]. This paper continues the discussion of some theoretical aspects that was started some years ago[3-6], to calculate the energies necessary to reach the even-point of energy production, i.e. where energy production equals energy input. Besides this concept of inertially confined plasma, we shall not discuss the aspects of laser ignated shock waves with stationarily driven fusion fronts[8] or implosion waves[9] driven by lasers[10]. In this paper the influence of fast ion losses in inertially confined plasma is discussed.

Being aware of the complexity of the whole subject, especially of nonlinear effects[11,12] at high laser intensities, we treat here a much simplified model, where we start from a spherical D-D or D-T plasma of volume V_o and density N_o to which an energy E_o has been transferred within 10^{-9} sec to attain thermal equilibrium of electrons and ions (Fig. 1). The equipartition time between electrons and ions based on two-body Coulomb-collisions is of the right order of magnitude. The ion temperature reached 9o % of the electron temperature (exceeding 7 keV) within 10^{-10} seconds according to Lubin's[13] numerical calculations. At high densities, however, the equipartition can be concluded to be even faster[12,14]. The following adiabatic free expansion and the created fusion reactions have been calculated before[6] without allowance for the possibility that the fast ions may leave the plasma without having produced a fusion reaction. We will treat here the influence of these losses.

Two models will be considered. In the first one the dynamics are described by a hydrodynamic theory as in previous work[3-6], and for fusion reactions, furthermore, certain assumptions about the loss of fast ions are made. These assumptions are probably too pessimistic as can be seen from our second model, in which the dynamics are described by a one species kinetic theory including collisions. The time dependent problem with initial Gaussian density profile and a Gaussian velocity distribution is solved exactly. The fast ion losses are then automatically taken into account. It turns out that the result for the fusion gains is exactly the same as found from a hydrodynamic model with the same initial density profile and temperature, but without assuming any fast ion losses.

HYDRODYNAMIC MODEL WITH LOSSES

An amount of energy E_o (Fig. 1) may be transferred within 10^{-9} sec into a pure deuterium or deuterium-tritium plasma of solid state density. The plasma will then expand adiabatically from an initial radius $R=R_o$ and an initial temperature T_o. This expansion is governed by the law[15]

$$\dot{R} = \left\{ \frac{5kT_o}{m_o} (1-(R_o/R)^2 + \dot{R}_o^2 \right\}^{1/2} \quad (1)$$

where $\dot{R}_o=0$ was used and k is Boltzmann's constant, m_o is the averaged mass of the plasma particles (electrons and ions), and $R=R(t)$ is the actual radius of the sphere at the time t. Using Eq.(1) together with the adiabatic law for the temperature T, we find for the fusion gain

$$G = \frac{\varepsilon_F}{E_o} \int_0^\infty dt \int_0^{R(t)} dx\, dy\, dz\, \frac{n(R(t))^2}{A} \langle \sigma v \rangle \quad (2)$$

where ε_F is the energy released by one fusion reaction, A is a factor equal to 2 for pure deuterium and equal to 4 for the D-T mixture. To determine the average value $\langle \sigma v \rangle$, where σ is the reacting cross section and v the relative velocity of two ions, we use for v a Maxwellian distribution modified in such a way as to take into account that the fast ions leave the plasma without reacting. This modification is found in the following way: The averaged distance $\langle \ell_F \rangle$ which an ion with a velocity v travels until it produces a fusion reaction is, because of numerous elastic collisions leading to a kind of Brownian motion of the ion

$$\langle \ell_F \rangle = \ell_c \sqrt{\nu_c / \nu_F} \quad (3)$$

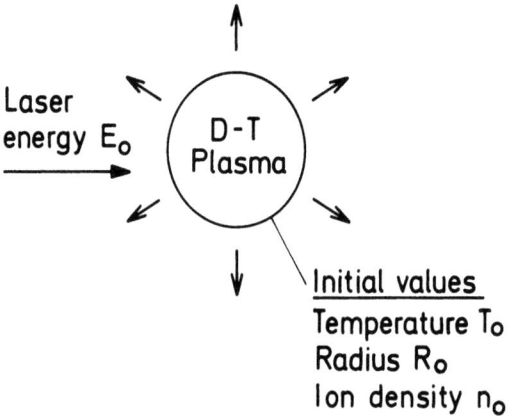

Fig. 1

Laser energy E_o is transferred to a D-T plasma producing a box-like density profile at given initial values for a following free expansion.

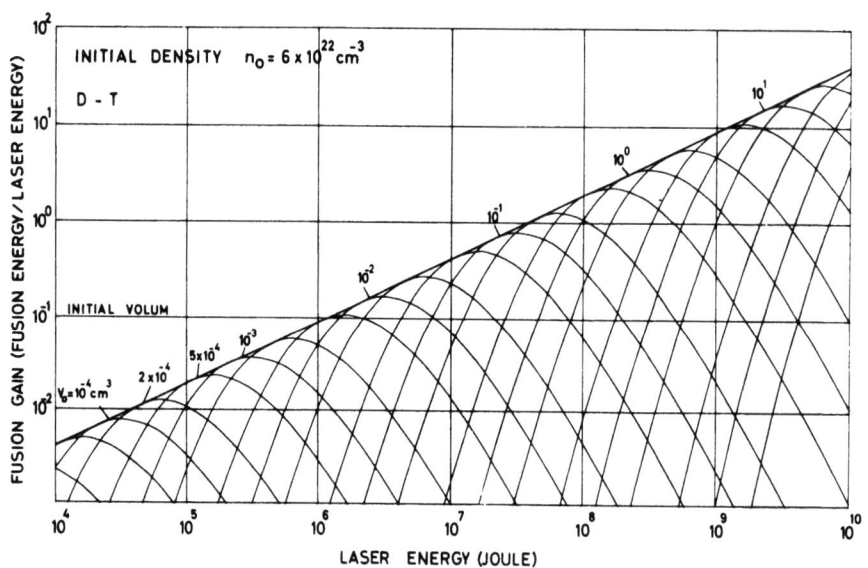

Fig. 2

Fusion gain G with ion losses at varying initial values of the laser energy E_o and volume V_o for an initial density of solid deuterium $n_o = 6 \times 10^{22}$ cm^{-3}.

where ν_F is the reaction frequency for fusion processes, ν_c the ion collision frequency, and ℓ_c the mean free path for elastic collisions. We have to compare $\langle \ell_F \rangle$ with the averaged mean distance $\bar{R}$ between the surface of the actual sphere and the considered point in the plasma. The averaged distance $r_m(r_1)$ of the surface to a point of radius r_1 within the sphere of radius R is

$$r_m = \frac{1}{2R} \int_0^\pi d\varphi_1 \, r(r_1, \varphi_1) \sin\varphi_1$$

where r is the actual distance between r_1 and the surface. We find

$$r_m = \frac{1}{2} + \frac{R^2 - r_1^2}{r_1} \ln \sqrt[4]{\frac{R+r_1}{R-r_1}}$$

because of identical vanishing of an elliptical integral at the given limits of the integral. The average value of r_m for all points within the sphere is

$$\langle r_m \rangle = \bar{R} = \int_0^R dr_1 \, 3\frac{r_1^2}{R^3} r_m = \frac{3}{4}R$$

If $\langle \ell_F \rangle$ is less than $\bar{R}$, which represents an effective radius in the sense of averaging over position and flight direction of the particles, the probability that the particle will react is unity, and if $\bar{R} \ll \langle \ell_F \rangle$ the probability is $\bar{R}/\langle \ell_F \rangle$. For fusion gain resulting from Eqs.(1) and (2) we obtain by methods similar to those used before[6]

$$G = \frac{\varepsilon_F}{E_o} n_o V_o \sqrt{\frac{m_o}{5kT_o}} \int_{R_o}^\infty \frac{dR}{(1-(R_o/R)^2)^{1/2}} \langle \sigma v \rangle \frac{n}{A} \quad (4)$$

with

$$\langle \sigma v \rangle = \left\{ \int_0^\infty Min\left(1; \frac{\bar{R}(R)}{\langle \ell_F(n,R,v) \rangle}\right) \cdot \sigma(v) v^3 \exp\left(-\frac{mv^2}{2kT}\right) dv \right\} /$$

$$/ \left\{ (2kT(R)/m)^{3/2} \pi^{1/2}/4 \right\} \quad (5)$$

where $\bar{m}$ is the reduced mass of the colliding nuclei. The variation of n and T is given by the adiabatic laws

$$n = n_o (R_o/R)^3, \qquad T = T_o (R_o/R)^2$$

The results of the calculation of G for varying laser energy E_o and different initial volumes V_o at the initial density $n_o = 6 \times 10^{22}$ cm^{-3} of the solid materials are shown in Fig. 2. The maximum gain for different values of V_o increases with very good approximation as $E_o^{2/3}$, while the corresponding law for the case without losses is $E_o^{1/3}$. The even-point, G = 1, is reached with losses at $E_o = 3.5 \times 10^7$ joules, while without losses G was 1.6×10^6 joules [6]. From a set of plots of the kind in Fig. 2 we derive a formula for the maximum G valid for 10^3 joules $< E_o < 10^{10}$ joules, and 10^{20} cm$^{-3} < n_o < 10^{25}$ cm^{-3}, for a D-T plasma

$$G = \begin{cases} 4.42 \times 10^{-36} E_o^{2/3} n_o^{4/3}, & \text{if } E_o n_o^2 \lesssim 3.7o \times 10^{55} \text{ j/cm}^6 \\ \\ 1.47 \times 10^{-17} E_o^{1/3} n_o^{2/3}, & \text{if } E_o n_o^2 \gtrsim 3.7o \times 10^{55} \text{ j/cm}^6 \end{cases} \qquad (6)$$

For a deuterium plasma we find for the same region of validity

$$G = 3 \times 10^{-39} E_o^{2/3} n_o^{4/3} \qquad (7)$$

The splitting of G into two ranges, Eq.(6), expresses immediately the influence of the fast ion losses. The upper line with a stronger increase of G on E_o and n_o, is the branch of losses, while the lower line is identic with the former case [6] with hardly any fast ion losses. At the limits of the range of validity the formulae may deviate from the numerical results by up to 1o %.

The maximum temperature T_o of the plasma at the time of beginning expansion can be derived very simply from the results of Fig. 2 with allowance for

$$T_o = \frac{1}{k} \frac{E_o}{2 n_o V_o}$$

where the factor 2 is due to the fact that electrons and ions gain the same energies. At the even-point for D-T in Fig. 2 we have, for example, $T_o = 12$ keV. This result also justifies neglecting the depletion of the fuel by burning because energy gained from each fusion process is more than a thousand times higher than the temperature and because the fusion gain G is of the order of one or less.

KINETIC MODEL

It is well known in hydrodynamics from numerical calculations that an initially boxlike density profile becomes Gaussian-like[16]. The same happens if the initial density profile is triangular[16]. Besides this numerically shown asymptotic behaviour, it was shown analytically[14] that only an initially Gaussian density profile of spatially constant temperature conserves this profile self-similarily in a strictly mathematical sense. Self-similarity is also conserved for elliptical targets[17] and the variation of the cross section of the target compared with that of the laser focus can be included by an iteration[14]. Because of these facts and because it is not completely known whatever the real initial state of a plasma produced by laser is, the approximation of the dynamics by a self-similarity expansion of Gaussian density profiles appears to be justified. In the following kinetic treatment, we therefore choose this type of initial conditions.

In order to get information about the effect of the losses of fast ions, we use a simplified model where the dynamics of only one particle species, namely the ions, is taken into account. From the results obtained in this way we refer to what might occur in a real plasma in which the electrons also contribute to the pressure. In the final numerical results, these electron pressure effects are taken into account in such a way that they act as a further gasdynamic component. Neglecting the dynamics of the electrons and ion-electron collisions is justified by the small electron mass.

We are dealing with a distribution $f(x,\underline{v},t)$ for the ions obeying a kinetic equation

$$\frac{\partial}{\partial t} f + \underline{v} \frac{\partial}{\partial x} f = \left(\frac{\partial f}{\partial t}\right)_{coll} \tag{8}$$

The collision term is assumed to be a local one, describing only ion-ion collisions. It is easy to find a solution which makes both sides of Eq.(8) equal to zero. To make the left-hand side vanish, f must depend only on the constants of motion $\underline{v}$ and $\underline{x}-\underline{v}t$. The right-hand side of the additive collisional invariants $\underline{v}$ and $\underline{v}^2$. Such a function is

$$f = \frac{n_o}{(2\pi kT_o/m_i)^{3/2}} \exp\left\{-\frac{1}{R_o^2}(\underline{x}-\underline{v}t)^2 - \frac{1}{kT_o}\frac{m_i}{2}v^2\right\} \tag{9}$$

where T, R and n_o are constants and m_i is the ion mass. Equation (9) can be rewritten in the form

$$f = \frac{n_o}{(2\pi k\theta/m_i)^{3/2}} \left(\frac{\theta}{T_o}\right)^{3/2} \exp\left\{-\frac{x^2}{R_o^2}\frac{\theta}{T_o}\right\} \exp\left\{-\frac{m_i}{2k\theta}\left(v - \frac{2k\theta}{m_i R_o^2}t\underline{x}\right)^2\right\}$$

(10)

$$\theta = \frac{T_o}{1 + \frac{2kT_o}{m_i R_o^2}t^2}$$

It can be shown that the dynamic expansion is like an adiabatic one with the ratio of the specific heats $\gamma = 5/3$. The local fusion rate is

$$\frac{1}{2}n^2\langle\sigma v\rangle = \frac{1}{2}\int d^3v_1 \int d^3v_2\, \sigma(|\underline{v}_1-\underline{v}_2|)|\underline{v}_1-\underline{v}_2| f(\underline{x},\underline{v}_1,t)f(\underline{x},\underline{v}_2,t)$$

(11)

Introducing new coordinates

$$\underline{v} = \underline{v}_1 - \underline{v}_2 = \left(\underline{v}_1 - \frac{2k\theta}{m_i R_o^2}t\underline{x}\right) - \left(\underline{v}_2 - \frac{2k\theta}{m_i R_o^2}t\underline{x}\right)$$

$$\underline{V} = \frac{1}{2}\left(\underline{v}_1 - \frac{2k\theta}{m_i R_o^2}t\underline{x} + \underline{v}_2 - \frac{2k\theta}{m_i R_o^2}t\underline{x}\right),$$

we obtain

$$\frac{1}{2}n^2\langle\sigma r\rangle = \frac{n_o^2}{2}\int d^3V\int d^3v\, v\, \frac{\exp\left[-\frac{m_i/2}{2k\theta}\underline{V}^2\right]}{(2\pi k\theta/(m_i/2))^{3/2}} \times$$

(12a)

$$\times \frac{\exp\left[-\frac{2m_i}{2k\theta}\underline{v}^2\right]}{(2\pi k\theta/(2m_i))^{3/2}} \exp\left[-\frac{2\underline{x}^2}{R_o^2}\frac{\theta}{T_o}\right]\left(\frac{\theta}{T_o}\right)^3$$

$$\frac{1}{2} n^2 \langle \sigma v \rangle = \frac{m_i^2}{2} \left(\frac{\theta}{T_o}\right)^3 \exp\left[-\frac{2x^2\theta}{R_o^2 T_o}\right] \int d^3v \sigma(v) \frac{\exp[-(m/2)\underline{v}^2/(2k\theta)]}{(2\pi k\theta/(m/2))^{3/2}}$$

(12b)

$$\frac{1}{2} n^2 \langle \sigma v \rangle = \frac{n_o^2}{2} \left(\frac{\theta}{T}\right)^3 \exp\left[-\frac{2\underline{x}^2}{R_o^2} \frac{\theta}{T}\right] \langle \sigma v \rangle_o = \frac{n^2(\underline{x},t)}{2} \langle \sigma v \rangle_o$$

(13)

the expression $\langle \sigma v \rangle_o$ is identical with the one in Eq.(5) if no attention is paid to losses of fast ions, i.e. the function Min is set equal to unity.

Using a normalization of n_o such that the total number of ions with a Gaussian profile is the same as these with a box profile, and such that the initial central densities are the same, the ratio of the corresponding fusion gains

$$\frac{G_{box}}{G_{Gauss}} = \frac{\int n_o^2 \, d\tau}{\int r_{Gauss}^2 \, d\tau} = 2^{3/2}$$

(14)

where G_{box} is calculated without allowance for the fast ion losses. This fusion gain G_{box} is therefore by definition the same as discussed earlier[6].

In Fig. 3 we report the results of $G = G_{Gauss}$ for a D-T plasma with an initial solid-state density. In Fig. 4 $G = G_{Gauss}$ is shown for pure deuterium. Interpolation formulae for the maximum fusion gains for 1 joule $< E_o <$ 10^{15} joule and 10^{16} cm^{-3} $< n_o <$ 10^{25} cm^{-3} are

$$G_{Gauss} = 3.93 \times 10^{-18} \, n_o^{2/3} \, E_o^{1/3} \quad (D-T)$$

$$G_{Gauss} = 1.79 \times 10^{-19} \, n_o^{2/3} \, E_o^{1/3} \quad (D-D)$$

Therefore, a D-T plasma will reach the even-point at $G_{Gauss} = 4.6$ MJ and for pure deuterium $G_{Gauss} = 2.4 \times 10^4$ MJ.

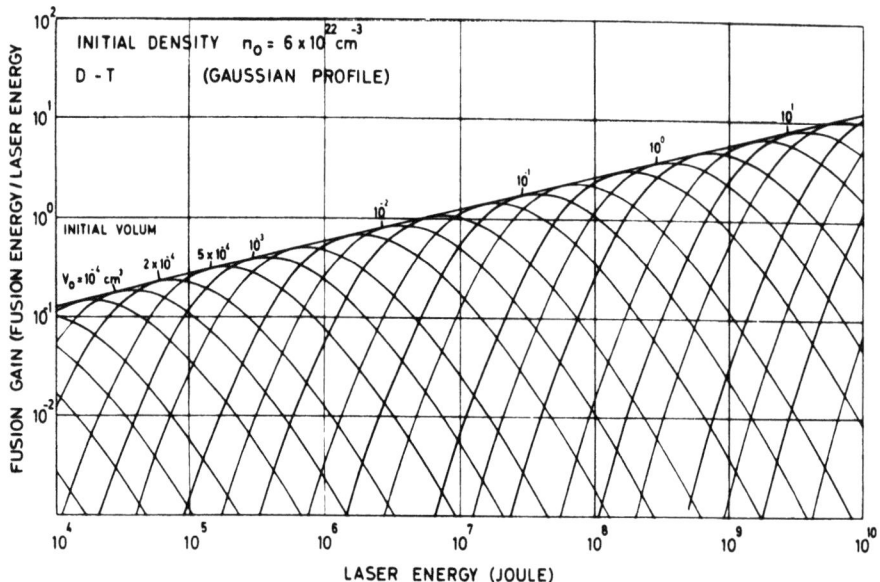

Fig. 3

Fusion gains $G = G_{Gauss}$ for a D-T plasma for varying laser energy E_o, volume V_o and for an initial density of $n_o = 6 \times 10^{22}$ cm^{-3}.

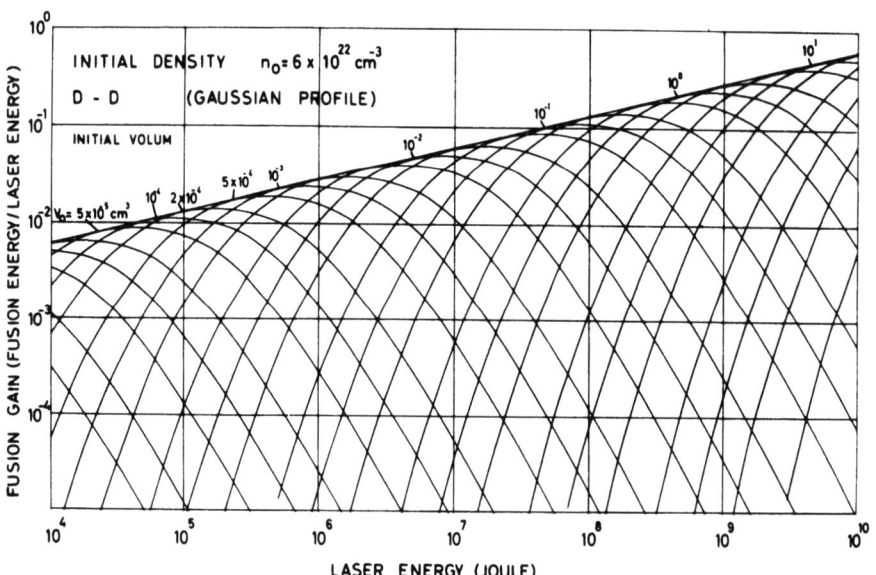

Fig. 4

Fusion gains $G = G_{Gauss}$ for pure deuterium for varying laser energy E_o and volume V_o and for an initial density of $n_o = 6 \times 10^{22}$ cm^{-3}.

CONCLUSIONS

We have shown that according to a rather pessimistic estimate laser energies of about 3×10^7 joules are necessary to reach the even-point, or according to a probably more realistic estimate only 4.6×10^6 joules. Any reactor concept would need not only to reach the even-point but also, for example, a minimum fusion gain G=3. Because of the dependence of G on the cubic root of E_o we find in the realistic case a E_o of at least 1.24×10^8 joules. But even such figures are perhaps not beyond future capabilities, especially as regards chemical lasers.

Comparing the results with those of other authors, we have to take into account the possibility of errors in using measured values of the fusion cross section $\sigma(v)$. We have used the values of σ from[18] and reproduced Tuck's values[19] of $\langle \sigma v \rangle$ from our program. In the dominant range (at ion energies of 40 keV), we can approximate $\sigma \sim v^6$. Therefore an error of 10 % for v results in an error of 60 % in σ and consequently also in G, and in an error of about factor 5 in E_o. The agreement of the computed energy result of G with that of Engelhardt[5] without considering the loss of fast ions, was pointed out[6], where the same initial conditions had to be presumed. Our higher fusion gains[6] at other initial conditions compared with Engelhardt[5] were due to optimation by the computer program. Therefore our much more general results presented here are a consistent continuation of the former treatments[5,6]. A comparison with the results of other treatments, including other details than considered here, see e.g. [8,13], was given by Spalding[20].

REFERENCES

1. F. Floux, D. Cognard, G. Denoeud, G. Piar, D. Parisot, J.L. Bobin, F. Delobeau, C. Fauquignon, Phys. Rev. $\underline{A1}$, 821 (1970); N.G. Basov, O.N. Krokhin and P.G. Kriukov, Paper 2.1, Sixth Int. Quantum Electronics Conf. Kyoto, Sept. 1970.
2. F. Winterberg, Phys. Rev. $\underline{174}$, 212 (1968); Bull. Amer. Phys. Soc. $\underline{15}$, 1453 (1970), W.H. Bennett, USA Pat. 3, 526, 575 (Sept. 1970).

3. N.G. Basov and O.N. Krokhin, Proc. <u>3rd Int. Quantum Electronics Conference, Paris 1963</u>; N. Bloembergen
4. J.M. Dawson, Phys. Fluids <u>7</u>, 981 (1964).
5. A.G. Engelhardt, Bull. Amer. Phys. Soc. <u>9</u>, 305 (1964).
6. H. Hora, Max-Planck-Institut für Plasmaphysik, Garching, Report 6/23 (1964); USAEC Rept. NRC-TT 1193 (1965); Application of Laser Produced Plasmas for Controlled Thermonuclear Fusion in <u>Laser Interaction and Related Plasma Phenomena</u>, H. Schwarz and H. Hora, Eds. (Plenum Press, New York, 1971, p. 427).
7. J.G. Linhart, Nuclear Fusion <u>1o</u>, 211 (1970).
8. Ming-sheng Chu, Dr. Thesis, Columbia University 1971; B. Ahlborn, Phys. Lett. J.L. Bobin, and G.F. Tonon, Proc. Europ. Conf. CTR, Rome, September 1970, p. 61.
9. G. Guderley, Luftfahrtforschung <u>19</u>, 3o2 (1942), C.F. von Weizsäcker, Z. Naturforsch. <u>9a</u>, 269 (1954).
10. J.W. Daiber, A. Hertzberg, and C.E. Wittliff, Phys. Fluids <u>9</u>, 617 (1966).
11. H. Hora, D. Pfirsch, and A. Schlüter, Z. Naturforsch. <u>22a</u>, 278 (1967); H. Hora, Phys. Fluids <u>12</u>, 181 (1969); Opto-Electronics <u>2</u>, 2o1 (1970); J.W. Shearer, R.E. Kidder, J.W. Zink, Bull. Amer. Phys. Soc. <u>15</u>, 1483 (1970).
12. P.K. Kaw, and J.M. Dawson, Phys. Fluids <u>12</u>, 2586 (1969).
13. M. Lubin, Astronautics and Aeronautics, Nov. 1970, p. 42.
14. H. Hora, Some Results of the Self-Similarity Model, in <u>Laser Interaction and Related Plasma Phenomena</u>, H. Schwarz and H. Hora Eds. (Plenum Press, New York, 1971, p. 365).
15. See Ref. 4, Eq.(3o) with adequate initial conditions.
16. W. Fader, Phys. Fluids <u>11</u>, 22oo (1968).
17. J.M. Dawson, P. Kaw, and B. Green, Phys. Fluids <u>12</u>, 875 (1969).
18. W.R. Arnold, J.A. Phillips, G.A. Sawyer, J.E. Stovall, and J.L. Tuck, Phys. Rev. <u>93</u>, 483 (1954).
19. J.L. Tuck, Nuclear Fusion <u>1</u>, 2o1 (1971).
20. I.J. Spalding, Culham Report CLM-R 1o9 (1970); Kvantova Elektronika (N.G. Basov) <u>1</u>, Nr. 3 (1971).

FUSION NEUTRON AND SOFT X-RAY GENERATION IN LASER ASSISTED DENSE PLASMA FOCUS*

R. A. Shatas, T. G. Roberts, H. C. Meyer, and J. D. Stettler
U.S. Army Missile Command, Redstone Arsenal, Alabama

ABSTRACT

The Langmuir frequency of the dense plasma focus matches closely the CO_2 laser frequency; consequently, a strong absorption is expected within the volume of contained plasma. For thermonuclear radiation effects studies, laser assisted plasma focus can be employed for fusion neutron or soft X-ray pulse generation. In the latter case, high-Z material is laser-injected into the nascent focus, and plasma cooling by enhanced X-ray radiation is compensated by laser heating. Laser to X-ray energy conversion efficiencies near 50% are feasible. Numerical calculations for ff, fb, bb and total radiation for 5% Cu or Fe injection into focus to enhance the X-ray yield are presented for electron temperatures from 0.5 to 10 keV. Contrary to the X-ray generation, estimates of increase in neutron yield depend upon model assumed for dense focus. Furthermore, at electron temperatures in excess of 1 keV, the critical electron density must be maintained and the anomalous absorption invoked to obtain the laser energy absorption in a single pass through focus. Estimates for the threshold of the anomalous absorption are presented. Most of the calculations were performed in the context of the single fluid boiler model, although estimates for the X-ray enhancement are model independent. The required laser energies of 10^2 to 10^3 J per 10^{-7} sec pulse are obtainable from a segmented cylindrical electron-beam preionized CO_2 laser device.

---

*Presented at the Second Workshop on "Laser Interaction and Related Plasma Phenomena" at Rensselaer Polytechnic Institute, Hartford Graduate Center, August 30-September 3, 1971.

1. INTRODUCTION

It is well known that laser heating of plasmas to keV temperatures is one of probable means in achieving thermonuclear reactors.[1] However, even the lowest estimates[2] of energy needed to achieve the break-even point require lasers that are beyond the current state of the art. Since primary considerations have been given to Nd^{+3}-glass and ruby lasers, the plasma electron concentration must be of the order of 10^{21} cm^{-3} to satisfy the condition $\omega_{pe} \approx \omega_\lambda$ needed to obtain a short absorption path. At the required plasma concentrations, practically only inertial confinement can be considered as feasible. Although implosion schemes[3] have recently been realized in spherically symmetric configurations,[4-6] a large fraction of laser energy deposited in the plasma goes into expansion. On the other hand, the critical plasma electron concentration for the 10.6 μm CO_2 laser radiation is of the order of 10^{19} cm^{-3}. Magnetic confinement of plasma at these densities and keV temperatures requires field intensities of the magnitude of megagauss. Small plasma volumes meeting these requirements are obtainable in dense focus[7,8]; enhancement of neutron[9] and X-ray[10] production of dense focus is of technological interest in designing pulsed radiation generators for testing of thermonuclear fusion reactor materials and devices.

2. DENSE PLASMA FOCUS

Although many designs have been employed in the neutron[7,8,11] and X-ray[12-14] generation by the dense focus without laser assistance, there is no generally accepted single process describing the operation. In all experimental arrangements, electrical energy stored in a bank of low-inductance capacitors is applied across coaxially-arranged electrodes filled with d-t gas at a few Torr pressures. The current sheet formed across the electrodes after the initial breakdown is driven along the electrodes by $\vec{j} \times \vec{B}$ force which snowplows the neutral gas. Since the magnetic field decreases with radius, plasma is driven axially and radially outward and stores a considerable amount of magnetic energy. At the end of the center electrode, this stored magnetic field causes a rapid and violent pinching, and a portion of electrons and ions is driven by the collapsing plasma into a hot dense focus of some 10^{-2} cm^3 volume. Densities of the order of 10^{19} cm^{-3} at temperatures of several keV may be attained for some 100 nsec, although not all neutrons and X-rays are of thermal origin.[15-24] The snowplow model[25] accounts for most of the features right after the initial breakdown, but neither a simple collapse-expansion mechanism nor a sausage type instability can account for the relatively long duration of the focus.[26] In the ion beam, beam-target, or acceleration model,[19-21,27] very large electric fields near the neck of the focus accelerate hydrogen isotope ions to energies of several tens of keV. These ions produce neutrons by striking the quasi-stationary dense plasma just ahead of the focus. The acceleration may come from the 500 kV cm^{-1} fields created by the large values of IdL/dt at the time of pinch, or from an anomalous enhancement of the plasma resistivity which causes ohmic heating and permits a rapid constriction of the current distribution. This creates enormous magnetic fields even if only a small fraction of the discharge current flows near the axis. In the flow through focus model,[16,28] a reservoir of plasma is created near the center electrode and the subsequent outflow along the axis of this plasma through the magnetic nozzle provides during the confinement a locus of hot plasma. The plasma is not composed of the same particles during the entire confinement and the time each particle spends in the focus is determined by the ratio ℓ/v_{ax} where ℓ is the effective length of the focus and v_{ax} is the axial plasma velocity. In the moving boiler model,[17,29] the collapse produces a relatively stable magnetohydrodynamic structure which moves axially away from the face of the center electrode. Experimental data do not fit any single model exactly,[30] although there is evidence that a large fraction of neutrons is of thermonuclear origin accounted for by the moving boiler model.[15]

3. LASER HEATING OF DENSE FOCUS

The lack of a unique model to fit dense focus restricts the laser heating calculations to order of magnitude estimates particularly in the case of the enhancement of neutron production. However, the calculations of the soft X-ray enhancement are model independent.[10] For the enhancement of neutron fluence by laser irradiation, both the increase of plasma density by laser-driven injection of deuterium into focus[31] and the raise of reaction temperature[9] have been proposed. While the first scheme could be carried out with Nd^{3+} or ruby Q-switched laser pulse, the second one requires CO_2 lasers of tens of nanosecond duration. In both cases, a strong containment in the dense focus is presupposed such that the obvious blowout[32,33] of a $\beta = 1$ Z-pinch does not take place. If one neglects the radiation by the d-t plasma, the laser pulse energy needed to raise the temperature of the plasma can be estimated from Table 1.

Thus, for $\langle 5\ keV\rangle$ temperature rise of the focus, an absorbed laser pulse energy of the order of 250 J would be needed. Fig. 1 shows the general configuration of the laser assisted plasma focus device. For the operation in the X-ray generation mode, the thin foil is made out of high-Z material which is vaporized by a laser pre-pulse and subsequently injected into the nascent focus with velocities[34] of the order of $10^7 \ldots 10^8$ cm sec^{-1} obtainable by the Linlor effect. The laser pulse is absorbed by a free-free electron transition in the field of positive ions (inverse

TABLE 1
Laser Heating of Pinched Plasmas

Laser	Nd^{3+}-glass	CO_2
Wavelength	1.06 μm	10.6 μm
Critical density	10^{21} cm^{-3}	10^{19} cm^{-3}
Thermal energy per 1 keV at critical density	$5 \cdot 10^5$ J cm^{-3}	$5 \cdot 10^3$ J cm^{-3}
Lawson's containment time	$\tau > 10^{-7}$ sec	$\tau > 10^{-5}$ sec
Confinement field at $\langle 10$ keV$\rangle$	$20 \cdot 10^6$ G	$2 \cdot 10^6$ G

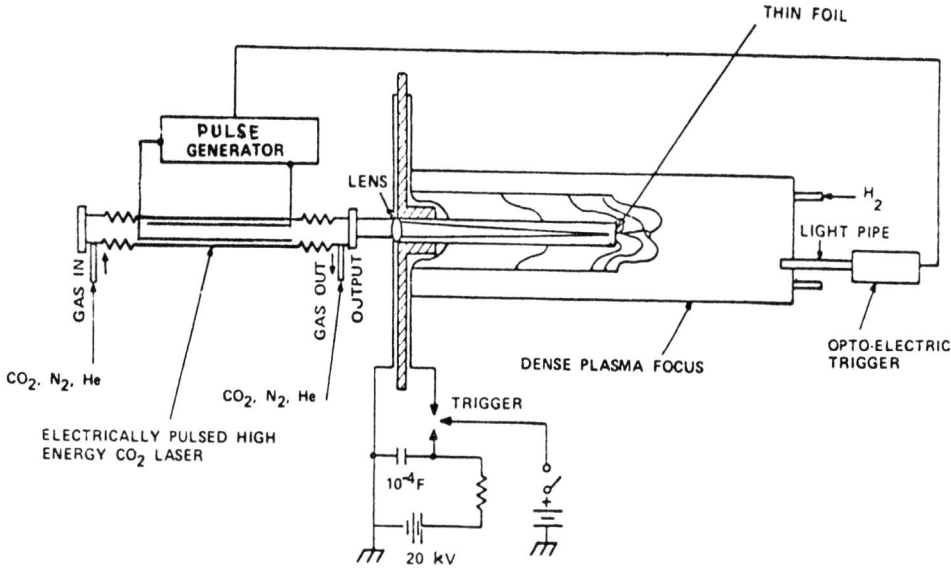

FIG. 1. Experimental Arrangement for the Laser-Assisted Plasma-Focus Device. For X-ray Enhancement, the Laser is Also Employed to Vaporize the High Atomic Number Foil Covering the Inner Coaxial Conductor of the Plasma Gun.

Bremsstrahlung) with a distance a few times the reciprocal of the absorption constant

$$\alpha_Z = \frac{N_e N_i}{n} Z^2 \frac{\bar{G}(Z)}{\bar{G}(1)} A [cm^{-1}] \qquad (1)$$

where $n = [1 - (\omega_{pe}/\omega_\lambda)^2]^{1/2}$, $A \approx T^{-3/2}\lambda^2 = 3 \cdot 10^{-39}$ for $T = 1$ keV $\lambda = 10.6$ μm, and $\bar{G}$ is the Gaunt factor.[35] When this device is operated in the X-ray generator mode both the absorption of the laser radiation and the production of X rays is strongly dependent on the presence of a small amount of a high-Z impurity introduced into the plasma. We illustrate this by calculating the absorption and reradiation of the energy for a pure hydrogen plasma and for a hydrogen plasma which contains 5% (atomic) of either Fe or Cu. All calculations are based on theories reviewed by Griem[36] unless otherwise explicitly stated.

Initial conditions taken for these calculations comprise singly ionized plasma at an electron concentration of 10^{19} cm^{-3} and a range of electron temperatures from 0.5 to 10 keV. Ionization or excitation by the electron impact governs the energy transfer of interest in this calculation. At these densities and temperatures, all thermalization times are of the order of 10^{-8} sec or less. Thus, a thermal distribution for the electrons will be assumed.[37] In Table 2 are given absorption coefficients calculated in the wings of the resonance absorption where n of Eq. (1) approaches unity. Thus, the hydrogen plasma with 5% impurity ions is optically thick to the CO_2 laser radiation even when n of Eq. (1) approaches unity at the wings of the resonance absorption seen at the critical density. Typical reflections of 20-30% have been observed in Nd^{3+}-glass laser heating of 10^{21} cm^{-3} density plasma.

TABLE 2
Absorbance of 10^{19} cm^{-3} Pure Hydrogen Plasma and 5% Fe or Cu Added Plasma

T[keV]	0.5	1	2.5	5	7.5	10
α_H[cm^{-1}]	0.93	0.33	$8.4 \cdot 10^{-2}$	$3 \cdot 10^{-2}$	$1.6 \cdot 10^{-2}$	10^{-2}
α_{Fe}[cm^{-1}]	17	5.9	3.2	1.2	0.64	0.44
α_{Cu}[cm^{-1}]	14	7.0	3.9	1.6	0.86	0.58

Although data are scare for CO_2 laser heating of 10^{19} cm^{-3} density plasma, we can estimate that the percentage of laser light reflected will be about the same; therefore it may be assumed that this impure plasma despite the small volume of the plasma focus device will absorb most of the incident laser radiation. However, the pure hydrogen isotope plasma in the dense plasma focus device considered here is optically thin to the CO_2 laser radiation unless the electron density is maintained such that the plasma Langmuir frequency remains quite close to the laser frequency where n is singular. Furthermore, the laser radiation is focused to an area of 10^{-2} cm^2; at the laser pulse energies in question, the incident laser flux may reach 10^{11} W cm^{-2}. Under these conditions, flux thresholds for the onset of the anomalous absorption may be exceeded. A simple process which excites ion waves is the inverse Landau damping, that is, the electron currents are dissipated, and the energy goes into ion acoustic waves. This process is analogous to the amplification (or damping) of acoustic waves in piezoelectric semiconductors. If we consider the process as the scattering of an electron from a phonon of the ion acoustic waves of frequency ν and velocity v_a, we have for conservation of energy and momentum:

$$h\nu = \frac{1}{2} m_e v_i^2 - \frac{1}{2} m_e v_f^2 \qquad (2)$$

$$h\nu/v_a = m_e v_f - m_e v_i \quad . \qquad (3)$$

For the generation of acoustic waves, we need $v_a < v_i$; so for a net amplification, let us take $v_a = v_f$. Solution of Eqs. (2) and (3) for ν yields

$$\nu = 4(m_e/h) v_a^2 \quad .$$

The velocity v_a is given[38] by

$$v_a^2 = k(T_i + ZT_e)/m_i \quad .$$

Therefore, we have for $T_e = T_i$

$$\nu = 4 \frac{m_e}{m_i} \frac{k}{h} T_e (1 + Z) \quad .$$

For $Z = 1$ (deuterium) and $T_e = 1$ keV, $\nu = 10^{15}$ sec^{-1} which is much higher than the ion Langmuir frequency

$$\omega_{pi}/2\pi = e(\pi m_i/n)^{-\frac{1}{2}} = 3 \cdot 10^{11} \; sec^{-1} \quad ,$$

hence unattainable. This simple process cannot conserve energy and momentum simultaneously so we consider the two-stream and parametric instabilities which Kaw and Dawson[39] and Nishikawa[40] have both treated. In these processes there are three particles interacting, the photon of the laser beam, the phonon of the ion acoustic wave, and the phonon of the electron or Langmuir wave. Both of these instabilities then involve the excitation of both ion acoustic waves and electron or Langmuir waves.

To obtain the two-stream instability threshold for 10.6 μm, Eq. (15) from Kaw and Dawson[39] (for $T_e > T_i$) is used:

$$eE_\lambda (m\omega_\lambda)^{-1} \simeq \sqrt{6}\, s_e \nu_e^{1/2} \omega_R^{-1/2} \quad . \tag{4}$$

Here, E_λ and ω_λ are the magnitude and frequency of the laser field, s_e is the electron thermal velocity, ν_e is the electron collision frequency, and ω_R, the frequency of the electron waves, is taken to be ω_o. We also have the relation $s_e^2 = \gamma_e T_e m_e^{-1}$, where γ_e is a numerical factor.

If we take the electron collision frequency ν_e to be the electron-ion collision frequency as Dawson does, ν_e is given[41] by

$$\nu_e = 2 \cdot 10^{-5}\, n_i Z_i^2 T_e^{-3/2} [\sec^{-1}] \tag{5}$$

where T_e is the electron temperature [eV]. The flux threshold P[W cm^{-2}] is then

$$P \simeq 10^{-16} (n_e^3/T_e)^{\frac{1}{2}} \tag{6}$$

or of the order of 10^{11} [W cm^{-2}] for our conditions.

The threshold for the parametric instability was obtained from the expression of Kaw and Dawson originally formulated by DuBois and Goldman[42]

$$P \simeq 5 \cdot 10^3 n_e T_e \gamma \omega_{pe}^{-1} \simeq (8.3) \cdot 10^{-8} n_e^{1/2} T_e \gamma \tag{7}$$

where units are same as in Eq. (6), γ is the linear plasma damping rate, and ω_{pe} is the plasma electron frequency. Taking $\gamma \omega_{pe}^{-1} \simeq 0.02$ according to Kaw and Dawson, we have obtained for our conditions the threshold of the order of 10^{12} W cm^{-2}.

We can estimate the minimum penetration depth ℓ after the two-stream instability has been excited and the anomalous collisional (and heating) effect is operating. We assume with Kaw and Dawson that the maximum anomalous collision frequency we can hope to get before the instability is turned off coincides with the plasma ion frequency. This gives us the expression $\ell \simeq 3 \cdot 10^{-2} T_e n_e^{1/2} p^{-1}$ [cm], which for 1 keV electron temperature and flux of 10^{12} W cm^{-2} yields a numerical value $\ell < 0.1$ cm. This is not quite an order of magnitude less than given in Table 2; however, the onset of instabilities may also affect the duration of the dense focus containment. Clearly, some experimental data are needed to pursue this point farther.

4. SOFT X-RAY ENHANCEMENT

Introduction of the high-Z impurities during the early formative stages of the discharge would limit the plasma temperature because of an intense line radiation and thus would be deleterious to the X-ray production. Therefore it must be accomplished preferably at the onset of the primary pinch. Because the spallated anode material is only of the order of 1% or less of the hydrogen density,[12] the required high-Z ions are externally injected as shown in Fig. 1.

During the 100 nsec confinement time the primary plasma cooling mechanism consists of radiation losses. These losses are compensated by the absorption of laser energy which is supplied at a rate tailored to the desired spectral shift of the X rays during the pulse. For the purposes of this discussion, it is taken that the laser power is equal to the total X-ray radiated power, and therefore the electron temperature remains nearly constant. The radiation losses consist of a Bremsstrahlung continuum which results from free-free (ff) transitions, a recombination continuum whcih results from free-bound (fb) transitions and line radiation which results from bound-bound (bb) transitions. In order to calculate these radiation losses, the degree of ionization expected for the Fe or Cu atoms has to be determined. To remove by electron impact one electron of r-equivalent ones with an ionization potential E_I from an ion of charge Z - 1, time of the order of

$$\tau_Z \approx 4 \cdot 10^6 \, E_I Z^2 [\exp(E_I/T)](rn_e)^{-1} T^{-1/2} \qquad (8)$$

is needed where T is in keV and n_e is the electron number density [cm^{-3}]. We use calculated ionization potentials[43] for Fe and Cu, and we impose the requirement that all ions be created in the early phase of the focus formation. The ions of interest and their time of creation τ calculated from Eq. (8) are given in Table 4.

The times of ionization in Table 4 are based on an electron

TABLE 4

Time of Ionization τ in Nanoseconds for the Z-th Degree of Ionization at a Given Electron Temperature T in keV

T	Fe			Cu		
	E_I[keV]	Z	τ	E_I[keV]	Z	τ
0.5	0.49	16	11	0.48	15	3
				0.52	16	4
1	0.49	16	5	0.67	19	12
2.5	1.95	23	24	2.3	25	31
	2.05	24	45	2.5	26	49
5.0	2.05	24	22	2.6	27	41
7.5	2.05	24	16	2.6	27	29
10	8.83	25	40	2.6	27	23
				10.1	28	66

number density of 10^{19} cm^{-3} and thus are upper limits because the density will actually increase with each successive degree of ionization.

The radiation in the form of Bremsstrahlung is given by

$$P_{ff} = 4.74 \cdot 10^{-31} Z^2 n_e n_i T^{1/2} [\text{W cm}^{-3}] , \qquad (9)$$

where T is in keV and n_e and n_i are respectively electron and ion densities. The Bremsstrahlung ratio of an impure to a pure hydrogen isotopic plasma is

$$P_{ffI}/P_{ffH} = fZ(1 + Z + Z^2 f) + 1 \qquad (10)$$

where f is the fraction of high-Z ions taken with respect to hydrogen. (Each of these high-Z ions is ionized to a positive charge of Z.) We use results given in Table 4 and Eqs. (9) and (10) to calculate the Bremsstrahlung radiation shown in Table 5.

In a similar manner we find that the recombination radiation is given by

$$P_{fbI} = fZ^2(1 + fZ) P_{ffH} E_I T^{-1} . \qquad (11)$$

TABLE 5

Power Density $P[\text{GW cm}^{-3}]$ Radiated at Temperature T in keV Free to Free Transitions in 10^{19} cm^{-3} Initially Pure Hydrogen Plasma to Which 5% Cu of Fe is Admixed

T	0.5	1	2.5	5	7.5	10
P_{ffH}	0.035	0.047	0.075	0.11	0.13	0.15
P_{ffFe}	0.83	1.2	4.7	7.0	8.5	10.5
P_{ffCu}	0.77	1.8	5.7	9.3	11.4	12.8

Numerical results are summarized in Table 6.

TABLE 6

Power Density $P[\text{GW cm}^{-3}]$ Radiated at Temperature T in keV by Free to Bound Transitions in 10^{19} cm^{-3} Initially Pure Hydrogen Plasma to Which 5% Cu or Fe is Admixed

T	0.5	1	2.5	5	7.5	10
P_{fbH}			- negligible -			
P_{fbFe}	0.76	0.54	3.6	2.8	2.3	9.0
P_{fbCu}	0.71	1.1	5.3	4.7	3.9	6.8

The line radiation has to be calculated from the relative populations of the excited levels of the ionic species given in Table 4. However, the plasma is too tenuous to assure a thermal distribution among such excited ionic levels. At the electron densities of interest, the lowest principal quantum number in local thermal equilibrium is of the order of $Z^{12/17}$. Thus, all the line radiation originates from levels populated only by the electron impact. Such radiation occurs principally at the energy E_2 of the resonance line of the ion. Since this quantity is not known for our highly ionized ions, a reasonable estimate is $E_2 \sim 1/2\, E_\infty = 1/2\, E_I$. Corrections to E_I for plasma effects and increase of the series limit vary as $1.5\, Z[\text{eV}]$ and $3\, Z^{3/5}[\text{eV}]$, respectively. For our conditions these corrections are negligible and therefore have not been included. Thus, the line radiation expressed by

$$P_{bbI} = 3 \cdot 10^4 \, f(1 + Zf) \, P_{ffH} \exp(-E_I/2T) \, T^{-1} \quad (20)$$

is calculated in Table 7.

TABLE 7

Power Density $P[\text{GW cm}^{-3}]$ Radiated at Temperature T in keV by Bound to Bound Transitions in 10^{19} cm^{-3} Initially Pure Hydrogen Plasma to Which 5% Cu or Fe is Admixed

T	0.5	1	2.5	5	7.5	10
P_{bbH}			- None -			
P_{bbFe}	111	101	66	57	50	32
P_{bbCu}	108	99	63	58	51	45

Table 8 shows the total X-ray power radiated in f to f, f to b and b to b radiation modes for the pure hydrogen isotopic and the impure plasmas obtained by adding data from Tables 5 to 7.

TABLE 8

Total Power P_T(f to f + f to b + b to b) Radiated at Temperature T in keV in the X-ray Form by 10^{19} cm^{-3} Hydrogenic Plasma (H) to Which 5% Fe or Cu is Admixed

T	0.5	1	2.5	5	7.5	10
$P_{TH}[\text{MW cm}^{-3}]$	35	47	75	10	130	150
$P_{TFe}[\text{GW cm}^{-3}]$	113	103	74	67	61	51
$P_{TCu}[\text{GW cm}^{-3}]$	109	102	74	72	67	61

Finally, the energy which must be supplied by the laser to multiply ionize the ions and to raise the temperature of both the ions and their electrons to that of the hydrogen plasma can be calculated from Table 9 by multiplying corresponding entries by the effective plasma volume in cm^3.

TABLE 9

Thermal and Ionization Energy Needed to Heat to Temp. T in keV 5% Fe and Cu Atoms Added to that of 10^{19} cm^{-3} Initially Pure Hydrogen Plasma

T	0.5	1	2.5	5	7.5	10
3/2 kT$_{ions}$ [J cm^{-3}]	60	120	300	600	900	1,200
3/2 kT$_e$'s(Fe) [kJ cm^{-3}]	1	2	6	14	22	30
3/2 kT$_e$'s(Cu) [kJ cm^{-3}]	1	2	8	16	24	32
E$^{Fe}_{Ionization}$ [kJ cm^{-3}]	0.3	0.3	1.2	1.3	1.3	2.0
E$^{Cu}_{Ionization}$ [kJ cm^{-3}]	0.3	0.4	1.5	1.8	1.8	2.0

5. CONCLUSIONS

Techniques for X-ray generation at low photon energies such as flash X-ray tubes, vacuum spark gaps, and coaxial discharges are characterized by their rather low conversion efficiency. Laser energy to X-ray energy conversion efficiencies of a few percent in laser created high-Z plasmas for kilovolt X-ray generation have been calculated by Bernstein and Comisar.[44]

This rapid expansion of a hot laser plasma restricts the effective X-ray radiation time to a few nanoseconds[45]; because of this, the X-ray fluence is limited. Consequently, the ratio of energy emitted in X rays to the energy expended in ionizing the high-Z atoms and heating the ions with their respective electrons to the required temperatures is low. Since dense focus possesses an inherent containment, a substantial increase in the laser energy to X-ray energy conversion efficiency can be expected. In Table 10 we show the estimated efficiency of laser energy to X-ray energy conversion for dense focus of 10^{19} cm^{-3} initial electron density and 10^{-2} cm^3 effective plasma volume at a temperature T in keV. Into this plasma we inject by a laser prepulse 5% Fe ionized and heated to the temperature T. A containment and X-ray radiation

TABLE 10

Laser to X-ray Conversion Efficiency at Electron Temperature T [keV] for 10^{19} cm^{-3} density, 10^{-2} cm^3 Volume and 10^{-7} sec Duration Dense Focus with 5% Fe Injected. $E_I + E_T$ Refer to Ionization and Heating of Added Fe; E_L is the Incident Laser Energy of Which 80% is Absorbed.

T	0.5	1	2.5	5	7.5	10
X rays [J]	113	103	74	67	61	51
$E_I + E_T$ [J]	13	23	72	153	233	320
E_L [J]	157	157	183	275	380	475
η	0.72	0.65	0.40	0.24	0.17	0.11

duration of 100 nsec is assumed. The energy radiated in the ff, fb and bb modes of the X-ray pulse is shown in the second row; this energy must be supplied by the absorbed laser pulse to maintain a constant temperature. The third row indicates the energy needed to ionize and to heat the 5% Fe. The fourth row pertains to the energy contained in the laser pulse incident on the foil and plasma of Fig. 1; reflection losses of 20% are already incorporated into this number. Finally, the overall conversion efficiency is given in the last row. It is seen that the softer the X-ray radiation the higher is the conversion efficiency. Therefore, to generate X rays of photon energies in excess of 10 keV, the currently available conventional techniques are still preferable. In our efficiency estimate, we have not included the electrical energy needed to operate the dense focus because, first, the electrical energy stored in the capacitor bank of the focus is easy to come by as compared with that of the laser output. Second, the increase in electrical energy to the dense focus alone is not necessarily the sure way in increasing the output.[23]

REFERENCES

1. J. G. Linhart, Very High Density Plasmas for Thermonuclear Fusion, Nucl. Fus. 10, 211-234 (1970).

2. Heinrich Hora, Application of Laser Produced Plasmas for Controlled Thermonuclear Fusion, in Laser Interaction and Related Plasma Phenomena, H. J. Schwarz and H. Hora, eds. (Plenum Press, N. Y., 1971), p. 427.

3. J. W. Daibler, A. Hertzberg, and C. E. Wittliff, Laser-Generated Implosions, Phys. Fluids $\underline{9}$, 617-619 (1966).

4. S. W. Mead, Plasma Production with a Multibeam Laser System, Phys. Fluids $\underline{13}$, 1510-1518 (1970).

5. N. G. Basov, O. N. Krokhin and G. V. Slizkov, Heating of Laser Plasmas for Thermonuclear Fusion, Preprint #132, Lebedev Physical Institute, Acad. Sci. USSR (1971).

6. N. G. Basov, O. N. Krokhin, G. V. Slizkov, S. I. Fedotov and A. S. Shikanov, Powerful Laser Installation with the Successive-Parallel Amplifying System for Plasma Heating, Preprint #123, Lebedev Physical Institute, Acad. Sci. USSR (1971).

7. Joseph W. Mather, Formation of a High Density Deuterium Plasma Focus, Phys. Fluids $\underline{8}$, 366-377 (1965).

8. Joseph W. Mather and Paul J. Bottoms, Characteristics of the Dense Plasma Focus Discharge, Phys. Fluids $\underline{11}$, 611-618 (1968).

9. R. A. Shatas, T. G. Roberts, H. C. Meyer, and J. D. Stettler, CO_2 Laser Assisted Neutron Generation, Bull. Amer. Phys. Soc. $\underline{15}$, 1308 (1970).

10. Romas A. Shatas, John D. Stettler, Harry C. Meyer, and Thomas G. Roberts, Soft X-Rays from a Laser-Heated Dense Plasma Focus, J. Appl. Phys. $\underline{42}$ 5884-5886 (1971).

11. M. H. Dazey, H. L. L. van Paassen and V. Josephson, Electrode Metal Effects in a Deuterium Plasma Z-Pinch Device, J. Appl. Phys. $\underline{41}$, 3545-3546 (1970).

12. Everet H. Beckner, Production and Diagnostic Measurements of KiloVolt High Density Deuterium, Helium and Neon Plasmas, J. Appl. Phys. $\underline{37}$, 4944-4952 (1966).

13. Everet H. Beckner, Pulsed, High Intensity Source of Soft X Rays, Rev. Sci. Instr. $\underline{38}$, 507-511 (1967).

14. H. L. L. van Paassen, R. H. Vandre and R. Stephen White, X Ray Spectra from Dense Plasma Focus Devices, Phys. Fluids $\underline{13}$, 2606-2612 (1970).

15. C. Patou, A. Simonet and J. P. Watteau, Measured Anisotropies of the Plasma Focus Neutron Emission Compared with Proposed Mechanisms, Phys. Lett. $\underline{29A}$, 1-2 (1969).

16. M. J. Bernstein, D. A. Meskan and H. L. L. van Paassen, Space, Time and Energy Distributions of Neutrons and X Rays from a Focused Plasma Discharge, Phys. Fluids $\underline{12}$, 2193-2202 (1969).

17. V. P. Dyachenko and V. S. Imshennik, Plasma Focus and the Neutron Emission Mechanism in a Z Pinch, Zh. Exper. Theor. Fiz. 56, 1766-1777 (1969); Sov. Phys. - JETP 29, 947-948 (1969).

18. M. J. Bernstein and F. Hai, Evidence for Nonthermonuclear Neutron Production in a Plasma Focus Discharge, Phys. Lett. 31A, 317-318 (1970).

19. M. J. Bernstein, Deuteron Acceleration and Neutron Production in Pinch Discharges, Phys. Rev. Lett. 24, 724-727 (1970).

20. M. J. Bernstein and F. Hai, Evidence for Neutron Production via Enhanced Resistivity in a Plasma Focus, Phys. Rev. Letters 25, 641-642 (1970).

21. M. J. Bernstein, Acceleration Mechanism for Neutron Production in Plasma Focus and Z Pinch Discharges, Phys. Fluids 13, 2858-2866 (1970).

22. J. H. Lee, D. S. Loebbaka and C. B. Roos, Hard X Ray Spectrum of a Plasma Focus, Plasma Phys. 13, 347-349 (1971).

23. A. Bernard, A. Coudeville and J. P. Watteau, Neutron Yield of a Focus Discharge in Various Experiments, Phys. Lett. 33A, 477-478 (1970).

24. M. J. Bernstein, C. M. Lee and F. Hai, Time Correlations of X Ray Spectra with Neutron Emission from a Plasma Focus Discharge, Phys. Rev. Lett. 27, 844-847 (1971).

25. T. D. Butler, I. Henins, F. C. Jahoda, J. Marshall and R. L. Morse, Coaxial Snowplow Discharge, Phys. Fluids 12, 1904-1916 (1969).

26. G. G. Comisar, Hydromagnetic Instabilities in the Dense Plasma Focus, Phys. Fluids 12, 1000-1007 (1969).

27. R. E. Dunway and J. A. Phillips, Neutron Generation from Straight Pinches, J. Appl. Phys. 29, 1137-1143 (1958).

28. P. O. Morgan et al., Proc. 3rd Europ. Conf Fusion and Plasma Physics, p. 118 (Utrecht, 1968).

29. N. V. Filippov and T. I. Filippova, Phenomena Associated with the Buildup of a Noncylindrical Focussed Z Pinch, Plasma Physics and Controlled Nuclear Fusion Research (Culham Conf. Proc.) Vol. 2, 405-416 (IAEA, Vienna, 1966).

30. D. E. Potter, Numerical Studies of the Plasma Focus, Phys. Fluids 14, 1911-1924 (1971).

31. P. Guillaneux, C. Patou and G. Tonon, Laser Driven Axial Flow Pinch in a Dense Plasma Focus, Phys. Lett. 32A, 370-371 (1970).

32. Thomas P. Wright, Early-Time Model of Plasma Expansion, Phys. Fluids 14, 1905-1910 (1971).

33. George C. Vlases, Heating of Pinch Devices with Lasers, Phys. Fluids 14, 1287-1289 (1971).

34. David W. Gregg and Scott J. Thomas, Kinetic Energies of Ions Produced by Laser Giant Pulses, J. Appl. Phys. 37, 4313-4316 (1966).

35. See, e.g., George Bekefi, Radiation Process in Plasmas (Wiley, N.Y., 1966), Ch. 3, Emission and Absorption from Binary Encounters.

36. Hans R. Griem, Plasma Spectroscopy (McGraw-Hill, N.Y., 1964), Ch. 6, Equilibrium Relations and Ch. 8, Radiative Energy Losses.

37. See, e.g., Lyman Spitzer, Jr., Physics of Fully Ionized Gases, 2nd ed. (Interscience, N.Y., 1962).

38. L. A. Artsimovich, Controlled Thermonuclear Reactions, transl. from 2nd edition in Russian (Gordon and Breach, N.Y., 1964), pg. 105.

39. P. K. Kaw and J. M. Dawson, Laser-induced Anomalous Heating of a Plasma, Phys. Fluids 12, 2586-2591 (1969).

40. K. Nishikawa, Parametric Excitation of Coupled Waves - I General Formulation, Jour. Phys. Soc. Japan 24 916-922 (1968); II Parametric Plasmon-Photon Interaction, ibid. 24, 1152-1158 (1968).

41. L. A. Artsimovich, op. cit. pg. 41.

42. D. J. DuBois and M. V. Goldman, Radiation-Induced Instability of Electron Plasma Oscillations, Phys. Rev. Letters 14, 544-546 (1960); Parametrically Excited Plasma Fluctuations, Phys. Rev. 164, 207-222 (1967).

43. Wolfgang Lotz, Ionization Potentials of Atoms and Ions from Hydrogen to Zinc, J. Opt. Soc. Amer. 57, 873-878 (1967).

44. M. J. Bernstein and G. G. Comisar, X Ray Production in Laser Heated Plasmas, J. Appl. Phys. $\underline{41}$, 729-732 (1970).

45. J. Bruneteau, E. Fabre, H. Lamain and P. Vasseur, Experimental Investigation of the Production and Containment of a Laser Produced Plasma, Phys. Fluids $\underline{13}$, 1795-1801 (1970).

SUMMARY OF DISCUSSION

(VI. Fusion Neutrons from Laser
Irradiated High Density Plasmas)

F. Floux strongly felt that the future application of the laser will have a decisive place in nuclear fusion. The next few years will be devoted to the 1 kilojoule lasers. Although most of the experiments will be performed with Nd-glass lasers, the CO_2 laser radiation will be of equal merit. To date no phenomena have been observed that might become seriously detrimental to nuclear fusion experiments. The instabilities or some yet unexplainable turbulence phenomena promote a fast transfer of the laser energy into the plasma.

It was brought out that there is a possibility to build a laser of 10^6 joule pulses in the nanosecond range in the U.S. within one year. The concensus of the attendees was that the physics of laser created plasma is much more complicated than was originally thought. After experiments with the kilojoule laser, one will be able to judge whether the few simple scaling laws observed at the lower energies are still valid for the kilojoule range. A possible problem may be the blowing off process of the plasma boundary which can prevent sufficient heating of the plasma interior and thus reduce the neutron yield.

Further optimism was expressed by G. V. Sklizkov for the future development of the laser for nuclear fusion. However, he feels that it is too early to tell whether controlled thermonuclear fusion will be possible with lasers. At this time one should concentrate on the investigation of high temperatures, nonlinear phenomena and dynamics of laser produced plasmas. We are dreaming of 1 to 10 Megajoule laser energies. In order to decrease the energy for the even-point, one has to increase the ion density up to $10^{24} cm^{-3}$.

The attendees felt better diagnostic techniques should be developed to further the study of the properties of the laser produced plasmas. If it were possible to study the plasma on a microscopic scale, problems may be found due to nonequilibrium electron distribution or one may find different temperatures at different points.

M. Goldman brought up the possibility of driving an imploding hollow sphere by laser. A comparison was made between the possible

power generators with laser produced fusion plasma with the large
machines applying magnetically confined plasmas of low density. A
point was made that the laser plasma needs only relatively compact
equipment[1] and much simpler service techniques, whereas the Tokamak
and Stellerator[2] systems are much more complicated and any material
defects are quite difficult to repair.

P. Harteck questioned why in the laser produced plasmas for fusion
purposes one does not also use for targets D-T carbon, or lithium
compounds of tritium. The objections of several participants was
their feeling of the high radioactive danger involved in handling
tritium. P. Harteck, who actually is the co-discoverer of tritium
and has worked over several decades with tritium, assured the
attendees that there is no serious danger involved.

References
1. H. Hora, <u>Laser Interaction and Related Plasma Phenomena</u> (H.
   Schwarz and H. Hora, eds.) Plenum Press, New York, 1971, p. 427.
2. L. Rothard, Kernenergetic <u>13</u>, 269 (1970).

LIST OF CONTRIBUTORS* AND ATTENDEES

+*A. J. Alcock
 National Research Council
 Division of Physics
 Ottawa 7, Canada

+*A. J. Beaulieu
 Gen Tec (1979) Inc.
 2625 Dalton Street
 Quebec 12, P.Q., Canada

*D. K. Bhadra
 Gulf General Atomic
 P. O. Box 608
 San Diego, California 92112

Jean P. Biscar
 Laser Research - Physics
 University of Wyoming
 Laramie, Wyoming 82070

*Laird P. Bradley
 Sandia Laboratories
 P. O. Box 5800
 Albuquerque, New Mexico 87115

E. L. Breig
 Space Physics Laboratory
 Aerospace Corporation
 2300 E. El Segundo Blvd.
 El Segundo, California 90245

+*B. R. Bronfin
 United Aircraft Research Lab.
 East Hartford, Connecticut 06108

Philip E. Cassady
 Lockheed Research Labs.
 3251 Hanover Street
 Palo Alto, California 94304

E. S. Cassedy
 Polytechnic Institute of Brooklyn
 333 Jay Street
 Brooklyn, New York 11201

N. M. Ceglio, Jr.
 Naval Postgraduate School
 Monterey, California 93940

Stewart G. Chapin
 Martin Marietta Corporation
 P. O. Box 179
 Denver, Colorado 80201

+*A. J. DeMaria
 United Aircraft Research Labs.
 East Hartford, Connecticut 06108

Thomas A. Dillon
 National Bureau of Standards
 Boulder, Colorado 80302

Sherman R. Farrell
 Spectromagnetic Industries
 25393 Huntwood Avenue
 Hayward, California 94544

Frank D. Feiock
 The KMS Technology Center
 11689 Sorrento Valley Road
 San Diego, California 92121

Heinz Fischer
 Technische Hochschule
 Darmstadt, Germany

+*Francis Floux
 Commissariat à l'Énergie Atomique
 Centre d'Études de Limeil
 B.P. n° 27-94-Villeneuve St. Georges
 France

+INVITED Contributors

Brendan Godfrey
Kirtland Air Force Base
Hdqtrs. Air Force Special
  Weapons Center (AFSC)
New Mexico 87117

*Robert P. Godwin
Los Alamos Scientific Laboratory
P. O. Box 1663
Los Alamos, New Mexico 87544

Mark Goldstein
International Business & Research
P. O. Box 9062
Coral Gables, Florida 33124

Lester K. Goodwin
The KMS Technology Center
11689 Sorrento Valley Road
San Diego, California 92121

B. J. Graham
U. S. Naval Academy
Physics Department
Annapolis, Maryland 21402

+*A. H. Guenther
Effects Branch, Research Division
Air Force Weapons Lab.
Kirtland Air Force Base, New Mexico

+*Paul Harteck
Rensselaer Polytechnic Institute
Troy, New York 12181

+*Alan F. Haught
United Aircraft Research Labs.
East Hartford, Connecticut 06108

+*Heinrich Hora
Max-Planck-Institut für Plasmaphysik
Garching, Germany
Rensselaer Polytechnic Institute
Hartford Graduate Center
275 Windsor Street
Hartford, Connecticut 06120

*Robert J. Hull
MIT Lincoln Laboratory
Box 73
Lexington, Massachusetts 02173

+*Eric Jones
Sandia Corporation
P. O. Box 5800
Albuquerque, New Mexico 87115

Alan F. Klein
Physics International Company
2700 Merced Street
San Leandro, California 94577

+*Benedikt Kronast
National Research Council
Division of Physics
Ottawa 7, Canada

+*William E. Kruer
Plasma Physics Laboratory
Princeton University
P. O. Box 451
Princeton, New Jersey 08540

*Donald E. Lencioni
MIT Lincoln Laboratory
Box 73
Lexington, Massachusetts 02173

+*M. Lubin
Mech. & Aerospace Science Dept.
University of Rochester
Rochester, New York 14627

Michael M. Mann
Northrop Corporate Labs.
3401 West Broadway
Hawthorne, California 90250

John Marburger
Physics & Elec. Engr. Dept.
University of Southern California
University Park
Los Angeles, California 90007

*George H. Miley
Nuclear Engr. & Elec. Engr. Dept.
University of Illinois
214 Nuclear Engineering Lab.
Urbana, Illinois 61801

+*P. Mulser
Max-Planck-Institut für Plasmaphysik
8046 Garching Bei München
Germany

David L. Murphree
Mississippi State University
Aerophysics & Aerospace Engr. Dept.
Drawer A., State College
Mississippi 39762

CONTRIBUTORS AND ATTENDEES

Isiah Nebenzahl
Dept. of Applied Physics
Cornell University
Clark Hall
Ithaca, New York 14850

M. B. Nicholson-Florence
Dept. of Physics
University of Essex
Wivenhoe Park, Colchester
Essex, England

Gabriel Otis
Departement de Physique
Universite Laval
Quebec 10, Canada

*A. J. Palmer
Dept. of Physics
Memorial University of Newfoundland
St. John's, Newfoundland, Canada

+*R. Papoular
Association Euratom-CEA
Fusion Controlée
Centre d'Études Nucléaires
Boite Postale n° 6
94 - Fontenay-aux-Roses, France

P. P. Pashinin
P. N. Lebedev Physical Institute
  of the Academy of Sciences USSR
Leninsky Prospect 53
Moscow, USSR

+*W. K. Pendleton
Air Force Weapons Lab.
Kirtland Air Force Base
New Mexico 87117

Ralph L. Phelps, Jr.
Hdqrs. Air Force Special Weapons
  Center (AFSC)
Kirtland Air Force Base, New Mexico
  87117

Anthony N. Pirri
AVCO-Everett Research Lab.
2385 Revere Beach Parkway
Everett, Massachusetts 02149

Krishnan Raman
Wesleyan University
Middletown, Connecticut 06457

*Robert Reeves, Jr.
Rensselaer Polytechnic Institute
Troy, New York 12181

Jerrold G. Rittmann
University of California
Lawrence Radiation Lab.
P. O. Box 808
Livermore, California 94550

+*Helmut Schwarz
Rensselaer Polytechnic Institute
275 Windsor Street
Hartford, Connecticut 06120

*Stephen Segall
Institute for Fluid Dynamics and
  Applied Mathematics'
University of Maryland
College Park, Maryland 20742

*Romas A. Shatas
PSD, RDE & MSL
U. S. Army Missile Command
Redstone Arsenal, Alabama 35809

+*G. V. Sklizkov
P. N. Lebedev Physical Institute
  of the Academy of Sciences USSR
Leninsky Prospect 53
Moscow, USSR

Earl W. Smith
Plasma Physics Section 271.06
National Bureau of Standards
Boulder, Colorado 80302

+*John A. Stamper
Plasma Physics Division
Naval Research Laboratories
Washington, D. C. 20390

+*R. G. Tomlinson
United Aircraft Research Labs.
East Hartford, Connecticut 06108

Real Tremblay
Laboratoire d'Optique et
  Hyperfréquences
Université Laval
Quebec 10, Canada

David B. van Hulsteyn
Elec. Engr. Dept.
University of Texas
Austin, Texas

+*George C. Vlases
 Aerospace Research Lab. PL-10
 University of Washington
 Seattle, Washington 98105

 Thomas P. Wright
 Sandia Corporation
 P. O. Box 5800
 Albuquerque, New Mexico 87115

*Eli Yablonovitch
 Div. of Engr. & Applied Physics
 Gordon McKay Laboratory
 Harvard University
 9 Oxford Street
 Cambridge, Massachusetts 02138

+*Chiyoe Yamanaka
 Faculty of Engineering
 Osaka University
 Osaka, Japan

 Gerold Yonas
 Electron Beam Research and
   Technology Dept.
 Physics International
 2700 Merced Street
 San Leandro, California 94577

AUTHOR INDEX

Numbers in parentheses following the text page numbers are reference numbers, and are included to assist in locating a reference at the end of each contribution when the author's name is not cited at the point of reference in the text.

A

Adelman, A. M. 254 (2)

Afanasiev, Yu. V. 245 (1); 390 (8); 401 (8); 444 (12); 434 (1)

Aglitzky, E. V. 395 (12); 399 (17)

Agostini, P. 92 (14); 94 (14)

Ahlborn, B. 516 (8); 525 (8)

Ahlstrom, H. G. 38 (11,12,13); 39 (11,13); 267 (22); 344 (25); 350 (25); 353 (25); 354 (25); 382 (5)

Ahmad, N. 159 (7,10)

Aihara, S. 322 (19)

Alcock, A. J. 101 (13); 156 (6); 157 (6); 159 (9); 163 (13); 164 (9,13); 168 (9,14); 169 (15); 171 (9); 179; 254 (3); 266 (3); 267 (23); 378 (17); 379

Ali, A. W. 276 (3); 277 (3)

Allario, F. 47 (19)

Allen, C. W. 345 (29)

Alpher, R. Z. 111 (25)

Aman, R. 430

Andelfinger, C. 254 (8); 263 (8)

Anderson, O. A. 186 (11)

Andriakhin, V. M. 45 (13,14); 47 (13); 48 (13,14)

Arai, T. 241 (9); 302 (3); 303 (3)

Arnold, W. R. 525 (18)

Artsimovich, L. A. 532 (38); 533 (41)

Ascoli-Bartoli, U. 194 (27); 291 (5)

Askar'Yan, G. A. 97 (2); 172 (19)

B

Babykin, M. V. 254 (10)

Baconnet, J. P. 197 (36)

Balkanski, M. 215 (9); 271 (1)

Barnes, W. S. 362 (37); 419 (10); 420 (16); 427 (10); 469 (6)

Bartell, L. S. 212 (4); 213 (4)

Basov, N. G. 26 (6); 77 (1); 155 (1); 291; 342 (5); 390 (1,3-5,7); 391 (9); 392 (4); 395 (13); 397 (7); 398 (16); 399 (13,19,26); 402 (19); 409; 461 (21); 469 (4); 470 (4); 482 (1); 503 (4); 516 (1,3); 528 (5,6)

Baumhacker, H. 505 (9)

Beaulieu, A. J. 77; 246 (5)

Becchi, C. 215 (17); 217 (17); 220 (17)

Beckner, E. H. 145; 528 (12,13); 534 (12)

Bekefi, G. 531 (35)

Belenov, E. M. 390 (8); 401 (8)

Belland, P. 379 (19)

Benard, J. F. 342 (7); 343 (7)

Benesch, R. 182 (5); 184 (5); 192 (24); 204 (45); 205 (45)

Benford, J. 186 (14); 187 (14)

Bennett, W. H. 516 (2)

Bensinger, D. L. 144

Berejetskaya, N. K. 92 (14); 94 (14); 99 (5)

Bernard, A. 528 (23); 539 (23)

Bernstein, I. B. 233 (4); 291 (9)

Bernstein, M. J. 461 (22); 528 (16,18-21,24); 529 (16); 538 (44)

Bershader, D. 113 (28)

Bertotti, B. 420 (12)

Bespalov, V. I. 373 (15)

Bethe, H. 264 (16)

Bettis, J. R. 153 (2); 401 (18)

Bhadra, D. K. 291 (7); 315

Billebize, A. 410 (7); 421 (7)

Birdsall, C. K. 318 (14); 319 (16)

Bitterman, S. 29 (8)

Blanc, A. 87 (7); 90 (7); 92 (7); 99 (8)

Bloembergen, N. 367 (7); 387

Bobin, J. L. 107 (22); 390 (6); 396 (14); 410 (1,2,6); 418 (6); 425 (6); 430; 469 (1); 503 (1); 516 (1,8); 525 (8)

AUTHOR INDEX

Bodner, S. 337

Boehmer, L. D. 144

Boiko, V. A. 155 (1); 390 (4,7);
392 (4); 395 (12,13);
397 (7) 398 (16);
399 (13,17); 503 (4)

Boivin, A. 441 (8)

Bordier, C. et al 94 (17)

Boris, J. P. 318 (14)

Bottoms, P. J. 528 (8)

Bourrabier, G. 254 (6);
266 (6)

Boyer, K. 55 (24)

Bradley, L. D. 153

Braginskii, S. I. 277 (8)

Brekhovskikh, L. M. 382 (6)

Brienza, M. J. 470 (8)

Bristow, T. C. 456 (23)

Bronfin, B. R. 59; 77

Brooks, R. E. 114 (29);
130 (29)

Brown, C. O. 9 (5); 13 (5)

Bruneteau, J. 538 (45)

Buchelt, E. 254 (7,8);
262 (12); 263 (8,12);
264 (12); 266 (7)

Büchl, K. P. 135 (35); 246 (6);
254 (7,8); 263 (8); 266 (7);
308 (8); 342 (8); 422 (17);
425 (17); 469 (2); 503 (3);
504 (3); 505 (9); 509 (3)

Buchsbaum, S. 367 (1)

Budker, G. I. 254 (9)

Bunkin, F. V. 92 (10); 95 (10);
170 (17); 172 (17)

Burnham, J. B. 9 (4)

Bushnell, J. C. 390 (5);
469 (5); 470 (5)

Butler, T. D. 528 (25)

Byers, J. 319 (15); 337

Byron, S. 40

C

Calvert, J. G. 68 (3)

Canavan, G. H. 99 (11)

Canto, C. 107 (22)

Carbone, R. J. 278 (9)

Carolan, P. G. 203 (44);
204 (44)

Caruso, A. 245 (2); 246 (2);
250 (2); 343 (17);
420 (12,15); 469 (3)

Cesari, G. 197 (36)

Chalmeton, V. 82 (3); 87 (6);
91 (6); 92 (9); 99 (7)

Chan, P. W. 191 (20)

Chang, C. S. 215 (37,38);
217 (37, 38)

Chen, C. J. 53 (23)

Cheung, A. 228 (3)

Chin, S. L. 94 (15)

Cho, C. W. 370 (10); 379

Christiansen, W. H. 25 (2,3); 27 (2); 32 (2); 33 (2)

Chu, M.-S. 516 (8); 525 (8)

Chu, T. K. 318 (7)

Churney, K. L. 70 (9)

Clayden, W. A. 228 (1)

Cognard, D. 342 (7); 343 (7); 390 (6); 410 (2,6); 418 (6); 425 (6) 469 (1); 503 (1); 516 (1)

Cohen, R. 318 (10)

Collin, R. E. 220 (39)

Comisar, G. G. 367 (6); 461 (22); 528 (26); 538 (44)

Consoli, T. 254 (6); 266 (6)

Cook, R. D. 45 (9)

Cool, T. A. 9 (2)

Coudeville, A. 197 (36); 528 (23); 539 (23)

Crozo, H. 430

Cutler, L. S. 215 (24); 216 (24); 220 (24)

D

Daehler, M. 196 (31,32)

Daiber, J. W. 391 (24); 516 (10); 528 (3)

Damon, E. K. 80 (1)

Daughney, C. C. 196 (33); 199 (37); 200 (37)

David, J. 498 (4)

Davidson, R. C. 280 (13)

Davies, R. E. 144

Davies, T. J. 153

Davies, W. E. R. 191 (21)

Davis, B. T. 45 (9)

Davis, J. W. 9 (4-6)

Davis, L. I., Jr. 170 (16)

Dawson, J. M. 38 (10,11); 273 (1); 276 (4,7); 280 (13); 281 (1); 284 (1); 286 (7); 291 (3); 318 (1,2,5,12,13); 319 (16); 324 (12,20); 328 (2); 329 (23) 340 (1); 343 (20,21); 345 (30); 359 (20,21); 362 (20); 434 (5); 448 (15); 457 (15); 494 (17) 516 (4,12); 517 (15); 521 (17); 533 (39); 534

Dazey, M. H. 528 (11)

Dean, S. O. 273 (1); 276 (3); 277 (3); 278 (10); 280 (10); 281 (1,10); 284 (1)

AUTHOR INDEX

deAngelis, A. 343 (17)

Decker, G. 194 (30); 195 (30)

Decoux, J. 498 (4)

deGiovanni, G. 410 (1)

DeGroot, J. 318 (3); 337

DeJuren, J. A. 49 (9)

Delobeau, F. 390 (6); 410 (1,2,6); 418 (6); 425 (6); 469 (1); 503 (1); 516 (1)

Delone, G. A. 99 (5)

DeMaria, A. J. 470 (8)

deMetz, J. 413 (8)

DeMichelis, C. 98 (3); 101 (13); 159 (9); 164 (9); 168 (9,14); 169 (15); 171 (9); 173; 291 (5); 378 (17); 379 (19)

Denavit, J. 319 (15)

Denisov, N. G. 358 (35)

Denoeud, L. G. 390 (6); 410 (6); 418 (6); 425 (6); 469 (1); 503 (1); 516 (1)

DeShong, J., Jr. 44 (4); 45 (4,15); 46 (4)

Detlefs, F. 269

Deutsch, T. F. 9 (3)

Dietz, D. R. 370 (10)

diFrancia, G. T. 215 (32)

Dirac, P. A. M. 210 (1); 211 (1); 220 (1)

Dixon, J. K. 70 (10)

Dolgov, G. G. 111 (24)

Donaldson, T. P. 163 (12); 172 (12); 173 (12)

Dreicer, H. 318 (8)

Drozhbin, Yu. A. 395 (13); 399 (13)

DuBois, D. F. 367 (3,5); 387 (1); 434 (4); 533 (42)

Dubé, G. 435 (6) 444 (6); 470 (10)

Dumanchin 3

Duncan, A.B.F. 76

Dunway, R. E. 528 (27)

Dyachenko, V. P. 528 (17); 529 (17)

D'Yakonov, M. I. 215 (14,16); 221 (16)

E

Eckbreth, A. C. 9 (6)

Edwards, D. F. 102 (18)

Eerkins, J. W. 45 (9)

Eidmann, K. 308 (8); 342 (4,8); 422 (17); 425 (17); 469 (2); 503 (3); 504 (3); 505 (9); 509 (3)

Elgin, R. L. 126 (34)

Elton, R. C. 458 (19); 461 (20)

Emerson, J. A. 68 (1)

Engelhardt, A. G. 291 (2); 308 (7); 315 (3); 342 (2); 516 (5); 525 (5)

Eubank, H. P. 318 (6)

Evans, D. E. 181 (2); 186 (13); 187 (13); 188 (13); 194 (29); 203 (44); 204 (44)

Evans, L. R. 99 (9); 101 (14); 160 (11)

Everhart, E. 278 (9)

F

Fabre, E. 308 (5); 538 (45)

Fader, W. J. 289; 291 (8,9); 315 (2); 521 (16)

Fainberg, Ya. B. 254 (9)

Fallon, H. 337

Faucheux, G. 430

Faugeras, P. E. 308 (5,9)

Fauquignon, C. 390 (6); 410 (1,2,5,6); 418 (6); 425 (6); 469 (1); 503 (1); 516 (1)

Favro, L. D. 215 (21,25-28,44); 217 (21,25-28,44); 222 (26); 223 (44)

Fedotov, S. I. 399 (26); 401 (18); 528 (6)

Fejer, J. A. 184 (8); 186 (8); 194 (8)

Fenstermacher, C. A. 55 (24)

Fettis, H. E. 118 (32)

Filippov, N. V. 391 (11); 529 (29)

Filippova, T. I. 529 (29)

Floux, G. F. 107 (22); 308 (5); 342 (6,7); 343 (7); 390 (6); 396 (14); 410 (1,2,4-7); 418 (6); 421 (7); 425 (6); 429 (21); 469 (1); 482 (2); 503 (1);.516 (1); 545

Foltz, N. D. 379

Forrest, M. J. 182 (4); 183 (4); 186 (13); 187 (13); 188 (13); 194 (29)

Forslund, D. W. 201 (40)

Foster, J. D. 9 (8)

Fradkin, D. M.
215 (21,25-28,44);
217 (21,25-28,44);
222 (26); 223 (44)

Frenkel, N. 444 (11)

Fünfer, E. 194 (28)

Fuss, D. 318 (14)

G

Gaertner, W. W. 210 (2); 211 (2); 220 (2)

Gale, B. C. 159 (7,10)

Ganley, T. 43 (1,2); 44 (8,11); 45 (16); 47 (11,20,22); 49 (11,22); 51 (11)

Gatti, G. 343 (17)

AUTHOR INDEX

Gautier, J. 498 (4)

Gekker, I. R. 318 (4)

Generalov, N.A. et al 94 (16)

George. T. V. 308 (7); 315 (4); 342 (2)

Gerry, E. T. 25 (1); 182 (3)

Gerstmayr, J. 67

Gilinsky, V. 367 (3)

Ginzburg, V. L. 354 (33); 358 (33)

Glenn, W. J., Jr. 470 (8)

Glennon, B. M. 448 (14); 457 (14)

Glock, E. 194 (28)

Gobeli, G. W. 390 (5); 469 (5); 470 (5)

Godwin, R. P. 340

Goldman, E. 444 (13); 450 (24)

Goldman, M. V. 367 (5); 434 (4); 533 (42)

Goldstein, M. 545

Golubev, S. A. 45 (13); 47 (13)

Gondhalekar, A. M. 182 (5); 184 (5)

Good, R. H., Jr. 215 (34); 217 (34)

Gorbunov, A. V. 343 (23)

Gorenflo, R. 179 (3)

Görlich, P. 264 (13)

Graf, A. K. 105 (20)

Gratton, R. 245 (2); 246 (2); 250 (2); 343 (17); 420 (15); 431

Gray, M. A. 173 (24); 372 (14); 373 (14); 374 (14)

Green, B. 344 (26); 359 (26); 360 (26); 363 (26); 521 (17)

Gregg, D. W. 307 (3); 529 (34)

Grewal, M. 319 (15)

Gribkov, V. A. 390 (3,4); 391 (10); 392 (4); 393 (21); 398 (16)

Griem, H. R. 278 (10); 280 (10); 281 (10); 457 (16); 458 (16); 531 (36)

Guderley, G. 516 (9)

Guenther, A. H. 87 (5); 105 (20); 153 (2); 179; 401 (18)

Guillaneux, P. 529 (31)

Guipponi, P. 420 (12)

Gush, H. P. 212 (8)

Gush, R. 212 (8)

Guyot, D. 43 (1,2); 44 (5,10); 45 (1,10); 87 (7); 90 (7); 92 (7); 99 (8)

H

Haas, G. 443 (10)

Haber, I. 280 (13)

Hadley, C. R. 215 (35); 217 (35); 223 (43)

Hai, F. 528 (18,20,24)

Haines, M. G. 298 (12,13)

Hammer, D. 280 (13)

Hardway, G. A. 105 (20)

Harrigan, F. A. 9 (3)

Harris, L. 70 (9)

Harris, R. L. 215 (23); 216 (23); 220 (23)

Harteck, Paul v; 67; 546

Haught, A. F. 80 (1); 289; 291 (6,8); 315 (2)

Hauser, G. C. 479

Heflinger, L.O. 114 (29); 130 (29)

Heinicke, W. 307 (3)

Hendel, H. 318 (7); 337

Henderson, D. 318 (8)

Henins, I. 528 (25)

Henke, B. L. 126 (34)

Hercher, M. 92 (11); 95 (11)

Herman, R. M. 173 (24); 372 (14); 373 (14); 374 (14)

Herring, C. 264 (16)

Hertzberg, A. 9 (1); 26 (4); 33 (9); 38 (11); 40 (16); 391 (24); 516 (10); 528 (3)

Herwig, L. O. 44 (6); 46 (6)

Hess, R. V. 47 (19)

Hildebrand, F. B. 118 (31)

Hill, A. 21

Hirono, M. 240 (8); 271 (3)

Hobby, M. G. 391 (11)

Hoffman, A. L. 27 (7)

Hohla, K. 135 (35)

Holmes, L. S. 196 (33); 199 (37); 200 (37)

Hongyo, M. 483 (7,8)

Honig, R. E. 254 (4,5); 266 (5)

Hora, H. 77 (1); 172 (22); 215 (11); 216 (11); 264 (13,14); 265 (17); 267 (20,21); 271 (1,2); 276 (5); 291 (4); 307 (1); 308 (1,6,7); 309 (6); 312 (1); 315 (4); 342 (2); 343 (11-13,18); 344 (12); 345 (11,12,31); 350 (12,13); 353 (12); 354 (13); 356 (12); 357 (12); 358 (11,12,36); 360 (36); 363 (12); 369 (9); 378 (9); 281 (1-4); 382 (2); 384 (4); 390 (2) 434 (3); 505 (9); 516 (6,11,14); 520 (6); 521 (14); 523 (6); 525 (6); 528 (2); 546 (1)

Houtermans, F. G. v

AUTHOR INDEX

Hsi, C. G. 318 (12); 324 (12)

Huddleston, R. H. 172 (20)

Hughes, T. B. 343 (15)

Hurle, I. R. 9 (1); 26 (4)

Hutson, A. R. 215 (13); 217 (13)

I

Iannucci, J. 337

Imshennik, V. S. 528 (17); 529 (17)

Ingraham, J. 318 (8)

Isaenko, V. I. 390 (27)

Isawa, Y. 486 (12)

Isenor, N. R. 94 (17)

Itoh, T. 241 (9); 302 (3); 303 (3)

Ivanov, A. A. 254 (10)

Ivey, H. F. 265 (18)

Iwamoto, I. 240 (8); 271 (3)

Izawa, Y. 186 (12); 191 (22)

J

Jackson, D. A. 387

Jahoda, F. C. 528 (25)

Jaynes, E. T. 215 (22)

Johnson, D. J. 144

Johnston, E. 40 (16)

Jones, E. D. 390 (5); 469 (5); 470 (5)

Josephson, V. 528 (11)

K

Kachen, G. 430 (23); 470 (9)

Kadomtsev, B. B. 200 (38)

Kaiser, W. 307 (3)

Kaminsky, M. 239 (7)

Kang, H. 481; 483 (6); 486 (13); 491 (14); 503 (5)

Kapitza, P. L. 210 (1); 211 (1); 220 (1)

Kasper, J. V. V. 62 (1)

Katz, J. 318 (3)

Katzenstein, J. 181 (2); 186 (13); 187 (13); 188 (13); 194 (27,29)

Kaufman, F. 74 (13)

Kaw, P. K. 276 (7); 286 (7); 318 (1,2,5,10); 328 (2); 337; 342 (24); 343 (19,20); 344 (24); 347 (24); 353 (24); 354 (24); 356 (24); 357 (24,34); 358 (24); 359 (20,24); 362 (20); 363 (24); 384 (8); 434 (5); 494 (17); 516 (12); 521 (17); 533 (39); 534

Kegel, W. H. 188 (16); 189 (16); 494

Keilhacker, M. 196 (34); 200 (34); 201 (34)

Keldysh, L. V. 92 (12); 99 (6)

Kellerer, L. 201 (41,42); 202 (42)

Kelley, P. L. 171 (18)

Key, M. H. 159 (7,10); 163 (12); 172 (12); 173 (12)

Khvostionov, V. E. 45 (14)

Kidder, R. E. 38 (11); 342 (9); 343 (10,16); 344 (10); 360 (10); 361 (10); 363 (10); 420 (14,16); 434 (2); 503 (6); 516 (11)

Kim, P. H. 241 (9); 302 (2,3); 303 (2,3); 307 (2); 308 (2); 309 (2)

Kimel, S. 69 (7)

Kleen, W. 265 (18); 266 (18)

Klose, W. 264 (15)

Kluge, R. 337

Kompaneets, A. S. 102 (16); 297 (11)

Kononov, E. Ya. 398 (16)

Konyukhov, V. K. 26 (5)

Koopman, D. W. 153 (1); 228; 241 (12,13)

Korobeinikov 403 (20)

Korobkin, V. V. 102 (15); 155 (2); 156 (6); 157 (6); 168 (14); 169 (15); 173; 267 (23)

Kovetz, Y. 179 (3)

Krall, N. A. 280 (13)

Krasil'Nikov, S. S. 45 (13,14); 47 (13)

Krasyuk, I. K. 99 (4); 170 (17); 172 (17)

Krause, H. 504 (7)

Kriukov, P. G. 342 (5); 390 (5); 421 (20); 427 (18); 428 (20); 430 (22); 469 (4); 470 (4); 516 (1)

Krokhin, O. N. 155 (1); 245 (1); 291 (1); 390 (1,3,4,7,8); 391 (9); 392 (4); 395 (13); 397 (7,15); 399 (13,19,26); 401 (8); 402 (19); 419 (11); 420 (13); 421 (20); 427 (11,18); 428 (20); 430 (22); 442 (9); 444 (12); 503 (4); 516 (1,3); 528 (5,6)

Kroll, N. M. 367 (2,4)

Kronast, B. 179; 182 (5); 184 (5); 188 (15); 192 (24); 194 (28); 198 (15); 204 (45); 205 (45); 271

Kruer, W. L. 276 (7); 286 (7); 318 (2,12,13); 319 (15); 324 (12); 328 (2); 340 (1); 343 (21); 359 (21);

Kryukov, D. N. 419 (11); 427 (11)

Kundo, J. 215 (32); 217 (32)

Kunze, H. J. 181 (1); 186 (10)

Kuo, P. K. 215 (21,25-28,44);
  217 (21,25-28,44); 222 (26);
  223 (44)

Kysilka, J. 470 (9)

L

Laflamme 3

LaFleur, P. L. 215 (36);
  217 (36)

LaFramboise, J. G. 233 (5)

Lamain, H. 308 (5); 538 (45)

Lamberton 3

Landau, L. D. 309 (10); 323;
  345 (28); 369 (8)

Landolt-Börnstein 113 (27)

Langer, P. 396 (14); 410 (3);
  411 (3)

Langmuir, I. 228 (6); 235 (6)

Lashmore-Davies, C. N. 201 (39)

Lawson, J. D. 38 (15)

LeBlanc, J. M. 299 (15)

Ledingham, R. B. 126 (34)

Lee, C. M. 528 (24)

Lee, J. H. 528 (22)

Lee, K. S. H. 184 (6)

Lehner, G. 202 (43)

Leighton, P. A. 68 (4)

Leising, W. 433

Leland, W. T. 55 (24)

Lent, R. E. 126 (34)

Leonard, S. L. 172 (20)

Leontovich, M. A. 277 (8)

Leteinturier-Laprise, A. 430

Lichtman, D. 254 (1)

Lifshitz, E. M. 309 (10);
  345 (28); 369 (8)

Lindl, J. 342 (24); 344 (24);
  347 (24); 353 (24);
  354 (24); 356 (24);
  357 (24); 358 (24);
  359 (24); 363 (24);
  384 (8)

Linhart, J. G. 77 (2); 516 (7);
  528 (1)

Linlor, W. I. 255 (11)

Lipkin, H. J. 215 (33,35);
  217 (33,35)

Litvak, A. G. 102 (18); 156 (4)

Loebbaka, D. S. 528 (22)

Lotz, W. 534 (43)

Lovisetto, L. 194 (27)

Lubin, M. 387; 427 (19);
  469 (7); 503 (2); 516 (13);
  525 (13)

Lucht, R. A. 47 (19)

Ludwig, D. 190 (19); 191 (19);

204 (19)

Lynch, D. W. 223 (43)

Mc

McCoy, J. W. 9 (10)

McLean, E. A. 273 (1); 276 (3); 277 (3); 278 (10); 280 (10); 281 (1,10); 284 (1)

McWhirter, R. W. P. 457 (17)

M

MacDonald, A. 179 (2)

Mack, M. E. 470 (8)

Macke, W. 264 (13)

Mahn, C. 190 (19); 191 (19); 204 (19)

Maker, P. D. 80 (1); 97 (1)

Mandel'Shtam, S. L. 102 (15); 111 (24); 155 (2); 398 (16)

Marchenko, V. M. 170 (17); 172 (17)

Marcuse, D. 215 (31); 217 (31)

Marlow, W. C. 113 (28)

Marshall, J. 528 (25)

Martellucci, S. 343 (17)

Mather, J. W. 528 (7,8)

Matsui, I. 212 (6); 213 (6)

Mattioli, M. 308 (5,9); 369 (9); 378 (9); 379 (19)

Mazzucato, E. 291 (5)

Mead, S. W. 342 (9); 435 (7); 503 (6); 528 (4)

Melnikov, N. S. 403 (20)

Mennicke, H. 505 (9)

Meskan, D. A. 528 (16); 529 (16)

Meyer, H. C. 528 (9,10); 529 (9,10)

Meyer, J. 193 (25)

Meyer, J. W. 215 (15); 216 (15)

Meyerand, R. G. 80 (1)

Meynial, D. 430

Miley, G. H. 43 (1,2); 44 (10-12); 45 (10); 46 (17,18); 47 (20,22); 49 (11,22); 51 (11)

Miller, M. A. 343 (23)

Minck, R. W. 107 (21); 342 (1)

Möller, H. G. 269

Morgan, C. G. 99 (9); 101 (14); 160 (11)

Morgan, F. 95 (18); 99 (9); 100 (9); 107 (9)

Morgan, P. O. 391 (11); 529 (28)

Morpurgo, G. 215 (17); 217 (17); 220 (17)

Morse, R. L. 201 (40); 318 (14); 528 (25)

AUTHOR INDEX

Mott-Smith, H. M. 228 (6)

Motz, H. 254 (3); 266 (3); 343 (22); 347 (22)

Mudko, R. I. 9 (3)

Müller, H. 264 (14)

Mulser, P. 245 (3); 308 (8); 342 (8); 344 (26,27); 359 (26); 360 (26); 362 (27); 363 (26,27); 382 (7); 384 (7); 387; 422 (17); 425 (17); 469 (2); 503 (3); 504 (3); 505 (9); 509 (3)

Myers, G. H. 74 (13)

N

Nagao, Y. 483 (7)

Naiman, C. S. 81 (2); 92 (2)

Nakanishi, Y. 186 (12)

Namba, S. 241 (9); 302 (2,3); 303 (2,3); 307 (2); 308 (2); 309 (2)

Neuberger, D. 76

Neusser, H. G. 391 (25)

Nicols, M. H. 264 (16)

Nicholson-Florence, M. B. 343 (15)

Nielsen, P. E. 99 (11)

Nielson, C. W. 201 (40); 318 (14)

Nikulin, V. Ya. 391 (10); 393 (21)

Nishikawa, K. 321 (18); 322; 387 (2); 491 (16); 533 (40)

Nodwell, R. A. 189 (17,18); 190 (18); 191 (20); 193 (25); 197 (17,18)

Nutter, M. J. 55 (24)

O

Oberman, C. R. 276 (4,7); 286 (7); 318 (2); 324 (20); 326 (21); 328 (2); 337; 345 (30); 448 (15); 457 (15); 494 (17)

Oliver, B. M. 215 (24); 216 (24); 220 (24)

Olsen, J. N. 469

Onishi, M. 486 (9)

Opower, H. 307 (3); 391 (25)

Oraevskii, A. N. 26 (6)

Ott, W. 254 (8); 263 (8)

Owen, C. S. 215 (20)

P

Pack, J. L. 308 (7); 315 (4); 342 (2)

Palmer, A. J. 173 (23); 367

Pao, Y. H. 68 (4,5)

Papadopoulos, K. 273 (1); 280 (13); 281 (1); 284 (1)

Papas, C. H. 184 (6)

Papoular, R. 82 (3,4); 87 (6);

91 (6); 179; 308 (5,9);
369 (9); 378 (9); 379 (19)

Pappert, R. A. 184 (6)

Parisot, D. 390 (6); 410 (6);
418 (6); 425 (6); 469 (1);
503 (1); 516 (1)

Pashinin, P. P. 99 (4);
102 (15); 155 (2); 170 (17);
172 (17); 390 (23,27)

Patou, C. 528 (15);
529 (15,31)

Paul, J. W. M. 196 (33);
199 (37); 200 (37)

Peacock, N. J. 182 (4);
391 (11)

Pechacek, R. E. 184 (6);
185 (7)

Peercy, P. S. 390 (5);
469 (5); 470 (5)

Pendelton, W. K. 87 (5); 179

Perkins, F. W. 318 (10)

Peshkin, M. 215 (33,35);
217 (33,35)

Pfeiffer, H. Chr. 212 (5);
213 (5)

Pfirsch, D. 77 (1); 267 (20);
343 (11); 344 (11);
358 (11); 381 (1); 505 (9);
516 (11)

Phelps, A. V. 92 (11); 94 (11);
95 (11)

Phillips, J. A. 525 (18);
528 (27)

Piar, G. 390 (6); 410 (6);
418 (6); 425 (6); 469 (1);
503 (1); 516 (1)

Pierce, J. R. 255; 261

Pietrzyk, Z. A. 188 (15);
197 (15)

Pignerol, M. 396 (14)

Pimentel, G. C. 62 (1)

Pinsley, E. A. 9 (6)

Piskova, G. K. 99 (5)

Pis'Mennyi, V. D. 45 (13,14);
47 (13)

Pitts, J. M., Jr. 68 (3)

Platzman, P. M. 367 (1)

Pohl, F. 202 (43)

Polk, D. H. 289; 291 (6,8);
315 (2)

Poluektov, I. A. 390 (8);
401 (8)

Porter, G. 69 (6,8); 76 (6)

Post, R. F. 458 (18)

Potter, D. E. 529 (30)

Preston, D. A. 163 (12);
172 (12); 173 (12)

Prokhindeev, A. V. 102 (15);
155 (2)

Prokhorov, A. M. 26 (5);
45 (13); 47 (13); 92 (10);
95 (10); 99 (4); 102 (15);
155 (2); 170 (17); 172 (17);
390 (23,27)

AUTHOR INDEX

Puell, H. 307 (3); 391 (25)

Pustovalov, V. V. 326 (22)

Q

Queffelec, A. 430

R

Rabinovich, M. S. 97 (2); 172 (19)

Rabinowitz, I. N. 233 (4)

Rado, W. G. 107 (21); 342 (1)

Raether, P. 88 (8)

Raizer, Yu. P. 92 (10); 95 (10); 99 (10); 101 (12); 102 (17); 149 (1); 179 (1); 274 (2); 278 (11); 295 (10); 371 (12,13)

Rakhimiov, A. T. 45 (13); 47 (13)

Ramsden, S. A. 191 (21)

Rand, S. 343 (14)

Rank, D. H. 370 (10)

Ready, J. F. 254 (1,3); 266 (3)

Reeves, R. 67

Rehm, R. G. ix

Rentzepis, P. M. 68 (4,5)

Reuss, J. 107 (22)

Rhoads, H. S. 47 (21); 50 (21)

Ribe, F. L. 196 (32)

Richardson, M. C. 94 (17); 101 (13); 159 (9); 163 (13); 164 (9,13); 168 (9,14); 173; 169 (15); 171 (9); 378 (17)

Ringler, H. 189 (17,18); 190 (18); 197 (17,18)

Roberts, K. V. 318 (14)

Roberts, T. G. 528 (9,10); 529 (9,10)

Robinson, D. C. 6; 182 (4)

Rockwood, S. D. 99 (11)

Röhr, H. 191 (23); 192 (23); 194 (28,30); 195 (30)

Rolnick, W. B. 215 (21,27,44); 217 (21,27,44); 223 (44)

Ron, A. 367 (2)

Roos, C. B. 528 (22)

Rosauer, E. A. 223 (43)

Rose, D. J. 182 (3)

Rosen, B. 318 (13); 337

Roskos, R. R. 212 (4); 213 (4)

Rostocker, N. 367 (2)

Rothard, L. 546 (2)

Rothe, H. 265 (18); 266 (18)

Rubin, P. L. 215 (19)

Rudakov, L.I. 254 (10)

Rusk, J. R. 44 (9); 45 (9)

Russell, D. A. 33 (9)

Russell, G. R. 44 (7)

Ryazanov, E. V. 403 (20)

S

Sakurai, 103 (19); 135 (19); 136

Salat, A. 215 (12); 217 (12)

Saleres, A. 342 (7); 343 (7) 415 (9)

Salpeter, E. E. 186 (9)

Salzmann, H. 308 (8); 342 (4,8); 422 (17); 425 (17); 469 (2); 503 (3); 504 (3); 505 (9); 509 (3)

Sanmartin, J. R. 321 (18)

Sannikov, V. V. 182 (4)

Sasaki, T. 315 (3); 481; 482 (4,5); 483 (6-8)

Savage, C. M. 80 (1); 97 (1)

Savchenko, M. M. 97 (2); 156 (5); 168 (5)

Sawyer, G. A. 196 (31); 525 (18)

Scherzer, O. 215 (45); 217 (45); 223 (45)

Schilling, H. B. 262 (12); 263 (12); 264 (12)

Schlüter, A. 267 (20); 268; 343 (11); 345 (11); 358 (11); 381 (1); 516 (11)

Schneider, R. T. 43 (3); 46 (3); 47 (19,21); 50 (21)

Schoenebeck, H. 212 (7)

Schwarz, H. J. 210 (2,3); 211 (2,3); 213 (3); 215 (9,10,11,29); 216 (11); 217 (29); 218 (29); 220 (2,3); 221 (40,42); 222 (40,42); 241 (9,10); 271 (1); 301; 302 (1-3); 303 (1-3); 307 (1,2); 308 (2,4,1); 309 (2); 312 (1); 315

Scorer, R. S. 298 (14)

Sedov, L. I. 297 (11)

Segall, S. B. 227; 271; 315

Senatsky, Yu. V. 342 (5); 390 (5); 469 (4); 470 (4)

Shanny, R. 280 (13); 318 (12); 324 (12); 329 (23)

Shatas, R. A. 528 (9,10); 529 (9,10)

Shcheglov, V. A. 26 (6)

Shchelev, M. Ya. 102 (15); 155 (2)

Shearer, J. W. 318 (9); 343 (10); 344 (10); 360 (10); 361 (10,38); 362 (37); 363 (10); 419 (10); 427 (10); 461 (21); 469 (6); 516 (11)

Shen, Y. R. 367 (7)

Shikanov, A. S. 399 (26); 401 (18); 528 (6)

Shimoda, K. 172 (21)

AUTHOR INDEX

Shirley, J. A. 9 (2)

Siemon, R. E. 144; 186 (14); 187 (14)

Sigel, R. 246 (4); 308 (8); 342 (3,4,8); 422 (17); 425 (17); 469 (2); 503 (3); 504 (3,7,8); 505 (9); 507 (8); 509 (3)

Silin, V. P. 321 (17); 326 (22); 387

Siller, G. 254 (7,8); 262 (12); 263 (8,12); 264 (12); 266 (7)

Silver, D. M. 74 (13)

Simonet, A. 528 (15); 529 (15)

Sizukhin, O. V. 318 (4)

Sklizkov, G. 155 (1); 245 (1); 390 (3,4,7,22); 391 (9,10); 392 (4); 393 (21); 395 (12,13); 397 (7); 398 (16); 399 (13,17,19,26); 401 (18); 402 (19); 503 (4); 528 (5,6); 545

Slama, L. 254 (6); 266 (6)

Smetna, F. O. 228 (2)

Smith, C. 135 (36)

Smith, D. C. 9 (10)

Smith, H. M. 114 (30)

Smith, M. W. 448 (14); 457 (14)

Smith, R. F. 215 (23); 216 (23); 220 (23)

Smith, W. H. 68 (2)

Sollid, J. E. 184 (6)

Sommerfeld, A. 264 (16)

Souli, M. 498 (4)

Soures, J. 433

Sower, V. K. 479

Spalding, I. J. 77 (1); 525 (20)

Speiser, S. 69 (7)

Spitzer, L., Jr. 111 (26); 120 (26); 143 (26); 276 (6); 286 (14,6); 293; 345 (32); 346; 370 (11); 371 (11); 531 (37)

Stamper, J. A. 273 (1); 276 (3); 277 (3); 278 (10); 280 (10); 281 (1,10); 284 (1); 315

Stanek, E. J. 215 (34); 217 (34); 223 (43)

Stansfield, B. L. 193 (25)

Stehle, P. 215 (37, 38)

Steinfeld, J. I. 69 (8)

Steinhauer, L. C. 38 (11,13); 39 (11,13); 267 (22); 344 (25); 350 (25); 353 (25); 354 (25); 382 (5)

Steinmetz, J. 470 (9)

Stepanov, V. K. 97 (2); 156 (5); 168 (5)

Stettler, J. D. 528 (9,10); 529 (9,10)

Steuer, K. H. 196 (34);

200 (34); 201 (34)

Stone, G. 278 (9)

Stovall, J. E. 525 (18)

Stringer, T. E. 280 (12)

Sudan, R. N. 273 (1); 281 (1); 284 (1)

Sukhodrev, N. K. 102 (15); 155 (2)

Sun, Y. C. 337

Swain, J. E. 342 (9); 343; 498 (2,3); 503 (6)

T

Takamura, S. 322 (19)

Takayama, K. 322 (19)

Takeda, Y. 212 (6); 213 (6)

Talanov, V. I. 373 (15)

Targ, R. 9 (7-9)

Taylor, R. L. 29 (8)

Terhune, R. W. 80 (1); 97 (1)

Theimer, O. 184 (6); 195 (26)

Thiess, P. E. 43 (2); 44 (12); 46 (17,18)

Thom, K. 43 (3); 46 (3)

Thomas, K. S. 196 (31)

Thomas, S. J. 307 (3); 529 (34)

Thompson, H. B. 212 (4); 213 (4)

Tidman, D. A. 153 (1)

Tiffany, W. B. 9 (7-9)

Tomlinson, R. G. 80 (1); 159 (8); 168 (8); 171 (8); 177; 179 (2); 378 (17)

Tonon, G. 308 (5); 516 (8); 525 (8); 529 (31)

Tourtellotte, H. 210 (2); 211 (2); 220 (2)

Trivelpiece, A. W. 184 (6); 185 (7)

Tschekalin, S. V. 342 (5); 390 (5); 469 (4); 470 (4)

Tschuchimori, N. 315 (3)

Tsongas, G. A. 25 (3)

Tuck, J. L. 525 (18,19)

Tyurin, E. L. 419 (11); 421 (20); 427 (11,18); 428 (20); 430 (22)

Tzoar, N. 367 (1)

U - V

Vail, J. R. 153 (1)

Valeo, E. 276 (7); 286 (7); 318 (5); 326 (21); 337; 494 (17)

Vandre, R. H. 528 (14)

Vaniukov, M. P. 390 (27)

Van Paassen, H. L. L. 528 (11,14,16); 529 (16)

Van Zandt, L. L. 215 (15,18);

AUTHOR INDEX

216 (15); 220 (18)

Varshalovich, A. D. 215 (14,16); 221 (16)

Vasil'nov, V. V. 45 (14)

Vasseur, P. 538 (45)

Veksler, V.I. 254 (9)

Velikhov, E. P. 45 (13); 47 (13)

Venchikov, V. 390 (27)

Verber, C. M. 254 (2)

Verdeyen, J. T. 43 (1,2); 44 (10,11); 45 (10); 47 (20,22); 49 (11,22); 51 (11)

Veyrie, P. 107 (22)

Vlases, G. C. 27 (7); 38 (11,12,14); 529 (33)

Voinov, Yu. P. 398 (16)

Volkov, T. F. 387

Von Ardenne, M. 263

Von Weizsäcker, C. G. 516 (9)

Voronov, G. S. 99 (5)

W

Waki, M. 481; 483 (6); 486 (13); 491 (14)

Walch, A. P. 9 (4)

Walsh, D. 254 (3); 266 (3)

Wang, C. C. 170 (16)

Ward, G. 185 (7)

Watson, C. J. H. 343 (22); 347 (22)

Watteau, J. P. 197 (36); 528 (15,23); 529 (15); 539 (23)

Waylonis, J. E. 443 (10)

Weaver, H. T. 479

Weibel, E. S. 336 (24)

Weise, W. L. 448 (14); 457 (14)

White, D. R. 111 (25)

White, R. S. 528 (14)

Whitehead, J. D. 318 (10)

Wick, R. V. 107 (23); 108 (23); 144

Wienecke, R. 135 (35)

Wiggins, T. A. 370 (10)

Wilcock, P. D. 182 (4)

Wilhelm, H. 345 (31)

Wilkerson, T. D. 153 (1)

Wilson, J. R. 299 (15); 315 (1)

Winterberg, F. 254 (10); 516 (2)

Witkowski, S. 135 (35); 308 (8); 342 (8); 382 (7); 384 (7); 422 (17); 425 (17); 469 (2); 503 (3); 504 (3,7); 505 (9); 509 (3)

Wittliff, C. E. 391 (24);

516 (10); 528 (3)

Wolf, E. 441 (8)

Wong, A. 318 (11)

Wright, T. P. 529 (32)

Wuerker, R. F. 114 (29); 130 (29)

Y

Yablonovitch, E. 243

Yakovlev, V. A. 395 (13); 399 (13)

Yamanaka, C. 77; 186 (12); 191 (22); 240 (11); 271; 308 (8); 315 (3); 387; 482 (3-5); 483 (3,6-8); 486 (9-13); 491 (11,14); 496 (1); 503 (5)

Yamanaka, T. 315 (3); 481; 482 (3-5); 483 (3,6); 486 (10,11,13); 491 (11,14); 503 (5)

Yokoyama, M. 186 (12); 191 (22)

Yonas, G. 271

Yoshida, K. 481; 483 (6); 486 (13); 491 (14); 503 (5)

Young, C. G. vii

Young, M. 92 (11); 95 (11)

Z

Zakharov, S. D. 342 (5); 390 (4,5,7); 392 (4); 395 (12,13); 397 (7); 399 (13,17); 419 (11); 421 (20); 427 (11,18); 428 (20); 430 (22); 469 (4); 470 (4); 503 (4)

Zavoiskii, E. K. 254 (10)

Zeldovich, Ya. B. 92 (10); 95 (10); 99 (10); 102 (16); 179 (1); 215 (30); 274 (2); 295 (10)

Zink, J. W. 343 (10); 344 (10); 360 (10); 361 (10); 363 (10); 516 (11)

Zorev, N. N. 401 (18)

Zwicker, H. 194 (28)

SUBJECT INDEX

A

Abel equation 118, 137

Absorption *(See also Nonlinear absorption)* 6, 39, 70, 72, 73, 79, 80, 101, 102, 109, 140, 141, 149, 161, 162, 318, 340, 343, 350, 359, 360, 370, 396, 410, 413, 418, 420, 434, 443, 444, 449, 455, 457, 461, 463, 464, 477, 482, 491, 512, 528, 532, 534

Acoustic frequency *(See also Ion-acoustic)* 493

Adiabatic expansion 297

Admixed metal vapor 537

Aerosol 149

Alkali-halides 243

Alpha radiation "pumping" 44

Aluminum target 228, 283, 300, 442, 446

Anti-Stokes shifts 162

Argon 159

Attachment 100

Avalanche 87, 95, 99

B

Background gas 511

Backward scattering *(See Scattering)*

Barium crown glass 499

Beam trapping 177, 378

Beer's Law 70, 72, 74, 76

Bernstein wave 190, 199

Blackbody radiation 264

Blown-off plasma *(See Nonlinear)* 545

Blueshift 184, 506

Blümlein cable 255

Bohm-Gross frequency 325, 331

Boron-lined 47, 48, 50, 53, 54

SUBJECT INDEX

Boundary interactions 109, 142

Bound-bound (bb) transitions 534

Bragg relationship 210

Breakdown 6, 7, 80, 84, 89, 94, 99, 100, 101, 102, 105, 109, 114, 117, 130, 135, 143, 145, 157, 156, 161, 162, 164, 169, 170, 177, 179, 243, 267
...*Blast wave* 101, 134, 149

Breakdown threshold 81, 102, 164, 172

Brightness 263, 400

Bremstrahlung *(See also Inverse Bremstrahlung)* 84, 123, 182, 196, 417, 531, 535
...*Continuum* 505, 534

Brillouin scattering 344, 363, 367, 368, 370, 373, 378

Bunching process 216

C

Carbon 392, 393, 477, 505

Cascade ionization 91, 92, 94, 99, 171, 172

"Castle" target 463

CD_2 target 477

CD_4 target 430

Caustic (focusing cone) 101

Channeling 153, 164

Chapman-Jouget theory 102

Characteristic distance 217

Characteristic wave length 220

Charge fluctuations 321

Chemical lasers 77

Chromium target 301

Clean air threshold 149

CO_2 1, 9, 25, 94, 243, 245, 246, 531, 532, 545

Coherence length of electrons 221, 222

Cold cathodes 271

Collective effects 275, 276, 321

Collision 12, 16, 28, 37, 72, 84, 85, 240, 275, 343, 457

Collision broadening 7

Collision frequency 28, 320, 343, 345, 353, 358, 371, 442, 443, 519, 533

Collisionless plasma 200, 201 275

Combustion driven gasdynamic laser 27

Complex targets 390

Compounds of tritium 546
(See also under name of compound)

Condition for thermonuclear fusion 391

SUBJECT INDEX

Convective cooling 14, 15, 18, 22

Copper 228, 536, 537

Coulomb forces 120, 278

Critical density 245, 512

Critical region 276

Cu added plasma 531

Cut-off density 421, 425, 444

Cutoff energy spread 222

D

Damage threshold for glass 497, 498

Damping 351, 352, 493

D-D plasma 516

D-D fusion 509

D-D reaction 409, 410, 420

Debye length 182, 233, 235, 239, 240, 265, 369, 493

Decay time 64

Deconfining acceleration 311, 352

Defocusing effect 172

Demodulation 215, 220

Dense focus 463, 527, 528, 529, 539

Dense plasma 442, 528, 529

Dense-plasma focus (DFS) 98

Detonation wave 102, 149, 278

Deuteration 469, 477

Deuterium 38, 98, 99, 112, 195, 409, 449, 461, 482, 485, 486, 488, 490, 491, 509, 529

Deuterium ice 413

Deuterium-lithium target 390, 430, 434, 455, 456, 461

Deuterium plasma 109, 410, 520

Deuterium pressure 130

Deuterium targets 504

Deuterium-tritium carbon 546

Deuterium-tritium plasma 464, 516, 517, 529

Deuterized polyethylene target 390, 397, 482

Diamagnetic 295

Dielectric strength 243

Dimerization 75

Doppler shift 163, 194, 195, 198, 280, 281, 371, 395, 399, 506

Double breakdown blast wave 139

Double discharge laser 3

Driving frequency 331

Dummy probe 230

Dye 69

Dynamic expansion 522

E

Eddies 299

Einstein's coefficient 211, 457

Electron beam 3, 68, 222, 516
...*Filamentary structure* 222

Electron conduction 411, 425, 428

Electron cyclotron drift 200

Electron cyclotron frequency 202, 204

Electron density *(See also Plasma density)* 84, 111, 114, 117, 119, 120-122, 129, 140-142, 167, 181, 182, 190, 216, 234, 238, 239, 271, 321, 357, 360, 368, 393, 394, 398, 443, 448, 450, 457, 458, 462, 463, 510, 528, 531, 534

Electron-density fluctuations 322

Electron drift 100

Electron emission 253, 254, 271

Electron-ion collision frequency 276, 382

Electron ion temperature 452

Electron monochromaticity 223

Electron monochromator 222

Electron multiplication 99

Electron pressure 286, 322

Electron probability density 217, 218

Electron radius 212

Electron temperature *(See also Temperature)* 120, 129, 130, 131, 132, 232, 238, 240, 265, 277, 344, 346, 352, 358, 398, 410, 421, 424, 425, 428, 450, 452, 458, 462, 482, 486, 489, 505, 513, 516, 534

Electron wave function 218, 221

Electron wave length 210

Electrostriction 171, 343

Emittance measurements 260

Energy analyzer 486

Energy density contours 441

Energy density profile 439

Energy per unit volume 4

Energy spread of electron 221, 263

Equilibrium 444

Even-point 525

Expansion *(See also Plasma expansion)* 419, 425, 470, 521

Expansion energies 477

Extinction 102

F

Faraday cell 411

Fast ion 241, 490, 517

Fe - admix 536, 537

Field emission 99

Filament 167, 168, 173, 267

Filamentary scattering 157

Filter factor 213

First electrons 88, 90, 147, 179

Flare interferogram 392

Flux threshold 533

Focusing 6
...*Cone (caustic)* 101

Foil transmission 126, 127, 129

Force density 309, 310

Forward-scattering *(See Scattering)*

Fracture 495

Free electrons 92, 93

Free-free electron transmissions (ff) 442, 529, 534

Freon 159

Frequency mixing 367

Fringe 119, 120, 130, 137, 164, 448
...*Shift* 119, 164, 448

Front velocities 278

Fusion 55, 516, 527, 545

Fusion energy gain 517, 519, 523-525

G

Gain coefficient 30

Gamma rays 56

Gas breakdown *(See Breakdown)*

Gaunt factor 531

Gladstone-Dale relation 135

Glan polarizers 470
... *Prisms* 483

Grid-laser 3

Growth of breakdown 99, 103, 109, 141, 145, 323, 324, 326, 371, 493

H

Heating 274, 277, 389, 390, 416, 419, 428, 450, 461, 463, 482, 494, 529, 530, 531, 534
...*Anomalous* 319, 321

Helium 4, 159, 274, 511

High Z impurities 457

Holography 11, 103, 114, 117, 118, 130, 135

Hydrodynamic instabilities 321

Hydrogen 107, 123, 410, 531, 536, 538

I

Implosion 450, 516

Impurity doped plasma 459, 460

Incident waves 384

Index of refraction *(See Refractive index)*

Initial electron *(See First Electrons)*

Initiation process *(See First Electrons)*

Interaction time *(Electron-photon)* 212, 213, 214

Interferogram 118, 164, 166, 168, 251

Interferometry 165, 182, 196, 392, 446

Inverse bremsstrahlung 99, 120, 179, 245, 276, 370, 421, 443, 444, 448, 457, 496, 529

Iodine laser 61, 64

Ion-acoustic oscillation 321, 323

Ion density 234, 312, 545

Ion energy due to magnetic field 301, 307, 315

Ion reflectivity 234

Ion temperature *(See also Temperature)* 233, 452, 464, 510

Ion velocity 477, 479

Ionization 49, 53, 80, 81, 84, 87, 102, 105, 135, 151, 153 159, 172, 250, 276, 444, 457, 458, 496, 498, 510, 534, 538

J-K

Kapitza-Dirac effect 209, 220

Kerr effect 171, 172

Kilojoule range 545

Kinetic model 521

Kofink refractive index *(See also Refractive index)* 382

Krypton 159

L

Lagrangian code 444

Landau damping 182, 191, 324, 331, 532

Langmuir frequency *(See also Plasma frequency)* 532

Langmuir probe *(See also Probe)* 228, 229, 232, 235, 240, 271

Langmuir wave 533

Lattice spacing 210

Lawson criterion 38, 39

LiD *(See Deuterium-lithium)*

Light pressure 381

LiH 482, 486, 487

Lithium ions 44

SUBJECT INDEX

Long sparks 155

Luboshez lens 485

M

Mach number 103, 135

Macromolecules 94

Magnetic confinement 38, 143, 429, 546

Magnetic field 65, 181, 190, 201, 202, 204, 241, 254, 273, 281, 289, 291, 296, 298, 303, 305, 308, 309, 315, 486, 516, 528

Magnetic permeability inside plasma 303

Magnetic pressure 297

Magnetic Reynold's number 284, 285

Magnetic stress tensor 298

Magnetic vector potential 217

Magnetrostrictive 343

Megajoule laser 545

Methane 159

Microcracks in glass 497

Microwave gas breakdown (See also Breakdown) 99

Microwaves 318

Minimum fusion gain (See also Fusion energy gain) 525

Modulation of electron beams
at optical frequencies 215

Momentum transfer 218, 219, 275, 278

Multibeam laser 399, 436

Multichain amplifier system 430

Multiphoton absorption 68, 99, 496

Multiphoton ionization 94, 179

N

"Natural time" operation 27

Navier-Stokes equation 370

Neon 159

Neutron 44, 50, 52, 55, 56, 196, 342, 363, 410, 415, 419, 422, 434, 448, 450, 453, 464, 509, 529

Neutron emission 397, 398, 410, 414, 415, 420, 425, 482, 503, 505, 509, 512

Neutron yield 390, 414, 418, 426, 434, 455, 486, 491, 492, 545

Nitrogen 11, 159, 274

Nitrogen addition 4

Nitrogen laser 51, 153

NMR analysis 477

Non-adiabatic effects 285

Nonequilibrium 545

Non-linear 31, 308, 343, 416, 420, 429, 545
...*absorption* 434

Non-linear acceleration 254, 267

Non-linear 163, 171, 172, 326, 367, 372, 444, 516, 545

Non-linear forces 267, 309, 341, 343, 344, 346, 350, 357, 387

Nuclear fusion *(See also Fusion and Thermonuclear fusion)* 77, 254, 410, 515, 545

Nuclear radiation 47, 49

Nuclear radiation pumping 43, 46

Nuclear effects simulation 107

O

Optical tunneling 99

Optimum amplifier length 5

Oscillation energy 350

Overdense plasma 342, 383, 422, 425, 433

P

Particle-pushing model 318

Particle trapping 326, 334

Penetration depth 533

Phase shift 321

Photochemical 72

Photoelectric emission 264

Photoionization *(See also Ionization and First electron)* 92, 172, 275, 276

Photolytic 64

Phthalocyanine 69

Pierce optical system 261

Pinch 182, 183, 187, 188, 192, 195-201, 204, 298, 528, 530

Plasma acceleration 394

Plasma boundary 111

Plasma-core 47

Plasma density *(See also Electron density and Ion density)* 309, 311, 359, 360, 361, 377, 394, 416, 445, 446, 448, 451

Plasma dynamics 103, 104

Plasma expansion 101, 102, 394, 419, 420, 504

Plasma filament 167, 173

Plasma focus 196, 391, 527, 530

Plasma frequency 111, 189, 197, 200, 325, 339, 369, 371, 382, 482, 512, 443

Plasma growth 101, 109

Plasma filaments 173

Plasma heating *(See Heating)*

Plasma implosion 391

Plasma luminosity 102

Plasma momentum 395

Plasma pressure 391, 395

Plasma sheath 234, 235, 303

Plasma time history 417

Plasmoid 291, 292

Platinum inclusion in glass 501

Pockels cell 66, 228, 411, 470, 483, 503

Polyethylene deuteride 430

Ponderomotive forces 172, 308, 309, 315, 343, 368, 372, 381, 383

Pondermotive force density 310, 345

Positioning of target 441

Prebreakdown ionization *(See also First electron)* 81

Pre-ionization 3

Prepulse 442, 448, 455, 529

Pressure profile 362

Priming electrons *(See First electron)*

Probe 228, 230, 249, 250, 251, 308, 315, 505, 506

Probe plasma interaction 234

Probe signals 228

Proton beam 47

Pulse compression 27

Q

Q-switch 6, 63, 69, 81, 82, 483, 496, 503, 529

Quantum mechanical bunching of electrons 215, 217

Quantum mechanical scattering 217

Quartz 501

Quenching 64, 74, 75

R

Radiation pressure 265, 286, 343, 358, 361

Radiative transport wave 101

Radio-frequency waves 318

Radioactive danger 546

Radioisotope 56

Raise damage threshold 498

Raman scattering 368

Rayleigh scattering 370, 374, 379, 486

Reaction frequency 519

Recoil electron momentum 216, 218, 220

Recombination 98, 100, 101, 123, 251, 457

Recombinative radiation 406, 535

Red shift *(See Doppler shift)*

Reflection *(See also Kapitza-Dirac effect)* 111, 142, 159, 177, 339, 342, 355, 362, 384, 411, 415, 420, 423, 470, 476, 482, 488, 489, 505, 507

Reflection coefficient 416, 421, 425, 513

Reflection probability of electron *(See also Kapitza-Dirac effect)* 211, 214

Refraction 39, 111, 113, 117, 142, 172

Refractive index 2, 6, 7, 111, 118, 130, 135, 137, 164, 167, 172, 266, 343, 345, 350, 354, 355, 373, 375, 382, 383, 442

Refractive index changes 171
...*Spacial variation* 355

Resistivity 324, 325, 330, 331, 332 ...*Anomalous* 318

Retarding field method 302

Reynold number *(See Magnetic Reynold number)*

Rogowski coil 261

Rotation of plasma 298

S

Saha equation 444

Scaling law *rule* 33, 427, 545

Scattering *(See also Raman and Rayleigh)* 2, 111, 142, 149, 177, 179, 181, 182, 185, 187, 189, 190, 191, 193, 196, 197, 199, 201, 202, 203, 204, 217, 271, 277, 368, 371, 470, 474, 491
...*backward* 372
...*forward* 158, 161, 170

Scattering losses 5, 6

Scattering potential 217

Schlieren system 164

Self-focusing 155, 157, 159, 163, 168, 169, 170, 171, 172, 177, 268, 343, 367, 368, 369, 377
...*in liquids* 157

Self-focusing filament 373

Self-focusing length 173

Self-similarity expansion 521

Self-trapping 497

Shadowgraphy 156, 278, 279

Shock front 140, 159, 199, 295

Shock plasma 197

Shock profile 137, 138, 140

Shock tube laser 27

Shock wave *(See also Detonation wave)* 26, 101, 130, 136, 137, 141, 245, 246, 249, 275, 278, 292, 295, 296, 297, 403, 404, 411, 425, 444, 450, 496, 497, 507

Shock wave expansion 98, 103, 404

SUBJECT INDEX

Shock-wave ionization 102

Shock wave radius 135, 137, 405

Similarity model 445

Skin depth 291, 295, 360, 443

Slow group *(See also Thermal group)* 241, 302, 490

Smokatron 254

Snowplow model 528

Soft X-rays 458, 461, 534

Solid deuterium *(See also Target)* 390, 410

Solid hydrogen *(See also Target)* 245, 382, 504

Solid target *(See also Target)* 228, 240, 254, 274, 289, 413, 434, 442, 464, 469, 477, 486, 487, 505

Sound wave 151

Space charge effects 264, 266

Spatial beating wave length for electron modulation 215, 219, 220

Spherical divergence 275

Standing light wave 209, 271

Standing wave 343, 359, 360, 362

Stark line 276, 398

Stellerator 546

Stimulated thermal Thomson scattering *(See also Thomson scattering)* 374, 377, 378

Stokes shift 372, 374

Stratified plasma 310

Streak photography 109, 110, 170, 171, 196, 248, 417, 425, 486, 504

Streamer 88, 153

Super-radiance 5, 6, 411

Supersonic 26, 237

Surface damage of glass laser *(See also Damage threshold)* 497

T

Tantalum target 301

Target 241, 441, 461, 464 *(See also under name of material)*
...Complex targets 390
...Plate target 486
...Particle target 289

Target core 464

Target dimensions 391

Taylor-Sedov blast wave theory 103, 135, 136

TEA-CO_2-Laser 247

Temperature *(See also Electron temperature and X-ray temperature determination)* 6, 13, 26, 31, 85, 120, 186, 203, 246, 286, 293, 295, 302, 342, 359, 361, 382,

391, 395, 410, 455, 456, 457, 488, 510, 531

Thermal conduction 102

Thermal equilibrium 536

Thermal force density 383

Thermal front 464

Thermal group 302, 308, 309, 490

Thermal radiation 102

Thermal self-focusing (See also Self-focusing) 375, 377

Thermalization times 120, 531

Thermionic electron emission 264

Thermoelectric force 283

Thermokinetic force 349

Thermonuclear fusion 7, 38, 40, 389, 433, 481, 482, 516, 528, 545

Thermonuclear initiation 107

Thermonuclear reactors 528

Thin films 339

Thomson scattering (See also Stimulated thermal Thomson scattering) 177, 179, 181, 201, 211, 486

Threshold (See also Breakdown) 105, 140, 149

Threshold intensity (See also Breakdown) 100, 171

Time of flight 238, 271, 308, 315, 410, 478, 479, 491, 505, 509

Time of ionization 535

Tokamak 546

Total reflection 355, 360

Transmission 103, 107, 108, 124, 126, 141, 339, 474, 476, 482

Transverse excitation 2

Transverse flow 19-21

Trapping 329

Traveling wave tube 216

Tritium 38, 47
...*lithium compound of* 546

Tungsten target 301

Turbulence 197, 200, 545

Two-dimensional simulations 334

Two groups of plasma 241, 302, 315, 490

Two-photon 68, 69, 72

Two-stream instabilities 321, 323, 343, 533

U

Ultraviolet 3, 398, 429

Underdense region 276

SUBJECT INDEX

V

Vaporization 442, 444, 445

Volterra integral equation 118

Vortex motion 299, 315, 326

W

Wazer computer program 343, 361, 420

WBK approximation 347, 352, 381

WBK condition 348

Weibel instability 336

X

X-rays *(See also Soft X-rays)* 126, 129, 396, 456, 461, 505, 507, 508, 512, 513, 528, 529, 531, 537, 539

X-ray temperature determination 434

Xenon 159

Y-Z

Zeeman shift 65

MIX
Papier aus verantwortungsvollen Quellen
Paper from responsible sources
FSC® C105338

If you have any concerns about our products,
you can contact us on
ProductSafety@springernature.com

In case Publisher is established outside the EU,
the EU authorized representative is:
**Springer Nature Customer Service Center GmbH
Europaplatz 3, 69115 Heidelberg, Germany**

Printed by Libri Plureos GmbH
in Hamburg, Germany